Solar Energy Conversion

Dynamics of Interfacial Electron and Excitation Transfer

RSC Energy and Environment Series

Editor-in-Chief:
Professor Laurence Peter, *University of Bath, UK*

Series Editors:
Professor Heinz Frei, *Lawrence Berkeley National Laboratory, USA*
Professor Ferdi Schüth, *Max Planck Institute for Coal Research, Germany*
Professor Tim S. Zhao, *The Hong Kong University of Science and Technology, Hong Kong*

Titles in the Series:
1: Thermochemical Conversion of Biomass to Liquid Fuels and Chemicals
2: Innovations in Fuel Cell Technologies
3: Energy Crops
4: Chemical and Biochemical Catalysis for Next Generation Biofuels
5: Molecular Solar Fuels
6: Catalysts for Alcohol-Fuelled Direct Oxidation Fuel Cells
7: Solid Oxide Fuel Cells: From Materials to System Modeling
8: Solar Energy Conversion: Dynamics of Interfacial Electron and Excitation Transfer

How to obtain future titles on publication:
A standing order plan is available for this series. A standing order will bring delivery of each new volume immediately on publication.

For further information please contact:
Book Sales Department, Royal Society of Chemistry, Thomas Graham House, Science Park, Milton Road, Cambridge, CB4 0WF, UK
Telephone: +44 (0)1223 420066, Fax: +44 (0)1223 420247
Email: booksales@rsc.org
Visit our website at www.rsc.org/books

Solar Energy Conversion
Dynamics of Interfacial Electron and Excitation Transfer

Edited by

Piotr Piotrowiak
Rutgers University, New Jersey, USA
Email: piotr@andromeda.rutgers.edu

RSCPublishing

RSC Energy and Environment Series No. 8

ISBN: 978-1-84973-387-8
ISSN: 2044-0774

A catalogue record for this book is available from the British Library

Published by The Royal Society of Chemistry,
Thomas Graham House, Science Park, Milton Road,
Cambridge CB4 0WF, UK

Registered Charity Number 207890

For further information see our web site at www.rsc.org

To the memory of Robin Hochstrasser, his curiosity and the inspiration
he gave to many of us

Preface

Solar energy research, from basic science to device engineering, is undergoing unprecedented growth which most certainly will continue in the foreseeable future. The broad boundary between chemistry, physics and materials science is particularly rich in new discoveries and holds the greatest potential for future breakthroughs. The design of superior photovoltaic materials, photocatalysts for water splitting and the reduction of carbon dioxide depends on a detailed, molecular level understanding of the dynamics of the interfacial electron and electronic energy transfer processes. In this monograph we have gathered a sample of state-of-the-art experimental and theoretical methods which were recently applied and, in many instances expressly developed, for the study of these fast interfacial processes that are crucial to solar energy conversion. All experimental techniques described in the monograph are capable of sub-picosecond temporal resolution, which is necessary in order to monitor the dynamics of photoinduced charge separation and exciton relaxation. The selected computational approaches are highly scalable enabling static and dynamic modelling of hybrid systems consisting of molecular chromophores and either semiconductor or metallic substrates.

Naturally, the overview presented is by no means comprehensive and at best represents a fleeting and highly subjective snapshot of this rapidly evolving area. Nevertheless, the experimental time-resolved methods discussed in the monograph span the frequency range from THz to X-ray and the corresponding instrumental setups range from those that can easily fit on a small table to those which require a national facility. The wide cross-section of techniques is meant to reflect the ease with which researchers in this field move between the bench top, laser table, synchrotron source and theory. This flexibility and interdisciplinary approach are the trademarks of modern solar photochemistry. Reflecting the current activity in the field, many of the applications of the described techniques are devoted to photoinduced electron

RSC Energy and Environment Series No. 8
Solar Energy Conversion: Dynamics of Interfacial Electron and Excitation Transfer
Edited by Piotr Piotrowiak
© The Royal Society of Chemistry 2013
Published by the Royal Society of Chemistry, www.rsc.org

transfer at the interface between a wide band gap semiconductor and a molecular redox active chromophore; however, all presented methods are general and not restricted to one type of a system or one process. Plasmonic photoelectrodes are among the most promising new approaches to solar energy conversion and photocatalysis. For this reason, a special effort has been made to introduce theoretical and experimental methodology used in the study of dynamic processes in chromophores coupled to nanostructured metallic surfaces.

It is worthwhile to note that when one takes a somewhat broader and longer perspective, one inevitably realizes that solar photochemistry research has become the leading driving force behind the development of new experimental and theoretical tools in physical chemistry and chemical physics. The primary targets and the most elegant realization of, for example, multidimensional femtosecond spectroscopy, time-dependent density functional theory and time-resolved X-ray diffraction, all revolve around the dynamics of electron and excitation transfer in either natural or synthetic systems that are relevant to solar energy conversion. It has been fascinating to witness the remarkable progression of what a few decades ago was merely a niche area to the very forefront of the chemical sciences. Photochemical energy conversion today captures the imagination of young scientists in a similar manner to the search for a cure for cancer or space exploration. The impact of solar photochemistry on the evolution of modern chemical sciences extends well beyond efficient conversion of photons into electrical current or electrochemical potential. In keeping with this spirit, our intent was not only to present a status report, but even more so to provide a starting point for new challenging exploratory research that has been always associated with this field.

In closing, I would like to express my most sincere gratitude to the authors and co-authors of all the chapters. A monograph can be only as good as the scientists who are willing to contribute and share their work with the readers. It is enormously satisfying to see how many leaders in their respective areas found the time to participate in this endeavour. Thank you!

Piotr Piotrowiak
New Jersey, USA

Contents

Chapter 1   **Computational Modeling of Photocatalytic Cells**   **1**
Steven J. Konezny and Victor S. Batista

1.1	Introduction	1
1.2	Photoelectrochemical Device Modeling	3
	1.2.1 Modeling Current-Voltage Characteristics	3
	1.2.2 Bioinspired High-Potential Porphyrin Photoanodes	5
1.3	Inverse Design of Photoabsorbers	10
1.4	Charge Transport in Nanoporous Metal Oxides	12
	1.4.1 Fluctuation-Induced Tunneling Conductivity	13
	1.4.2 Power Law Dependence of the Dark AC Conductivity	19
	1.4.3 Experimental Methods	20
1.5	Calculations of Redox Potentials: Reduction of Systematic Error	21
	1.5.1 Methodology and Benchmark Results	23
	1.5.2 Density Functional Theory Computational Methods	26
	1.5.3 Method Benchmark Results	26
	1.5.4 Choice of Reference Redox Couple	27
	1.5.5 Accounting for Solvent Polarity and Supporting Electrolyte	28
1.6	Conclusions and Outlook	29
	Acknowledgements	31
	References	31

RSC Energy and Environment Series No. 8
Solar Energy Conversion: Dynamics of Interfacial Electron and Excitation Transfer
Edited by Piotr Piotrowiak
© The Royal Society of Chemistry 2013
Published by the Royal Society of Chemistry, www.rsc.org

**Chapter 2   Charge and Exciton Dynamics in Semiconductor Quantum
             Dots: A Time Domain, *ab Initio* View   37**
             Amanda J. Neukirch and Oleg V. Prezhdo

2.1   Introduction   37
2.2   Theoretical Approaches   38
      2.2.1   Hartree–Fock Method   39
      2.2.2   Incorporation of Electron Correlation in
              Hartee–Fock with Configuration Interaction   39
      2.2.3   Density Functional Theory   40
      2.2.4   Time Domain Density Functional Theory   40
      2.2.5   Nonadiabatic Molecular Dynamics   40
2.3   Proposed Mechanisms for Multiple Exciton
      Generation   41
      2.3.1   Impact Ionization Process   41
      2.3.2   Direct Mechanism   42
      2.3.3   Dephasing Mechanism   42
2.4   Excited States and Symmetry Adapted
      Cluster-Configuration Interaction (SAC-CI)   42
      2.4.1   Multiexciton Generation (MEG)   43
      2.4.2   MEG with Dopants, Defects and
              Charging   46
2.5   Phonon Induced Dephasing   50
      2.5.1   Optical Response Function   50
      2.5.2   Phonon Dephasing in PbSe Quantum Dots   52
      2.5.3   Temperature Dependence of Phonon
              Dephasing   54
      2.5.4   Multiple Exciton Generation, Fission and
              Luminescence and Dephasing   54
2.6   Electron Phonon Relaxation   58
      2.6.1   Time Dependent Density
              Functional Theory   58
      2.6.2   Non-Adiabatic Molecular Dynamics   60
      2.6.3   Phonon-Assisted Relaxation of Charge
              Carriers in PbSe   61
      2.6.4   Temperature Dependence   63
      2.6.5   Ligands Saturate Dangling Bonds and
              Accelerate Electron–Phonon Relaxation   66
2.7   Time Domain *ab Initio* Study of Auger and
      Phonon-Assisted Auger Processes   67
      2.7.1   Auger Theory   68
      2.7.2   Results of Auger Studies   69
2.8   Conclusion   70
Acknowledgements   72
References   72

Chapter 3 Multiscale Modelling of Interfacial Electron Transfer   77
Petter Persson

  3.1 Introduction   77
  3.2 Materials Modelling   79
       3.2.1 Modelling Methods   79
       3.2.2 Multiscale Modelling   81
  3.3 Interfacial Electron Transfer   90
       3.3.1 Excitations at Interfaces   90
       3.3.2 Interfacial Interactions   94
       3.3.3 Surface Electron Transfer   96
       3.3.4 Modelling Complex Materials   101
  3.4 Conclusions and Outlook   105
  References   106

**Chapter 4 Plasmon-enhanced Solar Chemistry: Electrodynamics and
            Quantum Mechanics   111**
Hanning Chen, George C. Schatz and Mark A. Ratner

  4.1 Introduction   111
  4.2 Continuum Models   115
       4.2.1 Finite-difference Time-domain (FDTD)   116
       4.2.2 Finite-element Method (FEM)   117
       4.2.3 Discrete Dipole Approximation (DDA)   120
  4.3 Many-body Theories   123
       4.3.1 Linear Response Time-dependent Density
              Functional Theory (LR-TDDFT)   123
       4.3.2 Real-time Time-dependent Density Functional
              Theory (RT-TDDFT)   125
  4.4 Hybrid Approaches   127
       4.4.1 Multiscale Maxwell–Schrödinger Scheme
              (MMS)   128
       4.4.2 Hybrid RT-TDDFT/FDTD Approach (QM/ED)   129
  4.5 Conclusions and Future Direction   131
  Acknowledgments   132
  References   132

**Chapter 5 Dynamics of Interfacial Electron Transfer in Solar Energy
            Conversion As Viewed By Ultrafast Spectroscopy   135**
Villy Sundström and Arkady Yartsev

  5.1 Dye Sensitized Nanostructured Metal Oxides for
       Grätzel Solar Cells   135
  5.2 Electron Injection from Sensitizer to Semiconductor in
       Dye Sensitized Solar Cells   136

5.3   Electron–Cation Charge Recombination in Dye
      Sensitized Semiconductor Materials   143
5.4   Dye Sensitized Solar Cell Performance in Relation
      Electron–Cation Recombination and Sensitizer
      Binding Geometry   148
5.5   Electron–Cation Interactions as a Source of
      Fast Recombination and Slow Charge
      Transport in Dye Sensitized Nanostructured
      Semiconductor Films   151
5.6   Conclusions   157
Acknowledgements   158
References   158

Chapter 6   **Semiconductor Nanocrystals Studied by Two-Dimensional**
            Photon Echo Spectroscopy   **161**
            Cathy Y. Wong, Shun S. Lo and Gregory D. Scholes

6.1   Semiconductor Nanocrystals and Quantum
      Dot-Based Solar Cells   161
6.2   Two-Dimensional Photon Echo Spectroscopy:
      Background   164
      6.2.1   Fundamentals of 2DPE Spectroscopy   164
      6.2.2   Solvation Dynamics   165
      6.2.3   Electronic Coherences   167
6.3   Two-Dimensional Photon Echo Spectroscopy:
      The Experiment   168
      6.3.1   Wedge Calibration   169
      6.3.2   Spectral Interferometry   170
      6.3.3   Summary   182
6.4   Simulation of the 2D Spectra: The Case of
      CdSe NCs   183
      6.4.1   Fine Structure of the Lowest Exciton and
              Biexciton States in CdSe   183
      6.4.2   Determination of Exciton–biexciton
              Transitions   186
      6.4.3   Simulating the 2D spectra   187
6.5   CdTe/CdSe Core/Shell Nanocrystals Probed
      by 2DPE   189
      6.5.1   Simulating the 2D Spectra of CdTe/CdSe
              Core/Shell Nanocrystals   191
      6.5.2   Discussion   195
6.6   Conclusion   198
Acknowledgements   199
References   199

Chapter 7 Ultrafast Optical Imaging and Microspectroscopy   **203**
Piotr Piotrowiak, Libai Huang and Lars Gundlach

7.1   Introduction   203
7.2   Kerr-Gated Femtosecond Fluorescence Microscopy
      and Micro-Spectroscopy (KGFM)   205
      7.2.1   Background and Experimental
              Considerations   205
      7.2.2   Excitation Fluence, Orientation, Shape and
              Size Dependence of Carrier Dynamics in
              CdS_xSe_{1-x} Nanobelts   208
      7.2.3   Fluorescence Dynamics of Quantum Dots in
              Close Proximity to Metal Surfaces   212
7.3   Femtosecond Pump Probe Transient Absorption
      Microscopy (PPTAM)   213
      7.3.1   Background and Experimental
              Considerations   213
      7.3.2   Examples of Applications of PPTAM to
              the Study of Exciton and Charge Carrier
              Dynamics in Nanostructures and
              Heterogeneous Materials   215
7.4   Future Challenges: Single Molecule Detection, Higher
      Time Resolution and Spatial Super-Resolution in
      Femtosecond Microscopy   218
      7.4.1   Single Molecule Femtosecond
              Microscopy   218
      7.4.2   Sub 100 Femtosecond Time Resolution   220
      7.4.3   Spatial Super-Resolution   221
7.5   Conclusions   222
Acknowledgements   222
References   223

**Chapter 8 Ultrafast Multiphoton Photoemission Microscopy of Solid
            Surfaces in Real and Reciprocal Space**   **225**
*A. Winkelmann, C. Tusche, A.A. Ünal, C.-T. Chiang,
A. Kubo, L. Wang and H. Petek*

8.1   Introduction   225
8.2   Frequency *vs.* Time Domain Measurements of
      Interfacial Electron Dynamics   228
8.3   Time-Resolved Ultrafast Multiphoton Photoemission   229
8.4   Time-Resolved Photoemission Electron Microscopy   231
8.5   Time-Resolved Photoemission Electron Microscopy
      (TR-PEEM) Imaging of Plasmonic Phenomena   234

8.5.1 TR-PEEM Imaging of Localized SP Modes 237
8.5.2 TR-PEEM Imaging of Surface Plasmon
Polariton (SPP) Dynamics 241
8.5.3 Nanoplasmonic Optics 244
8.6 Photoelectron Momentum Mapping 246
8.6.1 Momentum Mapping of Delocalized Bands 246
8.6.2 Momentum Mapping of Localized
Electronic States 250
8.7 Summary 256
Acknowledgements 256
References 256

**Chapter 9 Light at the Tip: Hybrid Scanning Tunneling/Optical
Spectroscopy Microscopy 261**
Jao van de Lagemaat and Manuel J. Romero

9.1 Introduction 261
9.2 Light Emission from the STM 262
9.2.1 Mechanism 262
9.2.2 Role of Surface Plasmons in Scanning
Tunneling Luminescence 264
9.2.3 Experimental Considerations 265
9.3 Light Emission from Nanostructures 266
9.3.1 Individual Atoms, Molecules and
Surface States 266
9.3.2 Thin Film and Organic Semiconductors 268
9.3.3 Nanocrystals 271
9.4 Time-Resolved Studies of Photon Emission
in the STM 275
9.5 Conclusions and Outlook 276
Acknowledgment 277
References 277

**Chapter 10 Time Resolved Infrared Spectroscopy of Metal Oxides and
Interfaces 281**
Akihiro Furube

10.1 Introduction 281
10.2 Experimental Techniques 286
10.3 Mechanism of Interfacial Electron Injection into
Metal Oxides 287
10.3.1 Metal Oxide Semiconductor Dependence 288
10.3.2 Dye Dependence 289
10.3.3 Effect of Solvent 292
10.3.4 Effect of Ions 292
10.3.5 Metal Nanoparticle to Metal Oxides 294

10.4   Summary   296
References   297

Chapter 11 **Carrier Dynamics in Photovoltaic Structures and
Materials Studied by Time-Resolved Terahertz
Spectroscopy**   **301**
*Enrique Cánovas, Joep Pijpers, Ronald Ulbricht and
Mischa Bonn*

11.1   Introduction   301
11.2   Time-Resolved Terahertz (THz) Spectroscopy
(TRTS)   303
11.2.1   Terahertz Generation, Detection and
Time-Resolved Terahertz Spectroscopy
Setup   303
11.2.2   Characteristic Terahertz Responses in
Semiconductors   308
11.3   TRTS Carrier Dynamics Studies of High Efficiency
Photovoltaic Concepts   314
11.3.1   Relaxation of Hot Carriers in
Semiconductor Quantum Dots   315
11.3.2   Carrier Multiplication (CM) in
Semiconductors   316
11.4   Interfacial Electron Transfer in Photovoltaic
Structures Probed by Time-Resolved Terahertz
Spectroscopy   323
11.4.1   Dye Sensitizing Mesoporous Oxide Films   323
11.4.2   Quantum Dot Sensitizing Mesoporous
Oxide Films   327
11.4.3   Quantum Dot Superlattices   329
11.4.4   Exciton Dissociation in Semiconducting
Polymers   330
11.5   Summary   330
References   331

Chapter 12 **X-ray Transient Absorption Spectroscopy for Solar Energy
Research**   **337**
Lin X. Chen

12.1   Introduction   337
12.2   X-ray Transient Absorption Spectroscopy (XTA):
Capabilities and Development   338
12.2.1   Development of X-ray Transient
Absorption Spectroscopy   338

12.3 Applications of XTA in Solar Energy Conversion
     Research: Examples   347
     12.3.1 Metalloporphyrin Excited State Structural
            Dynamics (Homogeneous Electron/Energy
            Transfer)   347
     12.3.2 Photoinduced Interfacial Electron Transfer
            (Heterogeneous Electron Transfer)   352
12.4 Future Research and Development   356
     12.4.1 Visualization of Fundamental Events in
            Photon–Matter Interactions: Capturing the
            Transition States   356
     12.4.2 Detecting Transient Structures with
            Low Concentrations in Photocatalytic or
            Irreversible Processes and Rephasing the
            Coherence of Nuclear Motions in an
            Ensemble   358
     12.4.3 Theoretical Modeling   360
12.5 Summary   362
Acknowledgements   362
References   363

Subject Index   **371**

CHAPTER 1

Computational Modeling of Photocatalytic Cells

STEVEN J. KONEZNY AND VICTOR S. BATISTA*

Department of Chemistry, Yale University, P.O. Box 208107, New Haven, CT 06520-8107, USA
*Email: victor.batista@yale.edu

1.1 Introduction

Solar energy conversion into chemical fuels is one of the "holy grails" of the 21st century. Significant research efforts are currently underway toward understanding natural photosynthesis and artificial biomimetic systems. Photocatalytic cells absorb solar energy and use it to drive catalytic water oxidation at photoanodes:[1,2]

$$2H_2O \rightarrow 4H^+ + 4e^- + O_2(g) \qquad (1.1)$$

effectively extracting reducing equivalents from water (*i.e.*, protons and electrons) that can be used to generate fuel, for example $H_2(g)$ by proton reduction:

$$4H^+ + 4e^- \rightarrow 2H_2(g) \qquad (1.2)$$

The water oxidation half-reaction, introduced by Equation (1.1), is the most challenging obstacle for solar hydrogen production,[3–6] since it requires a four-electron transfer process coupled to the removal of four protons from water molecules to form the oxygen–oxygen bond. In Nature, this process is driven by solar light captured by chlorophyll pigments embedded in the protein antennas

RSC Energy and Environment Series No. 8
Solar Energy Conversion: Dynamics of Interfacial Electron and Excitation Transfer
Edited by Piotr Piotrowiak
© The Royal Society of Chemistry 2013
Published by the Royal Society of Chemistry, www.rsc.org

of photosystem II and the energy harvested is used to oxidize water in the oxygen-evolving complex.[3,7] The development of photocatalytic solar cells made of earth-abundant materials that mimic these mechanisms and photo-catalyze water oxidation has been a long-standing challenge in photoelec-trochemistry research, dating back to before the discovery of ultraviolet water oxidation on n-TiO$_2$ electrodes by Fujishima and Honda 40 years ago.[8] However, progress in the field has been hindered by a lack of fundamental understanding of the underlying elementary processes and the lack of reliable theoretical methods to model the photoconversion mechanisms.

The landscape of research in solar photocatalysis has been rapidly changing in recent years, with a flurry of activity in the development and analysis of catalysts for water oxidation[9–23] and fundamental studies of photocatalysis based on semiconductor surfaces.[24,25] Significant effort is currently focused on the development of more efficient catalysts based on earth-abundant materials and various strategies for the design of molecular assemblies that efficiently couple multielectron photoanodic processes to fuel production. The outstanding challenge is to identify robust materials that could catalyze the necessary multielectron transformations at energies and rates consistent with solar irradiance.

A promising approach for photocatalytic water oxidation involves designing high potential photoanodes based on surface-bound molecular complexes and coupling the anodic multielectron reactions to fuel production at the cathode with long-range free energy gradients. Such a design problem requires funda-mental understanding of the factors affecting photoabsorption, interfacial electron transfer to the photoanode, charge transport, storage of oxidizing equivalents for catalysis at low overpotentials and irreversible carrier collection by fuel-forming reactions at the cathode. The characterization of these processes by computational techniques clearly requires methods for modeling the complete photocatalytic mechanism as well as methods for understanding and characterizing the elementary steps at the detailed molecular level, including visible light sensitization of semiconductor surfaces by molecular adsorbates, charge transport and redox catalytic processes. This chapter reviews recent advances in the field, with emphasis on computational research focused on modeling photocatalytic cells and the elementary processes involved at the molecular level. The reviewed studies are part of an interdisciplinary research program including synthesis, electrochemistry and spectroscopy in a joint theoretical and experimental effort to advance our understanding of structure/function relations in high potential photoanodes based on functio-nalization of nanoporous TiO$_2$ thin films with transition metal catalysts.

The chapter is organized as follows. Section 1.2 reviews recent computational efforts focused on modeling current–voltage characteristics of functional dye-sensitized solar cells (DSSCs) under operational conditions, with emphasis on the effect of the nature of the molecular adsorbates and redox couple on the overall efficiency of photoconversion. Section 1.3 reviews recent developments of methods for inverse design of molecular adsorbers with suitable solar-light photoabsorption. Section 1.4 reviews the development of theoretical models of charge transport in nanoporous metal oxide thin films, with emphasis on the fluctuation-induced tunneling conduction (FITC) model as applied to the description of the temperature dependence of dc and ac conductivities and direct comparisons to experimental data. Section 1.5 is focused on reliable methods for modeling the redox properties of molecular adsorbates, with emphasis on the reduction of systematic errors introduced by either the level of theory (*i.e.*, the choice of density functional theory (DFT) functional, basis set and solvation model) or the electrochemical measurement conditions, including the nature of the solvent, electrolyte and working electrode. Section 1.6 presents a summary of the conclusions and outlook.

1.2 Photoelectrochemical Device Modeling

Modeling can provide a fundamental insight into the effects of individual system components on the overall device functionality.[26] A parameter-space analysis by systematic variation in device composition can lead to the discovery of assemblies with optimum performance. Insight into observed trends can be extracted from the parameters of equivalent-circuit current–voltage simulations. Here, we illustrate this systematic approach by analyzing of a series of high-potential porphyrin photoanodes, metal oxides and redox couples, suitable for photocatalytic cells. We show that DSSCs based on porphyrin dyes demonstrate superior performance owing to increased open-circuit voltage and the short-circuit current when using the relatively high-potential bromide–tribromide redox couple as a regenerative electron mediator rather than the standard iodide–triiodide couple. The resulting potentials progress toward sufficiently positive values suitable for water oxidation chemistry.

1.2.1 Modeling Current-Voltage Characteristics

The current–voltage characteristics, that is the current density J as a function of the applied voltage V, of photoelectrochemical cells are used to extract important parameters related to device performance. In the case of a DSSC, parameters of interest include the short-circuit current $J_{SC} \equiv J(V=0)$, open circuit voltage $V_{OC} \equiv V(J=0)$ and the power conversion efficiency. However, owing to the complex nature of DSSC operation, these data alone offer a limited understanding of the underlying physical/electrochemical processes. Device modeling of current–voltage characteristics is needed in order to develop this understanding and establish property–performance relationships. This kind of an analysis is useful not only to DSSCs but to photoelectrochemical cells in general since they share common system components.

The equivalent circuit diagram used to model solar cell current–voltage characteristics is shown at the top of Figure 1.1. The schematic energy level diagram of a DSSC at the bottom of Figure 1.1 shows the various charge transfer processes that occur in photoelectrochemical cells and relates these processes to current pathways *via* components of the model circuit. An illumination current density J_L is induced upon photoexcitation of the

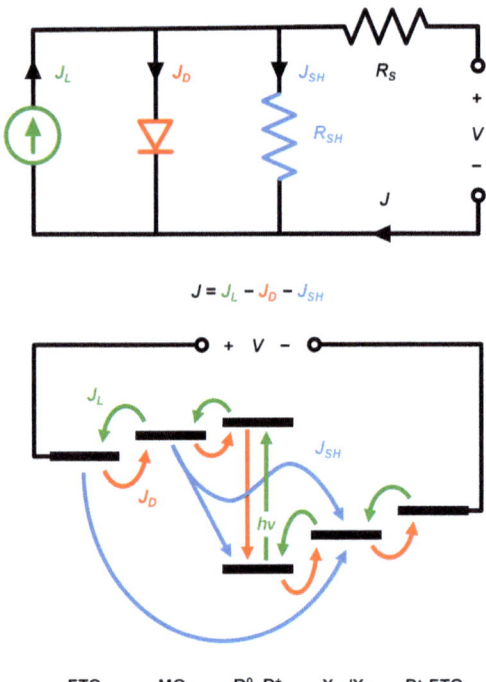

Figure 1.1 Simple equivalent circuit (top) for modeling solar cell current–voltage characteristics and energy level diagram (bottom) mapping the various charge transfer processes in a DSSC to the current pathways of the model circuit. The dominant mechanisms are described by a current density J_L induced upon photoexcitation and electron injection into the conduction band of the metal oxide semiconductor surface MO, linear (J_{SH}) and non-linear (J_D) reverse current densities in parallel with photocurrent source and a series resistance R_S to account for electrode and ionic resistances. In Section 1.2.2: MO = TiO$_2$, SnO$_2$, X = Br, I.

adsorbed dye molecule, followed by interfacial electron injection into the conduction band of the metal oxide semiconductor. Direct recombination of the photoinjected carriers with redox species in the electrolyte solution and the dark current along the edges of the cell contribute to the shunt resistance R_{SH} of the cell. Under applied bias, the cell can be modeled as a diode with a current density J_D that runs parallel to J_L and shunt current density J_{SH} that opposes J_L. The cell also has a net series resistance R_S, which includes the resistance of the metal oxide and the ionic resistance of the redox pair in the electrolyte.

The output current density J of the solar cell as a function of applied bias voltage in the equivalent circuit model is:[26,27]

$$J = J_L - J_0 \left\{ \exp\left[\frac{q(V + JAR_S)}{fkT} \right] - 1 \right\} - \frac{V + JAR_S}{AR_{SH}}, \qquad (1.3)$$

where J_0 is the reverse saturation current, k is the Boltzmann constant, T is the absolute temperature, q is the electronic charge, A is the device area and f is the ideality factor. Equation (1.3), which was originally developed to describe the non-ideal diode behavior of inorganic semiconductor *p–n* junctions, has been applied to a wide range of solar cell technologies including hydrogenated amorphous silicon p–i–n cells, Cu(In,Ga)Se$_2$ cells, organic bulk heterojunction cells and DSSCs.[26–29] In the case of photoelectrochemical cells, J_0 is related to the energy level alignment between the bottom of the conduction band of the metal oxide and the excited state donor level of the dye D*. This differs from J_L, which is related to the coupling between D* and states in the metal oxide that are closer in energy and therefore typically deeper in the conduction band. This means that a clear strategy for improving device performance is to adjust the energetic position of D* relative to the metal oxide density of states to achieve maximum J_L and minimum J_0. The benefits of this approach are apparent in Equation (1.3), which gives an estimate of the open circuit voltage of the cell in the regime of high shunt resistance:

$$V_{OC} \approx \frac{fkT}{q} \ln\left(\frac{J_L}{J_0} + 1\right) \tag{1.4}$$

We note that this model could be extended to incorporate electrochemical processes. For example, the electrochemical kinetics at the electrodes is more appropriately described by the Butler–Volmer equation with non-linear *J–V* characteristics rather than an ohmic resistor. However, as shown in Section 1.2.2, much insight can be gained when using the equivalent circuit model of Figure 1.1 and Equation (1.3) to do a comparative analysis of a series of device architectures in which the relevant energy levels are systematically varied.

1.2.2 Bioinspired High-Potential Porphyrin Photoanodes

Table 1.1 compares the equivalent circuit parameters that characterize the performance of solar cells based on sensitizers **1**, **2** and **3** (Figure 1.2) and N719. The analysis includes solar cells with either I_3^-/I^- or Br_3^-/Br^- redox couples as regenerative electron mediators and TiO$_2$ or SnO$_2$ nanocrystalline substrates. It is shown that the series resistance increases when the TiO$_2$ photoanode is replaced by SnO$_2$, for a given dye and electrolyte, consistent with lower *in vacuo* room temperature nanoporous-film dark conductivities in TiO$_2$ ($\sim 10^{-10}\,\Omega^{-1}\,cm^{-1}$; Figure 1.7) compared to SnO$_2$ ($\sim 10^{-6}\,\Omega^{-1}\,cm^{-1}$; Figure 1.10). In addition, the series resistance R_S decreases for iodide relative to bromide for dyes **1–3**, consistent with the higher conductivity of iodide when compared to bromide at low concentrations.[30] In addition to changes caused by the intrinsic properties of the ions, the substitution of I$^-$ by Br$^-$ leads to a reduction in the saturation recombination current J_0 and an increase in the open circuit voltage V_{OC}, as observed for solar cells based on sensitizers **1–3**. These changes in the current–voltage characteristics are due to a tighter binding of bromide to the porphyrin adsorbates, as shown by the analysis of the electrostatic potentials.

Table 1.1 Solar cell performance parameters including the parameters of the equivalent-circuit model, short-circuit current density J_{SC}, open-circuit voltage V_{OC}, fill factor FF and solar-to-electrical energy conversion efficiency η at $100\,mW\,cm^{-2}$ illumination (AM 1.5).

Dye	Oxide	Electrolyte	J_L (mA/cm²)	J_0 (mA/cm²)	AR_S (Ωcm²)	AR_{SH} (kΩcm²)	f	J_{SC} (mA/cm²)	V_{OC} (V)	FF	η (%)
1	TiO₂	I_3^-/I^-	0.15	3.81×10^{-6}	4.60	2.74	1.72	0.14	0.37	0.31	0.02
		Br_3^-/Br^-	0.13	1.59×10^{-7}	19.5	4.64	2.23	0.13	0.62	0.26	0.02
	SnO₂	I_3^-/I^-	0.97	1.47×10^{-2}	4.72	1.23	3.18	1.01	0.31	0.32	0.10
		Br_3^-/Br^-	2.58	3.35×10^{-4}	77.8	9.79	2.81	2.53	0.64	0.47	0.76
2	TiO₂	I_3^-/I^-	0.93	4.20×10^{-6}	5.65	1.12	1.84	0.90	0.55	0.51	0.25
		Br_3^-/Br^-	0.17	9.47×10^{-7}	27.9	4.15	2.43	0.21	0.69	0.42	0.06
	SnO₂	I_3^-/I^-	2.83	4.82×10^{-6}	9.31	0.16	1.34	2.51	0.37	0.41	0.39
		Br_3^-/Br^-	1.35	1.16×10^{-6}	138	24.7	1.62	1.32	0.59	0.51	0.40
3	TiO₂	I_3^-/I^-	0.26	2.15×10^{-3}	2.69	0.22	3.18	0.25	0.37	0.26	0.02
		Br_3^-/Br^-	0.17	5.46×10^{-9}	30.4	3.81	1.76	0.16	0.62	0.30	0.03
	SnO₂	I_3^-/I^-	0.76	1.85×10^{-2}	4.85	11.3	3.46	0.81	0.31	0.37	0.09
		Br_3^-/Br^-	3.48	3.13×10^{-5}	69.4	2.47	2.20	3.37	0.66	0.45	1.00
N719	TiO₂	I_3^-/I^-	9.62	5.53×10^{-3}	3.96	45.9	3.97	9.59	0.77	0.61	4.48
		Br_3^-/Br^-	0.16	7.09×10^{-5}	9.58	7.95	3.63	0.15	0.67	0.45	0.05
	SnO₂	I_3^-/I^-	11.1	3.26×10^{-2}	16.3	0.30	2.92	10.33	0.46	0.38	1.82
		Br_3^-/Br^-	0.25	6.95×10^{-5}	9.97	5.03	3.46	0.04	0.08	0.71	0.00

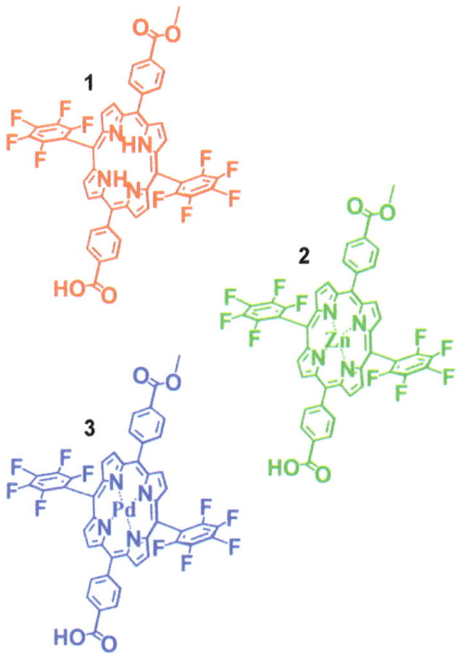

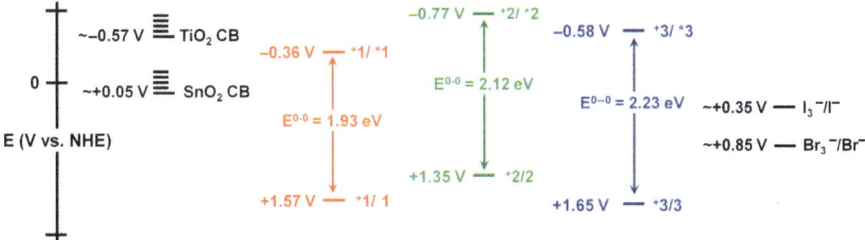

Figure 1.2   Compounds **1**, **2** and **3**.

Figure 1.3   Energy level diagrams illustrating reduction potentials of relevant half reactions. Ground state and excited state potentials of porphyrins **1** (red), **2** (green) and **3** (blue) were determined by cyclic voltammetry measurements together with absorption and emission spectra. Approximate potentials for the TiO$_2$ and SnO$_2$ conduction bands[31,32] and the I$_3^-$/I$^-$ and Br$_3^-$/Br$^-$ couples[33,34] are also shown. The O$_2$/H$_2$O couple is 0.82 V at pH = 7. All potentials are reported in V *vs.* the normal hydrogen electrode (NHE).

The analysis of electrostatic potentials suggest that Br$^-$ ions interact more strongly than I$^-$ and bind more closely to the aromatic rings owing to their smaller ionic radius (Br$^-$: 1.82 Å, I$^-$: 2.06 Å). The resulting stabilization includes anion π interactions with the permanent quadrupole moments of the aromatic rings.[35,36] Up to 4 Br$^-$ anions per porphyrin are predicted to bind to **1** and **3** with binding energies in the –5 to –13 kcal mol^{-1} range. Two Br$^-$ ions

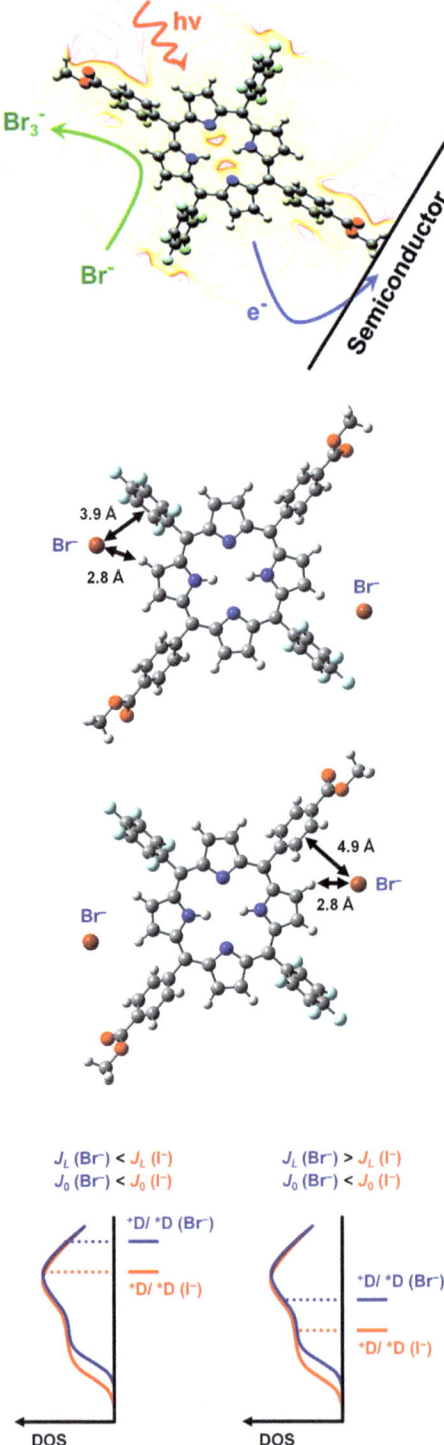

bind to the pentafluorophenyl groups while the other two bind to the carbo-methoxyphenyl groups (Figure 1.4). In addition, Br^- can bind to Zn in **2**, with a binding energy of -12.5 kcal mol^{-1}. However, metal–anion interactions for $Zn - I^-$ and $Pd - Br^-$ are much weaker (*i.e.*, comparable to thermal fluctuations).[26] In contrast to bromide, iodide ions show much weaker specific interactions with **1–3** probably due to the larger ionic radius.

The analysis of adsorbate ions interactions provided fundamental understanding of the origin of changes in the I–V characteristics induced by changes in the nature of the redox couple. The larger concentration of smaller ions shifts the electron donor and acceptor states and the edge of the conduction band to more negative potentials, preventing recombination and reducing J_0 (Figure 1.4). Therefore, the open circuit voltage V_{OC} is increased and the slope of the characteristic curve is reduced as $V \rightarrow 0$. The illumination current density increases (or decreases) when the donor state is poised where the conduction band has a larger (or smaller) density of states.

The comparative analysis of porphyrin-based photoanodes suggests that new high-potential sensitizers and device architectures can significantly expand the available parameter space of DSSC current–voltage simulation methods. In particular, a suitable choice of electron-withdrawing substitution groups, or transition metal, can shift the porphyrin redox potential and make it sufficiently positive to activate water oxidation catalysts.[26,34] However, the porphyrin excited state which injects the photoexcited electrons into the metal oxide conduction band typically experiences a similar stabilization when changing the metal center or the substitution groups, giving similar vertical excitation bands as seen in the absorption spectra (Figure 1.3). Therefore, it is natural to expect that adsorbates that generate deep positive holes by excitation with visible light would photoinject electrons only into metal oxides with conduction bands that are more positive than the NHE (*e.g.*, SnO_2). Achieving efficient water oxidation and proton reduction by visible light photocatalysis on sensitized metal oxide surfaces thus might require two-photon schemes.[37] A simple example is shown in Figure 1.5 where red photons inject electrons into the photocathode at potentials more negative than the NHE, while blue photons generate holes in

---

Figure 1.4   Top: Illustration of photoinduced electron transfer into the semiconductor conduction band. Middle: Bromide binding to the bis-pentafluorophenyl free-baseporphyrin sensitizer bearing linkers for attachment to metal oxide surfaces: two equivalent low-energy positions at 3.9 Å from the pentafluorophenyl groups, stabilized by π-anion interactions with a stabilization energy of -7.9 kcal mol^{-1} and two similar positions at 4.9 Å from the carbomethoxyphenyl groups stabilized by -4.9 kcal mol^{-1}. Bottom: Shift of the edge of the conduction band upon substitution of I_3^-/I^- (dotted lines) by Br_3^-/Br^- (solid lines) in solar cells based on **1**, **2**, or **3**, using TiO_2 or SnO_2 nanocrystalline substrates. Such shift leads to a reduction of J_0 since the barrier for back electron transfer, from the bottom of the metal oxide conduction band to the donor state, is increased. The illumination current density J_L is smaller (left) or larger (right) in Br_3^-/Br^-, according to the smaller or larger amplitude of the DOS at the energy of the donor state.

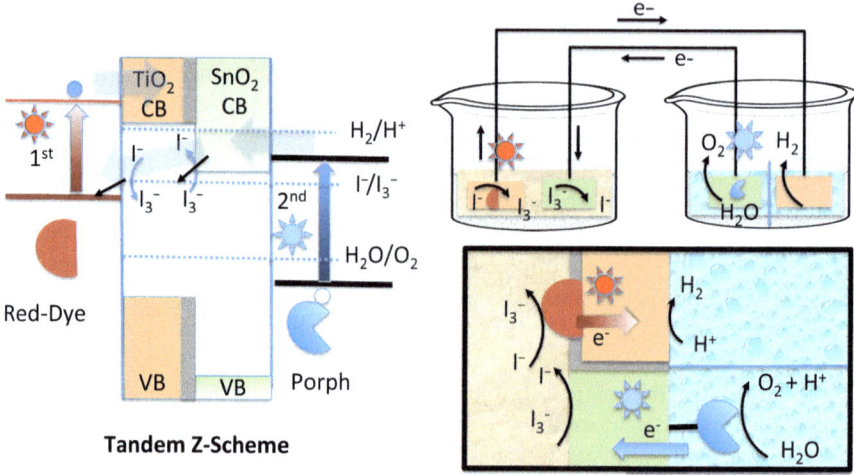

Figure 1.5 Energy diagram (left) and schematic representation (right) of a photo-catalytic cell based on a 2-photon tandem Z-scheme with high potential porphyrins covalently attached to SnO_2.

porphyrins adsorbed on the photoanode, sufficiently positive as to activate water oxidation catalysts.

1.3 Inverse Design of Photoabsorbers

The search for molecular adsorbates for high potential photoanodes with high photoconversion efficiency is challenging. Most of the work reported to date has been based on direct molecular design using empirical strategies, where typical "guess-and-check" procedures face the challenge of selecting suitable candidates from an immense number of accessible stable molecules. Therefore, there is significant interest in the development of inverse design methods that bypass the combinatorial problem in the development of materials for solar energy conversion.

In recent work, a systematic inverse design methodology suitable to assist the synthesis and optimization of molecular sensitizers for dye sensitized solar cells has been proposed.[38] The method searches for molecular adsorbates with suitable photoabsorption properties through continuous optimization of "alchemical" structures in the vicinity of a reference molecular framework, avoiding the exponential scaling problem of high throughput screening techniques. It has been as applied to the design and optimization of linker chromophores for TiO_2 sensitization using the recently developed phenylace-tylacetonate (*i.e.*, phenylacac) anchor[39] as a reference framework. A novel anchor (3-acac-pyran-2-one) was found to be a local optimum, with improved sensitization properties compared to phenyl-acac. Its molecular structure is re-lated to known coumarin dyes which could be used as lead chromophore anchors for practical applications in DSSCs. The findings are particularly relevant to the

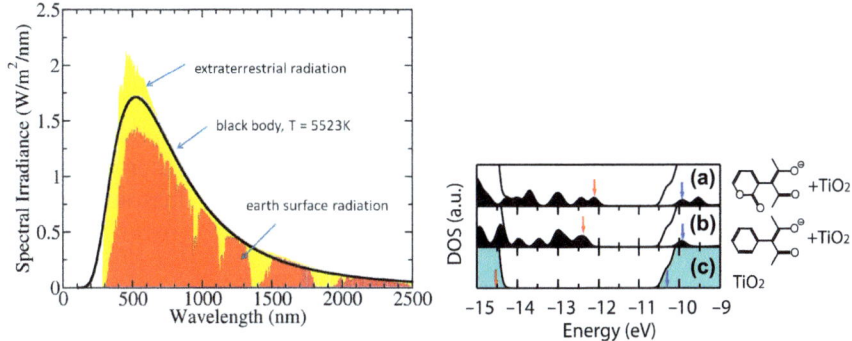

Figure 1.6 Right: Density of states (DOS) of a bare TiO_2 anatase nanostructure (a) and functionalized with phenyl-acac (b) and 3-acac-pyran-2-one (c). The solid-black areas in (b) and (c) are the projected DOS of the adsorbate molecules. Red and blue arrows point to electronic states with a predominant contributions to the absorption. Left: Solar spectrum model based on a black body at 5523 K.[38]

design of sensitizers for DSSCs because of the wide variety of structures that are possible and they should be equally useful for other applications such as ligand design for catalytic nanocrystalline structures (Figure 1.6).

The proposed methods use linear combination of atomic potential Hamiltonians

$$H_{i,j} = \sum_{A=1}^{N_{type}^i} \sum_{A'=1}^{N_{type}^j} b_i^{(A)} b_j^{(A')} h_{i,j}^{(A,A')},$$

defined in terms of tight-binding model Hamiltonians $h_{i,j}^{(A,A')}$ and participation coefficients $b_i^{(A)}$ that give the probability weight for the various possible atom types A at atomic sites i in the molecule. While for real molecules these coefficients $b_i^{(A)}$ are either 1 or 0, since they must define pure atom types, they can change continuously *in silico* (between 0 and 1) while the molecule undergoes transformations in the "alchemist" space of intermediate states. The evolution is typically steered by gradients that optimize the expectation value of the molecular properties of interest. Upon convergence, optimum structures typically have atomic sites with fractional $b_i^{(A)}$ that are readily rounded off to obtain the closest possible real molecule with optimized properties, assessed by using high-level quantum chemistry methods.

Photoanodes are designed for optimum activation with solar light, quantifying the photoabsorption intensity, $f_{sum} = \sum_{pq} = f_{pq} \times P(\lambda_{pq})$, as the sum of products of the solar power spectrum $P(\lambda_{pq})$ and the oscillator strengths $f_{pq} = 8\pi^2 v_{pq} m_w c |\mu_{pq}|^2/(3he^2)$ associated with electronic state transitions $q \leftarrow p$ with wavelengths λ_{pq}. The photoabsorption properties are computed as generalized expectation values,

$$\bar{O} = \sum_{i=1}^{N} \sum_{j=1}^{N} \sum_{A=1}^{N_{type}^{i}} \sum_{B=1}^{N_{type}^{j}} Q_i^{*(A)} Q_j^{(B)} O_{i,j}^{(A,B)}$$

in the basis of states obtained by solving the Schrödinger equation: $\mathbf{H\,Q} = \mathbf{E\,S\,Q}$, where $\mathbf{Q}$ is typically the matrix of eigenvectors in the basis of Slater-type atomic orbitals (STO), $\mathbf{E}$ is the diagonal matrix of energy eigenvalues and $\mathbf{S}$ is the overlap matrix

$$S_{i\alpha,j\beta} = \sum_{A=1}^{N_{type}^{i}} \sum_{B=1}^{N_{type}^{j}} b_i^{*(A)} b_j^{(B)} S_{i\alpha,j\beta}^{(A,B)}$$

While computationally demanding, these methods can be parallelized and scale only as $(NN_{type})^2$, with N and N_{type} the number of atoms and atom types, respectively. In contrast, assessing all possible N_{type}^N structures by brute force screening would obviously scale exponentially with N_{type}. Therefore, it is natural to anticipate that this kind of systematic methodology will provide powerful computational methods as well as guidelines for design of new materials for efficient solar energy conversion.

1.4   Charge Transport in Nanoporous Metal Oxides

The electronic mechanisms responsible for dark conductivity in nanoporous metal oxide materials control the efficiency of energy photoconversion in a wide range of applications, including photovoltaic devices,[40–42] photocatalysis and remediation of hazardous waste,[41–44] electrochromic windows and displays,[42,45,46] and chemical sensors.[42,47] Measurements in the 78–335 K temperature range show dc conductivity values spanning over four orders of magnitude, with a high-temperature Arrhenius dependence that gradually changes into a temperature-independent plateau at low temperatures. It has recently been shown that a FITC mechanism is fully consistent with TiO_2 and SnO_2 conductivity measurements over this entire temperature range.[48,49] These findings suggest that charge transport in nanoporous metal oxide films depends strongly on the properties of the junctions linking the constituent nanoparticles.

In recent years, there have been significant contributions towards understanding charge transport in nanoporous titania thin films.[50–60] The most popular models include variable range hopping (VRH) and/or multiple trap and release (MTR) of electrons in an electrically homogeneous medium containing a distribution of traps. These models have been particularly successful when applied to the conductivity of photogenerated electrons at temperatures experimentally accessible in a native device environment. However, they predict a temperature (T) dependence of the dark DC conductivity of the form $\ln \sigma \propto T^{-\alpha}$, where $\alpha = 1$ in the MTR model[61] and $a = 1/4$ for VRH.[62] As shown in Figure 1.7, this dependence does not account for the observed saturation of dark conductivity at low temperature, as reported here and elsewhere.[63–66] In contrast, the FITC model offers a quantitative description of conductivity over

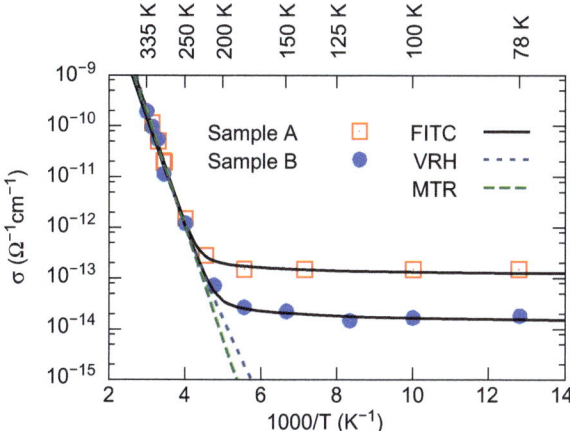

Figure 1.7 Comparison of the dark DC conductivities of nanoporous TiO_2 films, made by sintering Sigma-Aldrich (Sample A, filled circles) and Ishihara (Sample B, open squares) nanoparticles, as a function of inverse temperature (temperatures in K are shown on the top axis). Experimental data (solid circles and red boxes) are fitted by using the fluctuation-induced tunneling conduction (FITC) model (solid line), the variable-range hopping model (VRH, short-dashed line) and the multiple-trap-and-release model (MTR, long-dashed line).

the entire temperature range with a single set of structural parameters, predicting not only the Arrhenius high-temperature behavior but also the tunneling regime at low temperature. Since the model can be closely tied to the nanoporous film microstructure, it should provide valuable insight in the development of high performance electrode materials.

Most conductivity measurements reported to date have been limited to the high-temperature regime since the low-temperature conductivity is typically very small ($\sigma \approx 10^{-13} \Omega^{-1} cm^{-1}$) and, therefore, difficult to measure. In the limit, the Arrhenius-type behavior predicted by the FITC model is indistinguishable from models based solely on thermally activated processes. Thus, it is not surprising that, until recently, the FITC mechanism has been largely overlooked.[48,49] Recent work, however, has been focused on high sensitivity measurements that allow for reliable measurements at very low temperatures and on the characterization of the samples by scanning electron microscope (SEM) images, powder X-ray diffraction (XRD) measurements and atomistic modeling. The combined experimental and theoretical analysis provided insight into the nature of conductivity in sintered TiO_2 thin films, as characterized by the FITC mechanism, with rate-limiting processes determined by the barriers for electron transport through the nanoparticle contact junctions.[67]

1.4.1 Fluctuation-Induced Tunneling Conductivity

The FITC model for dc conductivity is described in our previous work[48,49,68, 69] and in other studies that apply a FITC model to systems with a comparable

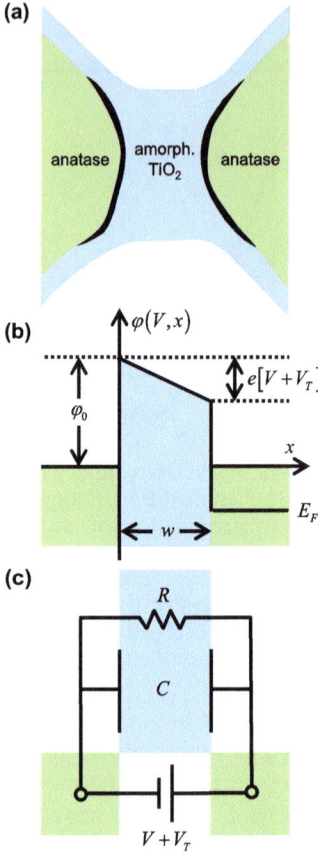

Figure 1.8   (a) Contact junction of two noncrystalline TiO_2 nanoparticles (blue) sintered by disordered TiO_2 (green). (b) Energy diagram for electron transport through the junction, showing the potential energy barrier under a bias $V + V_T$, where V is the applied voltage and V_T is the voltage fluctuation caused by thermal fluctuations in the density of free electrons at the junction. (c) Equivalent RC circuit of the contact junction.

microstructure.[70–73] A nanoporous metal oxide film can be modeled as a network of junctions that form where the nanoparticles come into contact [Figure 1.8(a)]. An important component of the FITC mechanism is that the junction is electrically inhomogeneous. Molecular dynamics simulation results for sintered TiO_2 nanoparticles show that anatase TiO_2 nanoparticles have a non-crystalline TiO_2 shell surrounding a crystalline core (Figure 1.9), which is consistent with reports of TiO_2 nanoparticle sizes, obtained from microscopy data, being larger than the size of the crystalline centers of the nanoparticles size, determined by XRD.[48,74] The conduction band offset between the crystalline and non-crystalline TiO_2 phases creates a barrier for electron transport. The fact that our data are consistent with the FITC model suggests that this

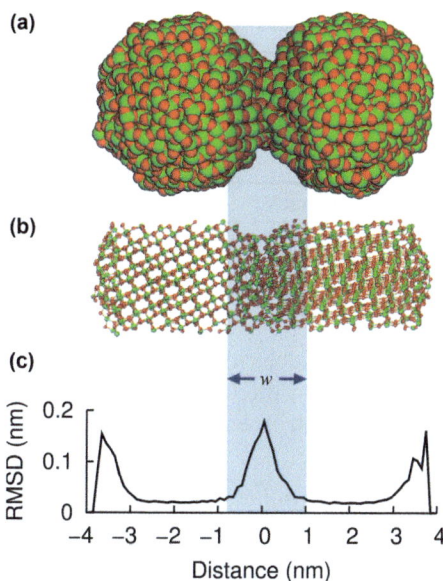

Figure 1.9 (a) Molecular dynamics simulation results of two 4-nm particles sintered at 450 °C. (b) Cylindrical sample cut from (a) showing the crystalline anatase core and noncrystalline shell of the nanoparticles. (c) Average atomic root-mean-squared deviation (RMSD) relative to bulk anatase.

core shell structure applies to SnO_2 nanoparticles and, by extension, other nanoporous metal oxides as well.

The FITC model describes the nanoparticle contact junction as an effective potential energy barrier for the transferring electron, owing to the intrinsic disorder of the semiconductor material between the nanocrystalline particles [Figure 1.8(b)].[48] At a given temperature, the electron transmission probability depends on the zero-field barrier height φ_0, effective width w and junction area A. In addition, fluctuations in the free-electron density on either side of the junction give rise to large voltage fluctuations that modulate the effective tunneling barrier. Typical parameters for TiO_2 nanoparticle junctions are given in Table 1.2. For a static dielectric constant of $\varepsilon_r \approx 20$ inside the junction, the effective junction capacitance $C = \varepsilon_r \varepsilon_0 A/w$ [Figure 1.8(c)] is on the order 10^{-6} pF. Therefore, small thermal fluctuations in the density of free electrons near the junction give rise to significant voltage fluctuations with $\langle V_T^2 \rangle = k_B T / C$.

As a function of the electric field $\mathscr{E}$, the dark dc conductivity can be expressed as:

$$\sigma_{dc}(\mathscr{E}) = \frac{\gamma \langle j_{dc}(\mathscr{E}) \rangle A}{tV} \tag{1.5}$$

where V is the voltage across the junction, t is the thickness of the sample, $\langle j_{dc}(\mathscr{E}) \rangle$ is the thermal average of the net dc current density and γ is a factor that relates the conductance of a single junction to the sheet conductance of the film.

Table 1.2   FITC model parameters used for samples A and B in Figure 1.7, including the effective area A, width w, zero-field barrier height φ_0 of the junction and the effective junction diameter given by $d_j \equiv 2\sqrt{A/\pi}$ compared to the nanocrystallite diameter d_{nc} measured by XRD.[48]

	d_{nc} (nm)	φ_0 (meV)	w (nm)	A (nm^2)	d_j (nm)
A	17.0	421	3.45	71.6	9.54
B	15.9	388	3.73	22.5	5.35

The tunneling current density is determined by the density of electrons per unit time incident on the junction and the transmission probability as a function of energy. Integrating over all electron energies one obtains:[48,68]

$$j_{dc}(\mathscr{E}) = \frac{mqk_B^2 T^2}{2\pi^2\hbar^3} \left(\left\{ \frac{e^{-2\chi w\xi}}{T'^2}(1 - e^{-T'q\mathscr{E}w/k_BT}) \right\} + \left\{ \frac{e^{-2\chi w\xi}}{1+T'}(1 - e^{-T'q\mathscr{E}w/k_BT}) \right\} \right)$$

$$+ \left\{ e^{-\varphi_m/k_BT}(1 - e^{-q\mathscr{E}w/k_BT}) \right\} + \left\{ \frac{e^{-2\chi w\xi}}{1+T'}[(1 - e^{-(1-T')\varphi_m/k_BT}) \right.$$

$$\left. - (1 - e^{-(1-T')(\varphi_m + q\mathscr{E}w)/k_BT})e^{-T'q\mathscr{E}w/k_BT}] \right\},$$

$$(1.6)$$

where m is the charge carrier mass, $\hbar$ is the reduced Planck constant, $\chi = (2m\varphi_0/\hbar)^{1/2}$ is the tunneling constant, q is the electron charge, the dimensionless temperature-dependent parameter T' is given by $2\chi w\eta k_B T/\varphi_0$ and φ_m is the maximum in the potential barrier. ξ and η are field-dependent dimensionless parameters that originate from the first two terms in a power series expansion of the exponent of the transmission coefficient under the WKB approximation.[75] Equations (1.5) and (1.6) provide a more explicit form of the dc conductivity than the approximation $\sigma \approx \sigma_0 \exp[-T/(T_0 + T)]$, where σ_0, T_0 and T_1 are temperature-independent parameters. The simplified expression has a limited range of validity and can lead to inaccurate estimations of the contact junction parameters.

Equation (1.6) accounts for both the Arrhenius regime and the temperature-independent low-temperature behavior, as described by the fluctuation-induced tunneling conductivity model. Each of the terms in curly brackets include a description of the forward current density component, in the direction of the applied electric field and a backflow current density in the opposite direction. The first term corresponds to the net current in the low-temperature limit, with an abrupt change in the density of states at the Fermi energy, while the other terms are corrections caused by expansion of the Fermi–Dirac distribution to first order in temperature.

Figure 1.7 compares the DC conductivity of nanoporous TiO$_2$ thin films and the corresponding conductivity curves obtained according to the FITC model with structural parameters given in Table 1.2. Note that the FITC model describes the conductivity data over the entire temperature range, including both the thermally activated high-temperature regime ($T > 250$ K) and the temperature-independent regime ($T < 150$ K) in quantitative agreement with experiments. In contrast, the VRH and MTR models can only account for one of the two regimes with a unique set of parameters. The high-temperature conductivity for sample B gives an activation energy of 439 meV and a pre-exponential factor of 7.4×10^{-4} $\Omega^{-1}cm^{-1}$, assuming a thermally activated process (MTR model), while the VRH model gives $\sigma = \sigma_0 \exp[-(T_0/T)^{1/4}]$ with $\sigma_0 = 1.3 \times 10^{20}$ $\Omega^{-1}cm^{-1}$ and $T_0 = 7.4 \times 10^9$ K. The effective zero-field barrier heights of 420 and 390 mV predicted by the FITC model for samples A and B, respectively, are similar to the activation energies E_a measured for the high-temperature data in various studies.[63–66] The slope of ln σ vs. $1/T$ in the high-temperature regime is largely determined by φ_0 and therefore $\varphi_0 \approx E_a$.

The value of φ_0 depends on the distribution of traps below the conduction band, typically attributed to oxygen vacancies, Ti_3^+ states and interface states.[63–66,76] Therefore, φ_0 depends on measurement and fabrication conditions such as the sintering time and exposure to ambient oxygen and N-doping.[64–66,76] In fact, the experimental data show that the low-temperature plateau in the conductivity of sample A is nearly an order of magnitude lower than that of sample B, reflecting both a larger barrier height and a wider tunneling width (see Table 1.3). Increasing φ_0 typically decreases the low-temperature plateau conductivity and increases the slope of ln σ vs. $1/T$ in the high-temperature region. Increasing the tunneling width w only reduces the former,[69] while increasing the effective junction area A increases the conductivity for the entire temperature range, with a negligible effect on the high-temperature slope. Therefore, it is clear that the parameters A and φ_0 determine the dark dc conductivity and can be tuned to optimize performance. The challenge ahead is thus to find materials that generate lower conduction band offsets, or fabrication conditions that optimize the defect distributions at the contact junctions.

Table 1.3 Power law exponent s for the frequency-dependent component of the ac conductivity measured on different days and the FITC model parameters extracted from the dc component for the sintered SnO_2 nanoparticle film. A, w and φ_0 are the effective area, width and zero-field barrier height of the junction, respectively and d_j is the effective junction diameter given by $d_j \equiv 2\sqrt{A/\pi}$.

Day	s	φ_0 (meV)	w (nm)	A (nm^2)	d_j (nm)
1	1.23	166	3.13	5.58	2.67
9	1.13	137	3.68	44.2	7.50
11	1.09	113	3.49	11.6	3.83
20	σ_{dc} only	102	3.43	29.7	6.14

The FITC model parameters have been compared to atomistic models of the contact junctions, obtained by sintering anatase nanoparticles using annealing molecular dynamics simulations, after thermalization of the system at 450 °C. Typical simulations of the sintering process were performed by using the LAMMPS package[77] in the NVT ensemble and the force field by Matsui and Akaogi[78] to describe the interactions between the atoms of the nanoparticle as in previous work.[79] The temperature was maintained at 450 °C using a Nose-Hoover thermostat with a damping time of 10 fs. The atomic positions and velocities were updated using the velocity-Verlet algorithm with a time step of 1 fs and a total run time of 250 ps. Atomic root mean square deviations (RMSD) were quantified to characterize the extension of the contact junctions by alignment of the anatase crystal structure to the crystalline core of each nanoparticle. Figure 1.9(a) shows a representative configuration of two sintered TiO_2 nanoparticles, obtained by annealing molecular dynamics simulations at 450 °C. A cylindrical sample extracted from these two nanoparticles, shown in Figure 1.9(b) reveals the internal atomic structure with non-crystalline TiO_2 at the contact junction and nanoparticle surface. The analysis of atomic RMSDs relative to the anatase crystal structure shown in [Figure 1.9(c)] quantifies the disorder observed in the non-crystalline phase at the junction and extremities of the cylindrical core. These RMSDs clearly indicate that the TiO_2 nanoparticles have a non-crystalline shell surrounding the crystalline core. This core shell structure is consistent with reports of the overall TiO_2 nanoparticle size, obtained from microscopy data, being larger than the nanocrystallite size, determined by XRD.[48,74] These results also agree with conductivity measurements of unsintered, spin-coated nanoparticle thin films that display both thermally activated and temperature-independent conductivity regimes,[63] suggesting fluctuation-induced carrier tunneling between crystalline cores through non-crystalline shells.

SEM and XRD data, as well as computational structural models, show that tunneling junctions are formed upon sintering TiO_2 nanoparticles, owing to the resulting multiphase composition at the contact.[48] The tunneling barrier height is determined by the energy difference between the conduction band of crystalline anatase and the electronic states of disordered TiO_2 at contact (see Figure 1.8). This picture is consistent with optical pump/THz probe measurements providing evidence that the AC conductivity is within a factor of 2 to 4 of the single crystal DC value, although the DC conductivity of photoexcited nanocrystalline TiO_2 and ZnO is suppressed relative to bulk values.[52, 80] These results are therefore consistent with conductivity limited by transport through disordered TiO_2 at the contact junctions, while carriers are relatively free within the crystalline nanoparticles. These observations are limited to vacuum conditions, while the presence of impurities, gas, or an electrolyte at the junctions might have a significant influence on the shape of the barriers and therefore on the overall conductivity through the nanoparticle network. It is therefore imperative to explore these effects since they are of great technological relevance.

1.4.2 Power Law Dependence of the Dark AC Conductivity

The total measured ac conductivity can often be separated into frequency-dependent and dc components and expressed as:

$$\sigma(\omega) = \sigma_{dc} + \sigma'(\omega) \qquad (1.7)$$

where $\sigma'(\omega)$ is the frequency-dependent component of the total conductivity σ. In a wide variety of disordered materials, including the systems studied in this work, Equation (1.7) is valid and $\sigma'(\omega)$ has a power law form:[81–83]

$$\sigma'(\omega) = c\omega^s \qquad (1.8)$$

where both the power law coefficient c and exponent s have temperature and frequency dependences that are indicative of the underlying charge transport mechanism. The functional forms of these parameters have been derived for many different models based on quantum mechanical tunneling or hopping. For example, the VRH model predicts a temperature-independent though frequency-dependent s, with a value of 0.8 for a typical characteristic relaxation time of $10^{-13}\,\mathrm{s}^{-1}$ and a frequency of 10 kHz.[82] The correlated barrier hopping (CBH) model also has a frequency-dependent s, though its value tends toward $s = 1$ as $T \rightarrow 0$ K.[83] Current efforts are focused on deriving functional forms of c and s based on FITC. The present study focuses on extractions of the dc component, which was discussed in the previous section.

Figure 1.10 shows the temperature dependence of the dc conductivity on several measurement days spanning a 20-day period. The conductivity of the sample steadily grows with time and saturates after 20–25 days.[49] Such aging of the sintered SnO_2 film is not surprising, as it is well known that the nanoporous semiconductor films are highly sensitive to the environment owing to their large

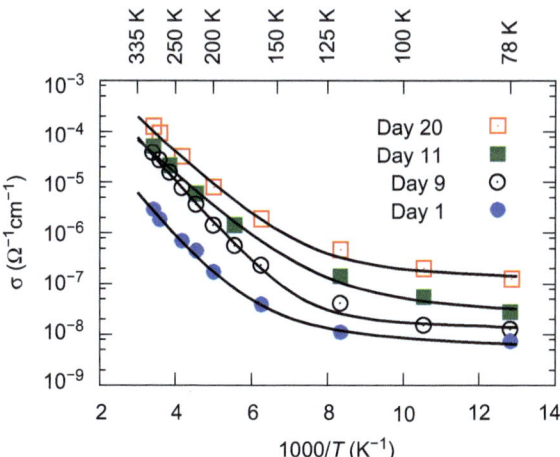

Figure 1.10 Temperature dependence of the dc conductivity of the SnO_2 sample. Solid lines illustrate the FITC model fitting. The corresponding fitting parameters are listed in Table 1.3.

surface areas, a fact widely exploited in chemical sensor applications.[42,47] We ascribe the aging to changes (under ambient conditions) in the concentration of trap states, which have been attributed to oxygen vacancies, Ti_3^+ states and interface states.[76] In the FITC model, φ_0 is affected by defect states since they can reduce the effective barrier height for tunneling, though it is important to note that the defect states at the contact junction are relevant.

The measured conductivity is well described by the model at all stages of sample aging. The results of FITC model fitting are shown by solid lines in Figure 1.10 and the contact junction parameters are listed in Table 1.3. The junction width w did not change appreciably as the sample aged and no systematic change in the junction area A was observed. This suggests that physical rearrangement of the nanoparticles was not the primary cause of the increased conductivity over time. The large spread in the obtained values of A is likely to be due to a low sensitivity of the FITC fitting procedure to this parameter compared to w and φ_0 for the range of junction parameters involved. There was, however, a systematic lowering of the tunneling barrier φ_0 with sample aging (Table 1.3), which is consistent with continuous contaminant adsorption to SnO_2 at the nanoparticle contact junctions. Contaminant adsorption can result in mid-barrier electronic states in the necking region that effectively lower φ_0. Learning how to design materials and fabrication conditions to minimize φ_0 is of particular interest for designing materials for solar energy conversion applications because the conductivity is most sensitive to this parameter for the entire range of temperatures studied, including room temperature.

1.4.3 Experimental Methods

Anatase nanoparticles were obtained from Sigma-Aldrich Corporation and Ishihara Corporation with diameters of <25 nm and 7 nm, respectively. 1.5 g of TiO_2 was added to 1.3 mL of water, followed by sonication for 15 minutes. For the conductivity measurements, the nanoparticle slurry was spread onto a bottom electrode between two fiberglass spacers, which were 200 ± 4 μm thick with measured resistance > 200 TΩ. Immediately thereafter, the top electrode was applied, the excess slurry was removed and the sample was dried for 24 h at ~ 70 °C. Dried samples were annealed at 450 °C for 1 h, the typical sintering conditions used for DSSCs. The sample electrodes were high chromium content stainless steel (grade 309). Samples for electrical measurements were lightly clamped with grade 309 bolts to maintain constant sample-to-electrode contact during the measurement and eliminate the effects of thermal expansion. The samples were loaded in a cryostat (Janis ST-100) and kept in the dark under vacuum (< 20 mTorr) for more than 24 h prior to measurement. A Stanford Research Systems SR570 low-noise current preamplifier was used to supply a 4 V bias and measure the current in a two-terminal measurement. To quantify the contact resistance in our devices, we varied the thickness of the TiO_2 nanoparticle films. We found that the use of chromium-rich stainless steel as the contact surface minimizes both the oxide layer formed on the electrode during sintering and the corresponding contact resistance, which is negligible compared to the resistance of the TiO_2 film.

SnO_2 nanoparticles with diameters of 22–43 nm where purchased from Alfa Aesar (tin (IV) oxide, NanoArc) for ac impedance and dc resistance measurements on sintered nanoparticle films. Nanoparticle slurry was made from 1.1 g of nanoparticle powder and 1 ml of distilled water by mixing with pestle and mortar. The sample was prepared with electrodes as described above for anatase samples, dried for 1 week at room temperature and then annealed in air at 450 °C for 40 minutes, the typical sintering conditions used for SnO_2 in DSSCs.[84] The applied bias voltage for the SnO_2 sample was 0.2 V and the sample displayed a linear *I–V* curve at room temperature at bias voltages up to 1 V. AC impedance was measured in the 0.3–30 000 Hz range using a Stanford Research Systems SR830 digital lock-in amplifier with ac biases of 0.25 and 1 V.

1.5 Calculations of Redox Potentials: Reduction of Systematic Error

We recently presented a practical, yet rigorous, methodology for quantitative predictions of redox potentials of transition metal complexes.[85] This is a particularly powerful tool in the context of solar energy conversion since it can give valuable insight into catalytic mechanisms and provide guidelines for the design of new electrocatalysts. In Section 1.2, we emphasized the importance of the potential of the regenerative electron mediators (*e.g.*, I_3^-/I^- *vs.* Br_3^-/Br^- redox couples), the position of the donor excited state level relative to the DOS of the metal oxide conduction band and the general investigation and optimization of the energy level alignment between of the redox active species in photoelectrochemical cells (Figure 1.1). Just as the ground and excited states can be estimated experimentally using cyclic voltammetry together with absorption and emission spectra, these data can be predicted using calculations of the ground state redox potential in parallel with the absorption spectra simulation methodologies discussed in Section 1.3. These predictive abilities enable the rational design of system components (*e.g.*, sensitizer, redox mediator, photocatalyst) with optimal energetics.

The method reduces systematic errors that result from the theoretical approach (*i.e.*, the choice of DFT functional, basis set and solvation model) as well as the electrochemical measurement conditions, including the nature of the solvent, electrolyte and working electrode. Therefore, this method is particularly reliable in correlating experimental and theoretical data, even for second- and third-row transition metal complexes for which larger deviations have been previously reported. Standard methods for computation of redox potentials are commonly applied in electrochemical studies,[86–101] although methodologies that could account for systematic uncertainties about experimental or computational origin had yet to be established. Earlier reports typically documented deviations between experimental and theoretical values of redox potentials in the 150–540 mV range for most of the available methodologies.[86-94] While these deviations continue to stimulate the development of more sophisticated DFT functionals, basis sets and solvation models, we note that deviations in the documented experimental data can often be a major factor in accounting for discrepancies of comparable magnitude. This is largely due to the fact that redox properties are typically quite sensitive to the particular choice of solvent, electrode, or electrolyte conditions. Therefore, identifying and reducing these sources of error is critical in establishing the capabilities and limitations of existing methods as well as for the design of new computational approaches.

In this section we outline a systematic methodology for removing uncertainties that are commonly included in comparisons between experimental and theoretical redox potentials, which are frequently reported relative to external reference couples, or reference electrodes. We study benchmark redox couples, including complexes that span three transition metal rows in various non-aqueous solvents. It is shown that the use of appropriate references, measured under the same conditions and calculated by using compatible computational frameworks, allows quantitative correlations between experimental and theoretical data. This approach leads to DFT redox potentials with standard deviations comparable to the experimental errors of cyclic voltammetry measurements, even at a rather modest level of theory (64 mV standard deviation for DFT/UB3LYP/LACVP/6-311G* level; see Figure 1.11).

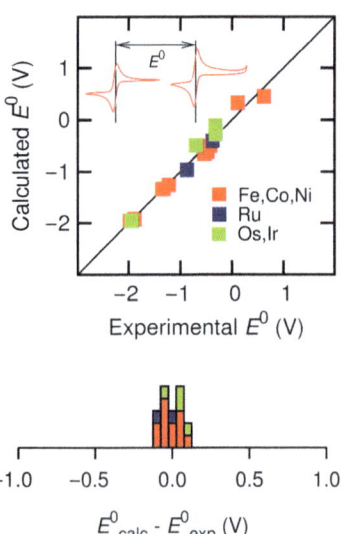

Figure 1.11   Top: Correlation between measured E_{exp}^0 and calculated E_{calc}^0 redox potentials in V *vs.* RC, where RC = $[FeCp_2]^{0/+}$, $[Ru(bpy)_3]^{2+/3+}$ and $[Ir(acac)_3]^{0/+}$ for first-, second-, and third-row transition-metal complexes, respectively, at the DFT/UB3LYP/LACVP/6-311G* level of theory. Bottom: Distribution in $E_{calc}^0 - E_{exp}^0$ (− 2 mV mean and 64 mV standard deviation).

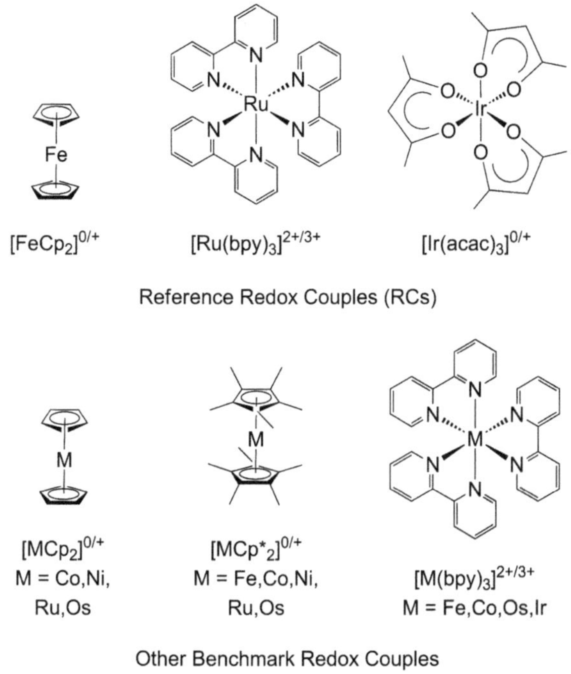

[FeCp$_2$]$^{0/+}$ [Ru(bpy)$_3$]$^{2+/3+}$ [Ir(acac)$_3$]$^{0/+}$

Reference Redox Couples (RCs)

[MCp$_2$]$^{0/+}$ [MCp*$_2$]$^{0/+}$

M = Co,Ni,   M = Fe,Co,Ni,   [M(bpy)$_3$]$^{2+/3+}$

Ru,Os   Ru,Os   M = Fe,Co,Os,Ir

Other Benchmark Redox Couples

Figure 1.12 Benchmark redox couples.

This is shown for a series of benchmark redox couples (Figures 1.12 and 1.13), including ([MCp$_2$]$^{0/+}$ (Cp = η^5-cyclopentadienyl), [MCp*$_2$]$^{0/+}$ (Cp* = η^5 −1,2,3,4,5-pentamethylcyclopentadienyl), [M(bpy)$_3$]$^{2+/3+}$ (bpy = 2,2′-bipyridine) and [Ir(a-cac)$_3$]$^{0/+}$ (acac = acetylacetonate), with M = Fe, Co, Ni, Ru, Os, or Ir) in various non-aqueous solvents [acetonitrile (MeCN), dimethyl sulfoxide (DMSO) and dichloromethane (DCM)].

1.5.1 Methodology and Benchmark Results

The Gibbs free energy change ΔG(soln), associated with the redox transition illustrated in Scheme 1.1, gives the absolute potential of a redox couple R/P, as follows:

$$E_{\text{calc}}^{\text{abs}} = \frac{\Delta G(\text{soln})}{nF} \tag{1.9}$$

where n is the number of moles of electrons involved in the redox reaction, F is the Faraday constant and $\Delta G(\text{soln}) = \Delta G(\text{g}) + \Delta G_{\text{solv}}^{\text{P}} - \Delta G_{\text{solv}}^{\text{R}}$. The resulting values are usually reported relative to a reference electrode (RE), such as the standard hydrogen electrode (SHE), or the silver-silver chloride saturated calomel electrode (SCE):

$$E_{\text{calc}}^0(\text{V } vs. \text{ RE}) = E_{\text{calc}}^{\text{abs}}(\text{V}) - E_{\text{exp,RE}}^{\text{abs}}(\text{V}) \tag{1.10}$$

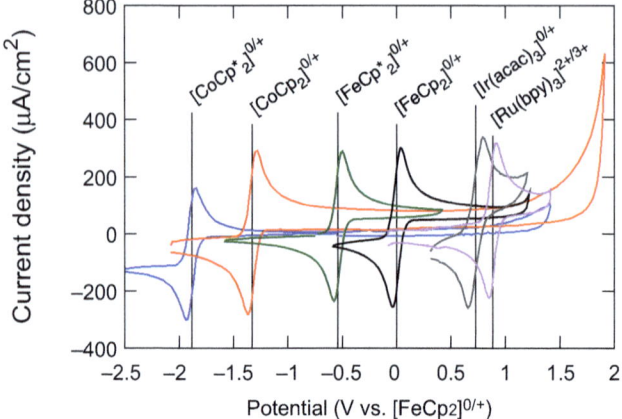

Figure 1.13 Cyclic voltammograms of [CoCp*$_2$]$^{0/+}$ (blue), [CoCp$_2$]$^{0/+}$ (red), [FeCp*$_2$]$^{0/+}$ (green), [FeCp$_2$]$^{0/+}$ (black), [Ir(acac)$_3$]$^{0/+}$ (grey) and [Ru(bpy)$_3$]$^{2+/3+}$ (purple) couples in 0.1 M [NBu$_4$][BF$_4$] in acetonitrile (100 mV s^{-1} scan rate, $T = 25\,°C$, current density of [Ir(acac)$_3$]$^{0/+}$ is reduced by a factor of 3).

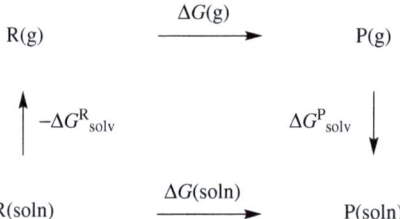

Scheme 1.1 Thermodynamic cycle used for the calculation of changes in free energy in solution $\Delta G(\text{soln})$ from states R to P based on gas-phase minimum-energy geometries and solvation free energies ΔG_{solv}.

Experimental redox potentials are typically reported with respect to a reference redox couple (RC), such as ferrocene/ferrocenium ([FeCp$_2$]$^{0/+}$):

$$E^0_{\text{exp}}(\text{V }vs.\text{ RC}) = E^0_{\text{exp}}(\text{V }vs.\text{ RE}) - E^0_{\text{exp,RC}}(\text{V }vs.\text{ RE}) \qquad (1.11)$$

Therefore, comparisons between calculated $E^0_{\text{calc}}(\text{V }vs.\text{ RE})$ and experimental $E^{0'}_{\text{exp}}(\text{V }vs.\text{ RE})$ values require accurate potentials of the redox couples relative to the reference electrodes. However, the experimental values of both the redox couples and reference electrodes vary by hundreds of millivolts from solvent to solvent, or from experiment to experiment with the same solvent but different electrolytes (Table 1.4). Furthermore, the SCE typically generates a liquid junction potential that may not be reproducible in a given experiment

Table 1.4 Experimental redox potentials reported for the SHE in various solvents.[102–105]

Solvent	$E_{\text{exp,SHE}}^{\text{abs}}(V)$
Water	4.24–4.44
Acetonitrile	4.56–4.66
Dimethylsulfoxide	3.83–4.04
Ethanol	4.20–4.24

Table 1.5 Redox potentials for the $[\text{FeCp}_2]^{0/+}$ couple (in V *vs.* SCE) in various solvents and electrolyte solutions.[107–110]

Solvent	$Li[ClO_4]^a$	$[NBu_4][ClO_4]$[108]	$[NEt_4][PF_6]$[109]	$[NBu_4][PF_6]$[110]
MeCN	0.31	0.40	0.38	0.40
DMSO		0.45	0.43	
DCM		0.48		0.46
DMF		0.47	0.46	0.45

^{a}Kuwana *et al.*,[107] quarter-wave potential.

and non-aqueous reference electrodes are prone to issues of reproducibility caused by electrode surface chemistry.[106] In addition, the reference redox couple (such as the ferrocene/ferrocenium pair) can change by tens to hundreds of mV with solvent or electrolyte (see Table 1.5 and Section 1.5.3).[107–110] All of these aspects therefore introduce systematic errors into the correlation between experimental and theoretical data. Removing this uncertainty requires comparison of experimental and calculated redox potentials relative to a redox couple measured under the same solvent and electrolyte conditions, rather than relative to a reference electrode.

An outstanding challenge is to identify a set of redox couples that could be used as internal references for a wide range of electrolytes, solvents and working electrode conditions. Calculated redox potentials would then be reported as follows:

$$E_{\text{calc}}^0(V \text{ } vs. \text{ RC}) = E_{\text{calc}}^{\text{abs}}(V) - E_{\text{calc,RC}}^{\text{abs}}(V) \quad (1.12)$$

where both $E_{\text{calc}}^{\text{abs}}$ and $E_{\text{calc,RC}}^{\text{abs}}$ are computed according to Equation (1.9). Analogous internal reference methodologies are common practice in other fields (*e.g.*, reports of NMR chemical shifts) and reduce the systematic uncertainties caused by the sensitivity of the measurements to experimental conditions. Recent studies have shown that errors are significantly reduced when $E_{\text{exp}}^0(V \text{ } vs. \text{ RC})$, obtained *via* Equation (1.11), is compared to $E_{\text{calc}}^0(V \text{ } vs. \text{ RC})$ calculated *via* Equation (1.12) and the following criteria are met:[85] (*i*) $E_{\text{calc}}^{\text{abs}}$ and $E_{\text{calc,RC}}^{\text{abs}}$ in Equation (1.12) are calculated using identical conditions (*e.g.*, same level of theory and solvent parameters); (*ii*) E_{exp}^0 and

$E^0_{exp,RC}$ in Equation (1.11) are measured under identical conditions, for example, the same solvent, electrolyte and working electrode; (*iii*) A reference transition metal complex is chosen for $E^0_{exp,RC}$ in Equation (1.11) and the calculation of $E^{abs}_{calc,RC}$ in Equation (1.12) is such that the metal lies in the same row of the periodic table as the complex used to calculate E^0_{calc}. These criteria are discussed in more detail in Sections 1.5.3 to 1.5.5.

1.5.2   Density Functional Theory Computational Methods

DFT calculations were performed using the B3LYP exchange correlation functional with unrestricted Kohn–Sham wave functions (UB3LYP) as implemented in the Jaguar electronic structure program.[111] Minimum energy configurations were obtained by using a mixed basis in which the metal centers are described by the non-relativistic effective core potentials (ECPs) of the LACVP basis set. Different levels of theory were used in an effort to compare the effect of the ligand basis set on the resulting correlation between calculated and experimental data. In order of increasing computational cost, the ligand basis sets investigated were 6-31G, 6-311G* and Dunning's correlation-consistent triple-ζ basis set[112–114] cc-pVTZ(-f), which includes a double set of polarization functions. In addition, the geometry optimizations based on 6-31G and 6-311G* basis sets were followed by UB3LYP single point energy calculations with the cc-pVTZ(-f) basis. The resulting correlations with experimental data were analyzed to assess the minimum computational effort necessary for quantitative prediction of redox potentials and the validity of the single-point approximation method, a commonly used method to save on computational cost, as it applies to calculations of redox potentials.

All reduction potentials were computed, according to Equation (1.9), by calculating the free energy changes $\Delta G(soln)$ associated with reduction of the complexes in solution, as follows:

$$\Delta G(soln) = \Delta G(g) + \Delta G^P_{solv} - \Delta G^R_{solv} \qquad (1.13)$$

where $\Delta G(g) = \Delta H(g) - T\Delta S(g)$ is the free energy change for the reduction reaction in the gas phase. Solvation free energies for reactants and products, ΔG^R_{solv} and ΔG^P_{solv}, respectively, were computed using the standard self-consistent reaction field approach for the gas phase minimum energy configurations with dielectric constants of $\varepsilon = 8.93$, 37.5 and 47.24 and solvent radii of 2.33, 2.19 and 2.41 Å for DCM, MeCN and DMSO, respectively.[111,112]

1.5.3   Method Benchmark Results

The correlation between the computational and experiment data *vs.* RC, where RC = [FeCp$_2$]$^{0/+}$, [Ru(bpy)$_3$]$^{2+/3+}$ and [Ir(acac)$_3$]$^{0/+}$ for first- (red), second- (navy) and third-row (green) metal complexes, respectively, is shown in

Figure 1.11 for the DFT/UB3LYP/LACVP/6-311G* level of theory. This level of theory, though not the most computationally expensive, is the best performing in terms of standard deviation with respect to the experimental data. This level gives a standard deviation of 56 mV for 12 of 18 couples that lie in the first row, compared to 150 and 90 mV when using 6-31G and cc-pVTZ(-f) basis sets, respectively. When the same $[FeCp_2]^{0/+}$ RC is used and extended to all couples, the standard deviation using 6-311G* rises to 148 mV. Although this standard deviation is an improvement over previous reports of method performance, which we attribute to reductions in systematic error in the experimental data, it still marks a sharp decrease in performance compared to the first-row couples alone. However, when a RC with a similar ECP is used, the values are comparable to the first-row statistics. The DFT/UB3LYP/LACVP/ 6-311G* level of theory yields a standard deviation of 64 mV and a mean of –2 mV for all 18 couples.

1.5.4 Choice of Reference Redox Couple

Internal reference redox couples can provide results that are consistent over a wide range of experimental conditions since they bypass problems common to reference electrodes Like SCE or SHE that might generate liquid junction potentials. Even non-aqueous reference electrodes are prone to issues of reproducibility owing to electrode surface chemistry.[106] These factors can contribute to discrepancies on the order of tens to hundreds of millivolts, even when comparing measurements reported under the same electrolyte and solvent conditions. Internal reference redox couples can remove these systematic uncertainties.

Earlier computational studies have been focused on finding appropriate functionals and basis sets to obtain accurate estimates of *absolute* redox potentials.[86–94] However, removing systematic errors caused by variations in the experimental conditions requires referencing the calculated potentials to redox couples calculated with the same functional and basis set, as in Equation (1.12). These systematic errors are partially canceled when the reference is calculated directly (*e.g.*, see Table 1.4). This is important since the accuracy of E_{calc}^0(Vvs.RC) is equally as dependent on the reference value as it is on the calculated absolute potential of the couple being studied [Equations (1.10) and (1.12)].

The results of Section 1.5.3 show that a suitable choice of a reference redox couple is period dependent.[85] For example, the $[FeCp_2]^{0/+}$ redox couple is a valuable reference for transition metal complexes of the first period in non-aqueous solutions and has been extensively used in a host of electrochemical studies under a wide variety of experimental conditions.[85,108,110,115–119] In addition, several studies have explored first principle methods for calculating redox potentials with $[FeCp_2]^{0/+}$ as a reference RC.[87,98] While comparisons for transition metal complexes of the first period have been successful, comparisons for second- and third-row transition metal complexes have proven to be more challenging. These discrepancies are likely to be due to systematic errors introduced by the choice of basis sets, pseudopotentials, or solvation models that are expected to be comparable for transition metals of the same period. Therefore, reference couples based on transition metals of the same period as the system of interest should partially cancel the resulting systematic deviations and provide satisfactory results.

$[FeCp_2]^{0/+}$ is the most common choice of RC. References for the second and third row, however, are a bit more challenging and have yet to be established. Electrochemical data is scarce and for most complexes (*e.g.*, the metallocene analogs) the measurements are reproducible only under limited experimental conditions. For example, it has been shown that ruthenocene and osmocene exhibit single, quasi-reversible oxidation waves in solutions of DCM and $[NBu_4][B(Ar^F)_4]$ ($Ar^F = 3,5$-bistrifluoromethylphenyl).[119] However, in MeCN, oxidation of these metallocenes involves irreversible processes.[115] In addition, the $[RhCp_2]^{0/+}$ couple can be highly reversible while the lifetime of rhodocene is on the order of seconds and is unstable at room temperature on the cyclic voltammetry timescale.[120] Therefore, instead of the metallocene analogs, reference couples for the second and third period have been based on $[Ru(bpy)_3]^{2+/3+}$, which is of interest in photoredox catalysis and artificial photosynthesis,[121–123] and $[Ir(acac)_3]^{0/+}$, which is often used as a precursor for complexes relevant to organic light-emitting diodes.[124–126] Both of these complexes show reversible or quasi-reversible peaks in both DCM and MeCN.

1.5.5 Accounting for Solvent Polarity and Supporting Electrolyte

The solvent and supporting electrolyte can significantly influence both the reference and the couple being studied. As an example, Figure 1.14 shows cyclic voltammograms of $[FeCp_2]^{0/+}$ measured in 0.1 M $[NBu_4][BF_4]$ in MeCN, DMSO and DCM solvents. The redox potential of such a couple is 84 (MeCN), 33 (DMSO) and 210 (DCM) mV *vs.* $Ag/AgNO_3$, with a significant shift as large as 177 mV when comparing the oxidation potential in DCM relative to DMSO (see Figure 1.14). Similarly, calculations predict a shift of 179 mV in the calculated absolute redox potential of the $[FeCp_2]^{0/+}$ couple in DMSO and DCM using the DFT/UB3LYP/LACVP/6-311G* level of theory. These results illustrate the capabilities of self-consistent reaction field methods to account for solvent effects in electrochemistry calculations which are essential for comparison to experimental data.

The supporting electrolyte also affects the values for the redox potentials obtained by cyclic voltammetry and should be accounted for when comparing calculated and experimental data. The electrolyte affects the ionic strength, the conductivity and reactivity as well as the effective dielectric constant of the medium.[127] In addition, ion pairing with the supporting electrolyte counterions can become favorable in low dielectric solvents or when the complex is highly charged, as in the case of the $[Ru(bpy)_3]^{2+/3+}$ couple. All of these effects can be partially removed by referring the measured potentials relative to an internal

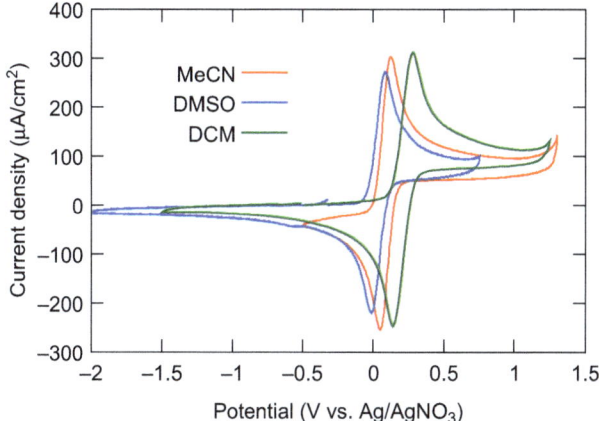

Figure 1.14 Cyclic voltammograms of the $[FeCp_2]^{0/+}$ couple in 0.1 M $[NBu_4][BF_4]$ in acetonitrile (red), dimethylsulfoxide (blue) and dichloromethane (green) solvents (100 mV s^{-1} scan rate, $T = 25$ °C).

reference to analogous electrolyte interactions. As an example, we consider the redox potential of the $[Ru(bpy)_3]^{2+/3+}$ couple which is higher in DCM than in MeCN by 174 mV *vs.* Ag/AgNO$_3$. However, when explicit $[BF_4]^-$ counterions are included in DCM, the calculated difference in absolute potential is 102 mV at the DFT/UB3LYP/LACVP/6-311G* level of theory. In contrast, a shift of 574 mV is obtained when ion pairing is neglected. Therefore, it is clear that ion pairing can account for significant shifts (*e.g.*, 472 mV in low-polar DCM, $\varepsilon = 8.93$). These results emphasize the importance of including explicit counterions for highly charged species in low-polarity solvents.

1.6 Conclusions and Outlook

Photocatalytic solar cells based on semiconductor materials functionalized with earth-abundant transition metal complexes are promising developments that should allow the sustainable production of chemical fuel from renewable resources (*e.g.*, water). In this chapter we have reviewed recent advances in computational modeling of fundamental aspects and mechanisms that affect the overall efficiency of solar light photoconversion, as determined by photoabsorption, charge transport and redox properties of systems with molecular adsorbates on metal oxide surfaces. While powerful on its own, when combined with structural and mechanistic characterization based on high-resolution spectroscopy and electrochemistry, the resulting computational methods are powerful tools for the design and characterization of new photocatalytic materials. These emerging methods are thus expected to continue to make significant contributions to the development of novel semiconductor materials for photocatalytic solar cell applications.

The chapter has focused on computational methods for analyzing complete solar cell assemblies using equivalent circuit modeling. A photoelectrochemical device model can be used to study current–voltage characteristics to provide a deeper understanding of the physical and electrochemical processes important to device performance. When used in conjunction with the methodologies discussed in order to calculate redox potentials, absorption spectra and the other relevant device parameters, these predictive abilities enable the rational design of photocatalytic solar cells, including tandem two-photon Z-schemes based on high potential photoanodes. Therefore, the computational methods discussed represent valuable tools for accurate prediction of redox potentials, photoabsorption and conductivity at the photoanode surface, which are necessary for optimizing the energetics of fundamental processes that determine the performance of photoelectrocatalytic devices.

The analysis of redox potentials in transition metal complexes shows that calculated and experimental data can be correlated with a standard deviation of 64 mV for 18 benchmark redox potentials and seven reference potentials calculated at the DFT/UB3LYP level of theory with LACVP and 6-311G* basis sets. This correlation is within the experimental error of cyclic voltammetry measurements and is achieved when using three important guidelines for reducing systematic error: (i) using an experimental reference redox couple measured under the same experimental conditions as the system of interest; (ii) using a computational reference redox couple calculated at the same level of theory (which serves the dual purpose of removing computational systematic error while preventing the use of an external measurement of the absolute potential of a reference electrode that can introduce additional experimental systematic error); (iii) using references with metal centers of the same row of the periodic table as the electrochemically active complex. The choice of reference is essential to ensure a reduction of computational systematic error since elements in the same row share a common set of core electrons and there is therefore better agreement of the pseudopotentials of the metal basis set. The resulting methodology leads to accurate calculations of redox potentials at a relatively low computational cost, even for transition metal complexes in the second and third rows for which poor performance was previously reported and assigned to deficiencies in the DFT functionals.

Charge carrier collection and series resistance can have a significant impact on the efficiency of photoelectrocatalytic devices. It is therefore important to understand the mechanisms of electrical conductivity in photoanodes and photocathodes. It has been shown that the FITC model can describe electron transport in nanoporous metal oxide thin films in quantitative agreement with experimental data over the entire range of temperature accessible to reliable measurements. The structural parameters extracted from the conductivity data are consistent with the characterization of sintered TiO_2 nanoparticle films by SEM, XRD and annealing molecular dynamics simulations. The analysis reveals that the sintering process forms disordered TiO_2, surrounding the crystalline core of nanoparticles and at the points of contact between the nanoparticle cores. The contact junctions determine rate-limiting energy barriers for electron transport from nanoparticle to nanoparticle that are significantly affected by the preparative procedures and by thermal fluctuations in the density of free carriers. Therefore, improvements in emerging technologies based on nanoporous TiO_2 electrodes could be focused on ways of optimizing the nanoparticle interconnectivity and on sintering conditions that would provide optimum particle size and shape, choice of electrolyte, ambient conditions that affect the level of doping and therefore the alignment of electronic energy levels at the interparticle contact junctions. AC conductivity measurements of sintered SnO_2 nanoparticle films exhibit a power law behavior with a dc component that is in quantitative agreement with the FITC description of electronic transport. This, coupled with the results for nanoporous TiO_2 films and other systems that share a similar inhomogeneous microstructure, suggests that the FITC mechanism may have broad applicability for the wide variety of materials used for solar energy conversion.

Acknowledgements

This work was supported by the Office of Basic Energy Sciences of the US Department of Energy (DE-FG02-07ER15909) and supercomputer time from NERSC and XSEDE. This work was also supported in part by the facilities and staff of the Yale University Faculty of Arts and Sciences High Performance Computing Center and by the National Science Foundation under grant #CNS 08-21132 that partially funded acquisition of the facilities.

References

1. N. S. Lewis and D. G. Nocera, *Proc. Natl. Acad. Sci. U.S.A.*, 2006, **103**, 15729–15735.
2. A. J. Esswein and D. G. Nocera, *Chem. Rev.*, 2007, **107**, 4022–4047.
3. J. P. McEvoy and G. W. Brudvig, *Chem. Rev.*, 2006, **106**, 4455–4483.
4. S. Karlsson, J. Boixel, Y. Pellegrin, E. Blart, H.-C. Becker, F. Odobel and L. Hammarstrom, *Faraday Discuss.*, 2012, **155**, 233–252.
5. R. Eisenberg and H. B. Gray, *Inorg. Chem.*, 2008, **47**, 1697–1699.
6. F. Liu, J. J. Concepcion, J. W. Jurss, T. Cardolaccia, J. L. Templeton and T. J. Meyer, *Inorg. Chem.*, 2008, **47**, 1727–1752.
7. Y. Umena, K. Kawakami, J.-R. Shen and N. Kamiya, *Nature*, 2011, **473**, 55–60.
8. A. Fujishima and K. Honda, *Nature*, 1972, **238**, 37–8.
9. N. D. McDaniel, F. J. Coughlin, L. L. Tinker and S. Bernhard, *J. Am. Chem. Soc.*, 2008, **130**, 210–217.
10. J. F. Hull, D. Balcells, J. D. Blakemore, C. D. Incarvito, O. Eisenstein, G. W. Brudvig and R. H. Crabtree, *J. Am. Chem. Soc.*, 2009, **131**, 8730–8731.
11. G. Li, E. M. Sproviero, R. C. Snoeberger III, N. Iguchi, J. D. Blakemore, R. H. Crabtree, G. W. Brudvig and V. S. Batista, *Energy Environ. Sci.*, 2009, **2**, 230–238.
12. J. Limburg, J. S. Vrettos, L. M. Liable-Sands, A. L. Rheingold, R. H. Crabtree and G. W. Brudvig, *Science*, 1999, **283**, 1524–1527.

13. E. M. Sproviero, J. A. Gascon, J. P. McEvoy, G. W. Brudvig and V. S. Batista, *Curr. Opin. Struct. Biol.*, 2007, **17**, 173–180.

14. E. M. Sproviero, J. A. Gascon, J. P. McEvoy, G. W. Brudvig and V. S. Batista, *Coord. Chem. Rev.*, 2008, **252**, 395–415.

15. W. J. Youngblood, S.-H. A. Lee, K. Maeda and T. E. Mallouk, *Acc. Chem. Res.*, 200, **42**, 1966–73.

16. M. W. Kanan and D. G. Nocera, *Science*, 2008, **321**, 1072–1075.

17. M. W. Kanan, Y. Surendranath and D. G. Nocera, *Chem. Soc. Rev.*, 200, **38**, 109–114.

18. Z. Chen, J. J. Concepcion, J. W. Jurss and T. J. Meyer, *J. Am. Chem. Soc.*, 2009, **131**, 15580–15581.

19. J. J. Concepcion, J. W. Jurss, M. K. Brennaman, P. G. Hoertz, A. O. T. Patrocinio, N. Y. Murakami Iha, J. L. Templeton and T. J. Meyer, *Acc. Chem. Res.*, 2009, **42**, 1954–1965.

20. J. J. Concepcion, J. W. Jurss, M. R. Norris, Z. Chen, J. L. Templeton and T. J. Meyer, *Inorg. Chem.*, 2010, **49**, 1277–1279.

21. S. W. Gersten, G. J. Samuels and T. J. Meyer, *J. Am. Chem. Soc.*, 1982, **104**, 4029–4030.

22. R. Brimblecombe, A. M. Bond, G. C. Dismukes, G. F. Swiegers and L. Spiccia, *Phys. Chem. Chem. Phys.*, 2009, **11**, 6441–6449.

23. R. Brimblecombe, G. F. Swiegers, G. C. Dismukes and L. Spiccia, *Angew. Chem., Int. Ed.*, 200, **47**, 7335–7338.

24. A. Mills and S. L. Hunte, *J. Photochem. Photobiol., A*, 1997, **108**, 1–35.

25. J. A. Treadway, J. A. Moss and T. J. Meyer, *Inorg. Chem.*, 1999, **38**, 4386–4387.

26. G. F. Moore, S. J. Konezny, H.-e. Song, R. L. Milot, J. D. Blakemore, M. L. Lee, V. S. Batista, C. A. Schmuttenmaer, R. H. Crabtree and G. W. Brudvig, *J. Phys. Chem. C*, 2012, **116**, 4892–4902.

27. S. M. Sze, *Physics of Semiconductor Devices*, John Wiley & Sons, 1981, p 868.

28. S. Dongaonkar, J. D. Servaites, G. M. Ford, S. Loser, J. Moore, R. M. Gelfand, H. Mohseni, H. W. Hillhouse, R. Agrawal, M. A. Ratner, T. J. Marks, M. S. Lundstrom and M. A. Alam, *J. Appl. Phys.*, 2010, **108**, 124509.

29. J. Halme, P. Vahermaa, K. Miettunen and P. Lund, *Adv. Mater.*, 2010, **22**, E210–E234.

30. Z.-S. Wang, K. Sayama and H. Sugihara, *J. Phys. Chem. B*, 2005, **109**, 22449–22455.

31. D. Rehm and A. Weller, *Ber. Bunsenges. Phys. Chem.*, 1969, **73**, 834–839.

32. M. Grätzel, *Nature*, 2001, **414**, 338–344.

33. G. Boschloo and A. Hagfeldt, *Acc. Chem. Res.*, 2009, **42**, 1819–1826.

34. G. F. Moore, J. D. Blakemore, R. L. Milot, J. F. Hull, H.-e. Song, L. Cai, C. A. Schmuttenmaer, R. H. Crabtree and G. W. Brudvig, *Energy Environ. Sci.*, 2011, **4**, 2389–2392.

35. C. Garau, A. Frontera, D. Quinonero, P. Ballester, A. Costa and P. M. Deya, *ChemPhysChem*, 2003, **4**, 1344–1348.
36. C. Garau, A. Frontera, D. Quinonero, P. Ballester, A. Costa and P. M. Deya, *J. Phys. Chem. A*, 2004, **108**, 9423–9427.
37. R. E. Blankenship, D. M. Tiede, J. Barber, G. W. Brudvig, G. Fleming, M. Ghirardi, M. R. Gunner, W. Junge, D. M. Kramer, A. Melis, T. A. Moore, C. C. Moser, D. G. Nocera, A. J. Nozik, D. R. Ort, W. W. Parson, R. C. Prince and R. T. Sayre, *Science*, 2, **332**, 805–809.
38. D. Xiao, L. A. Martini, R. C. Snoeberger, R. H. Crabtree and V. S. Batista, *J. Am. Chem. Soc.*, 2011, **133**, 9014–9022.
39. W. R. McNamara, R. C. Snoeberger, G. Li, J. M. Schleicher, C. W. Cady, M. Poyatos, C. A. Schmuttenmaer, R. H. Crabtree, G. W. Brudvig and V. S. Batista, *J., Am. Chem. Soc.*, 2008, **130**, 14329–14338.
40. C. J. Barbe, F. Arendse, P. Comte, M. Jirousek, F. Lenzmann, V. Shklover and M. Grätzel, *J. Am. Ceram. Soc.*, 1997, **80**, 3157–3171.
41. M. Grätzel, *Chem. Lett.*, 2005, **34**, 8–13.
42. X. Chen and S. S. Mao, *Chem. Rev.*, 2007, **107**, 2891–2959.
43. M. Hoffmann, S. Martin, W. Choi and D. Bahnemann, *Chem. Rev.*, 1995, **95**, 69–96.
44. O. Carp, C. L. Huisman and A. Reller, *Prog. Solid State Chem.*, 2004, **32**, 33–177.
45. C. Bechinger, S. Ferrere, A. Zaban, J. Sprague and B. A. Gregg, *Nature*, 1996, **383**, 608–610.
46. C. G. Granqvist, *Sol. Energy Mater. Sol. Cells*, 2007, **91**, 1529–1598.
47. M. Tiemann, *Chem. Eur. J.*, 2007, **13**, 8376–8388.
48. S. J. Konezny, C. Richter, R. C. Snoeberger III, A. R. Parent, G. W. Brudvig, C. A. Schmuttenmaer and V. S. Batista, *J. Phys. Chem. Lett.*, 2011, **2**, 1931–1936.
49. S. J. Konezny, D. Talbayev, I. E. Baggari, C. A. Schmuttenmaer and V. S. Batista, *Proc. SPIE*, 2007, **8098**, 809805.
50. P. E. Jongh and D. Vanmaekelbergh, *Phys. Rev. Lett.*, 1996, **77**, 3427–3430.
51. T. Dittrich, J. Weidmann, F. Koch, I. Uhlendorf and I. Lauermann, *Appl. Phys. Lett.*, 1999, **75**, 3980–3982.
52. G. M. Turner, M. C. Beard and C. A. Schmuttenmaer, *J. Phys. Chem. B*, 2002, **106**, 11716–11719.
53. C. O. Park and S. A. Akbar, *J. Mater. Sci.*, 2003, **38**, 4611–4637.
54. A. J. Frank, N. Kopidakis and J. van de Lagemaat, *Coord. Chem. Rev.*, 2004, **248**, 1165–1179.
55. J. Nelson and R. E. Chandler, *Coord. Chem. Rev.*, 2004, **248**, 1181–1194.
56. G. Boschloo and A. Hagfeldt, *J. Phys. Chem. B*, 2005, **109**, 12093–12098.
57. L. M. Peter, A. B. Walker, G. Boschloo and A. Hagfeldt, *J. Phys. Chem. B*, 2006, **110**, 13694–13699.
58. N. Kopidakis, K. D. Benkstein, J. van de Lagemaat, A. J. Frank, Q. Yuan and E. A. Schiff, *Phys. Rev. B*, 2006, **73**, 045326.
59. J. Bisquert, *J. Phys. Chem. C*, 2007, **111**, 17163–17168.

60. J. R. Jennings, A. Ghicov, L. M. Peter, P. Schmuki and A. B. Walker, *J. Am. Chem. Soc.*, 2008, **130**, 13364–13372.
61. T. Tiedje, J. M. Cebulka, D. L. Morel and B. Abeles, *Phys. Rev. Lett.*, 1981, **46**, 1425–1428.
62. N. F. Mott and E. A. Davis, *Electronic Properties in Non-Crystalline Materials*, Oxford University Press, New York, 1971.
63. A. K. Hassan, N. B. Chaure, A. K. Ray, A. V. Nabok and S. Habesch, *J. Phys. D-Appl. Phys.*, 2003, **36**, 1120–1125.
64. E. Barborini, G. Bongiorno, A. Forleo, L. Francioso, P. Milani, I. Kholmanov, P. Piseri, P. Siciliano, A. Taurino and S. Vinati, Sens. Actuators, *B*, 2005, **111–112**, 22–27.
65. K. Pomoni, A. Vomvas and C. Trapalis, *Thin Solid Films*, 2008, **516**, 1271–1278.
66. K. M. Reddy, S. V. Manorama and A. R. Reddy, *Mater. Chem. Phys.*, 2003, **78**, 239–245.
67. There are several studies that consider the effects of geometrical narrowing between nanoparticles when trapped in a homogeneous medium,[54, 128, 129] although they do not address the inhomogeneous nature of the contact junctions and the barriers to electron transport they create as described by the FITC model.
68. S. J. Konezny, M. N. Bussac and L. Zuppiroli, *Appl. Phys. Lett.*, 2008, **92**, 012107.
69. S. J. Konezny, M. N. Bussac, A. Geiser and L. Zuppiroli, *Proc. SPIE*, 2007, **6658**, 66580D.
70. P. Sheng, *Phys. Rev. B*, 1980, **21**, 2180–2195.
71. S. Paschen, M. N. Bussac, L. Zuppiroli, E. Minder and B. Hilti, *J. Appl. Phys.*, 1995, **78**, 3230–3237.
72. J. Ederth, P. Johnsson, G. A. Niklasson, A. Hoel, A. Hultåker, P. Heszler, C. G. Granqvist, A. R. van Doorn, M. J. Jongerius and D. Burgard, *Phys. Rev. B*, 2003, **68**, 155410.
73. M. Salvato, M. Cirillo, M. Lucci, S. Orlanducci, I. Ottaviani, M. L. Terranova and F. Toschi, *Phys. Rev. Lett.*, 2008, **101**, 246804.
74. H. B. Yin, Y. Wada, T. Kitamura, S. Kambe, S. Murasawa, H. Mori, T. Sakata and S. Yanagida, *J. Mater. Chem.*, 2001, **11**, 1694–1703.
75. J. J. Sakurai, *Quantum Mechanics*, Addison-Wesley, Massachusetts, 1994.
76. J. Nelson, A. M. Eppler and I. M. Ballard, *J. Photochem. Photobiol., A*, 2002, **148**, 25–31.
77. S. Plimpton, *J. Comput. Phys.*, 1995, **117**, 1–19.
78. M. Matsui and M. Akaogi, *Mol. Simul.*, 1991, **6**, 239–244.
79. V. N. Koparde and P. T. Cummings, *J. Phys. Chem. B*, 2005, **109**, 24280–24287.
80. J. B. Baxter and C. A. Schmuttenmaer, *J. Phys. Chem. B*, 2006, **110**, 25229–25239.
81. A. Jonscher, *Nature*, 1977, 673–679.
82. A. Long, *Adv. Phys.*, 1982, **31**, 553–637.
83. S. Elliott, *Adv. Phys.*, 1987, **36**, 135–217.

84. N. Dinh, M. Bernard, A. Hugot-Le Gof, T. Stergiopoulos and P. Falaras, *C. R. Chim.*, 2006, **9**, 676–683.
85. S. J. Konezny, M. D. Doherty, O. R. Luca, R. H. Crabtree, G. L. Soloveichik and V. S. Batista, *J. Phys. Chem. C*, 212, **116**, 6349–6356.
86. M.-H. Baik and R. A. Friesner, *J. Phys. Chem. A*, 2002, **106**, 7407–7412.
87. L. E. Roy, E. Jakubikova, M. G. Guthrie and E. R. Batista, *J. Phys. Chem. A*, 2009, **113**, 6745–6750.
88. I. Chiorescu, D. V. Deubel, V. B. Arion and B. K. Keppler, *J. Chem. Theory Comput.*, 2008, **4**, 499–506.
89. J. Li, C. L. Fisher, J. L. Chen, D. Bashford and L. Noodleman, *Inorg. Chem.*, 1996, **35**, 4694–4702.
90. J. Moens, P. Geerlings and G. Roos, *Chem.-Eur. J.*, 2007, **13**, 8174–8184.
91. J. Moens, P. Jaque, F. De Proft and P. Geerlings, *J. Phys. Chem. A*, 2008, **112**, 6023–6031.
92. M. Uudsemaa and T. Tamm, *J. Phys. Chem. A*, 2003, **107**, 9997–10003.
93. A. Galstyan and E.-W. Knapp, *J. Comput. Chem.*, 2009, **30**, 203–211.
94. M. T. Groot and M. T. M. Koper, *Phys. Chem. Chem. Phys.*, 2008, **10**, 1023–1031.
95. T. Wang, G. W. Brudvig and V. S. Batista, *J. Chem. Theory Comput.*, 2010, **6**, 2395–2401.
96. T. Wang, G. Brudvig and V. S. Batista, *J. Chem. Theory Comput.*, 2010, **6**, 755–760.
97. (a) J. Moens, G. Roos, P. Jaque, F. De Proft and P. Geerlings, *Chem.-Eur. J.*, 2007, **13**, 9331–9343; (b) N. Dinh, M. Bernard, A. Hugot-Le Gof, T. Stergiopoulos and P. Falaras, *C. R. Chimie*, 2006, **9**, 676–683.
98. L. E. Roy, E. R.v Batista and P. J. Hay, *Inorg. Chem.*, 2008, **47**, 9228–9237.
99. T. Wang and R. A. Friesner, *J. Phys. Chem. C*, 2009, **113**, 2553–2561.
100. M.-K. Tsai, J. Rochford, D. E. Polyansky, T. Wada, K. Tanaka, E. Fujita and J. T. Muckerman, *Inorg. Chem.*, 2009, **48**, 4372–4383.
101. R. Ayala and M. Sprik, *J. Chem. Theory Comput.*, 2006, **2**, 1403–1415.
102. S. Trasatti, *Pure Appl. Chem.*, 1986, **58**, 955–966.
103. C. P. Kelly, C. J. Cramer and D. G. Truhlar, *J. Phys. Chem. B*, 2006, **110**, 16066–16081.
104. C. P. Kelly, C. J. Cramer and D. G. Truhlar, *J. Phys. Chem. B*, 2007, **111**, 408–422.
105. W. R. Fawcett, *Langmuir*, 2008, **24**, 9868–9875.
106. G. Gritzner and J. Kuta, *Pure Appl. Chem.*, 1984, **56**, 461–466.
107. T. Kuwana, D. E. Bublitz and G. Hoh, *J. Am. Chem. Soc.*, 1960, **82**, 5811–5817.
108. D. Chang, T. Malinski, A. Ulman and K. M. Kadish, *Inorg. Chem.*, 1984, **23**, 817–824.
109. J. P. Chang, E. Y. Fung and J. C. Curtis, *Inorg. Chem.*, 1986, **25**, 4233–4241.
110. N. G. Connelly and W. E. Geiger, *Chem. Rev.*, 1996, **96**, 877–910.

111. *Jaguar, version* **7.7**, Schrodinger, LLC, New York, NY, 2010.
112. T. H. Dunning, *J. Chem. Phys.*, 1989, **90**, 1007–1023.
113. R. A. Kendall, T. H. Dunning and R. J. Harrison, *J. Chem. Phys.*, 1992, **96**, 6796–6806.
114. D. E. Woon and T. H. Dunning, *J. Chem. Phys.*, 1993, **98**, 1358–1371.
115. S. P. Gubin, S. A. Smirnova, L. I. Denisovich and A. A. Lubovich, *J. Organomet. Chem.*, 1971, **30**, 243–255.
116. S. P. Gubin, S. A. Smirnova and L. I. Denisovich, *J. Organomet. Chem.*, 1971, **30**, 257–265.
117. I. Noviandri, K. N. Brown, D. S. Fleming, P. T. Gulyas, P. A. Lay, A. F. Masters and L. Phillips, *J. Phys. Chem. B*, 1999, **103**, 6713–6722.
118. U. Kölle and F. Khouzami, *Angew. Chem., Int. Ed.*, 1980, **19**, 640–641.
119. M. G. Hill, W. M. Lamanna and K. R. Mann, *Inorg. Chem.*, 1991, **30**, 4687–4690.
120. N. El Murr, J. E. Sheats, W. E. Geiger and J. D. L. Holloway, *Inorg. Chem.*, 1979, **18**, 1443–1446.
121. K. Zeitler, *Angew. Chem., Int. Ed.*, 2009, **48**, 9785–9789.
122. A. J. Bard and M. A. Fox, *Acc. Chem. Res.*, 1995, **28**, 141–145.
123. A. Juris, V. Balzani, F. Barigelletti, S. Campagna, P. Belser and A. von Zelewsky, *Coord. Chem. Rev.*, 1988, **84**, 85–277.
124. A. Tsuboyama, H. Iwawaki, M. Furugori, T. Mukaide, J. Kamatani, S. Igawa, T. Moriyama, S. Miura, T. Takiguchi, S. Okada, M. Hoshino and K. Ueno, *J. Am. Chem. Soc.*, 2003, **125**, 12971–12979.
125. S. Lamansky, P. Djurovich, D. Murphy, F. Abdel-Razzaq, R. Kwong, I. Tsyba, M. Bortz, B. Mui, R. Bau and M. E. Thompson, *Inorg. Chem.*, 2001, **40**, 1704–1711.
126. K.-T. Wong, Y.-M. Chen, Y.-T. Lin, H.-C. Su and C.-c. Wu, *Org. Lett.*, 2005, **7**, 5361–5364.
127. D. Bao, B. Millare, W. Xia, B. G. Steyer, A. A. Gerasimenko, A. Ferreira, A. Contreras and V. I. Vullev, *J. Phys. Chem. A*, 2009, **113**, 1259–1267.
128. M. J. Cass, F. L. Qiu, A. B. Walker, A. C. Fisher and L. M. Peter, *J. Phys. Chem. B*, 2003, **107**, 113–119.
129. S. Nakade, Y. Saito, W. Kubo, T. Kitamura, Y. Wada and S. Yanagida, *J. Phys. Chem. B*, 2003, **107**, 8607–8611.

CHAPTER 2

Charge and Exciton Dynamics in Semiconductor Quantum Dots: A Time Domain, ab Initio View

AMANDA J. NEUKIRCH[a] AND OLEG V. PREZHDO*[a,b]

[a] Department of Physics and Astronomy, Bausch & Lomb Hall,
P.O. Box 270171, 500 Wilson Boulevard, University of Rochester, Rochester,
NY 14627-0171, USA; [b] Department of Chemistry, RC Box 270216,
Rochester, NY 14627-0216, USA
*Email: oleg.prezhdo@rochester.edu

2.1 Introduction

Quantum dots (QDs) are nanoscale clusters of bulk materials that exhibit the extraordinary property of quantum confinement. This feature allows the continuous tuning of various properties by changing cluster size and shape. QDs start exhibiting quantum confinement effects when they reach the size where electronic excitation takes place in bulk. This is known as the Bohr exciton radius and is different for every material. For a typical semiconductor it is in the range between a few nanometers and tens of nanometers, but is much larger for metals. Therefore, collective excitations in metallic nanoparticles, such as plasmons, can by tuned at a much larger particle size. On the sub-nanometer scale, clusters begin to behave like molecules. Their structures differ from the bulk and their properties change discontinuously with cluster size.

The dynamics of electronic excitations in semiconducting[1] and metallic[2] QDs is complex. High absorption cross sections, decreased electron–phonon

RSC Energy and Environment Series No. 8
Solar Energy Conversion: Dynamics of Interfacial Electron and Excitation Transfer
Edited by Piotr Piotrowiak

relaxation rates,[3,4] and generation of multiple electron (ME) hole pairs[5–7] make QDs excellent photovoltaic materials.[3,8] Electron–hole and charge–phonon interactions carry both fundamental and practical importance. For instance, QD photovoltaic efficiencies depend on the rates of electron–hole energy exchange and charge–phonon relaxation. Charge transfer processes are limited by phonon-induced dephasing of spin and electron states. Inelastic scattering is responsible for energy loss during charge tunneling though QDs.

QDs are studied by several scientific communities and are often modeled as either reduced-dimensional bulk or scaled-up molecular systems. QDs show discrete optical transitions like molecules, but consist of multiple unit cells and form electronic and vibrational bands like bulk materials. The charge–phonon and electron–hole interactions in QDs are stronger than those in bulk and weaker than those in the molecule. However, the molecular and bulk viewpoints can lead to contradictions. For example, the molecular view suggests that excitonic and charge–phonon interactions in QDs are strong and that relaxation is slow owing to mismatch between electron and phonon energies. In contrast, the bulk view indicates that the kinetic energy of quantum confinement is significantly greater than excitonic interactions and that charge–phonon relaxation through quasi-continuous bands is rapid. These qualitative differences generate debates that need to be resolved in order to advance both fundamental understanding and applications of QDs.

This chapter presents an *ab initio* description of the nature and dynamics of photoexcited states in semiconductor QDs, in the energy and time domains. By combining the bulk and molecular viewpoints, the analysis elucidates the controversies and provides a unified atomistic picture of the excited state processes. These *ab initio* methods are used to study excited state composition, evolution and relaxation, as well as electron phonon dephasing, all with an eye towards the incorporation of QDs in solar cells. For further reading on the work featured in this chapter see publications by the Prezhdo group.[8,9,93]

2.2   Theoretical Approaches

As mentioned above, the results discussed below are obtained using *Ab initio* methods. Other methods used to study QDs are effective mass theory[3,6] (EMT) and the pseudopotential techniques.[9–11] EMT uses a particle-in-a-box model where the electron and hole masses are given by their bulk values. EMT is an intuitive description that explains general trends seen in experiments. The atomistic pseudopotential technique can be applied to large systems, but requires careful parameterization for each material. *Ab initio* approaches use minimal parameterization and are applicable to most materials. This makes them particularly useful for studying dopants, defects, ligands, core/shell systems and QD synthesis. The Hartree–Fock (HF) method and density functional theory (DFT) have been around for many decades, while time domain (TD) DFT and non-adiabatic molecular dynamics (NAMD) are more recent areas of research. Currently, *ab initio* TDDFT/NAMD is the only technique that models QD dynamics in the time domain and at an atomistic level, directly mimicking time-resolved experiments. More information and on these theoretical approaches can be found in Craig *et al.*[12]

2.2.1 Hartree–Fock Method

The basic assumption behind the HF method is that the many-electron wave function can be written as a product of one-electron orbitals. HF approximates Schrodinger equation for a multidimensional wavefunction with coupled three-variable equations for single-electron orbitals. This allows for a great simplification of the many body problem. For example, the wavefunction describing valence electrons in a $Pb_{68}Se_{68}$ QD contains over 2000 variables; however single-particle HF equations depend on only three variables. The electrostatic field felt by an electron in an atom is due to the central potential of the nucleus together with the field created by the other electrons, Figure 2.1. HF achieves self-consistency, since electron orbitals are determined by the mean field, which depends on the orbitals. HF includes the Pauli Exclusion Principle by not allowing two identical electrons to occupy the same location through its exchange interaction, however it does not incorporate electron correlation.

2.2.2 Incorporation of Electron Correlation in Hartee–Fock with Configuration Interaction

Electronic correlation is responsible for excitonic effects and can be very strong. This correlation can be systematically added to HF using configuration

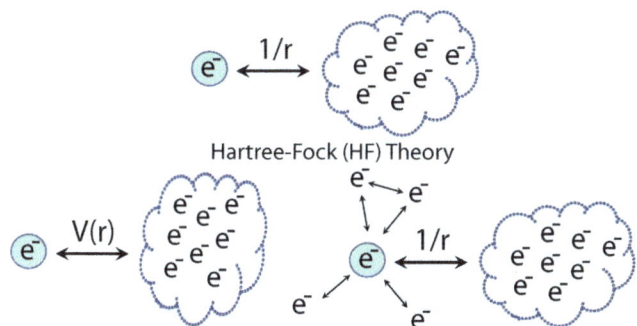

Figure 2.1 Illustration of single-particle HF theory and two alternative models that include electron-hole correlations. In HF, every single electron interacts with the mean-field created by the cloud of remaining electrons. CI adds correlation between electron pairs, triplets and so on. DFT builds an effective interaction potential that is different from $1/r$ which includes electron correlations within a modified single-particle description.[9] (Adapted from Ref. 9. Copyright 2009 American Chemical Society).

interaction (CI) and cluster expansions. This allows the exact many body wave function to be obtained, although at a very high computational expense. Configurations of two-body, three-body correlations and so on, Figure 2.1, represent multiple excitons, whose properties are particularly important for QD photochemistry. CI evaluates electron correlations explicitly but can be applied only to small clusters.

2.2.3 Density Functional Theory

DFT accounts for electron correlations indirectly. The Hohenberg–Kohn theorem proves that ground state properties can be obtained exactly from three-dimensional spatial density rather than multidimensional wavefunctions and DFT replaces coulomb interaction with a density functional, Figure 2.1. Kohn–Sham (KS) DFT ensures that density corresponds to a wavefunction. KS DFT is an effective single-particle theory where interacting electrons in a static external field are treated as non-interacting electrons moving in an effective potential. HF can be viewed as a special form of DFT that includes exact Pauli exchange with no electron correlation. As illustrated below, HF and DFT give different single-particle pictures. DFT is computationally inexpensive and works particularly well with extended systems that do not undergo complex chemical changes.

2.2.4 Time Domain Density Functional Theory

TDDFT describes a system's response to external perturbations, such as electromagnetic fields and phonons. Linear response TDDFT is frequently used to evaluate electronic excitation energies. The full TDDFT implemented in studies discussed throughout this chapter propagates electron density explicitly in time. Taking full computational advantage of time-independent DFT, its solutions are used as a basis for our TDDFT calculations.[12]

2.2.5 Nonadiabatic Molecular Dynamics

Classical mechanical prescriptions for phonon dynamics in response to changes in electron density constitute the quantum backreaction problem,[13] and the combined electron–phonon evolution is called NAMD. Traditional molecular dynamics (MD) is performed in a single, usually ground, electronic state. NAMD includes transitions between states. Most often, NAMD generates a quantum backreaction using surface hopping (SH), which stochastically correlates phonon dynamics with electronic states. Fewest switches SH (FSSH) minimized hops and satisfies a detailed balance between transitions up and down in energy as required by thermodynamic equilibrium.[14] FSSH was implemented in TDDFT by Prezhdo and Craig.[12]

2.3 Proposed Mechanisms for Multiple Exciton Generation

One of the most promising features of QDs for increased efficiency in solar cells is multiexciton generation (MEG). Predicted[3] several years before its discovery,[5] MEG from high energy photons avoids energy losses associated with electron–phonon relaxation to lower energy levels. MEG has drawn close attention owing to its potential for substantial improvement of photovoltaic device efficiencies.[3,5,6,8,15–23] The *Ab initio* analysis of the electronically excited states in the semiconductor QDs allows the different proposed mechanisms of MEG to be critically assessed.[3,6,11,20] Different proposals echo the rapidly shifting views on MEs and the phonon bottleneck and reflect the variety in the electronic structure of the materials exhibiting MEs. The *Ab initio* electronic structure calculations[22] discussed below show three types of photoexcited states in the semiconductor QDs, suggesting that three different mechanisms can be responsible for the ultrafast generation of MEs, a distinct mechanism for each type of the photoexcited state. The three mechanisms are impact ionization, the direct mechanism and the dephasing mechanism and are shown in Figure 2.2.

2.3.1 Impact Ionization Process

As in bulk materials, MEs in QDs can be created by an incoherent coulomb scattering mechanism, in which a high energy carrier relaxes to its ground state and excites valence electrons across the band gap, producing additional

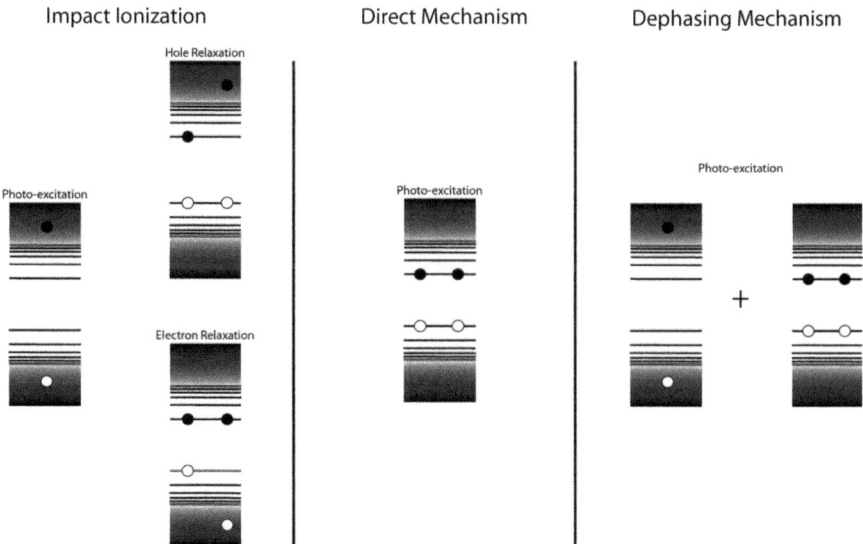

Figure 2.2 Illustration of the three distinct proposed mechanisms for MEG, impaction ionization, direct ME excitation and an initial superposition of SE and ME states that dephases into an ME state owing to increased phonon coupling.

electron–hole pairs. This effect is known as impact ionization (II)[3,11] and is the inverse of Auger recombination. The primary difference between exciton multiplication in bulk semiconductors and semiconductor quantum dots is the conservation of linear (crystal) momentum. In bulk semiconductors not only energy but also momentum (k) must be conserved, which increases the ideal energetic threshold of carrier multiplication beyond $2E_g$. In quantum dots this linear momentum constraint is lifted, reducing the carrier multiplication threshold. The efficiencies of II in semiconductor QDs depend on a couple of different things. First are the densities of single exciton (SE) and ME states. The lowest excitations, up to energies of twice the band gap, are single excitons. However, after that the ME density of states (DOS) increase significantly faster with energy than does the SE DOS. Thus if photoexcitation energy can be exchanged freely between SEs and MEs, most high energy excitations will be distributed among the MEs. However, many matrix elements coupling SE and ME states vanish by selections rules enforced by the two particle coulomb interactions, limiting II efficiency.

2.3.2 Direct Mechanism

Schaller *et al.*[5,8,19,20,22] proposed that a single absorbed photon generates a bi-exciton instantaneously by a second order perturbative process involving bi-exciton coupling to virtual single exciton states. Photoexcitation of multi-electron states is forbidden in the independent particle description. The cou-lomb interaction between independent electrons and holes couples singly and multiply excited states, generating non-vanishing oscillator strength for the multi-electron excitations.

2.3.3 Dephasing Mechanism

In the dephasing mechanism,[6,21,24] the photoexcited superposition of the single and ME states dephases by coupling to phonons. Immediately following the photoexcited superposition the electronic population oscillates between the two types of states coupled by the coulomb interaction. Then, the electronic energy decays into the phonon energy. Assuming that MEs couple to phonons more strongly than SEs, the energy decay is faster when the electronic population localizes in the ME state. Thus, the final surviving electronic state is a ME.[6] The *ab initio* methods discussed in the theory section above have suggested that all of these methods take place and that the dominant mechanism depends on the type of material, the temperature of the system, as well as defects and ligands. All of this will be reviewed and summarized in the remainder of this chapter.

2.4 Excited States and Symmetry Adapted Cluster-Configuration Interaction (SAC-CI)

Characterizing the excited states in the semiconductor QDs as single or MEs requires a proper single-particle description and a rigorous account of the electron–hole coulomb interaction. The non-interacting single-particle picture is provided by the HF approximation, which excludes electron correlation effects. These correlation effects are treated by symmetry adapted cluster (SAC) theory with configuration interactions (CI).[25,26] In contrast to DFT, SAC-CI can explicitly treat multiply excited electrons and holes allowing for the excited states to be quantified as single or MEs.[26–28] For example, the SAC-CI used below explicitly includes electron correlation through cluster expansion of the ground state wave function (WF):

$$|\Psi^{SAC}\rangle = \exp\left(\sum_{I=1}^{M} C_I \hat{S}_I\right)|\Phi_0\rangle = \left(1 + \sum_I C_I \hat{S}_I + \frac{1}{2}\sum_{I,J} C_I C_J \hat{S}_I \hat{S}_J + \cdots\right)|\Phi_0\rangle$$

(2.1)

Here, $|\Phi_0\rangle$ represents the HF WF, $\hat{S}_I$ are symmetry adapted excitation operators and C_I are variable coefficients. The excited state WFs $|\Psi^{SAC-CI}\rangle$ are calculated from the electron correlated ground state WF, $|\Psi^{SAC}\rangle$

$$|\Psi^{SAC-CI}\rangle = \sum_{K=1}^{N} d_K R_K |\Psi^{SAC}\rangle$$

(2.2)

where R_K represents an excitation operator and d_K is the SAC-CI coefficient. SAC-CI includes electron correlation for both the ground and excited electronic states and provides a very high-level description of electronic excitations with an explicit account of electron–hole coulomb interactions.

Using SAC-CI in the static, many-body picture, photoexcited states of quantum dots can be characterized by a superposition of electronic configurations in a molecular orbital basis, where the quantized band gap is computed as the difference between the ground state and the first excited state. A coefficient is associated with each configuration that denotes the configuration's significance or contribution to the electronic structure of the excited state. Large coefficients for singly excited configurations indicate that the excited state corresponds to the generation of a single exciton. Large coefficients for double excited configurations indicate non-trivial contributions from bi-excitonic excitation. Throughout this chapter SAC-CI is implemented within the Gaussian 03[29] computational package using the B3LYP functional. LANL2 relativistic effective core potentials were used for core electrons of all atoms and the basis set employed for the valence electrons was the corresponding LANL2DZ basis set.[30,31] More details on the theory and calculations presented in this section can be found in work by Prezhdo and co-workers.[22,32–36]

2.4.1 Multiexciton Generation (MEG)

Using a combination of the HF approximation and the SAC-CI, the nature of excited states in different systems can be determined. In Pb_4Se_4 almost all

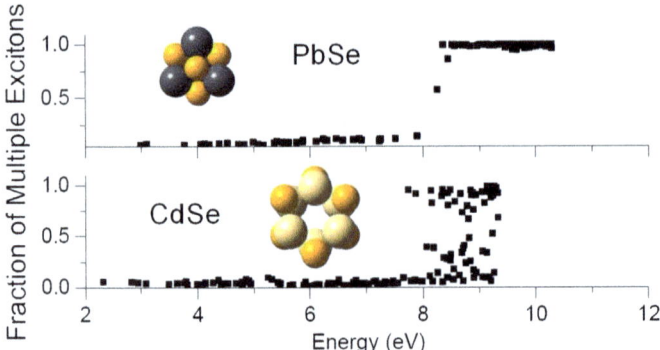

Figure 2.3   Fraction of MEs in the excited electronic states of Pb_4Se_4 and Cd_6Se_6 QDs, computed by summing the squares of the SAC-CI expansion coefficients corresponding to double and higher-order excitations. In all cases, MEs appear after a sharp threshold at a 2–3 times larger energy than the lowest excitation energy. Above the threshold, practically all PbSe states are MEs, while single excitons, MEs and their superpositions exist in CdSe.
(Adapted from Ref. 22. Copyright 2008 American Chemical Society).

optically excited states between 2.5 and 3 times the band gap become MEs, while both single and multiple excitons are seen in Cd_6Se_6, Figure 2.3. Figure 2.4 is a HF DOS of CdSe and PbSe QDs and demonstrates that small QDs provide good representation of large QD DOS, implying that conclusions derived from the small dots using computationally demanding CI should hold for large dots. Note that the excitation energies predicted by HF are three times larger than those computed by CI, indicating the importance of the electron–hole interaction. It is precisely this strong interaction that breaks selection rules and allows direct photoexcitation of MEs which explains the ultrafast generation of MEs without the need for phonon relaxation bottleneck.[22]

Silicon QDs deserve particular attention, since much of the present photovoltaic industry is already based on Si. Experimental measurements on Si QDs put the threshold of MEG at 2.4 times the first band gap excitation energy and found a quantum yield of 2.6 excitons per photon at 3.4 times E_g.[16] In order for Si QDs to become a viable option in solar devices, a theoretical understanding of MEG in Si QDs is necessary. First principles calculations show that at 2–3 times the lowest excitation energy the majority of the optically excited states in neutral Si_7 and Si_{10} take on ME character. The transition from SEs to MEs is not as sharp as in PbSe clusters, but it is much more pronounced than in CdSe. The closer similarity of Si to PbSe than CdSe is unexpected because Si clusters are less symmetric than PbSe clusters. The fraction of excited states that are MEs is plotted in Figure 2.5 as a function of the excitation energy for the neutral QDs. These data are obtained by summing the squares of the expansion coefficients corresponding to double and higher order excitations from the SAC-CI calculations. Both singles and

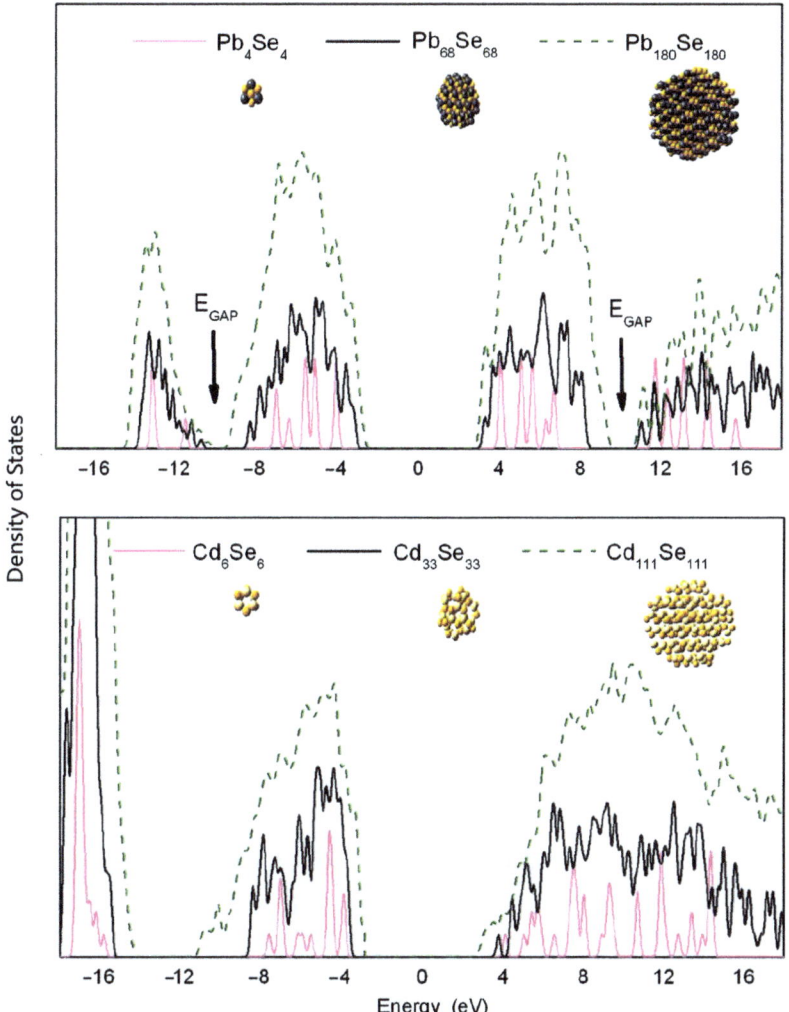

Figure 2.4 Single-particle DOS for Pb_4Se_4, $Pb_{68}Se_{68}$, $Pb_{180}Se_{180}$, Cd_6Se_6, $Cd_{33}Se_{33}$ and $Cd_{111}Se_{111}$ QDs (Figure 2.1) computed in the HF approximation and LANL2dz basis. The DOS of the smaller QDs provides a good approximation for the DOS of the larger QDs, indicating that the SAC-CI results for smaller dots hold for larger dots.
(Adapted from Ref. 22. Copyright 2008 American Chemical Society).

doubles (SD) and general (GEN) SAC-CI methods were employed in the calculations. The GEN calculations shows the onset of MEG to be approximately twice the band gap while the SD calculations put the threshold closer to three times the band gap. For both the SD and GEN calculations in Si_7, there is an extended region of superposition of SEs and MEs, while the Si_{10} cluster shows a much sharper transition from SEs to MEs.[33]

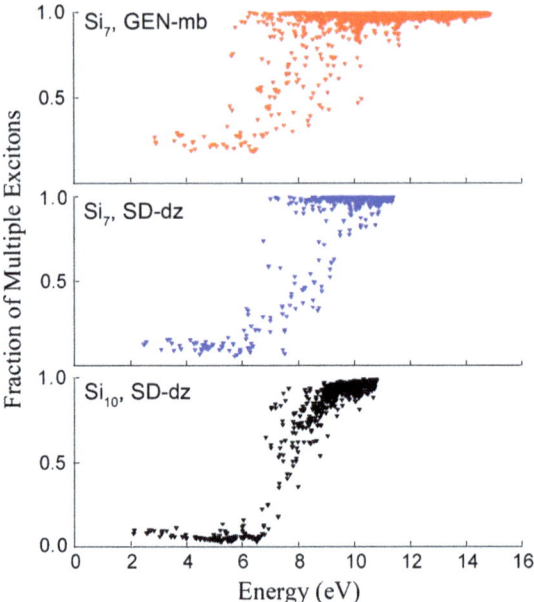

Figure 2.5   Fraction of MEs in the excited states of the neutral QDs calculated from the squares of the SAC-CI expansion coefficients corresponding to double and higher order excitations. MEs appear in both QDs at energies 2–3 times the first excitation energy and become predominant at higher energies.
(Adapted from Ref. 33. Copyright 2009 American Chemical Society).

2.4.2   MEG with Dopants, Defects and Charging

In order to realize QD photovoltaic cells successfully, detailed analysis of excitations in these materials is needed, extending beyond the ideal case and including various defects, such as photoionization, doping and surface defects.[35] In pure ideal semiconductors, excitations are relatively straight-forward; an absorbed photon excites an electron across the band gap from the valence band (**VB**) to the conduction band (**CB**). This produces an interacting electron–hole pair, or exciton, which is stabilized by the coulomb interaction of the two charges in the semiconducting material. Additionally, if the energy of the incident photon is at least twice the energy of the band gap then MEG can occur.

Semiconductor quantum dots often exhibit a variety of defects, such as missing or extra atoms, dopants and excess charges. Deviations from the ideal structure can occur either by design, such as doping Si with phosphorous or boron to create an n- or p-type material,[37] or by inherent surface defects[9] such as dangling bonds, or by inadvertent photoionization of the material. Materials containing defects exhibit transitions beyond SEs and MEs.[32,33] Figure 2.6 shows a diagram of the various transitions that can take place. If material becomes charged, an extra charge carrier appears in the VB or CB and

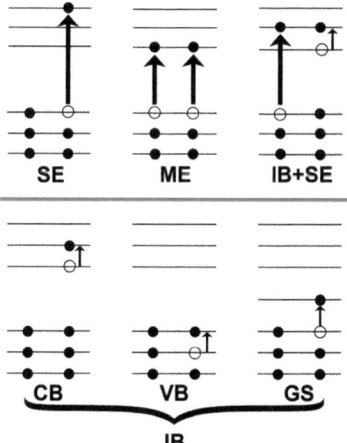

Figure 2.6 Possible electronic transitions in the semiconducting clusters. SE and ME transitions can occur in all the systems and conductions band (CB), valence band (VB) and gap-state (GS) transitions, collectively referred to as intra-band (IB) transitions, become possible when the material is modified from the ideal cluster. Combinations of IB and SE/ME transitions can also occur.
(Taken from Ref. 35. Copyright © 2011 Royal Society of Chemistry).

additional electronic transitions within these bands become possible. Dopants can introduce a state near the band edge and surface defects that create dangling bonds can create gap states, which can be occupied or unoccupied and are not necessarily near the band edge. For the remainder of this section VB, CB and gap state transitions will collectively be referred to as intra-band (IB) transitions.

IB transitions create excitons, but these excitons are contained within either the VB or CB. This renders them useless for photovoltaic applications, which need to separate the charges across the band gap. They also ensure that at sufficiently high energies excitations are complicated, since these IB transitions occur in conjunction with SEs. This results in a multi-electron excitation that only creates one new electron–hole pair spanning the band gap. While formally these transitions are MEs, in photovoltaic applications they are equivalent to SEs.

This section summarizes SAC-CI results found for positively and negatively charged clusters of PbSe[32] and Si,[33] respectively, as well as PbSe and Si clusters involving dopants and dangling bonds.[34] The calculations employed the same geometric structure for the neutral and charged clusters (inserts in Figure 2.7). The dopant case is represented by replacing a Si atom from the original cluster with a P atom. Dangling bonds are represented by removing a Pb atom from a PbSe cluster.

Figure 2.7 shows the optical spectra generated from the SAC-CI calculations for each of the systems studied. The spectra were obtained non-perturbatively, by explicitly computing transition dipole moment and oscillator strength for

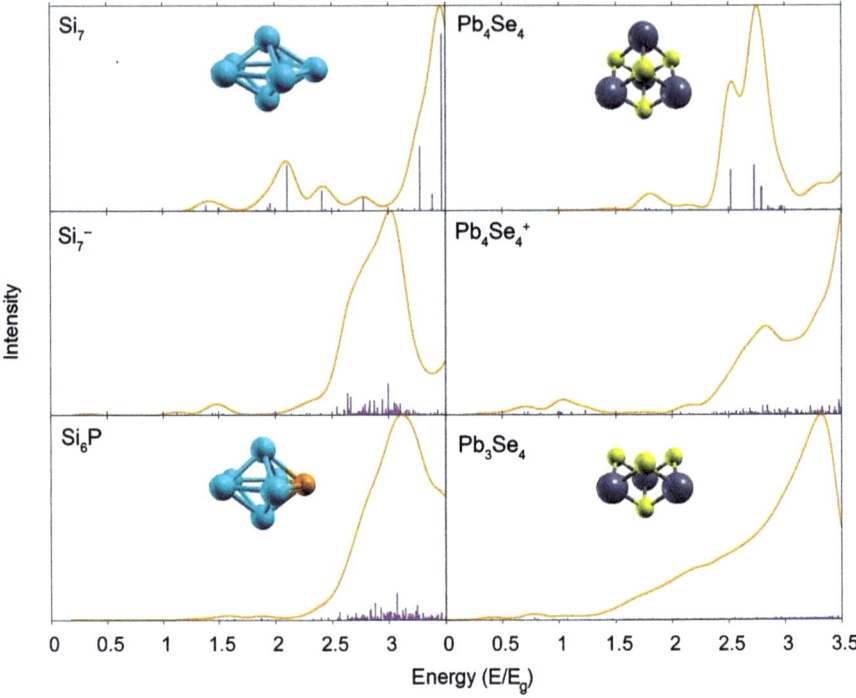

Figure 2.7   Optical absorption spectra. Modifications to the systems result in a blue
shift in the absorption spectra. Adding a dopant or introducing a dangling
bond causes a similar effect to charging the system. The energy scales for
the Se and PbSe clusters are shown relative to the lowest excitation
energies, *i.e.* band gaps E_g, of the corresponding neutral cluster, namely,
$E_g(Si_7) = 2.55$ eV, $E_g(Pb_4Si_4) = 2.72$. The inserts are the geometries of the
systems studied. The charged clusters have the same geometries as the
respective neutral clusters.
(Adapted from Ref. 35. Copyright © Royal Society of Chemistry 2011).

each transition. Generally, the bands seen in the spectra are composed of
multiple individual excitations. The calculations show that a modification of
the cluster from its ideal structure blue shifts the main absorption peaks in the
optical spectra; compare the top panels in Figure 2.7 with the middle and
bottom panels. This is seen with all types of defects including doping, charging
and dangling bonds. The effect has been observed experimentally and can be
rationalized by Pauli blocking.[38–40] The electron (hole) occupation of the lowest
(highest) energy CB (VB) state in an n-(p-) type cluster prevents a transition
into (from) this state from (to) a VB (CB) electron (state). While modifications
to the clusters open the possibility of much lower energy IB transitions, these
transitions do not make a significant contribution to the absorption spectra.

   The contributions of each excited state configuration to the overall character
of the excited state in the ideal and modified clusters are analyzed in Figure 2.8.
The data points shown in Figure 2.8 are the squares of the expansion

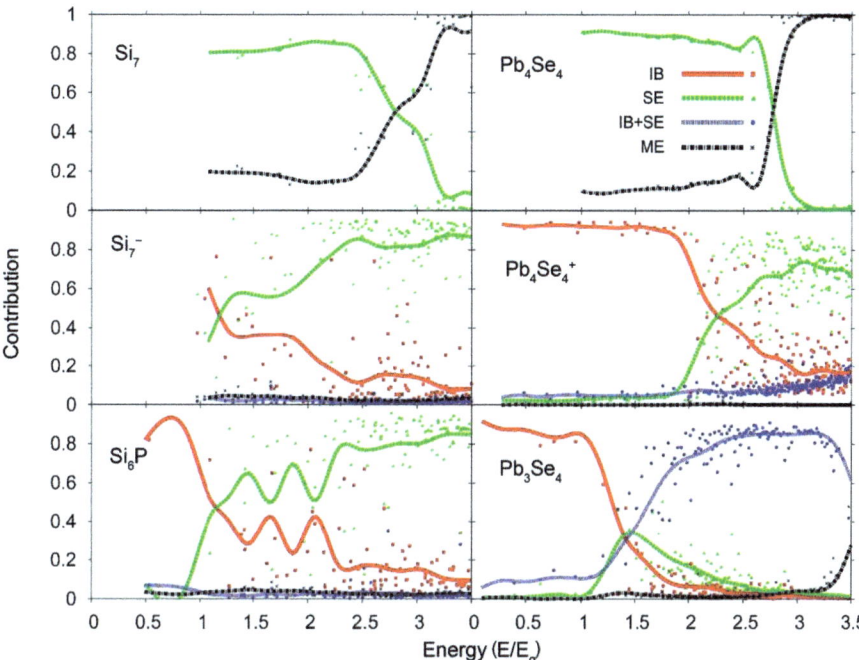

Figure 2.8 Contributions of the different electronic transitions to the excited state character for each of the studied systems. Red gives the contributions of intra-band (IB) transitions, green depicts single exciton (SE) transitions, blue corresponds to IB transitions combined with exciton transitions (IB + SE) and black shows ME transitions. The data points are the squares of the SAC-CI expansion coefficients for the corresponding transition types. Only every fifth data point is plotted for clarity. The lines give the average behavior of the respective transition type. (Taken from Ref. 35. Copyright © Royal Society of Chemistry 2011).

coefficients for a given excitation represented on the basis of the single- or multi-electron configurations defined in Figure 2.6. For both ideal Si_7 and Pb_4Se_4 clusters, the transformation from all predominantly SE to ME character occurs between 2.6 and 3 times the band gap energy. The transformation is sharper in the PbSe cluster. In both ideal cases the ME contribution is significant across the entire energy range.

The Si_7^- and Si_6P clusters show almost identical behavior, which is to be expected, since doping Si with P essentially amounts to adding an extra electron to the CB. Excitation energy in these clusters is smaller than that of the ideal cluster because the lowest excitations in the modified clusters arise from the IB transitions. As the energy increases, the IB contributions gradually decrease in magnitude and SEs become the dominant configuration. At energies where MEs overcome SEs in the ideal Si_7 cluster, SE configurations remain dominant in the Si_7^- and Si_6P clusters. The IB + SE and ME two-electron configurations begin contributing to the electronically excited states of the Si_7^- and Si_6P clusters at very high energies.[33]

The modified PbSe clusters show similar features to each other. IB configurations dominate at low energies in both modified PbSe clusters. At higher energies, the IB contributions fall and the SE contributions rise. SEs make significant contributions to the cationic $Pb_4Se_4^-$ cluster in a wider range, however, the onset of SEs is shifted to energies that are twice that of the SE energy onset in the neutral cluster. The ME transitions are also shifted to much higher energies. In contrast, in the Pb_3Se_4 cluster the SE contribution never gets above 50% and it is quickly overtaken by the IB + SE configurations. The MEs in the Pb_3Se_4 actually start around the same absolute energy as the ideal PbSe cluster. The fact that the MEs appear at lower energies in the NC, missing an atom, is due to reorganization of the local bonding pattern which partially eliminates the dangling bonds.[41]

2.5 Phonon Induced Dephasing

Electron–phonon interactions give rise to two related but qualitatively distinct phenomena: relaxation and dephasing. Compared to the electron–phonon energy relaxation, electron–phonon dephasing is a more subtle effect. Dephasing is an elastic process that conserves electronic energy. It destroys coherences between electronic states and converts them into ensembles of uncorrelated states. MEG is intimately related to phonon-induced dephasing. Of the three proposed mechanisms for MEG, two rely on the dephasing processes. Light absorption and emission create superposition of ground and excited states. Dephasing of this superposition determines optical line widths. Coulomb interactions generate superposition of single and multiple excited electron–hole pairs. Elastic dephasing of exciton superpositon plays a key role in the excited state dynamics in semiconductor QDs. Further reading on the theory and calculations discussed in this section can be found in work by Prezhdo and co-workers.[42–44]

2.5.1 Optical Response Function

Excluding inhomogeneous broadening associated with a distribution of optically active species, the intrinsic homogeneous line width, Γ, of an optical transition is inversely proportional to the dephasing time T_2.[45] The latter includes the excited state lifetime T_1 and pure dephasing time T_2^*:

$$\Gamma = \frac{1}{T_2} = \frac{1}{2T_1} + \frac{1}{T_2^*} \qquad (2.3)$$

For sufficiently long T_1, Γ is determined by T_2^*.

The pure dephasing time is associated with fluctuations and uncertainties in the energy levels owing to the electron phonon coupling in the semiconductor, ligands, and solvent, *etc.* Fluctuations in the energy levels are best characterized in terms of correlation functions. The unnormalized autocorrelation function (ACF), Cu(t), for a transition of energy E is defined as:

$$C_u(t) = \langle \Delta E(t) \Delta E(0) \rangle \tag{2.4}$$

where $\Delta E = E - \langle E \rangle$ and the angular brackets denote averaging over a statistical ensemble, in particular, canonical averaging in the present study. The initial value of the unnormalized ACF gives the average fluctuation in the transition energy, $C_u(0) = \langle \Delta E^2(0) \rangle$. Dividing $C_u(t)$ by $C_u(0)$ gives the normalized ACF:

$$C(t) = \frac{\langle \Delta E(t) \Delta E(0) \rangle}{\langle \Delta E^2(0) \rangle} \tag{2.5}$$

ACFs characterize periodicity and memory of the energy fluctuations. Rapid decay of an ACF indicates short memory and occurs if multiple phonon modes couple to the electronic transition and if these modes are anharmonic.

The Fourier transform (FT) of the ACF is known as the influence spectrum:

$$I(\omega) = \left| \frac{1}{\sqrt{2\pi}} \int_{-\infty}^{\infty} dt e^{-i\omega t} C(t) E \right|^2 \tag{2.6}$$

It identifies the frequencies, omega, of the phonon modes that efficiently couple to the electronic subsystem. The strength of the electron–phonon coupling for a particular mode is proportional to the intensity of the corresponding line in the influence spectrum.

The optical response functions characterizing the dephasing processes for a pair of states that are entangled in a coherent superposition can be obtained directly or *via* the second order cumulant expansion.[45] The cumulant expansion approximation together with the FT of the ACF provides additional information about the dephasing process. The cumulant dephasing function is obtained by double integration and exponentiation of the unnormalized ACF, Equation (2.4).

$$D(t) = \exp(-g(t)), \quad g(t) = \frac{1}{\hbar^2} \int_0^t d\tau_1 \int_0^{\tau_1} d\tau_2 C_u(\tau_2) \tag{2.7}$$

The above expression indicates that rapid dephasing is facilitated by a large fluctuation in the transition energy, (*i.e.* $C_u(0) = \langle \Delta E^2(0) \rangle$), as well as by a short memory of the fluctuation, (*i.e.* the rapidly decaying ACF).

Alternatively, the dephasing function can be computed directly as:

$$D(t) = \exp(i\omega t) \left\langle \exp\left(-\frac{i}{\hbar} \int_0^t \Delta E(\tau) d\tau \right) \right\rangle \tag{2.8}$$

Here, ω is the frequency corresponding to the thermally averaged transition energy $\langle \Delta E \rangle$. It is more difficult to achieve convergence with the direct expression since it involves averaging of a complex-valued oscillatory function,

$$\exp\left(-(i/\hbar) \int_0^t \Delta E(\tau)d\tau \right),$$

whose real and imaginary parts change signs. In comparison, the cumulant expression, Equations (2.4) and (2.7), involves averaging over a real and positively valued transition energy and its ACF.

2.5.2 Phonon Dephasing in PbSe Quantum Dots

The methods described above were initially used to study the electron–phonon dephasing in PbSe.[42] The simulations were performed with *ab initio* DFT as implemented in the Vienna *ab initio* Simulation Package (VASP).[46–48] Vanderbilt's ultrasoft pseudopotentials[49] and the PW91density functional[50] were used. Both $Pb_{16}Se_{16}$ (1 nm) and $Pb_{68}Se_{68}$ (1.4 nm) were studied. The initial geometries were generated based on the bulk structure of PbSe. Then the clusters were fully relaxed and optimized at zero temperature. Each cluster was brought up to 300 K by molecular dynamics with repeated velocity rescaling. A microcanonical trajectory was generated for each cluster using the Verlet algorithm with Hellmann/Feynmann forces. The theories are applied to a variety of electronic excitations, including high-energy exciton/biexciton superposition and the band gap excitation at room temperature. The exciton (biexciton) states were represented in the Kohn–Sham orbital picture by promoting one (two) electron(s) from occupied to unoccupied orbitals. The orbitals were optimized for the ground excited state. The excitation energies were estimated from the orbital energies and their occupation numbers. The methods described above were used in the remainder of this chapter.

Figure 2.9(a) shows the autocorrelation function for the E_g/ground and $3E_g$/biexciton pairs of states for the smaller and larger PbSe clusters. The amplitude of $C(t)$ is notably greater for the $3E_g$/biexciton pair, indicating that the energy difference between these states fluctuates much more than the energy difference between the band gap exciton and ground states. The autocorrelation functions for the larger dot oscillates much faster, implying that a much wider range of vibrational frequencies interacts with the larger dot. In both cases the correlation decays within several hundred femtoseconds.

The influence spectrum is shown in Figure 2.9(b) indicates that the dephasing takes place primarily owing to low-frequency acoustic modes. As expected from the autocorrelation function, the vibrations from the larger dot show a wider range of frequencies; however, each frequency contributes smaller amplitude compared to the smaller dot.

The dephasing functions obtained directly and with the cumulant expansion are plotted in Figure 2.9(c). The direct and cumulant dephasing functions show very good agreement. The minor differences seen at longer times indicate that

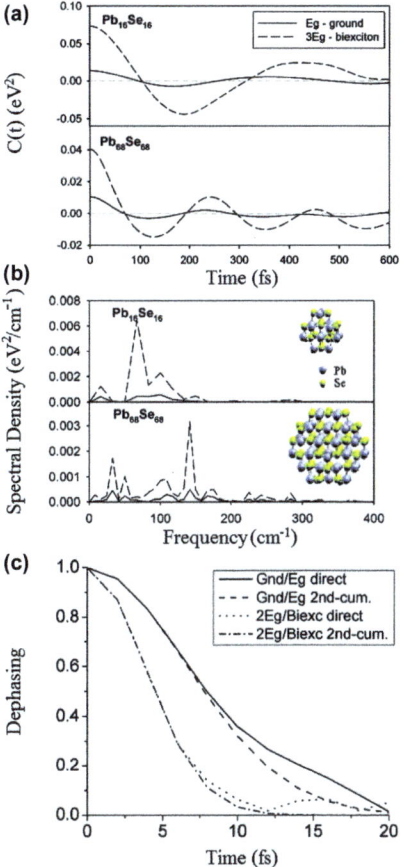

Figure 2.9 (a) Autocorrelation of the band gap (solid line) and of the energy difference between the triple gap exciton ($3E_g$) and biexciton state (dashed line). Both functions show similar decay times, on the order of hundreds of femtoseconds. The $3E_g$/biexciton energy difference fluctuates more. (b) Spectral density of the autocorrelation functions shown in part a. The larger dot shows a wider range of frequencies, which contribute smaller amplitudes individually. The QD size dependence of the phonon frequency modes indicates that dephasing occurs *via* acoustic phonons. The inserts are typical geometry of $Pb_{16}Se_{16}$ and $Pb_{68}Se_{68}$ clusters at room temperature. (c) Dephasing function for the band gap/ground state pair and triple gap exciton ($3E_g$)/biexciton pair obtained directly from Equation (2.8) and using the second-order cumulant expansion, Equations (2.4) and (2.7). The direct and cumulant expansion results agree very well. The minor differences seen at long times are due to memory effects.[42]
(Adapted from Ref. 42. Copyright 2006 American Chemical Society).

higher order correlations do play some role. The minor differences seen at longer times indicate that indeed higher order cumulants play a role in the current system.

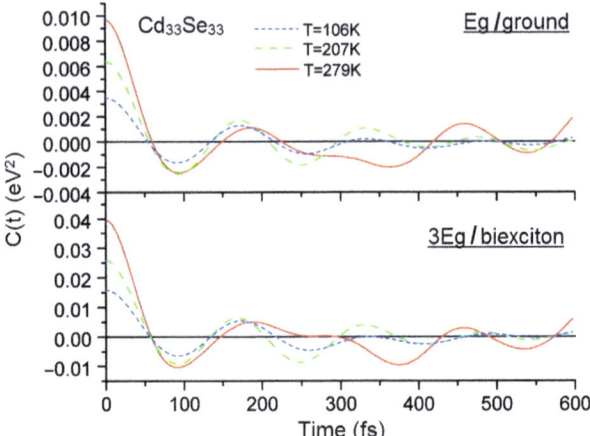

Figure 2.10   Autocorrelation functions $C(t)$ for the gaps between the lowest energy exciton and the ground state and a high energy exciton and biexciton in the $Cd_{33}Se_{33}$ QD.[43]
(Adapted from Ref. 43. Copyright 2008 American Chemical Society).

2.5.3   Temperature Dependence of Phonon Dephasing

The temperature dependence of decoherence was studied for $Cd_{33}Se_{33}$ and $Pb_{68}Se_{68}$.[43] Figure 2.10 shows the autocorrelation functions between E_g/ground and $3E_g$/biexciton for the CdSe dot at three different temperatures. The initial values of $C(t)$ differ considerably at the three different temperatures. As expected, the amplitude of the autocorrelation function is higher at higher temperatures indicating that the energy difference between states fluctuates much more at higher temperatures. However, $C(t)$ are quite close after the first oscillation for all temperatures. Since the decoherence time is an order of magnitude shorter than the oscillation period, the decoherence between the electronic states is dominated by the initial value of $C(0)$. This is just the square of the vibrationally induced fluctuation of the electronic energy gap. In either scheme a smooth Gaussian profile of the decay was observed. The pure dephasing/decoherence time was determined from Gaussian decay: $t = \lambda^{-1/2}, D(\tau) \propto -\lambda\tau^2$. The values are shown in Table 2.1. The decoherence time changed almost linearly with temperature.

2.5.4   Multiple Exciton Generation, Fission and Luminescence and Dephasing

MEG is intimately related to phonon-induced dephasing. Impact ionization[3] assumes incoherent transitions from high-energy SE states to MEs. The loss of electronic coherence occurs by coupling to phonons and, in this case, should be faster than MEG. The dephasing mechanism[6,21,24] starts with a coherent superposition of single and ME states and associates MEG with dephasing of the

Table 2.1 Phonon-induced pure dephasing/decoherence time for pairs of electronic states. Adapted from Ref. 43 (Copyright 2008 American Chemical Society).

Cluster	State pair	Pure-dephasing time/fs		
$Cd_{33}Se_{33}$		$T = 106$ K	$T = 207$ K	$T = 279$ K
	E_g/ground	16.2	11.2	9.6
	$2E_g/E_g$	17.8	12.1	11.0
	$3E_g/E_g$	13.3	10.7	8.9
	$2E_g$/biexciton	8.9	6.0	5.4
	$3E_g$/biexciton	7.5	5.6	4.8
$Pb_{68}Se_{68}$		$T = 128$ K	$T = 197$ K	$T = 347$ K
	E_g/ground	22.6	16.3	9.3
	$2E_g/E_g$	19.0	17.0	10.6
	$3E_g/E_g$	22.4	15.2	8.6
	$2E_g$/biexciton	10.7	8.6	5.3
	$3E_g$/biexciton	11.6	8.1	4.7

superposition. Finally, no matter how MEs are created they must dissociate into uncorrelated excitons that coexist in the same QD and luminesce independently. This loss of correlation within the coherent superposition of MEs has been named ME fission (MEF) and can be explained by phonon-induced dephasing.[44] Dephasing in the context of luminescence, MEG and MEF were studied for Si quantum dots because of Si's current importance in the solar cell industry.

Figure 2.11(a) presents the normalized ACFs, Equation (2.5), for the luminescence (E_g/ground), MEG ($3E_g$/triexciton) and MEF ($E_g/E_g^{e,h}$) processes at 300 K and 80 K. The first two ACFs track each other quite closely. In contrast, the ACFs describing MEF oscillate with much smaller amplitudes. Since the higher temperature excites a wider range of vibrations and makes them less harmonic, the ACFs oscillate more randomly at 300 K than 80 K.

The Fourier transforms (FT) of the ACFs, Equation (2.6), are shown in Figure 2.11(b). They indicate that both acoustic and optical modes contribute to luminescence linewidths and MEG. As with the ACFs, Figure 2.11a, the E_g/ground and $3E_g$/triexciton data closely match. In contrast, only low-frequency acoustic modes are involved in MEF. The majority of the modes contributing to the dephasing processes involved in luminescence and MEG lie within the 200–500 cm^{-1} frequency range. Additional peaks appear at 600 cm^{-1} and at the very low frequencies of less than 100 cm^{-1}. The phonon around 350–360 cm^{-1} is particularly important at the lower temperature, bottom panel of Figure 2.11(b). Thermal fluctuations break the symmetry of the QD and distort its geometry. As a result, a larger number of modes contribute to the luminescence and MEG dephasing processes at the higher temperature. This is particularly well demonstrated with the low frequency modes, which appear only at $T = 300$ K and arise from QD distortions. Compared to the PbSe and CdSe QDs of similar size, dephasing of luminescence and MEG is caused by higher frequency modes in the Si QD, as should be expected, since Si is a lighter atom. MEF is dominated by low frequencies at both temperatures. MEF

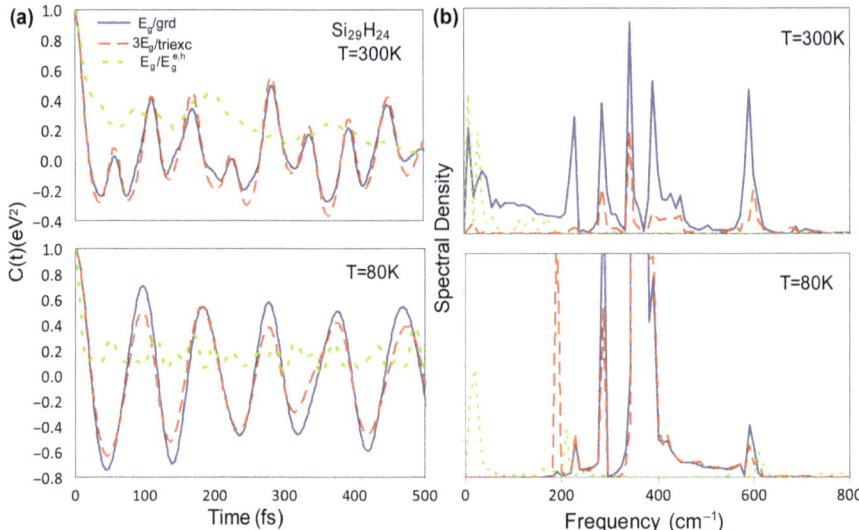

Figure 2.11   (a) ACFs for the luminescence (E_g/grd), MEG ($3E_g$/triexc) and MEF
($E_g/E_g^{e,h}$) dephasing processes. The top and bottom frames correspond to
ambient and low temperature as indicated. The luminescence and MEG
ACFs differ from MEF ACFs because MEF couples to a different set of
phonons. (b) Fourier transforms of the ACF shown in part (a).[44]
(Adapted from Ref. 44. Copyright 2009 American Chemical Society).

involves orbitals that are close in energy and therefore, have similar densities,
number of nodes, and so on. As a result, the energy gaps between pairs of low
energy excitons oscillate very slowly, even though the energies of individual
excitons oscillate at higher frequencies, as evidenced by the E_g/ground data.

The direct dephasing functions, Equation 2.8, corresponding to Figure 2.11
are plotted in Figure 2.12. The dephasing times shown in Table 2.2 were obtained
by fitting the dephasing functions with Gaussians. The dephasing times involved
in luminescence and MEG are all sub-10 fs. The dephasing involving a triexciton
occurs faster than the dephasing involving a biexciton or a single exciton.

The dephasing of MEF is significantly slower than the dephasing responsible
for luminescence linewidths and MEG. The order of magnitude difference
becomes more pronounced at the lower temperature. At first glance the drastic
disparity in the decay of the dephasing functions shown in Figure 2.12 con-
tradicts, the similarity of the ACF decay rates, Figure 2.11. In the cumulant
approximation, Equation 2.7, the dephasing rate can be attributed to a combination of decay time and amplitude of unnormalized ACF. It is the amplitude
that is responsible for the large difference between the dephasing times of MEF
and MEG/luminescence. The fluctuation of the MEG and luminescence energy
gaps determines the initial value of the unnormalized ACF, Equation 2.4.
Larger gap fluctuations result in faster dephasing. The states involved in MEF
are very close in energy. The state densities are similar and the energy gaps

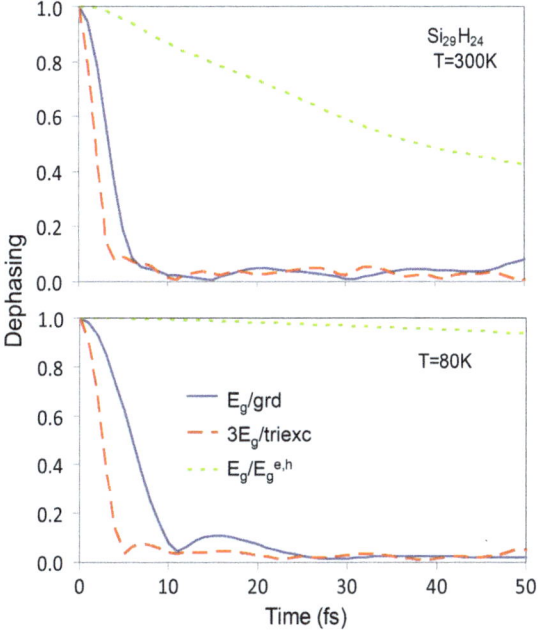

Figure 2.12 Dephasing functions for the luminescence, MEG and MEF processes. Dephasing times obtained by Gaussian fitting are shown in Table 2.2.[44] (Adapted from Ref. 44. Copyright 2009 American Chemical Society).

Table 2.2 Dephasing times (fs) for luminescence, MEG and MEF.

Process	$T = 80\,K\ <E_g> = 2.55\,eV$	$T = 300\,K\ <E_g> = 2.55\,eV$
Luminescence: E_g/grd	6.98 ± 0.21	3.96 ± 0.10
MEG: $2E_g$/biexc	7.33 ± 0.17	3.95 ± 0.13
$3E_g$/biexc	4.88 ± 0.11	2.63 ± 0.07
$3E_g$/triexc	2.96 ± 0.09	1.60 ± 0.05
MEF: $E_g/E_g^{e,h}$	205.33 ± 0.87	54.43 ± 1.06
E_g^h/E_g^e	334.78 ± 5.29	86.25 ± 1.58
E_g/E_g^h	337.52 ± 3.02	103.14 ± 1.63
E_g/E_g^e	367.01 ± 1.94	78.73 ± 0.96

between these states fluctuate negligibly. Density changes involved in MEG and luminescence create more significant perturbations to the vibrational lattice, involve a broader range of phonon motions, Figure 2.11 and accelerate dephasing, Figure 2.12 and Table 2.2. Since the MEF dephasing times arise from small fluctuations in the energy gaps, they should be sensitive to QD size,[2,3] surface ligands, solvent, and so on. Nevertheless, it is quite clear that, generally, MEF dephasing should take much longer than that of MEG and luminescence. While the MEF time has not been measured yet in semiconductor QDs, the MEF time calculated here for Si QD agrees with the time scale of singlet fission observed in molecular systems.[51]

2.6   Electron Phonon Relaxation

Owing to the broad bandwidth of the solar spectrum, electrons in semi-conductor photovoltaic materials can be excited into different energy levels in the conduction band. These excited electrons usually quickly lose excess energy to the lattice *via* a series of electron–phonon scattering events. To reduce this thermalization loss for possible solar efficiency enhancement, the electron relaxation rate should be minimized.[3,8,52] The intrinsic electron–phonon coupling in a bulk material cannot be modified. However, making the semiconductors into nanostructures such as QDs provides possibilities for decoupling electrons and phonons. Because of the quantum confinement effect, the electronic structure of quantum dots exhibits discrete energy levels and the spacing between adjacent levels may become larger than the energy of a single phonon. In this case only multiphonon processes will be able to relax the electrons, potentially lowering the relaxation rate compared to bulk materials.[3,53,54]

Experiments have also shown that both electron–phonon and Auger-type processes can coexist in the electron relaxation process in nanocrystals (NCs).[55] Depending on type of the material, status of surface passivation and type of surface ligands, either the electron–phonon process or the Auger process can dominate the charge carrier relaxation. However, the Auger process can be avoided if electrons and holes are separated or if electrons and holes have a similar DOS. Current experimental methods can enable effective separation of electrons and holes, making it possible to decouple the Auger and electron–phonon relaxation processes in NCs.[55,56] Slowing down electron–phonon coupling is favored for MEG efficiency and enhanced solar cell efficiency in general.

Owing to the fundamental and practical importance of electron–phonon coupling in photovoltaic nanomaterials, it is crucial to understand how the electron–phonon relaxation depends on a variety of factors including material, temperature, nanoparticle size and shape, surface terminations, surfactants, and so on.[9] Recently, a non-adiabatic molecular dynamics approach has been developed[12,57] to simulate the electron relaxation process in QDs[58–60] and this process has been used to investigate several different systems and is discussed below; for further reading consult Prezhdo and co-workers.[58–63,89–92]

2.6.1   Time Dependent Density Functional Theory

The electron density in the time dependent density functional theory (TDDFT) is written in the KS representation as:[64]

$$\rho(x, t) = \sum_{p=1}^{N_e} \left| \varphi_p(x, t) \right|^2 \qquad (2.9)$$

where N_e is the number of electrons and $\varphi_p(x, t)$ are single electron KS orbitals. The evolution of $\varphi_p(x, t)$ is determined by the application of the TD vibrational principle to the KS energy:[64]

$$
E\{\varphi_p\} = \sum_{p=1}^{Ne} \langle \varphi_p | K | \varphi_p \rangle + \sum_{p=1}^{Ne} \langle \varphi_p | V | \varphi_p \rangle
$$
$$
+ \frac{e^2}{2} \iint \frac{\rho(x', t)\rho(x, t)}{|x - x'|} d^3 x d^3 x' + E_{xc}\{\rho\}, \qquad (2.10)
$$

which contains the kinetic energy of non-interacting electrons K, the electron–nuclear attraction V which depends on the phonon coordinates R, the coulomb repulsion of density $\rho(x, t)$ and the exchange-correlation energy functional E_{xc} which accounts for the residual many-body interactions. The KS formulation of DFT uses a single-particle representation that effectively accounts for electron correlation and incorporates electron–hole interactions and excitonic effects.

Based on the Runge–Gross theorem, it is shown that the three-dimensional density of a many body quantum system is sufficient to describe the TD response of the system to an external perturbation, such as electromagnetic field and vibrational motion, $R(t)$. This is known as time dependent density functional theory (TDDFT). In the linear response approximation, TDDFT is frequently used to evaluate electronic excitation energies. Here, we use full TDDFT where the density is explicitly propagated in time. Application of the TD (Dirac) vibrational principle to KS energy generates density evolution:

$$
i\hbar \frac{\partial \varphi_p(x, t)}{\partial t} = H\{\varphi(x, t)\}\varphi_p(x, t), p = 1, \ldots, N_e. \qquad (2.11)
$$

The TDKS equations are coupled, since the Hamiltonian H depends on the overall density, Equation (2.9), and hence, all occupied KS orbitals. We solve the TDKS equation by expanding the TDKS orbitals $\varphi_p(x, t)$ in the adiabatic KS orbital $\tilde{\varphi}_k(x; R)$ basis obtained in the ground state DFT calculation:

$$
\varphi_p(x, t) = \sum_{k}^{Ne} c_{pk}(t) | \tilde{\varphi}_k(x; R) \rangle \qquad (2.12)
$$

Then, Equation (2.11) transforms into the equation for the expansion coefficients:

$$
i\hbar \frac{\partial c_{pk}(t)}{\partial t} = \sum_{m=1}^{Ne} c_{pm}(t) \left(\epsilon_m \delta_{km} - i\hbar d_{km} \cdot \dot{R} \right) \qquad (2.13)
$$

The non-adiabatic (NA) coupling describes the electron–phonon interaction:

$$
d_{km} \cdot \dot{R} = \langle \tilde{\varphi}_k(x; R) | \nabla_R | \tilde{\varphi}_m(x, R) \rangle \cdot \dot{R} = \left\langle \tilde{\varphi}_k(x, R) \left| \frac{\partial}{\partial t} \right| \tilde{\varphi}_m(x, R) \right\rangle \qquad (2.14)
$$

arising from the dependence of the adiabatic KS orbitals on the phonon coordinates $R(t)$.[65]

2.6.2   Non-Adiabatic Molecular Dynamics

The prescription for phonon dynamics $R(t)$ constitutes the quantum backreation problem and the combined evolution of the electrons and phonons is known as non-adiabatic molecular dynamics (NAMD). Traditionally, MD calculations are performed in the ground electronic state. NAMD includes transitions between different electronic states. In order to define the quantum back reaction, FSSH uses a stochastic algorithm that generates trajectory branching[66] and detailed balance.[14] Trajectory branching mimics the splitting of the quantum mechanical wavepackets describing phonons in correlation with different electronic states. Detailed balance ensures that transitions to higher energy are less likely than transitions to lower energy by the Boltzmann factor. This is essential when studying electron–phonon relaxation and achieving thermodynamic equilibrium.

FSSH is typically performed in the adiabatic basis obtained in the static calculation.[14,65,66] While the adiabatic forces in the ground and excited electronic states as well as the NA couplings between them can be calculated in TDDFT,[64] the NA coupling between excited electronic states remains an open question.[67] We use the zeroth-order adiabatic basis formed by the adiabatic KS orbitals. Compared to the original implementation of the TDKS–FSSH method,[12] in the work discussed in the remainder of the chapter, a further approximation is used by going from the many-particle Slater determinant basis to the single-particle representation. The FSSH simulation is performed separately for electrons and holes on the basis of single-particle adiabatic KS orbitals. The single-particle representation is appropriate for studies of QDs since their electronic structure is well represented by the independent electron and hole picture. Quantum confinement effects in QDs ensure that the electron and hole kinetic energies dominate the electrostatic interaction. As a result, even the basic effective mass theory provides a good description of the QD electronic structure.[68]

FSSH prescribes a probability for hopping between electronic states. The probability is explicitly time dependent and is correlated with the nuclear evolution. The probability of hopping between states k and m within the time interval Δt depends explicitly on phonon dynamics and equals:[66]

$$g_{km}(t, \Delta t) = \max\left(0, \frac{b_{km}\Delta t}{a_{kk}(t)}\right) \tag{2.15}$$

where

$$b_{km} = -2Re\left(a^*_{km}d_{km} \cdot \dot{R}\right); \quad a_{km} = c_k c^*_m \tag{2.16}$$

The coefficients c_k and c_m evolve according to Equation (2.13) and $d_{km} \cdot \dot{R}$ is the NA coupling in Equation (2.14). If the calculated g_{km} is negative, the hopping probability is set to zero; a hop from state k to state m can only occur if the electronic occupation of state k decreases and the occupation of state m increases. This feature of the algorithm minimizes the number of hops and is responsible for the name of the technique "fewest switches". To conserve the total electron nuclear energy after a hop, the original FSSH technique rescales the nuclear velocities along the direction of the electronic component, d_{km}, of the NA coupling.[65,66] If the kinetic energy available to the nuclei along the direction of the NA coupling is insufficient to accommodate an increase in the electronic energy, the hop is rejected. The hop rejection creates a detailed balance between upward and downward transitions.[14]

In the simplified implementation of FSSH, it is assumed that the energy exchange between the electronic and vibrational degrees of freedom during a hop is rapidly redistributed between all vibrational modes and the velocity rescaling and hop rejection is replaced by multiplying the probability (Equation 2.15) for transitions upward in energy by the Boltzmann factor. This simplification leads to significant improvements in the computational efficiency of FSSH, since it permits the use of the ground state nuclear trajectory to determine the TD potential that drives the electron dynamics.

2.6.2.1 Simulation Details

In all of the simulations discussed in this chapter the TDDFT–FSSH theory was implemented with the VASP[47] DFT package. The simulations were performed with either Perdew and Wang[50] or PBE exchange-correlation functions, Vanderbilt pseudopotentials[49] and a converged plane wave basis. The simulations were carried out in a cubic cell periodically replicated in three dimensions, as stipulated by the plane wave basis. To prevent spurious interactions between periodic images of the QD, the cell was constructed to have at least 8 Å of vacuum between the QD replicas.

The QDs are initially constructed from the bulk structures with the bond lengths at zero temperature and then relaxed to their lowest energy configuration. This relaxed configuration is then heated to a temperature, usually 300 K with repeated velocity rescaling. After the system is heated, a ground–electronic state microcanonical trajectory is performed for several picoseconds. The transition dipole moments and oscillator strengths for excitations between the KS orbitals were computed and used both to generate the optical absorption spectrum and to pick the most optically active excitations for the initial conditions of the NA runs. A 1 fs nuclear and a 10^{-3} fs electronic time steps were used for the dynamics calculations. The data shown in the figures below are converged by averaging over at least 500 runs. Further details can be found in many references.[12,58–63,69,70]

2.6.3 Phonon-Assisted Relaxation of Charge Carriers in PbSe

Of all semiconducting nanomaterials studied, lead salts, such as PbS and PbSe, show some of the most unique electronic and transport properties.[71–75] Their conduction and valence bands are more symmetric than other semiconductor materials, such as CdSe, potentially rendering Auger relaxation processes inefficient.[76] Also the effective mass of electrons and holes in lead salts are similar and small,[77] promising strong quantum confinement effects and inducing quantization of bulk electronic bands.[73] This discretization was expected to produce a mismatch between the electronic and vibrational energy quanta and is expected to produce a phonon bottleneck in the charge carrier relaxation.[3] Both of these attributes favor MEG and indeed high quantum yields in this material were observed.[5,6,19–21] However, ultrafast intraband charge–phonon relaxation in PbSe QDs was observed, suggesting that no phonon bottleneck exsisted.[78,79] Taking a step further beyond static EMT and pseudopotential calculations, time domain theoretical studies where able to directly mimic time-resolved experiments and provide a detailed atomistic description of the relaxation process.

Phonon motions induce fluctuations in electronic DOS, as illustrated by $Pb_{68}Se_{68}$ in the top panel of Figure 2.13. The fluctuations are minor: CB and VB edges change by less than 0.1 eV and are significantly smoothed, Figure 2.13. Phonon motions mix states of different symmetries and reduce gaps between states. The bottom panel of Figure 2.13 presents NA dynamics of electron and hole relaxation mediated by coupling to phonons.[8,58,60] Charge carriers visit multiple states during relaxation and no intermediate states play special roles. The population peaks created by photoexcitation spread to re-appear near the band gap. Relaxation is nearly complete within a picosecond in agreement with experiment.[78,80] No phonon bottleneck is observed at high photoexcitation energies. Most likely transitions involve small amounts of energy that are close to a phonon energy of 100–200 cm^{-1} (12–25 meV). This indicates that electronic energy gaps are small at high energies, explaining the

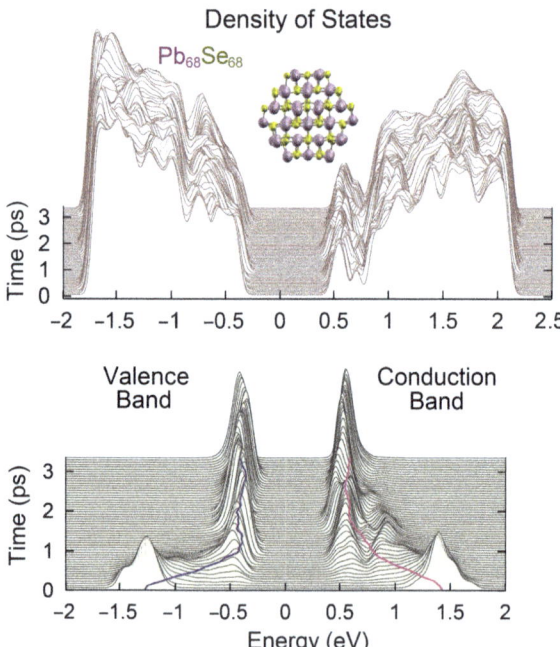

Figure 2.13   Evolution of the DOS (top panel) and the electron/hole energy relaxation (bottom panel) in $Pb_{68}Se_{68}$ at room temperature. The DOS fluctuates due to thermal atomic motions. Both electrons and holes relax within a picosecond. Holes decays faster than electrons due to higher DOS. The relaxation involves all starts at energies in the range of the photoexcited and band gap states.[60]

(Adapted from Ref. 60. Copyright 2008 American Chemical Society).

absence of the phonon bottleneck. Occasionally, up to 0.3–0.6 eV of electronic energy can be lost to phonons in single events.[8,58,60] Multiphonon relaxation was initially proposed to rationalize ultrafast experimental data[77] and is indeed seen in our simulation; however, faster resonant electron–phonon energy exchange is also efficient, particularly at higher energies.

2.6.4 Temperature Dependence

Temperature dependence of carrier relaxation distinguishes phonon-induced processes from other channels, such as Auger scattering.[68] Experiments have shown[56,78] that relaxation is much more temperature-dependent in PbSe than in CdSe QDs, where Auger processes are more effective. Different temperature dependencies suggest that phonons play an important role in most, but not all, QDs. Temperature dependence of electron–phonon relaxation was investigated in $Pb_{16}Se_{16}$[62] and $Cd_{33}Se_{33}$[63] quantum dots. Several things were ascertained from these studies, many ubiquitous for both systems.

The electron–phonon coupling, d_{km} in Equations (2.13) and (2.14), is directly related to the second derivative of the energy along the nuclear trajectory and therefore those vibrational modes that most strongly modulate the energy levels create the largest coupling.[81] For PbSe and CdSe both electrons and holes couple more strongly to lower acoustic frequency modes, which is consistent with co-herent phonon measurements.[81,82] This is unlike carrier relaxation dynamics in bulk semiconductors where only high-frequency optical modes are important.[83]

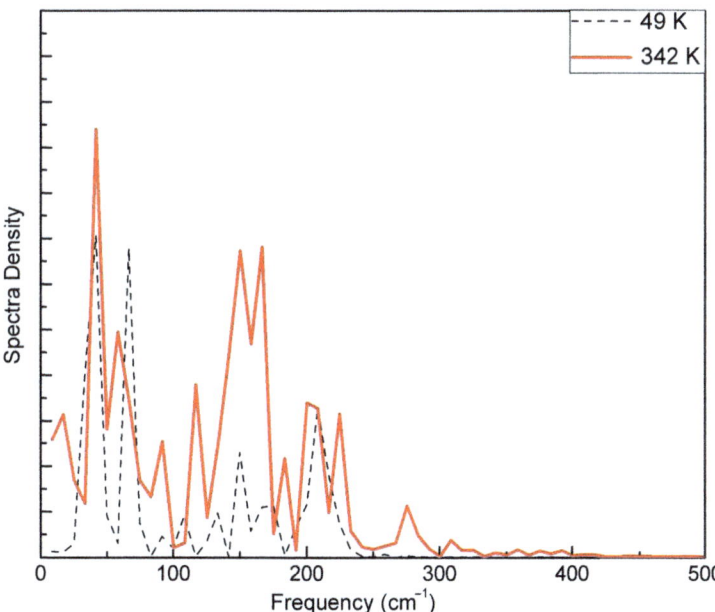

Figure 2.14 Fourier transform of the LUMO vibrations for the CdSe QD at high and low temperatures.[63]
(Adapted from Ref. 63. Copyright 2011 American Chemical Society).

Figure 2.14 shows the Fourier transform of the LUMO vibrations for the CdSe QD.[63] All of the conclusions drawn from Figure 2.14 hold for the spectral density of PbSe as well. Temperature affects the coupling in two ways. First, at higher temperature, the Fourier transforms show a high-frequency tail, indicating that a larger fraction of high-frequency surface modes are involved in the carrier decay dynamics. This can be understood by considering that increased temperature can activate higher frequency modes. Second, the Fourier transform curves at higher temperature are broadened, indicating that more modes are modulating the energy levels. Both the increase in the number of participating modes and the increased coupling to high-frequency modes would suggest an increase in relaxation rate with temperature. This is indeed what was found and can be seen for CdSe and PbSe in Figure 2.15 (a) and (b), respectively.

These studies allowed the development of a quantitative model on the temperature dependence of hot-carrier decay in NCs.[62,63] Notice that from Equation (2.14), if we assume d_{km} is only implicitly dependent on temperature, the NA coupling strength is proportional to the ion velocities $\dot{R}$, and in turn to

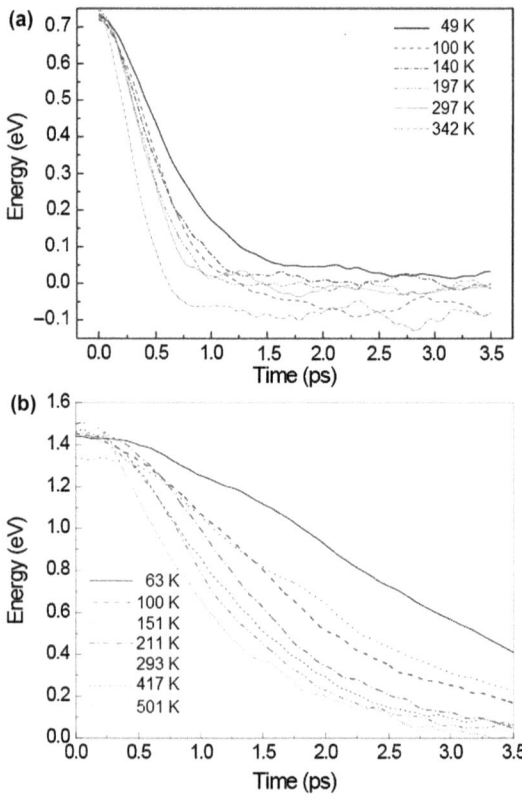

Figure 2.15   The average energy decay for different temperatures during 3.5 ps trajectories for CdSe (a)[63] and PbSe (b)[62] QDs.
(Part (a) of this figure was adapted from Ref. 63. Copyright 2011 American Chemical Society and Part (b) of this figure was adapted from Ref. 62. Copyright 2009 American Physical Society).

the square root of kinetic energy E_K. Since the MD process treats the ions classically, MD temperature T_{MD} is proportional to E_K. From time-dependent perturbation theory, a simple estimation is that the transition probability is proportional to the square of the off-diagonal element of the perturbation matrix, which is $|NA|^2$ in our case. Based on the above considerations the temperature dependence of the hot carrier relaxation rate could be simply written as:

$$\gamma \sim |NA|^2 \sim |d_{km}|^2 |\dot{R}|^2 \sim |d_{km}|^2 T_{MD} \qquad (2.18)$$

where, γ, NA, $\boldsymbol{d_{km}}$, $\dot{\boldsymbol{R}}$ and T_{MD} represent the relaxation rate, non-adiabatic coupling, electron–phonon coupling, ion velocity and temperature, respectively.

However with both PbSe and CdSe, calculated results deviated significantly from this expected trend. For both systems relaxation rates were better fitted to $T_{MD}^{0.4}$ than to T_{MD}, Figure 2.16 shows this for PbSe.[62] The weaker temperature dependence can only be attributed to the temperature dependence of electron–phonon coupling, d_{km}, strength. To investigate this term, we calculated the NA coupling strength between pairs of states for each system and found that it decreased with temperature as $T^{-0.3}$. Although QD geometry is weakly temperature dependent, the coupling can strongly depend on expansion.

Experimental results show no temperature dependence for CdSe QDs and a much weaker temperature dependence at low temperatures for PbSe.[78] In PbSe quantum dots, at low temperatures the Auger channel is more efficient than the multiphonon channel, which gives weak temperature dependence. At higher temperatures, when the electron–phonon channel becomes more efficient, it will

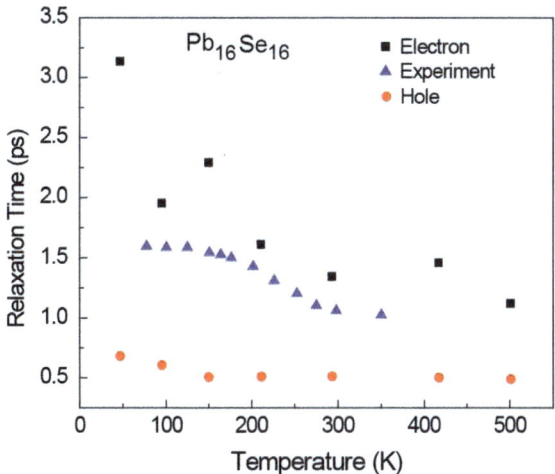

Figure 2.16   Calculated electron and hole relaxation times at different temperatures. Also shown is the temperature-dependent hot-carrier relaxation for NC with an average diameter of 3.8 nm.[62]
(Adapted from Ref. 62. Copyright 2009 American Physical Society).

contribute to the overall relaxation rate and then stronger temperature dependence is observed. This confirms that CdSe QDs relax predominantly from Auger relaxation channels. However current experimental methods can enable effective separation of electrons and holes, making it possible to decouple the Auger and electron–phonon relaxation processes in NCs. For CdSe NCs, with a surfactant of deep hole trapping molecule, such as pyridine, the electron–phonon interaction is dominant and a slow relaxation of ~ 200 ps is observed.[56,57] This is orders of magnitude longer than that in the bulk system.

2.6.5 Ligands Saturate Dangling Bonds and Accelerate Electron–Phonon Relaxation

Ligand contribution to QD excitation dynamics is exemplified by Si and Ge QDs in Figure 2.17. *Ab initio* calculations showed[59] that quantum confinement makes electron and hole DOS more symmetric in Si and Ge QDs compared with bulk. Despite symmetric DOS, electrons decay faster than holes, Figure 2.17. Asymmetric relaxation can be rationalized by stronger electron–phonon coupling in the CB, due to larger contributions of high-frequency phonons associated with Ge-H and Si-H surface bonds. As can be seen in the inserts it the right panel of Figure 2.18 the CB states for Ge are more delocalized on the surface than the VB states. Figure 2.18 shows spectral densities of the phonon modes that couple to CB and VB states of $Ge_{29}H_{24}$. Low-frequency motions correspond to Ge-Ge bond vibrations and show little difference between VB and CB, left panels. Asymmetry is clearly seen in high-frequency components that originate from Ge-H surface bonds, right panels of Figure 2.18. Even though low-frequency modes influence electronic energies more than high frequency modes (compare amplitudes in left and right panels

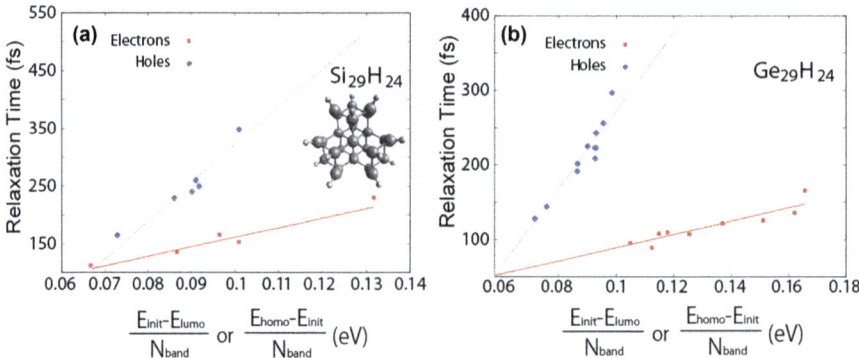

Figure 2.17   Electron and hole relaxation times in (a) $Si_{29}H_{24}$ and (b) $Ge_{29}H24$ QDs. Calculated relaxation times are correlated with state energy spacing averaged over the relaxation energy range. Electrons decay faster than holes in Si and Ge QDs owing to higher DOS and involvement of surface modes. See Figure 2.18.[59]
(Adapted from Ref. 59. Royal Society of Chemistry © Copyright 2009).

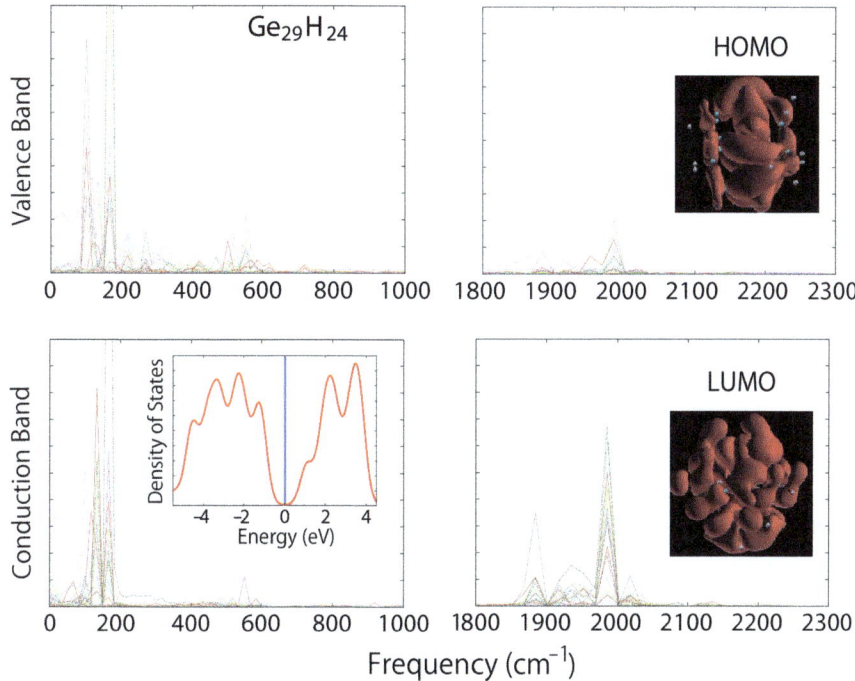

Figure 2.18 Frequencies of phonon modes that couple to CB and VB states in Ge29H24. The inserts show QD geometries, DOS and HOMO/LUMO orbital densities. In Ge and Si QDs, electrons decay faster than in holes, in contrast to PbSe and CdSe, because electrons have higher DOS and couple more strongly to high-frequency surface modes. Coupling to surface modes is rationalized by larger surface delocalization of CB orbitals, for example, LUMO, than VB orbitals, for example, HOMO. (Adapted from Ref. 59. Royal Society of Chemistry © Copyright 2009).

of Figure 2.18), higher frequency modes have higher velocities and accelerate charge relaxation through the velocity dependence of NA coupling.[12]

2.7 Time Domain *ab Initio* Study of Auger and Phonon-Assisted Auger Processes

Auger dynamics in semiconductor QDs is particularly important for photovoltaics. Photons absorbed at the blue end of the solar spectrum create high-energy excitations that relax to the lowest energy harvested in the red. Significant amounts of solar power are lost to heat, limiting photovoltaic efficiency to 31%.[3] Multiple exciton recombination (MER) by the Auger mechanism can further accelerate energy losses by transferring photon energy to the charge carrier that is more strongly coupled to phonons.[68] In contrast, the inverse process, known as impact ionization,[3,10] uses the excess energy to excite additional electrons across the band gap, resulting in carrier multiplication and

MEG. MEG can raise the photovoltaic efficiency to 44%.[3,5,6,15,84] The TDDFT–NAMD method described in Section 2.6.1 was used to simulate Auger dynamics in nanoscale systems. To simulate Auger dynamics, electron correlation is taken into account in the adiabatic basis in which all Coulomb interactions are described in the Hamiltonian for fixed nuclear coordinates. The eigenstates are coupled through NAC which arise from nuclear motion during molecular dynamics. This approach allows us to include phonon assisted Auger dynamics. This picture is complimentary to the traditional Auger model which employs diabatic initial and final states and is independent of nuclear coordinates.[85,86]

2.7.1 Auger Theory

Our simulation method including the ground, SE and double electron (DE) states, $|\phi_g(x; R)\rangle$, $|\phi_{SE}^{ij}(x; R)\rangle$, $|\phi_{DE}^{i,k,k,l}(x; R)\rangle$, respectively, was formulated using second quantization with the ground state as a reference.[86] SEs and DEs are obtained as:

$$|\phi_{SE}^{ij}\rangle = \hat{a}_i^\dagger \hat{a}_j |\phi_g\rangle, \quad |\phi_{DE}^{i,j,k,l}\rangle = \hat{a}_i^\dagger \hat{a}_j \hat{a}_k^\dagger \hat{a}_l |\phi_g\rangle \qquad (2.17)$$

where the electron creation and annihilation operators, $\hat{a}_i^\dagger$ and $\hat{a}_j$, generate and destroy an electron in the ith and jth adiabatic KS orbitals, respectively. The time-evolving wave function is then expressed by:

$$|\Psi(t)\rangle = C_g(t)|\phi_g\rangle + \sum_{i,j} C_{SE}^{ij}(t)|\phi_{SE}^{ij}\rangle + \sum_{i,j,k,l} C_{DE}^{i,j,k,l}(t)|\phi_{DE}^{i,j,k,l}\rangle \qquad (2.18)$$

Similar to Equation (2.13), the expansion coefficient appearing in Equation (2.18), evolves by the first-order differential equations:

$$i\hbar \frac{\partial C_X(t)}{\partial t} = C_X(t)E_X - i\hbar C_g(t)d_{X;g} \cdot \dot{R} - i\hbar \sum_{i'j'} C_{SE}^{i'j'}(t)d_{X;SE,i'j'} \cdot \dot{R}$$
$$- i\hbar \sum_{i',j',k',l'} C_{DE}^{i',j',k',l'}(t)d_{X;DE,i'j',k',l'} \cdot \dot{R} \qquad (2.19)$$

where X and Y now correspond to either ground, SE of DE state, E_X is the state energy and the NA couplings are defined by:

$$d_{X;Y} \cdot \dot{R} \equiv \langle \phi_X |\nabla_R|\phi_Y\rangle \cdot \dot{R} = \left\langle \phi_X \left| \frac{\partial}{\partial t} \right| \phi_Y \right\rangle \qquad (2.20)$$

The atomistic simulation of SE/DE generation and dynamics was performed by directly solving Equation (2.19) with time-dependent NA couplings and energies.[59] The ground, SE and DE states form the two-particle electronic basis in our method and each state can transit to another state due to the NA coupling. The energies appear in the diagonal parts of the Hamiltonian, while the NA couplings are embedded in the corresponding off-diagonal components. The simulations are extremely large scale, too large for direct numerical simulation; however, the Hamiltonian is sparse, since the NA couplings connect states that differ only in a single electron or hole.[12] Based on this fact, we developed an efficient simulation code that can remove all zero components from the sparse Hamiltonian and solve Equation (2.19) using only the extracted non-zero components of the Hamiltonian.[86]

2.7.2 Results of Auger Studies

These studies focus on Ge QDs. Ge belongs to the same group as Si which is currently the most common solar cell material. Compared to Si, bulk Ge has a significantly lower band gap: 0.67 *vs.* 1.1. While MEG has been observed in Si QDs,[15] the absolute excitation energies required for efficient MEG are rather high. Owing to its lower band gap, Ge is a promising alternative to Si.[87,88] It was established that MEG can be assisted by phonons at energies below the electronic threshold.[86]

Figure 2.19(a) describes MEG. It shows the population of all SE sates as a function of time for different excitation energies. The decay of SEs is accompanied by a corresponding growth of MEs, since the total population change remains negligible for the duration of the simulation, in agreement with nanosecond QD fluorescence lifetimes. The data are fitted to a linear combination of a Gaussian and exponential components shown in the figure caption. The dynamics show both Gaussian and exponential components. The initial decay of a quantum state is always Gaussian.[13] At long times the dynamics involve a large ensemble of final states and become exponential, as described for instance by Fermi's golden rule. The transition from Gaussian to exponential decay occurs when the SE population drops to 0.92–0.96, indicating that the short time Gaussian dynamics is responsible for 5–10%, validating the rate theory models that are limited to exponential decay. Note that our study shows that phonon assisted Auger processes are present.

Electron–phonon relaxation competes successfully with MER. MER becomes efficient only at low energies, after a significant amount of energy has been deposited into phonon modes by the relaxation. Figure 2.19(b) shows the population of SEs created starting from MEs with three different energies. The simulations clearly indicate that MER is much slower than MEG. The simulations also show the MER is preceded by electron–phonon relaxation and occurs only from low energy DEs. Since the coupling between SEs and DEs promotes both forward and backward processes, the directionality is determined by the relative DOS for SEs and MEs. At high energies SEs generate MEs, while at low energies MEs annihilate to form SEs.

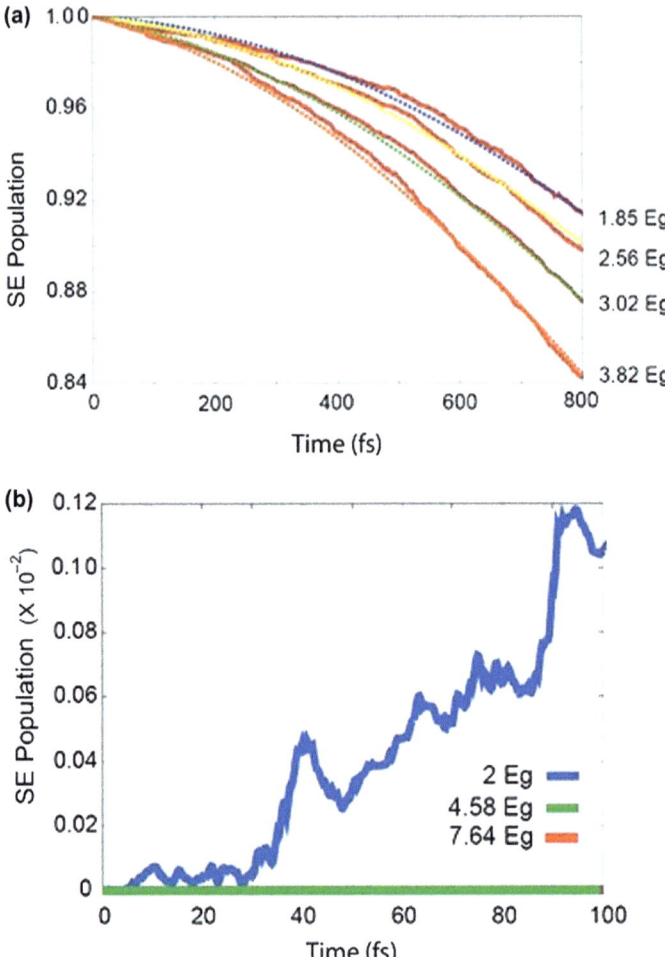

Figure 2.19   (a) Decay of the combined population of all SEs from ME generation, starting from the initially excited 158th (1.85E_g), 314th (2.56E_g), 418th (3.02E_g) and 600th (3.82E_g) SEs. The ground state population remains negligible within the shown time scale. The DE generation accelerates with increasing excitation energy. The data are fitted by $0.75e^{-t^2/\tau_g^2} + 0.25e^{-t/\tau_e}$ with τ_g/τ_e equal to 2.45/20.0, 2.39/10, 2.32/5 and $\frac{1}{2}$ ps, from lowest to highest energy, respectively. (b) Generation of SEs caused by ME recombination, starting from the initially excited 602nd (2E_g), 5929th (4.58E_g) and 98099th (7.64E_g) DEs. MER is preceded by electron–phonon relaxation and only low energy DEs lead to SEs.[86] (Adapted from Ref. 86. Copyright 2011 American Chemical Society).

2.8   Conclusion

"*Ab initio*" modeling of excited state dynamics in the energy and time domains generates valuable insights into semiconductor QD properties. TDDFT combined with NAMD simulates the complex evolution of coupled electronic and vibrational degrees of freedom as it occurs in nature. Compared with EMT and tight-binding calculations that predict highly degenerate energy levels in QDs, *ab initio* calculations demonstrate that atomic structure, surface effects, core/shell interactions, thermal fluctuations and coulomb coupling break degeneracies and create complicated distributions of electronic levels. *Ab initio* descriptions directly illustrate non-ideal processes and generate input for modification and extension of phenomenological models. Using SAC-CI it was found that QD charging gives rise to intraband transitions that exhibit little optical activity but dramatically modify the nature and energy of the excited electronic states. Charging also blue shifts absorption spectra and increases ME thresholds. Surface reconstructions significantly changes QD band gaps. Ligand and shell layers saturate surface dangling bonds and alter high-energy regions of CB and VB, while ligand high-frequency modes create strong electron–phonon coupling.

Electron–phonon interactions induce two distinct processes in QDs: dephasing and relaxation. Superposition of electronic states, created by the coulomb interaction during photoexcitation and subsequent time evolution, dephases into incoherent mixtures of states. Dephasing is ultrafast if it involves electronic states with substantially different energies and spatial densities. Examples of this include superposition of SEs and MEs and ground and excited states. ME fission into independent SEs occurs by dephasing that is much slower. This is because MEs are typically formed by SEs that are close in energy.

Time domain modeling of electron–phonon relaxation unified two seemingly contradicting experimental observations: despite large spacing between optical lines, phonon bottleneck to electron–phonon relaxation exists only under very special conditions. QD spectra are composed of multiple individual excitations that combine into distinct bands according to optical selection rules. Selection rules are much less stringent for electron–phonon transitions. Even though relatively few excitations are strongly optically active, most excited electronic states participate in phonon relaxation. Through these studies it was found that ligand and shell layers saturate dangling bonds and alter high-energy regions of the CB and VB. High-frequency ligand mode create strong electron-phonon coupling which speed up electron-phonon relaxation. It was also discovered that while overall relaxation rates increase at higher temperatures the non-adiabatic coupling constants actually decrease at higher temperatures.

Auger processes are particularly important in nanoscale materials. Until very recently they had only been modeled in the time-independent approach. The newly developed sparse matrix techniques discussed in this chapter allow the study of time domain simulations of Auger processes that inherently involve many electronic states. These studies have helped to validate standard rate theory models that only incorporate exponential decay. The electron–phonon relaxation pathway is dominant until lower energies. Since the coupling between SEs and MEs promotes both forward and backward processes, the directionality is determined by the relative DOS for SEs and MEs. At high energies SEs generate MEs, while at low energies MEs annihilate to form SEs.

With increased computational power, current methods will be applied to larger QDs and more diverse materials involving realistic ligands, surface defects and viable core/shell compositions. Longer scale simulations of luminescence quenching and the phonon bottleneck will become possible. Photovoltaic assemblies of QDs with other materials, such as molecular chromophores, polymers and inorganic semiconductors present new theoretical questions. The issues raised here with electronic states in semiconductor QDs find close similarities with spin states of metallic QDs. *Ab initio* approaches in time and energy domains will greatly advance our understanding of QD properties that govern solar energy harvesting and many other applications.

Acknowledgements

The research was supported by the National Science Foundation, grant CHE-13001 18, and US Department of Energy Grant DE-SC0006527. We thank colleagues and collaborators: Sean A. Fischer, Angeline B. Madrid, Christine M. Isborn, Svetlana V. Kilina, Xiaosong Li, Hideyuki Kamisaka, Koichi Yamashita, Kim Hyeon-Deuk and Bradley F. Habenicht, Hua Bao, Xiulin Ruan, Dimitri S. Kilin, Colleen F. Craig, Liangliang Chen, Heather Jaeger, Run Long and Taizhi Tan for their contributions.

References

1. D. J. Milliron, S. M. Hughes, Y. Cui, L. Manna, J. Li, L.-W. Wang and A. Paul Alivisatos, *Nature*, 2004, **430**, 190–195.
2. D. S. Kilin, O. V. Prezhdo and Y. Xia, *Chemical Physics Letters*, 2008, **458**, 113–116.
3. A. J. Nozik, *Annual Review of Physical Chemistry*, 2001, **52**, 193–231.
4. A. Pandey and P. Guyot-Sionnest, *Science*, 2008, **322**, 929–932.
5. R. D. Schaller and V. I. Klimov, *Physical Review Letters*, 2004, **92**, 186601.
6. R. J. Ellingson, M. C. Beard, J. C. Johnson, P. Yu, O. I. Micic, A. J. Nozik, A. Shabaev and A. L. Efros, *Nano Letters*, 2005, **5**, 865–871.
7. J. A. McGuire, J. Joo, J. M. Pietryga, R. D. Schaller and V. I. Klimov, *Accounts of Chemical Research*, 2008, **41**, 1810–1819.
8. O. V. Prezdho, *Chemical Physics Letters*, 2008, **460**, 1–9.
9. O. V. Prezhdo, *Accounts of Chemical Research*, 2009, **42**, 2005–2016.
10. L.-W. Wang, M. Califano, A. Zunger and A. Franceschetti, *Physical Review Letters*, 2003, **91**, 056404.
11. A. Franceschetti, J. M. An and A. Zunger, *Nano Letters*, 2006, **6**, 2191–2195.
12. C. F. Craig, W. R. Duncan and O. V. Prezhdo, *Physical Review Letters*, 2005, **95**, 163001.
13. O. V. Prezhdo and V. V. Kisil, *Physical Review A*, 1997, **56**, 162–175.
14. P. V. Parahdekar and J. C. Tully, *J. Chem. Phys.*, 2005, **122**.
15. M. C. Beard, K. P. Knutsen, P. Yu, J. M. Luther, Q. Song, W. K. Metzger, R. J. Ellingson and A. J. Nozik, *Nano Letters*, 2007, **7**, 2506–2512.

16. B.-R. Hyun, Y.-W. Zhong, A. C. Bartnik, L. Sun, H. D. Abruña, F. W. Wise, J. D. Goodreau, J. R. Matthews, T. M. Leslie and N. F. Borrelli, *ACS Nano*, 2008, **2**, 2206–2212.
17. G. I. Koleilat, L. Levina, H. Shukla, S. H. Myrskog, S. Hinds, A. G. Pattantyus-Abraham and E. H. Sargent, *ACS Nano*, 2008, **2**, 833–840.
18. E. Rabani and R. Baer, *Nano Letters*, 2008, **8**, 4488–4492.
19. R. D. Schaller, M. Sykora, J. M. Pietryga and V. I. Klimov, *Nano Letters*, 2006, **6**, 424–429.
20. R. D. Schaller, V. M. Agranovich and V. I. Klimov, *Nature Physics*, 2005, **1**, 189–194.
21. J. E. Murphy, M. C. Beard, A. G. Norman, S. P. Ahrenkiel, J. C. Johnson, P. Yu, O. I. Mićić, R. J. Ellingson and A. J. Nozik, *Journal of the American Chemical Society*, 2006, **128**, 3241–3247.
22. C. M. Isborn, S. V. Kilina, X. Li and O. V. Prezhdo, *The Journal of Physical Chemistry C*, 2008, **128**, 18291.
23. J. M. An, M. Califano, A. Franceschetti and A. Zunger, *The Journal of Chemical Physics*, 2008, **128**, 164720–164727.
24. A. Shabaev, A. L. Efros and A. J. Nozik, *Nano Letters*, 2006, **6**, 2856–2863.
25. H. Nakatsuji and K. Hirao, *The Journal of Chemical Physics*, 1978, **68**, 2053–2065.
26. H. Nakatsuji and M. Ehara, *The Journal of Chemical Physics*, 1994, **101**, 7658–7671.
27. A. Dreuw and M. Head-Gordon, *Chemical Reviews*, 2005, **105**, 4009–4037.
28. B. G. Levine, C. Ko, J. Quenneville and T. J. Martinez, *Molecular Physics*, 2006, **104**, 1039–1051.
29. M. J. Frisch, G. W. Trucks, H. B. Schlegel, G. E. Scuseria, M. A. Robb, J. R. Cheeseman, J. A. Montgomery, T. Vreven, K. N. Kudin, J. C. Burant, J. M. Millam, S. S. Iyengar, J. Tomasi, V. Barone, B. Mennucci, M. Cossi, G. Scalmani, N. Rega, G. A. Petersson, H. Nakatsuji, M. Hada, M. Ehara, K. Toyota, R. Fukuda, J. Hasegawa, M. Ishida, T. Nakajima, Y. Honda, O. Kitao, H. Nakai, M. Klene, X. Li, J. E. Knox, H. P. Hratchian, J. B. Cross, V. Bakken, C. Adamo, J. Jaramillo, R. Gomperts, R. E. Stratmann, O. Yazyev, A. J. Austin, R. Cammi, C. Pomelli, J. W. Ochterski, P. Y. Ayala, K. Morokuma, G. A. Voth, P. Salvador, J. J. Dannenberg, V. G. Zakrzewski, S. Dapprich, A. D. Daniels, M. C. Strain, O. Farkas, D. K. Malick, A. D. Rabuck, K. Raghavachari, J. B. Foresman, J. V. Ortiz, Q. Cui, A. G. Baboul, S. Clifford, J. Cioslowski, B. B. Stefanov, G. Liu, A. Liashenko, P. Piskorz, I. Komaromi, R. L. Martin, D. J. Fox, T. Keith, A. Laham, C. Y. Peng, A. Nanayakkara, M. Challacombe, P. M. W. Gill, B. Johnson, W. Chen, M. W. Wong, C. Gonzalez and J. A. Pople, *Gaussian 03, Revision C.02*, Gaussian, Inc., Wallingford CT, 2004.
30. P. J. Hay and W. R. Wadt, *The Journal of Chemical Physics*, 1985, **82**, 270–283.
31. W. R. Wadt and P. J. Hay, *The Journal of Chemical Physics*, 1985, **82**, 284–298.

32. C. M. Isborn and O. V. Prezhdo, *The Journal of Physical Chemistry C*, 2009, **113**, 12617–12621.
33. S. A. Fischer, A. B. Madrid, C. M. Isborn and O. V. Prezhdo, *The Journal of Physical Chemistry Letters*, 2010, **1**, 232–237.
34. S. A. Fischer and O. V. Prezhdo, *The Journal of Physical Chemistry C*, 2011, **115**, 10006–10011.
35. S. A. Fischer, C. M. Isborn and O. V. Prezhdo, *Chemical Science*, 2011, **2**, 400–406.
36. H. M. Jaeger, S. Fischer, and O. V. Prezhdo, *J. Chem. Phy.*, 2012, **136**, 064701–064710.
37. X. D. Pi, R. Gresback, R. W. Liptak, S. A. Campbell and U. Kortshagen, *Applied Physics Letters*, 2008, **92**, 123102–123103.
38. C. J. W. M. Shim, D. J. Norris and P. Guyot-Sionnest, *MRS Bulletin*, 2001, **26**, 1005–1008.
39. M. Shim and P. Guyot-Sionnest, *Nature*, 2000, **407**, 981–983.
40. D. Yu, C. Wang and P. Guyot-Sionnest, *Science*, 2003, **300**, 1277–1280.
41. A. Puzder, A. J. Williamson, F. Gygi and G. Galli, *Physical Review Letters*, 2004, **92**, 217401.
42. H. Kamisaka, S. V. Kilina, K. Yamashita and O. V. Prezhdo, *Nano Letters*, 2006, **6**, 2295–2300.
43. H. Kamisaka, S. V. Kilina, K. Yamashita and O. V. Prezhdo, *The Journal of Physical Chemistry C*, 2008, **112**, 7800–7808.
44. A. B. Madrid, K. Hyeon-Deuk, B. F. Habenicht and O. V. Prezhdo, *ACS Nano*, 2009, **3**, 2487–2494.
45. S. Mukamel, *Principles of Nonlinear Optical Spectroscopy*, Oxford University Press, 1995.
46. G. Kresse and J. Hafner, *Physical Review B*, 1993, **47**, 558–561.
47. G. Kresse and J. Furthmüller, *Physical Review B*, 1996, **54**, 11169–11186.
48. G. Kresse and J. Hafner, *Physical Review B*, 1994, **49**, 14251–14269.
49. D. Vanderbilt, *Physical Review B*, 1990, **41**, 7892–7895.
50. J. P. Perdew and Y. Wang, *Physical Review B*, 1992, **45**, 13244–13249.
51. G. Lanzani, G. Cerullo, M. Zavelani-Rossi, S. De Silvestri, D. Comoretto, G. Musso and G. Dellepiane, *Physical Review Letters*, 2001, **87**, 187402.
52. J. Nelson, *The Physics of Solar Cells*, Imperial College Press, London, 2003.
53. J. Urayama, T. B. Norris, J. Singh and P. Bhattacharya, *Physical Review Letters*, 2001, **86**, 4930–4933.
54. A. D. Yoffe, *Advances in Physics*, 2001, **50**, 1–208.
55. P. Guyot-Sionnest, M. Shim, C. Matranga and M. Hines, *Physical Review B*, 1999, **60**, R2181–R2184.
56. P. Guyot-Sionnest, B. Wehrenberg and D. Yu, *The Journal of Chemical Physics*, 2005, **123**, 074709–074707.
57. S. A. Fischer, B. F. Habenicht, A. B. Madrid, W. R. Duncan and O. V. Prezhdo, *The Journal of Chemical Physics*, 2011, **134**, 024102–024109.
58. S. V. Kilina, C. F. Craig, D. S. Kilin and O. V. Prezhdo, *The Journal of Physical Chemistry C*, 2007, **111**, 4871–4878.

59. K. Hyeon-Deuk, A. B. Madrid and O. V. Prezhdo, *Dalton Transactions*, 2009, 10069–10077.
60. S. V. Kilina, D. S. Kilin and O. V. Prezhdo, *ACS Nano*, 2009, **3**, 93–99.
61. B. F. Habenicht and O. V. Prezhdo, *Physical Review Letters*, 2008, **100**, 197402.
62. H. Bao, B. F. Habenicht, O. V. Prezhdo and X. Ruan, *Physical Review B*, 2009, **79**, 235306.
63. L. Chen, H. Bao, T. Tan, O. V. Prezhdo and X. Ruan, *The Journal of Physical Chemistry C*, 2011, **115**, 11400–11406.
64. M. A. L. Marques and E. K. U. Gross, *Time-dependent Density Functional Theory*, Annual Reviews, Palo Alto, CA, 2004.
65. S. Hammes-Schiffer and J. C. Tully, *The Journal of Chemical Physics*, 1994, **101**, 4657–4667.
66. J. C. Tully, *The Journal of Chemical Physics*, 1990, **93**, 1061–1071.
67. E. Tapavicza, I. Tavernelli and U. Rothlisberger, *Physical Review Letters*, 2007, **98**, 023001.
68. A. L. Efros, V. A. Kharchenko and M. Rosen, *Solid State Communications*, 1995, **93**, 281–284.
69. B. F. Habenicht, C. F. Craig and O. V. Prezhdo, *Physical Review Letters*, 2006, **96**, 187401.
70. W. R. Duncan, C. F. Craig and O. V. Prezhdo, *Journal of the American Chemical Society*, 2007, **129**, 8528–8543.
71. J. J. Peterson and T. D. Krauss, *Physical Chemistry Chemical Physics*, 2006, **8**, 3851–3856.
72. W. Lu, J. Fang, Y. Ding and Z. L. Wang, *The Journal of Physical Chemistry B*, 2005, **109**, 19219–19222.
73. F. W. Wise, *Accounts of Chemical Research*, 2000, **33**, 773–780.
74. B. L. Wehrenberg, C. Wang and P. Guyot-Sionnest, *The Journal of Physical Chemistry B*, 2002, **106**, 10634–10640.
75. H. Zeng, Z. A. Schelly, K. Ueno-Noto and D. S. Marynick, *The Journal of Physical Chemistry A*, 2005, **109**, 1616–1620.
76. V. I. Klimov, A. A. Mikhailovsky, D. W. McBranch, C. A. Leatherdale and M. G. Bawendi, *Science*, 2000, **287**, 1011–1013.
77. P. Liljeroth, P. A. Z. van Emmichoven, S. G. Hickey, H. Weller, B. Grandidier, G. Allan and D. Vanmaekelbergh, *Physical Review Letters*, 2005, **95**, 086801.
78. R. D. Schaller, J. M. Pietryga, S. V. Goupalov, M. A. Petruska, S. A. Ivanov and V. I. Klimov, *Physical Review Letters*, 2005, **95** 196401.
79. J. M. Harbold, H. Du, T. D. Krauss, K.-S. Cho, C. B. Murray and F. W. Wise, *Physical Review B*, 2005, **72**, 195312.
80. R. R. Cooney, S. L. Sewall, K. E. H. Anderson, E. A. Dias and P. Kambhampati, *Physical Review Letters*, 2007, **98**, 177403.
81. W. H. Miller and T. F. George, *The Journal of Chemical Physics*, 1972, **56**, 5637–5652.
82. T. D. Krauss and F. W. Wise, *Physical Review Letters*, 1997, **79**, 5102–5105.

83. U. Woggon, H. Giessen, F. Gindele, O. Wind, B. Fluegel and N. Peyghambarian, *Physical Review B*, 1996, **54**, 17681–17690.
84. J. A. McGuire, M. Sykora, J. Joo, J. M. Pietryga and V. I. Klimov, *Nano Letters*, 2010, **10**, 2049–2057.
85. K. Hyeon-Deuk and O.V. Prezhdo, *ACS Nano*, 2012, **6**, 1239–1250.
86. K. Hyeon-Deuk and O. V. Prezhdo, *Nano Letters*, 2011, **11**, 1845–1850.
87. J.-W. Luo, A. Franceschetti and A. Zunger, *Nano Letters*, 2008, **8**, 3174–3181.
88. S. Chatterjee, *Solar Energy*, 2008, **82**, 95–99.
89. A. J. Neukirch, Z. Guo and O. V. Prezhdo, *J. Phys. Chem. C*, 2012, **116**, 15034–15040.
90. S. V. Kilina, K. A. Velizhanin, S. Ivanov, O. V. Prezhdo, and S. Tretiak, *ACS Nano*, 2012, **6**, 6515–6524.
91. Long and O. V. Prezhdo, *J. Am. Chem. Soc.*, 2011, **133**, 19240–19249.
92. V. Kilina, D. S. Kilin, V. V. Prezhdo, and O. V. Prezhdo, *J. Phys. Chem. C*, 2011, **115**, 21641–21651.
93. K. P. Hyeon-Deuk, O. V. Prezhdo, *J. Phys.: Condensed Matter*, 2012, 24.

CHAPTER 3

Multiscale Modelling of Interfacial Electron Transfer

PETTER PERSSON*

Department of Theoretical Chemistry, Lund University, Box 124,
SE - 221 00 Lund, Sweden
Email: Petter.Persson@teokem.lu.se

3.1 Introduction

The development of theories and models for electron transfer (ET) in molecular and materials systems is an important and successful area of fundamental research. It is of great interest to both understand and control ET in a wide range of complex molecular and materials systems because of its importance in many physical, chemical and biological processes.[1,2]

Interfacial electron transfer (IET) constitutes a particularly challenging branch in this field owing to the inherent system complexity that is typically associated with any interface.[3–5] There is, however, currently strong international interest in developing new advanced applications based on IET processes utilizing advances in the design of new molecular and nanostructured materials. Promising technologies include chemical and biological sensors, as well as a range of electronic and optoelectronic devices. One flagship application that has attracted particular attention in the last 20 years is the development of new efficient, inexpensive and environmentally friendly energy conversion devices, such as solar energy conversion devices driven by light-harvesting and subsequent photoinduced ET processes.[6] Relevant photoinduced ET processes often take place either in supramolecular systems, or at molecular and material interfaces.[7,8] The fields of surface electron transfer and

RSC Energy and Environment Series No. 8
Solar Energy Conversion: Dynamics of Interfacial Electron and Excitation Transfer
Edited by Piotr Piotrowiak
Published by the Royal Society of Chemistry, www.rsc.org

photoinduced electron transfer are therefore often, though not always, intimately connected and a large part of the discussion in this chapter concerns photoinduced interfacial electron transfer processes.[9,10] Current research interests include emerging technologies for both solar electricity, using a variety of molecular, polymer and nanostructured materials,[11–16] as well as direct photocatalytic fuel production through so called artificial photosynthesis.[17]

Many fundamental properties of IET, such as dependence on driving force and electronic interactions between donor and acceptor states, have long been studied in considerable detail, including a wide range of phenomenological models and theoretical perspectives.[4,10] Accurate modelling of IET processes that provide an atomistically accurate perspective on specific IET interactions in molecular and materials systems of technological importance have, in contrast, been largely limited to comparatively simple systems. IET processes at molecule–metal oxide interfaces provide a good example because of their central importance to the dye-sensitized solar cell (DSSC) which has served as an important prototype nanodevice for approximately the last 20 years.[11–13] Adsorption of small molecular adsorbates, for example H_2O and simple organic molecules like methanol and formic acid, on surfaces of metal oxides such as TiO_2 have been studied theoretically for a long time.[18] In contrast, first principles electronic structure calculations aiming to provide a detailed molecular perspective on the atomistic structure and electronic interactions in DSSCs, including realistic models of IET processes in model systems for molecular dyes anchored to TiO_2, only started to emerge approximately ten years ago.[19] Since approximately 2000, there has been rapid computational progress with many advanced theoretical investigations performed for IET processes for a wide range of advanced molecular and materials interfaces,[10] notably including many important theoretical investigations of IET processes for example for DSSCs,[19–22] as well as polymer electronics that include bulk hetero-junction (BHJ) solar cells.[14–16]

Many IET processes in heterogeneous systems and mesoscopic materials continue to provide a serious challenge for accurate theoretical modelling both because of inherent structural complexity and the wide range of time- and length-scales on which electron transfer processes take place. Modern computational techniques that combine aspects of calculations for different length- and time-scales, so called multiscale modelling techniques, offer a viable approach to perform comprehensive modelling of complex molecular and materials systems,[23] including significant opportunities to investigate IET in nanoscale materials such as dye-sensitized nano-TiO_2 interfaces in DSSCs.[24]

Recent progress and on-going developments in modelling IET in selected complex molecular material interfaces of technological interest using atomistic first principles and multiscale methods are discussed in this chapter. Following a brief general overview over widely available computational tools in Section 3.2, modelling of IET processes will be discussed in Section 3.3 with a focus on molecule–metal oxide interfaces as a good example of a widely studied prototype application of broader general significance. This is followed by a brief description of recent and emerging applications of IET modelling to other technologically important systems, including nanoscale hetero-structures and polymer-based systems. The chapter ends with a discussion of some general conclusions and the outlook for continued progress of computational modelling of IET processes in heterogeneous systems and nanostructured materials.

3.2 Materials Modelling

The rapid rise of computing resources witnessed in the last 50 years has provided significant opportunities for accurate computational modelling of a broad range of molecular and materials systems in physics, chemistry and biology.[25,26] Computational material modelling today spans a wide range of levels of sophistication from accurate but time-consuming calculations based on a quantum mechanical (QM) treatment of molecules and materials, to essentially phenomenological treatments that can be many orders of magnitude faster than quantum calculations but instead typically lack the rigor, generality and accuracy of the former.[27] Figure 3.1 illustrates the typical hierarchy of approximations that is common for much of the research conducted in the field of computational materials science. The distinguishing features of three standard levels – quantum, molecular dynamics and coarse grained/finite element – are briefly outlined in the following paragraphs in order to provide a framework for the subsequent discussion of computational opportunities to model IET.

3.2.1 Modelling Methods

Computational materials science is a very wide field with many manifestations in different research areas. This section can only provide a very brief overview of standard computational methods to highlight pertinent capabilities and limitations of these levels in order to provide a framework for more specific

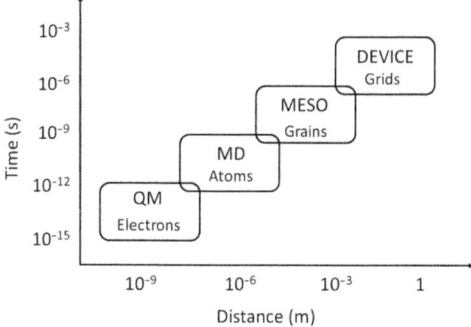

Figure 3.1 Many nanoscale energy conversion devices, like dye-sensitized and polymer-based solar cells, rely on efficient interfacial electron transfer processes. Multiscale modelling of such devices ultimately involves length- and time-scales that scan many orders of magnitude, ranging from ultrafast molecular-scale quantum phenomena such as photoinduced electron transfer, *via* mesoscopic properties such as charge transport, to macroscopic device performance such as *I–V* characteristics.

aspects of IET modelling to be discussed below. Further details of the many computational methods currently available can be found in many excellent textbooks.[25–27] Many valuable studies of IET have also been performed using non-adiabatic dynamic simulations and other advanced simulation methods; these are presented in detail elsewhere and are therefore not discussed in detail in this chapter.[10,20,21]

3.2.1.1 *Quantum Chemistry*

The current standard for quantum calculations are so called electronic structure calculations that focus on the quantum nature of the electrons, largely justified by the so-called Born–Oppenheimer approximation which separates the motion of the light electrons from the heavier nuclei that are assumed to behave essentially as point charges around which the electrons arrange in quantized levels, in other words what in a molecular language are known as molecular orbitals (MOs). This is routinely applied with great success to the study of the structural and electronic properties of a wide range of molecules and materials.[25,26,28]

From a computational point of view there is a natural distinction between the molecular and the materials perspective which lies chiefly in the treatment of the surroundings; a molecule is treated as a finite object in space, while spatial periodicity is imposed on the material description in order to describe its extensive nature. Today, so-called density functional theory (DFT) methods prevail for ground state calculations of both molecules and materials comprised of up to several hundred atoms.[29] For example, bond lengths can typically be calculated with standard DFT methodologies to accuracies within a few hundredths of an angstrom. Excitations can also be calculated with significant accuracy using an extension of DFT known as time-dependent DFT (TD-DFT). Optical excitations, calculated at the TD-DFT level, often match experiments with an accuracy of *ca* 0.1 eV for well-behaved systems.

The accuracy of a DFT calculation is largely determined by two factors: first the choice of functional with a wide range of widely available DFT functionals being used, such as the well-known B3LYP hybrid functional, and second the quality of the basis set used to describe the electronic distribution where more accurate descriptions generally come at a significantly increased computational cost. Because of the large number of atoms needed to describe most molecular and material interfaces, it typically becomes a difficult but important challenge to find a good balance between accuracy and computational feasibility. Some examples that illustrate current capabilities of DFT-based modelling of molecular, nanomaterials and interfacial properties of relevance for IET processes are discussed in Section 3.3. There is also significant current development of more accurate and efficient DFT methods that provides increasing opportunities to investigate advanced materials applications such as IET, including for example recent progress with new density functionals that include dispersion and long-range corrections.[30–33]

More sophisticated electronic structure treatments can be achieved for particularly challenging systems such as unconventional excited states using *ab initio* methods based on a more complete description of the many-electron wavefunction.[34] This includes many accurate investigations of electronic structure and excited states in small to medium sized molecules by multi-reference *ab initio* methods, such as CASPT2.[35] The drawback is mainly in the limitation of such methods to moderately sized molecular systems, typically a few tens of atoms.

Conversely, larger problems for example including thousands of atoms, can be addressed using a more approximate quantum method relying on parameterizations of the electronic structure problem, for example using semi-empirical methods. Semi-empirical methods typically suffer from significantly more limited applicability compared to first principles methods. Different parameterization schemes are, however, widely available to describe, for example, molecular structure, for example AM1 and PM6, or molecular electronic properties including electronic excitations, such as Zerner's Intermediate Neglect of Differential Overlap (ZINDO).[26,27]

3.2.1.2 Molecular Dynamics Simulations

In many situations, such as studying the melting of a frozen solid or describing how molecules move in a liquid, it is more interesting to describe the dynamics of a large ensemble of atoms or molecules than to focus on the internal electronic structure molecule by molecule. This can be done by so-called classical molecular dynamics (MD) simulations by introducing parameterized potentials that describe the intermolecular interactions.[27] There is a range of widely used implementations for materials, molecular and biological simulations, for example the AMBER force field. Such classical MD methods are generally characterized by a reduction in computational cost by several orders of magnitude compared to first principles methods. This currently allows systems of many thousands of molecules to be followed over picoseconds or longer timescales.[27] The loss of explicit information about electronic properties, however, makes classical MD methods less valuable for studies of electronic phenomena in general and IET processes in particular.

3.2.1.3 Coarse Grained and Finite Element Methods

A more severe step towards more phenomenological modelling of larger systems and longer timescales involves abandoning the atomistic perspective in favour of coarse graining or finite element methods where molecular or materials sections are lumped together in larger structural units. This is useful for studying a variety of systems ranging from protein interactions to transport phenomena that occur over size and timescales that go beyond what can realistically be studied by MD simulations, although this comes at the cost of concomitant loss in atomistic information that includes many fundamental quantum and electronic properties.[27]

3.2.2 Multiscale Modelling

In order to bridge the gap between accurate quantum methods that can only be applied to comparatively small systems and large-scale simulations which typically lack the capability to treat phenomena such as chemical reactions and ET processes that are fundamentally quantum in their nature, there is significant interest in developing methods that combine aspects of the different levels of calculation. These methods are referred to as multiscale modelling methods,[23] and some challenges and emerging opportunities for treating IET processes in nanostructured and heterogeneous materials are discussed in the following.

3.2.2.1 Multiscale Challenges

The main focus here is on computational approaches that take an atomistic perspective as a starting point for accurate descriptions of materials structure and function relevant for IET. Interactions of molecular adsorbates with solid surfaces, in particular semiconductor surfaces, constitute a recurring theme throughout much of the discussion as a good illustration of many key points. Computations in this field have in recent years shown emerging capabilities to contribute to a detailed understanding of several important properties at the molecular level, not least including interfacial charge transfer processes, where ultrafast time-resolved experiments are performed to understand basic mechanisms of interfacial charge transfer.[36] Advanced charge transfer processes taking place in complex mesoscopic systems comprising both molecular and materials components highlight the particular modelling challenges outlined in the textbox shown in Figure 3.2.

Despite the obstacles of computational complexity, an increasing number of theoretical and computational studies have been conducted in recent years, testing various strategies to capture some essential features relevant for IET processes also in complex materials such as dye-sensitized and bulk hetero-junction solar cells.[16,19–21]

A particularly successful strategy for many computational investigations has been to use reliable quantum calculations to investigate particular aspects of the full problem in detail. On-going progress in computational modelling capabilities gradually expands the realm of such investigations so that more and more realistic models can be constructed and DFT and TD-DFT calculations can today be routinely performed on molecular, supramolecular and heterogeneous donor–acceptor systems comprising many hundreds of atoms that were simply beyond treatment ten years ago. As a case in point, the development of new light-harvesting systems for solar energy conversion is today routinely complemented by spectral calculations at the TD-DFT level for both dye-sensitized[37] and polymer-based bulk hetero-junction solar cells.[16] Particularly fruitful in this regard have been many recent computational investigations that relate to detailed experimental investigations of fundamental molecular processes, including in particular ultrafast photoinduced IET at dye-sensitized semiconductor interfaces.[36] In several cases, exemplified by various molecule–metal oxide systems in Section 3.3.3, such a strategy allows a step-by-step build-up of knowledge about key phenomena, which can eventually be combined into a comprehensive picture of functioning systems.

- **System Size** – Interfaces often have complicated structures that can only be described at the molecular/material level using thousands of atoms. The problem is compounded by an inherent reduction of symmetry that prevents computational simplifications.

- **Time-scales** – Charge transfer processes at interfaces span many orders of magnitude in time, ranging from ultrafast electronic processes taking place in just a few femtoseconds, to slow long-range transport processes taking milliseconds or longer.

- **Computational accuracy** – The structural and electronic interactions between different materials at interfaces governing charge transfer processes at interfaces puts significant demands of high accuracy in computational modelling in order to avoid a mismatch between the different materials involved.

- **Electronic structure** – Interfaces with solid substrates include bands of electronic levels that need to be considered, and photoinduced charge transfer processes in particular involve manifolds of excited states.

Figure 3.2 Overview over multiscale modelling challenges in interfacial electron transfer processes.

It is, however, clear that the computational tools that currently find widespread use in much material and molecular modelling are insufficient to capture the full interfacial complexity of mesoscopic systems. In particular, many charge transfer processes, particularly photoinduced redox reactions, can only be described adequately at a quantum level, while the size of the full heterosupramolecular systems typically comprise many thousands of atoms, that is involving system sizes that can normally only be modelled computationally at a classical or phenomenological level. Similarly, the demand for accurate quantum calculations of femtosecond or picosecond reaction steps during photoexcitation or redox reactions is a poor match for studying long-time evolution accompanying charge transport that eventually leads to current generation on a microsecond or millisecond timescale in the complete devices.[13] This provides a strong incentive for the development of multiscale methods for IET applications.

3.2.2.2 Multiscale Modelling Strategies

An ambitious approach to overcome simultaneous demands for accurate treatment of quantum effects on an atomistic level, with a desire to model long timescale processes in complex systems is to devise methods to combine several levels of computational modelling in a unified framework.[23] Developing such *multiscale* approaches is an active field of method development for a variety of scientific and technologically important problems such as heterogeneous catalysis.[38]

The ability to model a system at distinctly different levels of theory can be readily achieved in cases where there is a natural structural division between components that, while all needed for a good system description, can be treated with different degrees of accuracy. A well-known example is the frequent need to include an approximate, polarizable continuum model to describe the environment by a quantum chemical description of a solvated molecule,[26] for example to find realistic redox or excitation energies for light-harvesting molecules used for molecular solar energy conversion devices.[39] In a more sophisticated treatment, the molecule that is treated at the full quantum mechanical (QM) level can be surrounded by a region of explicit solvent molecules, for example at a MD or molecular mechanics (MM) level. Such combinations allow dynamical processes to be modelled, for example capturing charge transfer processes within the QM region that are influenced by thermal fluctuations of the surrounding medium. Referred to as QM/MM methods, such approaches can in principle include several layers of gradually simpler treatments and have been extensively applied, for example in biochemical investigations of enzyme-catalysed reactions.[27]

The appealing simplicity of the approach for systems that naturally split in weakly interacting components, however, is complicated in many cases that would involve the splitting of a chemically continuous system. For example, QM/MM approaches are popular for studies of enzyme catalysis where there is a rather small active site that needs to be treated at the QM level, while large portions of the protein mainly have a structural role to play despite being chemically connected to the active centre. Much effort goes into devising working strategies to make systematic QM/MM investigations that are sufficiently accurate for chemical insight. In particular, the boundary between the levels must be treated properly, for example through inclusion of appropriate saturating groups for the broken bonds at the interfaces.[26,27]

The possibilities for such structural division very much depend on the type of system. In a polymer blend, a mesoscopic system can be easily divided without the need to break chemical bonds, whereas an obstacle for structural division of a heterogeneous system involving a solid substrate on one side lies in the difficulty of creating substrate interfaces without simultaneously introducing significant edge effects that influence the electronic structure of the substrate model. This relates to widely studied cluster termination and embedding problems. It is also more difficult to retain essential quasi-bulk properties of a nanostructured material using a small finite cluster model. Regardless of whether a periodic or non-periodic approach is adopted, it often becomes necessary to use a large quantum region to describe the surface charge transfer processes, even if this only can be achieved at significant computational cost. Exactly how large such a system needs to be is primarily determined by the relevant length-scales of interactions in the system at hand, for example screening lengths and quantum size effects, but will typically require at least several hundred atoms to be included.

An alternative deconstruction scheme can be applied to systems where chemical or electronic processes can be separated into fundamentally different timescales. For example, to model exciton dynamics in a molecular crystal one can, on the one hand, consider various quantum processes such as exciton generation, hopping and recombination at a QM level, while modelling macroscopic exciton distribution and kinetics at a phenomenological level that includes significantly larger spatial regions and time steps using parameters obtained from the QM calculations. Kinetic modelling of natural light-harvesting photosynthetic systems offers a relevant example where such a division of computational levels provides an appealing way to treat electron and energy transfer processes in a complex system.[40] This kinetic division of dynamical electron transfer processes taking place on different timescales offers opportunities to connect a detailed model of fundamental processes such as ultrafast IET (for example femto- or picoseconds) with device performance on longer timescales (millisecond or longer) in an integrated fashion.

3.2.2.3 Bridging the Quantum Mechanical– Molecular Dynamics (QM-MD) Gap

The introduction of advanced classical MD simulation techniques utilizing new force field methods designed to bridge the gap between QM and classical MD methods at the atomistic scale have recently shown notable promise for a wide range of complex materials modelling. Emerging capabilities for treating nanostructured materials in IET applications is discussed here in terms of so-called reactive force fields (as implemented in ReaxFF) – one of the most actively developed approaches for this purpose in the last few years.[41,42]

As outlined above, atomistic descriptions of matter can be made at either a QM or a classical MD level of theory. MD simulations are several orders of magnitude faster and demand much less memory per atom compared to calculations based on QM calculations. MD simulations therefore allow treatment of system size ranges to tens of thousands of atoms and picoseconds or longer simulation times, well beyond the reach of size and complexity of standard QM calculations. A large number of MD force fields have been developed to study many materials properties, but traditional classical MD simulations suffer from a significant shortcoming in that they cannot treat chemical reactions in a realistic manner. Instead bonding arrangements are typically predefined and the dynamics limited to problems where reactivity is not of essence, such as molecular motion within a solvent.[26,27] While QM/MM approaches can work well for separable problems as discussed above, is not always desirable to limit atomistic simulations to reactive and unreactive regions in a static fashion. For example, simulations of surface reactions in heterogeneous catalysis need to include dynamical descriptions of the substrates that can restructure during the catalytic cycle at the same time as adsorbate–substrate reactions take place at many different surface sites.[38] Similarly, it is not *a priori* obvious which bonds will break during a simulation of stress fracture in a solid material.

A modern approach to bridge the QM-MD gap is provided by the reactive force fields which employ a force field that is capable of describing chemical reactivity.[41,42] Such a force field needs to be much more sophisticated in terms of how atom–atom interactions are treated in order to allow breaking and formation of chemical bonds in a continuous fashion. This means that these advanced MD simulations are significantly slower than traditional MD simulations. However, as QM calculations are several orders of magnitude slower than MD calculations, even with a force field that is 10–100 times slower compared to a traditional MD simulation, new size and time regimes that are not accessible to standard QM calculations can be studied theoretically. Given the relative novelty of this multiscale methodology, it is necessary to evaluate the functionality, scope and limitations of this approach carefully, although it is firmly grounded in a first principles atomistic perspective.

Describing a chemical bond that can be formed or broken by a force field at first sight does not appear much more advanced compared to describing bonded interatomic interactions and intermolecular forces, as is done consistently in MD simulations. The typical interaction profile between two species includes a short-range repulsion, an energy minimum at some optimal separation and an attractive region that levels off at large separations. With appropriate adjustments of the scales for energy and separation distance, this could represent the interaction between two noble gas atoms or two solvent species in an MD simulation, or a molecular dimer such as H_2, in a QM study of a bond dissociation profile. For anything but the most trivial atomistic cases there are, however, difficult considerations that need to be addressed to formulate a fully successful MD strategy that incorporates bond making and bond breaking processes.

The first hurdle compared to traditional MD simulation methods is that, in order to accommodate the possibility of structural and chemical rearrangements, a generalized force field must be able to describe elements in different molecular environments in a consistent manner. For example, an oxygen atom forms C–O single and C=O double bonds in many organic compounds, such as alcohols, esters and carboxylic acids, it forms a rather special triplet homonuclear diatomic dioxygen molecule $O_2(g)$ in the gas phase and is a common anion in ionic oxides, such as magnesium oxide crystals, MgO(s). This is not a serious concern in classical MD simulations where different oxygen parameters can be used to describe oxygen in the different bonding environments, but such an approach is not viable when any given oxygen atom can transform dynamically from one type to another during a reactive simulation. A reactive force field therefore needs to be sufficiently flexible to describe various bonding environments and there can be no discontinuities in the force field description during the course of a dynamical transformation from one bonding situation to another. Most significantly, the simulation technique must also be able to assign an energy contribution that is in accordance with the particular bonding configuration at that particular point of the simulation.

This calls for a procedure to evaluate the bonding environment as the basis for calculating the forces that drive the dynamics. In ReaxFF, the energy of the system under investigation is calculated through a number of partial energy contributions according to:

$$E_{\text{system}} = E_{\text{bond}} + E_{\text{over}} + E_{\text{under}} + E_{\text{val}} + E_{\text{pen}} + E_{\text{tors}} + E_{\text{conj}}$$
$$+ E_{\text{vdWaals}} + E_{\text{coulomb}}$$

This includes two-, three- and four-atom terms and some fundamental properties of these are discussed below. A full account of the equations governing the reactive force field MD simulations can be found in the original literature.[41,42]

Making several of the energy terms depend critically on calculations of *bond orders* for each atom is of central importance to describe variable atomic bonding. It is assumed that the bond order of any given atom can be calculated by evaluation of interatomic distances to neighbouring atoms and that all such contributions add up to a net bond order for each atom. In particular, this prevents otherwise pathological tendencies of the atoms to make bonds beyond their natural valency capabilities by inclusion of an overcoordination penalty.

Angle and torsion energy terms are used to give appropriate energy penalties for deviations from normal angles and torsions, consider for example ethene with two trigonal planar carbon atomic centres and with a C=C π-bond that favours a planar arrangement of all atoms in the molecule. The description of this bonding situation is complemented by the inclusion of a conjugation term.

- *Non-bonded interactions.* A distance-corrected Morse potential is used to describe atom-pair interactions, with excessively high repulsions between bonded atoms avoided by including a shielded interaction.
- *Electrostatic interactions.* In order to take into account electrostatic interactions and system polarization effects arising from residual charges on atoms, not least important for consideration of charge transfer processes, interactions are calculated between all pairs of atoms in the system using a shielded Coulomb potential.

A so-called electron equilibration method (EEM), finally, is used to calculate the atomic charges in a consistent manner.[41,42]

The practical implementation of the reactive force field scheme in materials simulations requires extensive parameterization of the various terms entering the force field description. This parameterization is necessary for any classical MD method, but owing to the comparatively large number of adjustable parameters needed to achieve the added reactive flexibility, this is much more extensive for the reactive force field. There is a minimum of around 30 parameters per element that need to be determined prior to production type simulations, with increasing numbers of cross-terms arising from the many-body interactions in systems with many elements.

The strategy specifically adopted to obtain ReaxFF parameters has been to rely on extensive data obtained from state of the art density functional theory calculations performed at a consistent level, mainly B3LYP calculations with valence double-zeta quality or better basis sets. Parameters are fitted to both structural and energetic properties of a large data set typically comprising at least several hundred different results. Both materials properties, for example relating to structures and energies of solid phases, as well as molecular properties such as molecular energies and structures are included in the parameterization on an equal footing. The quality of the ReaxFF MD simulations can therefore approach that of a first principles DFT simulation, in particular as long as the bonding environments encountered during the simulation are compatible with those incorporated into the force field fitting by careful selection of model systems for the training set. The force field can be iteratively improved, if the simulations lead to new bonding situations not originally foreseen in post-MD testing of results by DFT calculations, with the possibility to readjust the parameters until a stable force field has been achieved when complete simulations can be performed and the results double-checked at the end without need for further recalibration. This approach has led to successful modelling of many different molecular properties and materials processes in recent years, convincingly demonstrating the good versatility and reliability of the basic methodology.

The success of the method comes at the expense of much work needed to construct extensive training sets and the devotion of considerable effort to tuning all parameters into a working system. Such parameterizations are motivated for systems that are of general interest for large number of simulations, rather than extensions to just a narrow set of problems. With parameters already available for a significant proportion of the common elements, it is also becoming possible to tackle new problems through more limited extensions of the parameterizations to just include the necessary new cross-terms between elements for which many of the atomic parameters already exist.

3.2.2.4   Surface Reactions

Surface science is a broad field of research in which many cases of heterogeneous catalysis and redox reactions have been investigated over the past decades. A fruitful merger between advanced spectroscopy and first principles calculations has provided significant insight into a number of important model interfaces for adsorbate–surface interactions, including both the properties of stable adsorption layers and catalytic reactions on metal, semiconductor and insulator surfaces.[43,28] It still remains a significant challenge, however, for traditional first principles (for example DFT) and first principles MD methods, such as Car-Parrinello MD, to model complete catalytic cycles of mesoscopic materials under realistic physical conditions, for example organic catalysis by mixed metal oxide catalysts. Development of multiscale methods capable of simulating systems with tens of thousands of atoms for microseconds or longer is therefore of great interest for complex heterogeneous structures and reactions.[38]

Using the reactive force field as an example of such on-going development, the basic functionality of the reactive force field approach was initially demonstrated for hydrocarbons[41] and silicon oxide systems.[42] It has subsequently been expanded to investigations of many different problems in materials science. Some examples of relevance for the modelling of complex interfaces include oxidative dehydrogenation of organic molecules in heterogeneous catalysis involving mixed metal oxides,[38,44] oxygen ion transport in solid oxide fuel cell systems,[45] proton transfer reactivity at metal oxide–water interfaces,[46] as well as doped $BaZrO_3$ proton conductors with applications to diffusion rates for multigranular systems.[47] The approach is illustrated in Figure 3.3 with an MD snapshot from a recent investigation of surface catalysis involving complex transition metal oxides.[38,44]

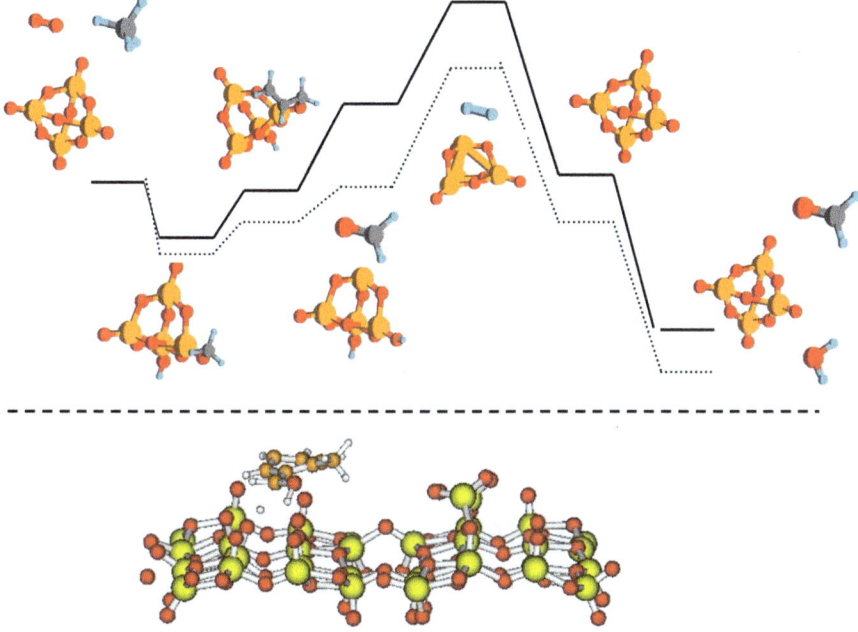

Figure 3.3 ReaxFF reactive force fields are developed to reproduce DFT data for complex chemical reactions, such as heterogeneous catalysis. In the top panel, a comparison of a calculated ReaxFF reaction energy profile (dashed line) is compared to the corresponding DFT results (solid line) for a model catalytic reaction, $CH_4 + O_2 \rightarrow CH_2O + H_2O$ on a small V_4O_{10} cluster. The method allows modelling of complex chemical reactions on picosecond or longer time-scales to be carried out for large simulation cells involving thousands of atoms or more under realistic temperature and pressure conditions. The lower panel shows a snapshot from a simulation of VO_x catalysis of complex hydrocarbons using a force field for hydrocarbon catalysis on vanadium oxide surfaces.[38,44]

Extensions to sensitization of nanostructured materials for solar energy conversion relevant to many of the IET applications discussed in this chapter are also in progress.[24,48]

3.2.2.5 Multi-hierarchy Methods

Using reactive force fields to enable more extensive atomistic simulations than are possible by QM methods alone is a good illustration of one step along a path towards a comprehensive multiscale description of materials properties and processes. Further steps towards modelling macroscopic phenomena and long timescales can be taken by interfacing with subsequent levels of approximations according to the scheme shown in Figure 3.1, for example coarse grained or finite element simulations. Such multiscale implementations have been successfully implemented, for example using scripting procedures to connect various levels of modelling into sophisticated multi-hierarchical frameworks.[49] With recent method developments and sufficient computing power becoming available, million-to-billion atom simulations of chemical reactions can also be demonstrated within so-called divide-and-conquer/cellular-decomposition frameworks.[50]

3.3 Interfacial Electron Transfer

Many recent computational investigations that address IET processes demonstrate current limitations and capabilities of first principles modelling. In this section, some specific examples are used to illustrate results, particularly for computational studies of photoinduced surface electron transfer at molecule–metal oxide interfaces. The examples that are discussed start from accurate quantum chemical descriptions of the initially photoexcited state of typical sensitizers, proceed to well-characterized interfacial systems used to investigate fundamental IET processes and end with a discussion of efforts to expand further the system complexity in the modelling.

3.3.1 Excitations at Interfaces

Photoinduced IET research is strongly promoted by a substantial synergy between the global technological interest in developing molecular devices for solar energy conversion, of necessity a photoinduced process, and the development of ultrafast laser spectroscopy to probe ultrafast processes that naturally are also photoinduced.[10] Computational modelling plays an increasingly important role in this development, for example providing insight into molecular properties that are hard to measure experimentally, such as structures and properties of transient species. First principles calculations can increasingly also be used for accurate predictions, for example of optical excitations, and thus serve as a valuable guide for experimental efforts to develop new molecular and materials components such as light-harvesting complexes with favourable photochemical properties.

3.3.1.1 Molecular Excitations

Light-harvesting by a designated molecular chromophore provides the first step in many photoinduced electron and energy transfer applications that include both supramolecular and heterogeneous systems.[8] Accurate modelling of excited states and subsequent excited state evolution is, however, generally much more demanding compared to ground state calculations both in terms of demand on computational resources and achievable accuracy.[26,27] Therefore, there are still many important questions to address computationally even for the initial excitation steps of many photoinduced IET processes.

Electronic structure calculations, in particular time-dependent DFT, have been successfully applied to studies of the optical absorption properties of a wide range of molecules,[51,52] including many dyes and molecular sensitizers for IET applications.[37,53] These calculations demonstrate significant predictive capabilities and provide insight into strategies for achieving efficient light-harvesting by tuning absorption spectra. They can also provide information about the influence of environmental effects, for example of solvents, counter ions and substrates, on excited state properties.

Calculations of excited state potential energy surfaces of light-harvesting chromophores beyond the initially excited Franck-Condon region are currently emerging as an important area for further theoretical work to guide development of improved molecular systems for photoinduced energy and electron transfer applications.[54,55]

Figure 3.4 provides a recent illustration of a promising molecular light-harvesting system, first developed for photoinduced electron transfer in linear molecular arrays for artificial photosynthesis, where a combination of DFT and TD-DFT calculations have provided theoretical insight guiding the design of new Ru^{II}-bistridentate complexes with improved excited state properties.[56–58] Computational predictions were made for new light-harvesting complexes that facilitated the search for complexes that overcame previous problems with short excited state room temperature lifetimes in Ru^{II}-bistridentate complexes. In particular, the $Ru(dqp)_2^{2+}$ complex, which was first characterized computationally, has subsequently been found to have a microsecond room temperature excited state lifetime – an improvement of several orders of magnitude compared to the standard $Ru(tpy)_2^{2+}$ complex.

The initial search criteria for the calculations were based on a consideration of promoting favourable ligand field splitting of the electronic structure *via* optimization of the octahedral metal coordination cage geometry. The complexes then served as a good prototype case for further in-depth computational studies aimed at investigating how multidimensional excited state potential energy surfaces play an important role in preventing activated decay and ground state recovery. This development makes the class of Ru^{II}-bistridentate complexes, with their useful structural capability to form linear arrays, useful for many supramolecular and interfacial charge transfer applications taking place on a picosecond timescale. These examples suggest that accurate first principles calculations that deal with the excited state evolution of

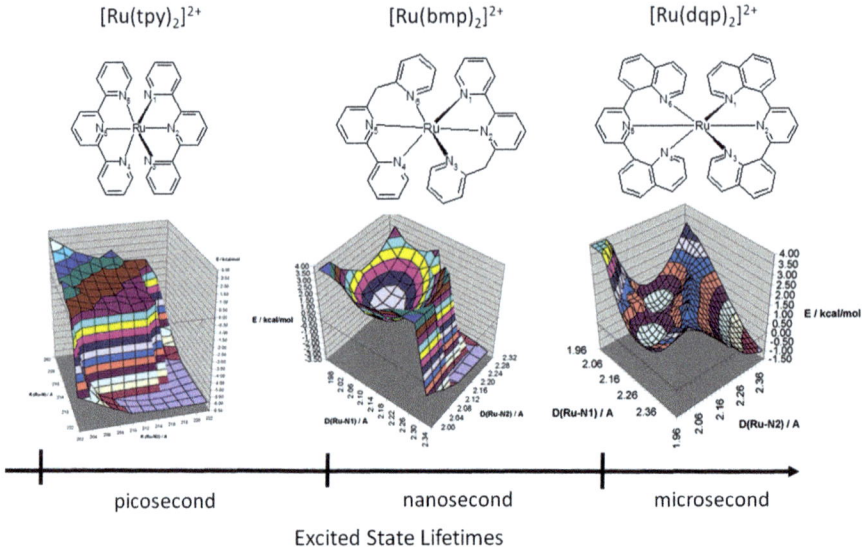

Figure 3.4 Quantum chemical calculations of multidimensional potential energy surfaces of three Ru-bistridentate complexes ($Ru(tpy)_2^{2+}$, $Ru(bmp)_2^{2+}$ and $Ru(dqp)_2^{2+}$) show significantly different barriers for activated decay,[55] consistent with experimental observation of variations in room temperature excited state lifetimes by several orders of magnitude.[57]

light-harvesting molecules beyond the initial Franck–Condon excitation region are set to gain prominence over the next few years.

3.3.1.2   Optoelectronic Properties of Nanomaterials

Electronic properties of solid materials have been extensively studied from a condensed matter theory perspective,[59] including calculation of the bulk and surface properties of many metals and semiconductors.[25] There has also been rapid development of strategies to synthesize, characterize and utilize a wide variety of metal and semiconductor clusters and nanoparticles.[60–62] Excitations in nanoparticles are of significant current interest for example optoelectronic and photovoltaic devices, as well as photocatalytic conversions such as water splitting at large exposed surface areas in such materials. Comprehensive modelling of nanoparticles in the quantum size regime, that is where the electronic properties differ from those of the corresponding bulk material, is more challenging compared to studies of ordered bulk solids and surfaces owing to an increased number of surface defects and the greatly increased possibility of different surface defects compared to high-quality crystals and their clean surfaces. Electronic properties and excitations can, however, be calculated for atomistic cluster models of increasing size, for example with similar TD-DFT methods which have been used extensively for molecular systems as described in the previous section. The ability to construct and investigate clusters of different size, shape and defect content provides a good testing ground to understand what surface defects arise in various nanomaterials that are not always readily accessible experimentally.

Computational cluster studies of nanostructured TiO_2, a prototype example of interest for IET processes studied in subsequent sections, show a gradual convergence of electronic properties, including excitation threshold, to the bulk regime for nanoclusters in the low-nanometre size regime, see Figure 3.5. The cluster models require realistic bonding throughout, essentially consistent with predictions from quantum confinement arguments, but with significant absorption threshold oscillations owing to local surface effects caused by a significant sensitivity of atomistic properties.[63,64]

There are also interesting examples of dynamical simulations and wider sampling of the ground state potential energy surfaces of nanoparticles that look promising for more general multiscale modelling for TiO_2 and related materials.[65,66]

These advances in the capability to model electronic structure properties in materials with realistic mesoscopic morphologies, for example by combining large-scale structural dynamics simulations with accurate excited state calculations of local defect sites, holds promise to provide much valuable

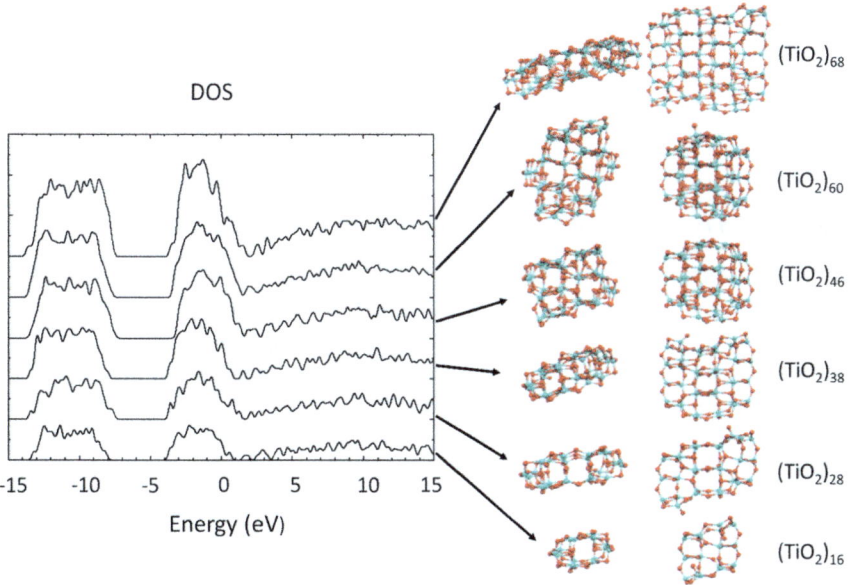

Figure 3.5 Quantum chemical calculations of TiO_2 nanoparticles with diameters up to *ca.* 2 nm. The optimized structures show significant surface relaxations and reconstructions for the small clusters, while the density of states (DOS) shows gradual development of a quasi-bulk band structure with increasing cluster size, including a filled valence band (below *ca.* −8 eV), a clean band gap at intermediate energies and a en empty conduction band (above *ca.* −4 eV).[63,64]

information for surface catalysis and interfacial charge transfer processes in the near future.

3.3.2 Interfacial Interactions

First principles based modelling, including extensions to multiscale methods, generally require the construction of a realistic atomistic model of the interface including both adsorbate and substrate. This can either be a computer-generated structure (either an optimized structure, or a simulated trajectory), or an experimentally known structure (common for example for biological systems such as enzymes). Based on such a model it is then possible to obtain information about the interfacial electronic structure and excitations at the interface, as well as providing a starting point for investigations of dynamical processes such as IET. The structural requirement is both a strength and a weakness compared to more phenomenological models that include advanced kinetic and dynamical modelling on parameterized systems.[67] The strength of the atomistic perspective is that a more detailed perspective can be gained without the need to invoke parameters that may, for example, only be obtainable by fitting to experiment. This gives atomistic first principle- based methods much greater predictive power for new systems. The weakness is that, with a few notable exceptions, little is known about the detailed structure of most experimentally investigated systems.[19,36] Many detailed time-resolved investigations of IET processes have been performed on complex systems, such as DSSCs, organic electronics and biological light-harvesting systems, where the atomistic structure is either unknown, or too complex to handle by first principles methods.

3.3.2.1 Dye-Sensitized Semiconductor Surfaces

The dye-sensitized nanostructured semiconductor interface constitutes a prominent test case for IET studies that is both of wide experimental interest and a good system to study fundamentally interesting phenomena, such as few-femtosecond IET processes.[7] However, the complex heterogeneous structure of working DSSC devices has in many cases provided little information about the detailed interfacial structure which is needed for atomistic modelling. Two paths for first principles and multiscale modelling are in focus here. First, there are only a limited number of prototype adsorbate–semiconductor systems that are of interest for IET studies and where the interfacial structure can be established to such a degree that they can be carefully studied both experimentally and computationally. While not always very good for device applications, these systems provide an invaluable testing ground for bench-marking emerging IET modelling approaches. Second, interfaces of significant experimental interest, for example the interaction of champion DSSC dyes with nano-TiO_2, are worth subjecting to large-scale computational investigations that aim to provide new insight into more complex systems, for example by evaluation of the efficiency of various IET processes.

3.3.2.2　Adsorbate Anchoring

Numerous computational investigations of small molecules on relevant metal oxide surfaces have been carried out in the broad area of surface science over the last decades, for example in studying heterogeneous catalysis.[19,36,68–72]

There is also a need to address larger systems by calculations in order to meet a growing experimental interest in spectroscopic studies of interfacial charge transfer in heterogeneous materials such as dye-sensitized nanostructured semiconductors. An important stepping stone towards atomistic models of experimentally interesting systems for spectroscopic studies of fundamental IET and DSSC applications is the determination of basic binding features of designated anchor groups such as carboxylic acids.[19,36,68–72]

Investigations of designated anchor units, typically comprising two or more anchor groups that combine together to form a strong interfacial link, have provided an important stepping stone towards systems of greater interest for IET applications. This includes in particular bi-isonicotinic acid binding to TiO_2, as this molecule is the binding ligand of champion RuN3-type dyes, see Figure 3.6. This model system illustrates much of the structural complexity that arises when more than one anchor group is interacting with the surface. Combined experimental and computational surface characterization under ordered conditions, including the use of single crystal substrates, provided an early example of detailed atomistic characterization of the interfacial interactions of a molecule with this complexity. Strong surface binding was found to be characterized by a compromise between strong local surface binding of the individual anchor groups together with utilization of some limited structural flexibility in the molecular backbone.[73,74] Figure 3.6 illustrates how essential surface binding features are captured by this system, although the interfacial structural complexity increases further when complete dye molecules, such

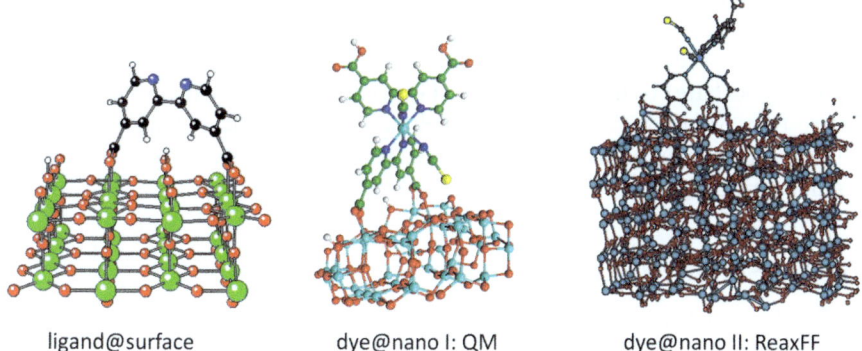

ligand@surface　　　　dye@nano I: QM　　　　dye@nano II: ReaxFF

Figure 3.6　Molecule–TiO_2 interfaces used to investigate ultrafast interfacial electron transfer processes. Increased system interaction complexity is illustrated for periodic surface QM calculations of a binding ligand on a clean rutile (110) surface (left), QM cluster calculations of a complete dye molecule (RuN3) in a nanocrystalline environment (middle) and a multiscale MD simulation (right), highlighting both differences and similarities.[24,73,82,]

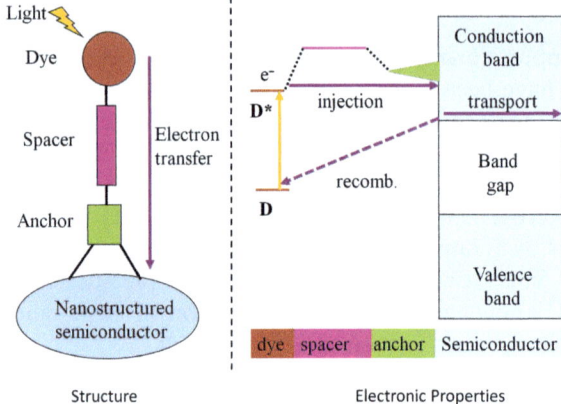

Structure   Electronic Properties

Figure 3.7 Schematic illustration of the relationship between structure and interfacial electron transfer properties in dye-sensitized nanostructured semiconductor materials. Control of the balance between electron injection, recombination and transport depends crucially on the physical separation and electronic coupling capabilities between the molecular donor and semiconductor acceptor states mediated by designated anchor and spacer groups.

as the widely used DSSC dye RuN3, are attached to more realistic nanostructured models.

3.3.2.3   *Interfacial Electronic Interactions*

Characterization of the electronic properties of interfaces of bi-isonicotinic acid (and close relatives including isonicotinic acid and benzoic acid) with TiO_2 single crystal surfaces, which combined quantum chemical calculations with X-ray spectroscopy, also provided early insight into significant interfacial electronic interactions, including surprising evidence of both adsorbate level shifting and of few-femtosecond electron injection (following X-ray absorption by the adsorbate) in this system.[73–76] This has spurred considerable continued interest in providing further experimental and theoretical details about the fundamental structural and electronic interactions of such well-behaved model surface systems.[19,21,22,71,77]

3.3.3   **Surface Electron Transfer**

3.3.3.1   *Surface Electron Transfer Mechanisms*

Almost all photoinduced surface electron transfer processes in sensitizer-semiconductor hetero-structures are most naturally characterized as two-step processes with initial excitation of the adsorbate, followed by interfacial electron transfer into a band of an acceptor state, as shown schematically in Figure 3.7.[7] There is, however, also a more unusual case in which the IET is caused by a direct interfacial charge transfer excitation which is described in the following section.[78] A wide range of advanced computational investigations have been performed in order to investigate these types of surface electron transfer processes in recent years and relevant computational modelling studies are discussed in the rest of Section 3.3.3.

In addition to the computational considerations presented in this chapter, there have been many important advances, for example employing sophisticated dynamical simulations of IET processes by several research groups. Some of these methods are highlighted elsewhere in this book and in several excellent recent review papers and are thus left outside the scope of the present chapter.[20,21,79]

3.3.3.2 *Interfacial Charge Transfer Excitations*

So-called interfacial charge transfer excitations have been observed for some chromophores attached to TiO_2 substrates, including catechol@TiO_2 which is a prominent example that is closely linked to IET in other adsorbed organic molecules. Charge transfer systems can often be recognized by clear changes in the absorption spectrum. In the case of catechol-sensitized TiO_2, a new low-energy shoulder appears below the TiO_2 fundamental band excitation threshold around 370 nm, that is, in a spectral region where neither TiO_2 nor catechol normally absorbs light. This represents an unusually strong adsorbate-substrate electronic interaction in which the charge transfer excitation can be seen from calculations to be dominated by direct excitation of an electron in a level mainly on the donor side to a level mainly on the acceptor side. This charge transfer transition typically gains intensity from orbital mixing or configuration interaction with higher excitations and is in this sense different from the level-shift discussed above for bi-isonicotinic acid.[26,71,75]

3.3.3.3 *Interfacial Electronic Coupling*

From a theoretical perspective, photoinduced IET across molecule–semiconductor interfaces can be broadly categorized into different regimes depending on what factors govern the transfer rate. For standard injection from a single donor state on the molecular side, coupling with a manifold of acceptor states in the substrate the *driving force* and *electronic coupling strength* are two key parameters. The driving force for molecule–semiconductor electron injection is closely related to the energetic alignment of the molecular donor level relative to the band of substrate acceptor levels. In particular, the position of the injecting state relative to the bottom of the acceptor band is crucial for the electron injection process.[4,36] For electron injection from a molecular donor state that is situated high above the bottom of the acceptor band (typically some hundred meV), the donor level is in constant resonance with the substrate band in which the so-called wide band limit is realized.[36] This case promotes ultrafast photoinduced electron injection, as is evident in many experimental systems.[7]

Electronic structure calculations of the interfacial electronic interactions can provide much valuable information about this type of system, including calculated interfacial level alignments giving information about the driving force and thus the potential realization of the wide band limit, as well as good predictions of interfacial coupling strengths.[80] In cases where the driving force for electron injection is small from the initially populated excited donor state, excited state vibrations and relaxation mechanisms may bring the injecting state below the bottom of the acceptor band during part of the excited state dynamics. The injection dynamics can then be expected to depend more strongly on vibrational aspects such as activation. In this case simulations taking explicit excited state dynamics into account, for example first principles non-adiabatic MD simulations, provide an appealing dynamical perspective.[20]

In the wide band limit where surface electron transfer rates are likely to be largely governed by interfacial electronic coupling, the strength of this coupling represents an important physical parameter which can be evaluated from calculations of interfacial electronic structures of dye-sensitized semiconductor systems from either periodic or cluster calculations.[77,80] From a qualitative point of view, weak and strong coupling cases can typically be distinguished by visualization of the involved molecular orbital levels in electronic structure calculations. In a further step, the interfacial electronic coupling can be evaluated more accurately by determination of the width, Γ, characterizing the spread in energy of the injecting level interacting with a quasi-continuum of acceptor states.[77,80] Estimates of injection rates can then be obtained assuming an inverse relationship between injection time, τ and the electronic coupling Γ as $\tau = k/\Gamma$, with a simple proportionality constant k.

This simple approach offers the possibility to gain fundamental insight into surface electron transfer processes from standard quantum chemical and condensed matter theory calculations,[77,80] and comparisons with experiments on perylenes with different spacer groups indicate that this simple approximation often reproduces essential experimental features well, such as the influence of the chemical nature of the anchor and spacer groups involved.[80]

The situation where there is minimal driving force for electron injection is experimentally important as a strategy to minimize losses in efficient solar energy conversion and this situation is likely to be often realized in champion DSSC systems, such as RuN3–TiO$_2$.[81] Calculations of champion DSSC systems such as the RuN3–TiO$_2$ interface are naturally of significant interest to understanding in detail why they work so well and how they may be improved.[82,83] In such subtly balanced IET systems, small variations in the predicted donor–acceptor alignment can, however, have a strong influence on any computational predictions, for example through strong variations in calculated interfacial electronic coupling strengths for small changes in donor–acceptor alignment. Significant caution is needed as the computational accuracy can be limited by both fundamental method limitations, for example the accuracy of the DFT functionals and modelling restrictions, for example using cluster models of limited size or periodic models that do not include surface defect states. There are, moreover, only a few systems that have been characterized in sufficient detail experimentally to be able to test thoroughly the underlying theoretical assumptions of different modelling approaches[84] and further computational progress can be expected as new interfacial systems that are amenable to detailed investigation can be identified and subjected to direct comparisons between theory and experiment.

In terms of the forward photoinduced IET, one interesting area of continued theoretical interest concerns explicit vibrational relaxation and vibrational coupling of the initially excited molecular state to product states reached immediately after IET.[36] Another area that is likely to profit from further computational investigation concerns the possible presence of intermediate states characterized by a bound electron–hole pair prior to release of electrons into a conduction band characterized by an experimentally discernible increase in electron mobility, for example from THz measurements.[85]

Calculations of interfacial electronic interactions can also be performed for snapshots from MD simulations. This can capture structurally induced fluctuations in electronic coupling over time, but explicit excited state dynamical effects, such as excited state relaxation *via* internal conversion and intersystem crossing process, require special attention. As mentioned above, more sophisticated treatments, in particular explicit excited state dynamics calculations, can then be useful for more sophisticated analysis of particularly interesting systems, as discussed elsewhere in this book and in reviews referred to above.

Another area that is subject to significant interest concerns substrate variations, for example comparisons of IET processes for TiO_2 and ZnO substrates, which have been carried out both experimentally[7,85] and theoretically.[77] Calculations suggest that the interfacial electronic coupling can also be sufficiently strong to mediate ultrafast IET for ZnO, but there may be other differences that are important experimentally for the ZnO performance, such as surface stability and screening of injected charges.[85]

3.3.3.4 Spacer Groups – The Weakest Link

Dye-sensitized semiconductors in general, and realizations of the wide band limit in particular, offer interesting opportunities to investigate the distance dependence of electron transfer in heterogeneous systems. Many ambitious experimental investigations have been carried out, although surprisingly few systems have displayed behaviour that can be rationalized based on the distance dependence of electron transfer alone.[7,36] In many DSSC-type systems studied experimentally using ultrafast spectroscopic techniques, structural disorder at the molecular level may contribute to influencing the IET kinetics. Thus even if IET rates have been measured accurately for many experimental systems,[7] it may often be difficult to correlate such information with molecular level donor–acceptor levels in a similar fashion as would have been done for chemically well-defined donor–acceptor systems.[86]

Perylene–TiO_2 interactions studied under highly controlled experimental conditions, including vacuum and single-crystal surface conditions, by Willig and co-workers presented some of the first and most convincing experimental results showing the significant influence of various anchor (for example carb-oxylic acid *vs.* phosphonic acid) and spacer groups (saturated *vs.* unsaturated) on surface electron transfer rates.[36] Together with first principles calculations on comparable systems, this provided direct evidence that the influence of the chemical nature of the bridge could largely override spatial distance depend-ence.[80] Figure 3.8 shows results from calculations of another closely related system comprising a pyrene chromophore bound to TiO_2 *via* a so-called rigid rod anchor and spacer designed to probe long-range IET.[87]

Interfacial systems of this size and complexity have only become amenable to calculations in the last few years, but increasing computer power and im-provements in efficient computational chemistry approaches, in particular in-volving DFT and TD-DFT methods, are providing good opportunities for an increasing number of studies that can be expected to flourish greatly in the

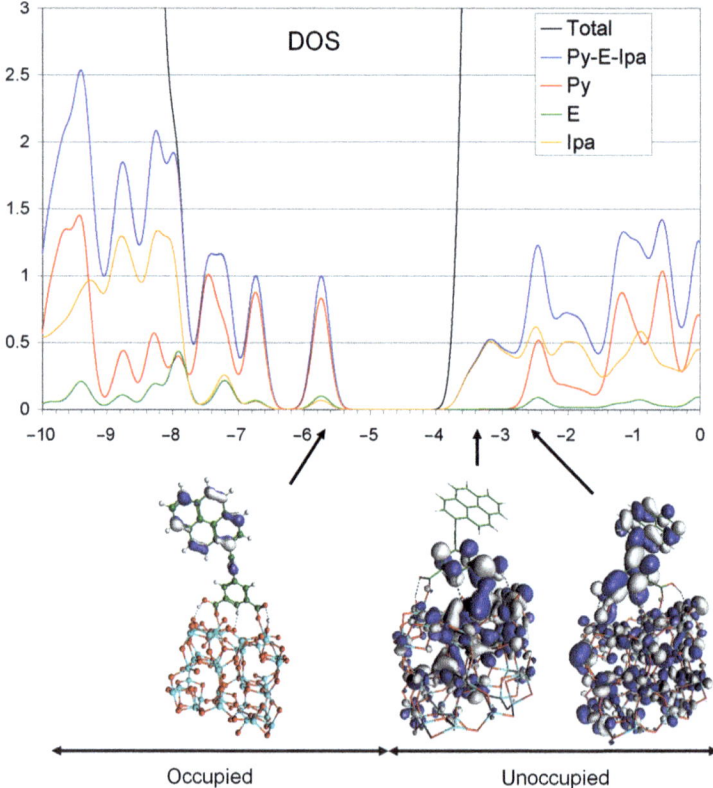

Figure 3.8   Variations in interfacial electronic coupling of different electronic levels of a pyrene chromophore interacting with a TiO_2 nanoparticle *via* an ethynylene-IPA spacer + anchor unit. The ground state HOMO level shows weak interaction with the TiO_2 substrate bands, while there is strong interfacial electronic interaction of the unoccupied adsorbate levels involved in photoinduced electron injection.[87]

dye-semiconductor injection, but also recombination processes and interactions with other surface species and redox mediators that open up a more comprehensive understanding of the basic electron transfer processes in advanced solar energy systems.[20]

3.3.3.5 Electron Dynamics

As discussed above, interfacial electron transfer at molecule–semiconductor interfaces may in many cases be limited mainly by the interfacial electronic coupling, rather than requiring structural activation processes. From a multiscale perspective that attempts to split complex materials processes into separable problems, this can be taken advantage of by performing electron dynamics simulations on structures that have been previously determined from either quantum chemical calculations or MD simulations.

It has been shown that such electron dynamics can be performed in good agreement with time-resolved experiments for perylene–TiO_2 interfaces with different anchor and spacer groups.[88,89] In particular, the use of electronic structure information from previous atomistic quantum chemical calculations provides an illustration of a successful combination of different levels of theory into a first principles-based multiscale approach. Theoretical developments of such electron dynamics simulations have recently also included vibrational effects that can be incorporated from separate calculations of molecular vibrational modes.[89]

3.3.4 Modelling Complex Materials

Detailed investigations of structurally well-characterized prototype systems, as discussed in the previous section, provide a solid basis for understanding fundamental charge transfer interactions at heterogeneous interfaces. One of the main challenges for developing working technology, however, is to be able to achieve a similar level of control in structurally disordered materials. Dye-sensitized solar cells provide a good example: the nanostructured metal oxide substrate provides a drastic and necessary increase in surface area at a highly competitive production cost, but the surfaces typically display significant structural disorder at the molecular level. This disorder makes it hard to control interfacial interactions. In particular, failure to control surface binding arrangements of many sensitizers precludes full functional control, often resulting in degraded functional performance both at the molecular level, for example in terms of lack of surface electron transfer control and at the device level, for example decreased maximum power output in working devices.[83]

Modelling individual sensitizer–substrate interactions in champion systems such as the RuN3 dye interaction, with more and more realistic descriptions of nanostructured TiO_2 substrates, provides one way of investigating how well current systems perform. For example, quantum chemical investigations of the RuN3-TiO_2 interface have provided insight into alternative binding configurations and how they influence interfacial properties.[82,83] Developing multiscale methods provides significant possibilities for expanding the reach for advance simulations of more complex situations, as illustrated in Figure 3.6.[24]

3.3.4.1   *Functional Control of Disordered Interfaces*

An alternative route to understanding the complications of disordered interfaces is to achieve functional control of intrinsically disordered materials interfaces as well. Such enhanced control can, for example, be achieved by construction of molecular sensitizers capable of binding in a functionally equivalent manner even to disordered nanostructured substrates. An important step in this direction is to find molecular sensitizers that can form interfaces with nanostructured substrates that display consistent interfacial electronic contacting capabilities, enabling systematic investigations of the distance dependence of surface electron transfer processes.

Several examples of advanced anchor motifs designed to facilitate uniform surface anchoring have been synthesized in recent years, including, for example, so-called tripods which can be used to anchor a wide range of different chromophores.[90] Owing to the structural complexity of these interfaces it has, however, become clear that it is often hard to predict binding capabilities with any certainty without resorting to some form of computational modelling. For example, many surface electron transfer measurements with the original *para*-anchored tripod unexpectedly indicated remaining problems in controlling the chromophore–substrate distance. Simple structural modelling recently provided a clue that the binding of the third leg of the tripod is likely to be less than ideal on typical substrate surfaces and suggested that a modified anchoring motif involving *meta*-substituted anchor groups could alleviate this problem. Two important aspects of the interfacial match that relatively simple atomistic modelling can provide information about concern first the simultaneous matching of several anchor groups to suitable surface attachment positions and second whether the anchor groups approach the surface at favourable orientations. For the tripods, in particular, the latter aspect could be improved by a switch from *para*- to *meta*-tripods. When tested experimentally, *meta*-tripods were indeed found to give enhanced two-dimensional surface control on favourable substrate morphologies even if other problems of interacting chromophores in so-called necking regions remained a problem in nanostructured materials that had small cavities compared to the size of the sensitizer, see Figure 3.9.[86]

Further improvement in interfacial control was recently demonstrated experimentally using a so-called Ru-star strategy in which homoleptic octahedral complexes with a central Ru^{II} metal centre provide rigid spacer groups pointing out in all six orthogonal directions, thus effectively preventing the metal core accidentally approaching the substrate surface, see Figure 3.9.[91,92] With improved 3D structural control over the interface, it is now possible to provide a better understanding of interfacial properties, in particular including electronic contacting capabilities, which could be extracted for several model sensitizers from a molecular perspective.[93] This provides an illustration of the emerging

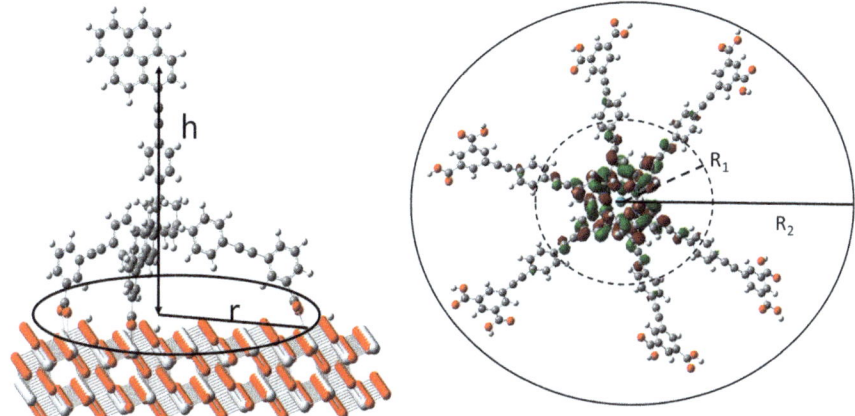

Figure 3.9   Advanced molecular functionalization strategies for enhanced interfacial electron transfer control. Tripods (left) achieve two-dimensional (2D) structural control on metal oxide surfaces, *i.e.* fixing chromophore–surface (h) and chromophore–chromophore (r) interaction distances.[86] Ru-star complexes (right) also provide three-dimensional (3D) control of the chromophore–substrate interaction distances in nanostructured systems.[93]

capabilities of first principles calculations to play an integral part in understanding the enhanced capabilities to control interfacial electronic properties and to guide further experimental developments to optimize IET processes.

3.3.4.2   Mesoscopic Charge Transport

Initial photoinduced surface electron transfer is of significant fundamental interest, but is typically only the first step in a complicated series of charge transport events in working devices, for example for achieving a working dye-sensitized solar cell device.[13] On one hand the injected electrons must be transported through the nanostructured semiconductor network to a back electrode and on the other hand the sensitized must be regenerated through a series of redox steps taking place in the electrolyte. Both processes involve modelling of charge transport over significantly larger distances (hundreds of nanometres to micrometres) and times (for example milliseconds) compared to the initial surface electron transfer processes that typically proceed in pico-seconds on a nanometre scale. Significant efforts have been made to model long-range charge transport properties as well, for example using random-walk models,[94,95] under alternative assumptions regarding electron transfer and transport mechanisms, substrate morphology, defect distributions and so on.

More sophisticated multiscale models capable of integrating the various length- and timescales to achieve a fully comprehensive description in which device performance, such *I–V* characteristics,[96] can be traced all the way down to fundamental phenomena and elementary electron transfer processes can hopefully play an increasingly important role over the next few years.

Investigations of photoinduced surface electron transfer in dye-sensitized semiconductor systems, as discussed extensively above, have provided a good testing ground for many calculations in the last decade. One advantage of these systems is that they offer significant opportunities to construct realistic atomistic models for a variety of first principles approaches because realistic surface binding modes can often be deduced from consideration of the involved anchor groups. Several other systems involving highly interesting surface electron transfer processes are, however, also emerging where computational modelling is becoming an increasingly valuable tool for understanding basic surface electron transfer mechanisms and efficiencies. Some of these modelling applications and opportunities are outlined more briefly below with the aim of highlighting research areas where multiscale modelling of IET is likely to be of significant interest in the near future.

3.3.4.3 *Polymers and Bulk Hetero-junctions*

As mentioned already in the introduction to this chapter, organic and polymer bulk hetero-junctions have shown very significant potential for electronic, optoelectronic, display and photovoltaic applications over approximately the last ten years.[97] Many electron and energy transfer processes superficially mirror those in the DSSCs, with a particular need to optimize interfacial charge separation and minimize recombination losses. Calculations of complex interfacial structure and functionality of such bulk hetero-junctions have been published for some systems,[15] but remain a challenging task for first principles calculations and multiscale modelling, as the functionality is often driven by electron transfer across an interface that is formed between weak non-bonding interactions, such as in polymer solar cells consisting of polymer–fullerene blends.[98] Predicting accurate atomistic structures for weakly interacting and flexible systems is particularly challenging for first principles DFT modelling as many standard DFT functionals are not able to describe weak interactions (in particular so-called dispersion) accurately, while calculated excited state properties are very sensitive to structural changes. The problem is compounded by the fact that polymer blends often do not have well-defined crystal structures, but display a significant degree of structural disorder. One way forward is to use cutting-edge DFT functionals that improve the description of the energy landscape. Another approach is to study idealized donor–acceptor interactions and rely closely on comparisons with experimental results, for example regarding spectral properties, which provide indirect confirmation of structural assumptions. In some cases such as, for example Poly(3,4-ethylenedioxythiophene) poly(styrenesulphonate) (PEDOT:PSS) blends, it has recently been possible to conduct first principles investigations of charge transfer properties using blend models with tractable periodic supercells.[99] Yet another strategy is to rely on highly parameterized methods, such as semi-empirical ZINDO spectral calculations and MD simulations for structural predictions. Calculations can currently serve an important role in rationalizing experimental observations, for example regarding driving energy and delocalization.[100] Given the technological importance of these systems, a lot of computational effort has been initiated and more can be expected in the near future, to provide accurate computational modelling of these elusive interfaces.[16]

Interesting comparisons between DSSCs and polymer-based solar cells can be expected to emerge over the next few years regarding fundamental IET processes and properties such as driving force requirements, as well as fundamental limitations on rates of interfacial electron transfer and charge transport.

3.3.4.4 Semiconductor Hetero-structures and Hybrid Systems

A number of novel nanostructured semiconductor hetero-structures have been explored extensively for device applications for several years.[101] Increasing computational capabilities to model IET processes in nanostructured materials in many ways provide similar opportunities as for the dye-sensitized solar cell applications discussed as a prototype example above. Moreover, recent progress in the growth of complex nanostructures based on new materials, such as III–V semiconductors like GaN opens up significant new opportunities for nanoscale energy conversion applications. This provides a good example of an emerging research area that is likely to offer an active field for computational materials research over the next fewyears.[102]

Other new types of molecular interface that can be expected to become interesting in the near future include soft materials and bio-interfaces, as well as molecular and polymer interactions with new types of nanostructured substrates, such as III–V nanowires. The novelty of such applications means that this is set to continue to offer new computational challenges for many years to come.

3.4 Conclusions and Outlook

Current computational modelling capabilities for IET systems are typically characterized to a high degree by successful applications of accurate calculations of particular properties of interfacial charge transfer systems. This includes atomistic models of interfaces of increasing complexity, excited states of many light-harvesting complexes and electron transfer dynamics in an increasing number of prototype systems of gradually increasing structural complexity. Calculations are now at a point where they frequently contribute valuable theoretical insight into the analysis of materials performance, for example for dye-sensitized solar cells.

Although much less abundant, there are also encouraging examples of calculations being used for predictive purposes, that is, carried out as an integral preliminary step in materials development to guide experimental efforts to synthesize new materials with specified properties. Recent successful examples include both the development of improved light-harvesting molecular systems and the development of new anchor motifs for molecular sensitization of nanostructured semiconductor substrates with improved surface binding and electronic contacting capabilities. Such predictive modelling has the potential to play a transformative role in a field where it is notoriously hard to predict from chemical and physical intuition how new materials will behave.

The combination of a complex materials structure with (i) realistic structural models containing thousands or more atoms, (ii) a manifold of electronic states and multidimensional potential energy surfaces and (iii) dynamical processes spanning many orders of magnitude in time from ultrafast elemental steps (femtoseconds) to macroscopic charge transport (milliseconds) still presents a formidable challenge for comprehensive modelling of many complex interfaces where important electron transfer processes take place. Significant inroads to overcome modelling of interfacial charge transfer have been made in recent years using high performance first principles simulations as well as emerging multiscale modelling techniques. In particular, significant progress has been achieved towards studying increasingly realistic interfaces. While early studies typically focused on simple model systems, such as the structural and electronic inter-actions of a small organic adsorbate on a single crystal surface, calculations are now performed on state-of-the art sensitizers in explicitly nanocrystalline en-vironments. Several of the issues adding to the structural complexity could be addressed along this path, including the effect of including designated spacer groups on interfacial charge transfer, solvent effects, effects of surface orientation and ordering. Nevertheless, much work is still needed to bridge significant gaps in time and size remaining before truly comprehensive multiscale system mod-elling can be routinely carried out on such complicated systems.

The prospect of complementing crucial technological developments in many new materials for interfacial charge transfer by fast and reliable computational modeling – capable of realistically capturing the full complexity of the materials and processes involved – is great. Developing and applying a range of first prin-ciples calculations and multiscale modelling approaches therefore presents a field of computational research where many interesting developments and applications of advanced computational modelling can be foreseen for coming years.

References

1. R. A. Marcus and N. Sutin, *Biochim. Biophys. Acta*, 1985, **811**, 265.
2. D. M. Adams, L. Brus, C. E. D. Chidsey, S. Creager, C. Creutz, C. R. Kagan, P. V. Kamat, M. Lieberman, S. Lindsay, R. A. Marcus, R. M. Metzger, M. E. Michel-Beyerle, J. R. Miller, M. D. Newton, D. R. Rolison, O. Sankey, K. S. Schanze, J. Yardley, and X. Y. Zhu, *J. Phys. Chem. B*, 2003, 107, 6668.
3. E. Ilisca and K. Makoshi (eds), *Electronic Processes at Solid Surfaces*, World Scientific Publishing, Singapore, 1996.
4. R. J. D. Miller, G. L. McLendon, A. J. Nozik, W. Schmickler and F. Willig, *Surface Electron Transfer Processes*, VCH Publishers, New York, 1995.
5. J. Stähler, U. Bovensiepen, M. Meyer and M. Wolf, *Chem. Soc. Rev.*, 2008, **37**, 2180.
6. N. S. Lewis and D. G. Nocera, *Proc. Nat. Acad. Sci.*, 2006, **103**, 15729.

7. J. B. Asbury, E. C. Hao, Y. Q. Wang, H. N. Ghosh and T. Q. Lian, *J. Phys. Chem. B*, 2001, **105**, 4545.
8. R. J. Forster, T. E. Keyes and J. G. Vos, *Interfacial Supramolecular Assemblies*, John Wiley & Sons, Chichester, 2003.
9. G. J. Kavarnos, *Fundamentals of Photoinduced Electron Transfer*, VCH Publishers, New York, Weinheim, Cambridge, 1993.
10. O. Kuhn and L. Wöste (eds), *Analysis and Control of Ultrafast Photoinduced Reactions*, Springer Series in Chemical Physics, Springer, Berlin, Heidelberg, **87**, 2007.
11. M. Grätzel, *Inorg. Chem.*, 2005, **44**, 6841.
12. S. Ardo and G. J. Meyer, *Chem. Soc. Rev.*, 2009, **38**, 115.
13. A. Hagfeldt, G. Boschloo, L. Sun, L. Kloo and H. Pettersson, *Chem. Rev.*, 2010, **110**, 6595.
14. J.-L. Brédas, D. Beljonne, V. Coropceanu and Cornil, *Chem. Rev.*, 2004, **104**, 4971.
15. N. Vukomirovic and L.-W. Wang, *NANO Lett.*, 2009, **9**, 3996.
16. C. Risko, M. D. McGehee and J.-L. Brédas, *Chem. Sci.*, 2011, **2**, 1200.
17. A. Magnusson, M. Anderlund, O. Johansson, P. Lindblad, R. Lomoth, T. Polivka, S. Ott, K. Stensjö, S. Styring, V. Sundström and L. Hammarström, *Acc. Chem. Res.*, 2009, **42**, 1899.
18. U. Diebold, *Surf. Sci. Rep.*, 2003, **48**, 53.
19. P. Persson, R. Bergström, L. Ojamäe and S. Lunell, *Adv. Quantum Chem.*, 2002, **41**, 203.
20. O. V. Prezhdo, W. R. Duncan and V. V. Prezhdo, *Prog. Surf. Sci.*, 2009, **84**, 30.
21. N. Martsinovich and A. Troisi, *Energy Environ. Sci.*, 2011, **4**, 4473–4495.
22. F. Labat, T. Le Bahers, I. Ciofini and C. Adamo, *Acc. Chem. Res.*, 2012, **45**, 1268.
23. S. Raimondeau and D. G. Vlachos, *Chem. Eng. J.*, 2002, **90**, 3.
24. P. Persson, M. J. Lundqvist, M. Nilsing, A. C. T. van Duin and W. A. Goddard III, *Proceedings of SPIE (The International Society for Optical Engineering): Physical Chemistry of Interfaces and Nanomaterials V*, ed. M. Spitler and F. Willig, 2006, 6325, 63250P.
25. K. Ohno, K. Esfarjani and Y. Kawazoe, *Computational Materials Science*, Springer Solid-State Series, Heidelberg, 1999.
26. C. J. Cramer, Essentials of Computational Chemistry, John Wiley & Sons, Chichester, 2002.
27. A. R. Leach, Molecular Modelling – Principles and Applications, Prentice Hall, 2nd edn, Harlow, 2001.
28. A. Gross, *Theoretical Surface Science*, Springer-Verlag, Berlin Heidelberg, 2003.
29. W. Koch and M. C. Holthausen, A Chemist's Guide to Density Functional Theory, Wiley-VCH, Weinheim, 2000.
30. Y. Zhao and D. G. Truhlar, *J. Chem. Theory Comput.*, 2011, **7**, 669.
31. S. Schenker, C. Schneider, S. B. Tsogoeva and T. Clark, *J. Chem. Theory Comput.*, 2011, **7**, 3586.

32. K. E. Riley, M. Pitonák, P. Jurecka and P. Hobza, *Chem. Rev.*, 2010, **110**, 5023.
33. L. A. Burns, Á. Vázquez-Mayagoitia, B. G. Sumpter and C. D. Sherrill, *J. Chem. Phys.*, 2011, **134**, 084107.
34. A. Szabo and N. S. Ostlund, *Modern Quantum Chemistry*, revised 1st edn, McGraw-Hill, New York, 1989.
35. G. Karlström, R. Lindh, P.-Å. Malmqvist, B. O. Roos, U. Ryde, V. Veryazov, P.-O. Widmark, M. Cossi, B. Schimmelpfenning, P. Neogrady and L. Sijo, *Comput. Mater. Sci.*, 2003, **28**, 222.
36. L. Gundlach, R. Ernstorfer and F. Willig, *Prog. Surf. Sci.*, 2007, **82**, 355.
37. S. Fantacci and F. De Angelis, *Coord. Chem. Rev.*, 2011, **255**, 2704.
38. W. A. Goddard III, A. van Duin, K. Chenoweth, M.-J. Cheng, S. Pudar, J. Oxgaard, B. Merinov, Y. H. Jang and P. Persson, *Topics in Catalysis*, 2006, **38**, 93.
39. M. J. Lundqvist, E. Galoppini, G. J. Meyer and P. Persson, *J. Phys. Chem. A*, 2007, **111**, 1487.
40. V. Sundström (ed.), *Nobel Symposium 101. Femtochemistry and Femtobiology: Ultrafast reaction dynamics at atomic scale resolution*, World Scientific, Imperial College Press, London, 1997.
41. A. C. T. van Duin, S. Dasgupta, F. Lorant and W. A. Goddard III, *J. Phys. Chem. A*, 2001, **105**, 9396.
42. A. C. T. van Duin, A. Strachan, S. Stewman, Q. Zhang, X. Xu and W. A. Goddard III, *J. Phys. Chem. A*, 2003, **107**, 3803.
43. G. A. Somorjai, Introduction to Surface Chemistry and Catalysis, John Wiley & Sons, New York, 1994.
44. K. Chenoweth, A. C. T. van Duin, P. Persson, M.-J. Cheng, J. Oxgaard and W. A. Goddard III, *J. Phys. Chem. C*, 2008, **112**, 14645.
45. A. C. T. van Duin, B. V. Merinov, S. J. Jang and W. A. Goddard III, *J. Phys. Chem. A*, 2008, **112**, 3133.
46. D. Raymand, A. C. T. van Duin, W. A. Goddard III, K. Hermansson and D. Spangberg, *J. Phys. Chem. C*, 2011, **115**, 8573.
47. A. C. T. van Duin, B. V. Merinov, S. S. Han, C. O. Dorso and W. A. Goddard III, *J. Phys. Chem. A*, 2008, **112**, 11414.
48. S. Monti, A. C. T. van Duin, S.-Y. Kim and V. Barone, *J. Phys. Chem. C*, 2012, **116**, 5141.
49. M. J. Buhler, J. Dodson, A. C. T. van Duin, P. Meulborek and W. A. Goddard III, *Materials Research Society Symposium Proceedings*, 2006, **894**, 327.
50. A. Nakano, R. K. Kalia, K. Nomura, A. Sharma, P. Vashishta, F. Shimojo, A. C. T. van Duin, W. A. Goddard III, R. Biswas and D. Srivasta, *Comput. Maert. Sci.*, 2007, **38**, 642.
51. A. Vlcek and S. Zalis, *Coord. Chem. Rev.*, 2007, **251**, 258.
52. C. J. Cramer and D. G. Truhlar, *Phys. Chem. Chem. Phys.*, 2009, **11**, 10757.
53. D. Jacquemin, E. A. Perpète, I. Ciofini and C. Adamo, *Acc. Chem. Res.*, 2009, **42**, 326.

54. M. Abrahamsson, M. J. Lundqvist, H. Wolpher, O. Johansson, L. Eriksson, J. Bergquist, T. Rasmussen, H.-C. Becker, L. Hammarström, P.-O. Norrby, B. Åkermark and P. Persson, *Inorg. Chem.*, 2008, **47**, 3540.
55. T. Österman, M. Abrahamsson, H.-C. Becker, L. Hammarström and P. Persson, *J. Phys. Chem. A*, 2012, **116**, 1041.
56. M. Abrahamsson, M. Jäger, T Österman, L. Eriksson, P. Persson, O. Johansson, H.-C. Becker and L. Hammarström, *J. Am. Chem. Soc.*, 2006, **128**, 12616.
57. M. Abrahamsson, M. Jäger, J. Kumar, T. Österman, P. Persson, H.-C. Becker, O. Johansson and L. Hammarström, *J. Am. Chem. Soc.*, 2008, **130**, 15533.
58. T. Österman and P. Persson, *Chem. Phys.*, 2012, **407**, 76.
59. D. Pettifor, *Bonding and Structure of Molecules and Solids*, Clarendon Press, Oxford, 1995.
60. R. L. Johnston, *Atomic and Molecular Clusters*, Taylor and Frances, London and New York, 2002.
61. C. Burda, X. Chen, R. Narayanan and M. A. El-Sayed, *Chem. Rev.*, 2005, **105**, 1025.
62. G. D. Scholes, *Adv. Funct. Mater.*, 2008, **18**, 1157.
63. P. Persson, J. C. M. Gebhardt and S. Lunell, *J. Phys. Chem. B*, 2003, **107**, 3336.
64. M. J. Lundqvist, M. Nilsing, P. Persson and S. Lunell, *Int. J. Quantum Chem.*, 2006, **106**, 3214.
65. S. Hamad, C. R. A. Catlow, S. M. Woodley, S. Lago and J. A. Mejias, *J. Phys. Chem. B*, 2005, **109**, 15741.
66. V. N. Koparde and P. T. Cummings, *J. Phys. Chem. B*, 2005, **109**, 24280.
67. D. V. Tsivlin, F. Willig and V. May, *Phys. Rev. B*, 2008, **77**, 035319.
68. R. Lindsay and G. Thornton, *J. Phys.: Condens. Matter*, 2001, **13**, 11207.
69. Diebold, *Surf. Sci. Rep.*, 2003, **48**, 53.
70. A. Vittadini, M. Sasarin and A. Selloni, *Theor. Chem. Acc.*, 2007, **117**, 663.
71. P. Zapol and L. A. Curtiss, *J. Comput. Theor. Nanosci.*, 2007, **4**, 222.
72. C. L. Pang, R. Lindsay and G. Thornton, *Chem. Soc. Rev.*, 2008, **37**, 2328.
73. P. Persson, A. Stashans, R. Bergström and S. Lunell, *Int. J. Quantum Chem.*, 1998, **70**, 1055.
74. L. Patthey, H. Rensmo, P. Persson, K. Westermark, L. Vayssieres, A. Stashans, A. Petersson, P. A. Bruhwiler, H. Siegbahn, S. Lunell and N. Mårtensson, *J. Chem. Phys.*, 1999, **110**, 5913.
75. P. Persson, S. Lunell, P. A. Bruhwiler, J. Schnadt, S. Södergren, J. N. O'Shea, O. Karis, H. Siegbahn, N. Mårtensson, M. Bässler and L. Patthey, *J. Chem. Phys.*, 2000, **112**, 3945.
76. J. Schnadt, P. A. Bruhwiler, L. Patthey, J. O'Shea, M. Odelius, R. Ahuja, O. Karis, M. Bässler, P. Persson, H. Siegbahn, S. Lunell and N. Mårtensson, *Nature*, 2002, **418**, 620.
77. P. Persson, S. Lunell and L. Ojamäe, *Chem. Phys. Lett.*, 2002, **364**, 469.

78. P. Persson, R. Bergström and S. Lunell, *J. Phys. Chem. B*, 2000, **104**, 10348.
79. L. G. C. Rego and V. S. Batista, *J. Am. Chem. Soc.*, 2003, **125**, 7989.
80. P. Persson, M. J. Lundqvist, R. Ernstorfer, W. A. Goddard III and F. Willig, *J. Chem. Theory Comp.*, 2006, **2**, 441.
81. G. Benko, J. Kallioinen, J. E. I. Korppi-Tommola, A. P. Yartsev and V. Sundström, *J. Am. Chem. Soc.*, 2002, **124**, 489.
82. P. Persson and M. J. Lundqvist, *J. Phys. Chem. B*, 2005, **109**, 11918.
83. F. De Angelis, S. Fantacci, A. Selloni, M. K. Nazeeruddin and K. Grätzel, *J. Am. Chem. Soc.*, 2007, **129**, 14156.
84. L. Gundlach, T. Letzig and F. Willig, *J. Chem. Sci.*, 2009, **121**, 561.
85. H. Nemec, J. Rochford, O. Taratula, E. Galoppini, P. Kuzel, T. Polivka, A. Yartsev and V. Sundström, *Phys. Rev. Lett.*, 2010, **104**, 197401.
86. S. Thyagarajan, E. Galoppini, P. Persson, J. M. Giaimuccio and G. J. Meyer, *Langmuir*, 2009, **25**, 9219.
87. S. Pal, E. Galoppini, V. Sundström and P. Persson, *Dalton Transactions*, 2009, **45**, 10021.
88. J. Li, M. Nilsing, I. Kondov, H. Wang, P. Persson, S. Lunell and M. Thoss, *J. Phys. Chem. C*, 2008, **112**, 12326.
89. J. Li, H. Wang, P. Persson and M. Thoss, *J. Chem. Phys. 22A529*, 2012, **137**.
90. E. Galoppini, *Coord. Chem. Rev.*, 2004, **248**, 1283.
91. P. G. Johansson, Y. Zhang, M. Abrahamsson, G. J. Meyer and E. Galoppini, *Chem. Commun.*, 2011, **47**, 6410.
92. Y. Zhang, E. Galoppini, P. G. Johansson and G. J. Meyer, *Pure Appl. Chem.*, 2011, **83**, 861–868.
93. P. Persson, M. Knitter and E. Galoppini, *RSC Adv.*, 2012, **2**, 7868.
94. J. van de Lagemaat and A. J. Frank, *J. Phys. Chem. B*, 2001, **105**, 11194.
95. J. Nelson and R. E. Chandler, *Coord. Chem. Rev.*, 2004, **248**, 1181.
96. R. Stangl, J. Ferber and J. Luther, *Sol. Energy Mater. Sol. Cells*, 1998, **54**, 255.
97. C. Deibel and V. Dyakonov, *Rep. Prog. Phys.*, 2010, **73**, 096401.
98. J. L. Brédas, J. E. Norton, J. Cornil and V. Coropceanu, *Acc. Chem. Res.*, 2009, **42**, 1691.
99. A. Lenz, H. Kariis, A. Pohl, P. Persson and L. Ojamäe, *Chem. Phys.*, 2011, **384**, 44.
100. A. A. Bakulin, A. Rao, V. G. Pavelyev, P. H. M. van Loosdrecht, M. S. Pshenichnikov, D. Niedialek, J. Cornil, D. Beljonne and R. H. Friend, *Science*, 2012, **335**, 1340.
101. M. T. Björk, B. J. Ohlsson, T. Sass, A. I. Persson, C. Thelander, M. H. Magnusson, K. Deppert, L. R. Wallenberg and L. Samuelson, *Appl. Phys. Lett.*, 2002, **80**, 1058.
102. M. Kula and L. Ojamäe, *Int. J. Quantum Chem.*, 2012, **112**, 1796.

CHAPTER 4

Plasmon-enhanced Solar Chemistry: Electrodynamics and Quantum Mechanics

HANNING CHEN, GEORGE C. SCHATZ* AND
MARK A. RATNER

Argonne-Northwestern Solar Energy Research Center, Department of
Chemistry, Northwestern University, 2145 Sheridan Road, Evanston,
IL 60208
*Email: schatz@chem.northwestern.edu

4.1 Introduction

According to the a recent annual energy outlook released by the United States
Department of Energy,[1] a 50% increase in global energy production rate needs
to be realized in 25 years to meet the needs of the rapidly growing economies
worldwide, particularly in the newly industrialized countries with massive
populations, such as China, India and Brazil.[2] However, the most urgent
concern about this impending energy shortage is not triggered by the larger
overall demands, instead it is the dominant share contributed by carbon-based
fossil fuels, leading to enhanced production of greenhouse gases.[3] Indeed, ex-
ploitation of fossil fuels threatens the ecological balance in many areas.[4] Given
the current world annual energy consumption rate of 15 TW (terawatts),[1] all
the world's readily extractable coal, oil and natural gas are expected to be
depleted in approximately 75 years.[5] Therefore, it has become an imperative

RSC Energy and Environment Series No. 8
Solar Energy Conversion: Dynamics of Interfacial Electron and Excitation Transfer
Edited by Piotr Piotrowiak
© The Royal Society of Chemistry 2013
Published by the Royal Society of Chemistry, www.rsc.org

task to seek alternative energy sources that are environmentally friendly, economically affordable and straightforwardly renewable.

There are many naturally usable forms of energy, including solar, wind, geothermal, biofuel and hydroelectric. Among them, solar cells stand out as a promising solution to the sustainable energy challenge as the total solar irradiance is 17 400 TW per year,[6] over 1000 times current demand. Also, there is no significant geographical variation in the solar radiation intensity across the world, alleviating the reliance on foreign energy sources for most developed and developing countries. Furthermore, it is much more cost effective to harness solar energy in remote regions than to lay the high voltage wires required to link to an electrical grid, making it an ideal portable power generator. In spite of these advantages of solar energy, its application has been severely hindered by high manufacturing and maintenance costs, short product lifetime, limited rechargeable storage capacity and especially to low light-to-electricity conversion efficiency.[7]

There are many different types of photovoltaic devices, including: semiconductor-based crystalline solar cells (CSC),[8] organic-based dye sensitized solar cell (DSSC) and bulk hetero-junction organic photovoltaics (BHJ).[9] Despite the importance of BHJ, we focus here on CSC and DSSC. In general, CSC achieve a standard incident photon conversion efficiency (IPCE) of $\sim 30\%$[10] at the expense of ultra-clean fabrication equipment and toxic heavy metal ingredients, making the disposal and recycling of the crystalline panels difficult. By contrast, most of the environmental issues associated with CSC are well resolved by DSSC, which utilize inexpensive and innocuous organic molecules as the light harvester, typically yielding an IPCE of $\sim 10\%$. The lower photovoltaic efficiency in DSSC compared to CSC can be ascribed to several limitations, including the narrow absorption window, the relatively weak extinction coefficient and energy mismatches in the electron transfer process that goes from the excited molecule to the semiconductor electrode.[11] In recent years, a so-called "plastic" solar cell has been developed by depositing thin-film semi-conductors on lightweight and disposable polymer chains.[12,13] By combining the flexibility and customizability of organic substrates with the efficiency and modularity of inorganic compounds, this new type of solar cell may circumvent the drawbacks of both materials and achieve a balanced compromise between expense and performance.[14]

In 1970s, it was discovered that Raman scattering intensity could be dramatically amplified by placing the analyte on a rough surface composed of a noble metal substrate, for example, silver, gold or copper.[15,16] It is widely accepted that the many orders of magnitude enhancement factor that results is due to the locally enhanced electric field near the substrate surface, resulting in an enormously intensified coupling between the analyte's excitations and the large collective motion of the substrate conduction electrons, the so-called surface plasmon polaritons.[17] Inspired by the elegant mechanism of surface enhanced Raman spectroscopy (SERS), extensive efforts have been devoted to increasing light absorption and expediting electron injection in solar cells using metal nanoparticles.[18]

Pioneering work in the area of plasmonic solar cells was performed by Stuart and Hall,[19] who observed a 20-fold photocurrent enhancement for a commercial silicon-on-insulator wafer at an incident wavelength of 800 nm when the average diameter of the silver islands reaches 108 nm. A subsequent experiment conducted by Pillai *et al.* demonstrated a 16-fold enhancement in light absorption for both thin-film and wafer-based silicon structures with the aid of 18-nm diameter silver particles when the incident photon energy is close to the band gap of silicon at 1200 nm.[20] The results of both studies are surprising because the locations of their enhancement peaks are distinct from the resonance wavelength of bulk silver at ~ 400 nm. However, according to Maxwell-Garnett theory,[21] the notable red shifting of the resonance modes in silicon-coated silver islands can be well explained by the much higher dielectric function of silicon. In fact, the tunability of the polariton peak is so good that a variation up to 200 nm has been recently reported on Si_3N_4-coated silver particles in concert with a two-fold transmission loss.[22]

Similar effects are seen for semiconductors other than bulk silicon. For example, the deposition of a layer of CdSe quantum dots on a silver substrate greatly attenuates the magnitude of the surface plasmon polariton (SPP) within a propagation length of 1.2 µm.[23] That study showed that the extent of transmission signal reduction is very sensitive to the incident wavelength, exhibiting a rapid decay at 514.5 nm, a moderate reduction at 687.9 nm and little change at 1426 nm. The absorption efficiency of the CdSe quantum dots was accurately measured through constructive or destructive interferences between multiple light sources, which can be readily modulated by changing the slit-groove distance. In another study which used the concept of all-optical modulation, a metal-insulator-metal type plasmon waveguide was fabricated by "sandwiching" a layer of alumina containing silicon quantum dots between two thin films of gold.[24] After the excitation of the embedded silicon nano-crystals by an impact ionization process under a sufficiently large bias voltage, plasmonic modes along the gold–silicon interface are created as a result of the radiative decay. Here the reverse of the usual photovoltaic effect lends support to the idea of direct near-field coupling between the quantum dots and metal film, as the resonant modes of the quantum dots are relatively diffuse and are not completely confined within the insulator layer. As an added benefit, some quantum dots can efficiently generate multiple excitons,[25] this can be exploited to facilitate electron–hole separation further. However, the relatively short lifetime of plasmon modes may become another limiting factor for multiple exciton generation that already faces stiff competition from the phonon scattering and exciton–exciton annihilation.[26]

In 1997, a remarkable absorption enhancement factor of 149 was reported by Ihara *et al.* upon binding *cis*-bis(isothiocyanato)-bis(2,2'-bipyridyl-4,4'-di-carboxylato)-ruthenium(II) molecules, known as N3 dye, directly to silver islands to form a Ag/N3 film.[27] Using field emission scanning electron microscopy, the average diameter and the average surface area of the silver islands were measured and a high surface coverage ratio of over 40% was found for all sizes of silver islands. Consequently, the light absorption reached its maximum when the concentration of N3 dye increased to a rather high value of 2×10^{-9} mol cm^{-2} which is sufficient to saturate the surface area of the silver islands. With further increasing concentration of the N3 dye, a reduction in the absorption efficiency per molecule was eventually observed owing to the rapidly diminished local electric fields on unbound dye molecules that are separated from the silver plasmonic surfaces by bound dye molecules. From a molecular orbital perspective, the enhanced absorption peak of the N3 dye at ~ 540 nm corresponds to two $t_2 \rightarrow \pi^*$ metal-ligand charge-transfer (MLCT) transitions, which happen to have excellent spectral overlap with the red-shifted (and broad) plasmonic peak of the N3-coated silver island centered at ~ 500 nm. By contrast, another important absorption peak of the N3 dye at ~ 400 nm, corresponding to two $\pi \rightarrow \pi^*$ intra-ligand transitions, is only moderately enhanced owing to its poor overlap with the plasmon resonance. Although the Ag/N3 film[27] is widely regarded as the first practical implementation of the plasmonic dye-sensitized solar cell (PDSSC), the direct incorporation of bare metal nanoparticles into dye molecules is usually detrimental to the particles owing to the presence of corrosive liquid electrolytes.[28] Moreover, the metal particles may act as charge recombination centers, reducing overall photovoltaic efficiency, as they are a natural electron donor to the excited dye molecule.[29]

To circumvent the direct contact between dye molecule and metal nanoparticle, a thin layer of TiO_2 was added between them in work by Standridge *et al.* using atomic layer deposition (ALD).[28] The TiO_2 layer not only serves as a robust corrosion protector but also functions as an ultrafast electron harvester, achieving excited state electron injection within several femtoseconds.[30] The resulting composite $Ag/TiO_2/N_3$ PDSSC gave a nine-fold photocurrent enhancement at an incident wavelength of 560 nm for a 2.0 nm TiO_2 layer,[31] the minimum thickness needed to prevent electrolyte-based corrosion. In comparison with the IPCE value of dye-only photoanodes, an enhancement factor of 6.0 was found with an extinction peak at ~ 560 nm when the thickness of the TiO_2 was chosen to be 4.8 nm. Very recently, a combination of SnO_2/Al_2O_3 was proposed to replace the TiO_2/N_3 for higher open circuit voltage and slower charge recombination.[32] Even without the presence of the plasmonic nanoparticles, a roughly five-fold IPCE enhancement was achieved thanks to the faster charge carrier transport in the SnO_2 and the better redox inactivity of Al_2O_3. Therefore, it is not surprising to see a 220-fold absorption enhancement for a 13-nm thick silver layer coated by erbium-doped Al_2O_3 at an incident wavelength of 520 nm.[33] In addition, Al_2O_3 apparently has better resistance to the electrolyte than TiO_2, making a thinner film down to 1.0 nm feasible.[31] Of course, the requirement of a protective layer can be entirely eliminated by the development of solid-state dye-sensitized solar cells (SS-DSSC) fabricated by nano-imprint lithography as shown in a two-dimensionally patterned plasmonic back reflector sintered with Z907 and C220 dyes.[34]

Regardless of the various implementations of PDSSC, their significantly improved photocurrent conversion efficiency can be ascribed to three light-trapping processes that utilize surface plasmon polaritons: light scattering, light concentration and light propagation. Upon placing large plasmonic nanoparticles at the air/dielectric interface, preferential scattering of the incident light into the dielectric layer takes place because of its higher permittivity.[35]

Subsequently, the scattered light can be elastically and internally reflected if the scattering angle is larger than the critical angle, effectively trapping the incident light by extending the optical path length in the dielectric layer.

Alternatively, small nanoparticles instead of large ones, can be anchored between the photoactive thin film and the dielectric semiconductor to exploit the near-field enhancement of plasmon resonance modes which concentrates the incident light into sub-wavelength volumes. This mechanism is particularly suitable for small nanoparticles with diameters ranging from 5–20 nm, owing to a scattering-to-absorption ratio that grows rapidly with increasing particle size, as can be seen using Mie theory in the Rayleigh regime, that is, $d_{particle} = \lambda_{inc}$.

The third scheme for light trapping is to flip the light propagation direction by 90° using a corrugated metal back surface to guide the evanescent SPP fields to travel along the lateral direction of the solar cell, which is much longer than the thickness of the light absorption layer. However, there is a substantial waste of incident photon energy if the absorption of the metal backing is stronger than that of the photoactive components. Therefore, an oxide coating of the metal back reflector is sometimes desired for raising the fraction of the injected energy which is finally converted into the in-plane SPP.[36]

To account for the above-mentioned plasmonic contributions to photovoltaic efficiency and to aid in the systematic design of PDSSC, several theories and computational methods have been developed and applied, including classical electrodynamics methods such as the finite-difference time-domain (FDTD) method,[37] the finite element method (FEM),[38] the discrete dipole approximation (DDA)[39] and the dyadic Green's function (DGF) method.[40] Quantum (many-body) approaches have also been developed, including a steady-state transport equation (SSTE) method,[41] linear response time-dependent density functional theory (LR-TDDFT),[42] real-time time-dependent density functional theory (RT-TDDFT),[43] tight-binding density functional theory (DFTB),[44] random phase approximation Maxwell–Schrodinger modeling (RPA-MS)[45] and many-body Green's function theory.[46] And very recently a hybrid quantum mechanics/classical electrodynamics (QM/ED) has been developed.[47] The remainder of this paper will describe these methods and their applications to solar cell problems.

4.2   Continuum Models

In continuum electrodynamics, an optical system is represented as a collection of discretization grids, each of which is characterized by its electric permittivity and magnetic permeability which are uniquely determined by the material properties. By solving Maxwell's equations or the coupled dipole-field equations in either the time or frequency domain, any macroscopic optical property of interest can be numerically determined subject to the desired boundary conditions. Since the continuum models are typically scale invariant, they are applicable to arbitrarily large systems. However a limitation in the description of metal nanoparticles is that grid sizes on the order of a few nanometers are necessary for convergence of the numerical methods, so this places an important limitation on how large a system can be treated (usually not larger than a few micrometers).

4.2.1 Finite-difference Time-domain (FDTD)

The basic idea of FDTD is to solve Maxwell equations:

$$\varepsilon(r)\frac{\partial}{\partial t}\vec{E}(r,t) = \nabla \times \vec{H}(r,t) - \vec{J}(r,t)$$

$$\mu(r)\frac{\partial}{\partial t}\vec{H}(r,t) = -\nabla \times \vec{E}(r,t)$$

$$(4.1)$$

to determine the time evolution of the electromagnetic fields, $\vec{E}(r,t)$ and $\vec{H}(r,t)$, subject to an assumed current density, $\vec{J}(r,t)$. Here ε is the electric permittivity and μ is the magnetic permeability. Once the fields are determined, the transmission spectra, $P_{tran}(\omega)$, across a plane on the far side of the optical system can be evaluated by:

$$P_{tran}(\omega) = \text{Re}\,\hat{n} \times \int E^*(\omega,r) \times H(\omega,r)dr \qquad (4.2)$$

where Re denotes the real part, $\hat{n}$ is the vector normal to the detection plane and the asterisk indicates the complex conjugate. The frequency dependent electromagnetic fields, $\vec{E}(r,\omega)$ and $\vec{H}(r,\omega)$, are obtained using a Fourier transform:

$$\vec{E}(\omega,r) = \int dt e^{-i\omega t}\vec{E}(t,r)$$

$$\vec{H}(\omega,r) = \int dt e^{-i\omega t}\vec{H}(t,r)$$

$$(4.3)$$

According to Poynting's theorem, the extinction spectra, $P_{ext}(\omega)$, the scattering spectra, $P_{sca}'(\omega)$ and the absorption spectra, $P_{abs}(\omega)$, are given respectively by:

$$P_{ext}(\omega) = \text{Re}\,\hat{n} \times \int \left[E_0^*(\omega,r) \times H_0(\omega,r) - E^*(\omega,r) \times H(\omega,r)\right]dr$$

$$P_{sca}(\omega) = \text{Re}\,\hat{n} \times \int \left[E^*(\omega,r) - E_0^*(\omega,r)\right] \times \left[H^*(\omega,r) - H_0^*(\omega,r)\right]dr \qquad (4.4)$$

$$P_{abs}(\omega) = P_{ext}(\omega) - P_{sca}(\omega)$$

where $\vec{E}_0(r,\omega)$ and $\vec{H}_0(r,\omega)$ are the incident electromagnetic fields.

As an application of FDTD, a silver cylinder with a diameter of 200 nm and a height of 50 nm was placed on a 20-nm thick SiO_2 dielectric spacer layer, under which there is a semi-infinite crystalline Si substrate (Figure 4.1).[37] In the case of rear illumination (the incident light comes from the substrate side) with a wavelength of 500 nm, an intensified electric field was found in the dielectric layer, particularly on the bottom edges of the silver cylinder, owing to the sharp change of dielectric environment at the Ag/SiO_2 interface (Figure 4.2). For front illumination (the incident light comes from the air) and taking the cylinder height to be 500 nm, two areas with amplified electric fields along the cylinder's height were observed with a separation of $\sim$210 nm (Figure 4.2).

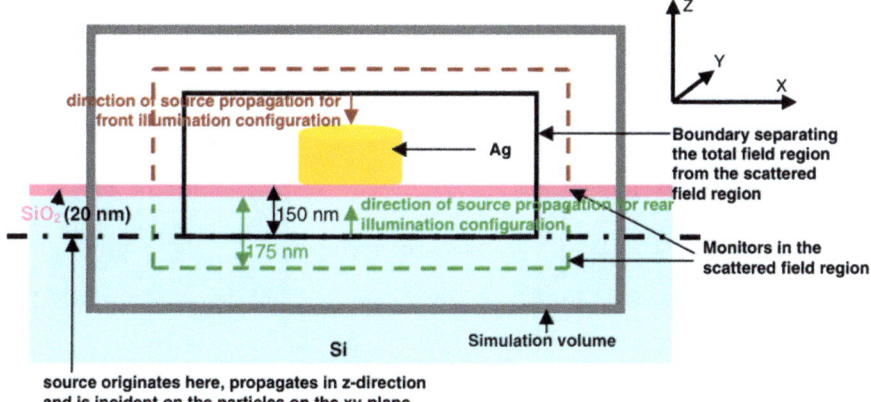

Figure 4.1   Silver cylinder on a crystalline Si substrate spaced by a SiO₂ layer (reprinted with permission from Mokkapati *et al.*,[37] Copyright 2011 IOP Science).

From this, it is easy to recognize that the wavelength of the SPP wave is twice the separation, that is, ~ 420 nm. It is also interesting to note that front illumination results in a considerably stronger local electric field within the dielectric layer than rear illumination, thanks to constructive interference between the incident field and the reflected field. Such interference may be easily modulated by the cylinder height or the dielectric thickness.

In terms of the normalized scattering cross section, $\bar{Q}_{sca}$, which provides a measure of scattered power normalized to the cross-sectional area of the silver cylinder, a taller cylinder is expected to be better at scattering than a shorter cylinder owing to simple geometrical effects. However, as indicated by the fraction, f_{sub}, of scattered light that couples into the substrate, shorter particles can have much better coupling efficiencies at wavelengths close to the plasmon wavelength, for both front and rear illumination. Also, when the particle's height is increased, the effective dipole moment is moved away from the Ag/SiO₂ surface, leading to reduced near-field coupling. Thus, better overall light-trapping efficiency is more likely using shorter silver cylinders on SiO₂-coated silicon substrates than taller ones unless their surface coverage is so high, that is, $\gg 1/Q_{sca}$, so that each particle's availability for scattering light is significantly diminished.

4.2.2   Finite-element Method (FEM)

As a frequency-domain alternative to the FDTD method, FEM employs the well-known time-harmonic Maxwell equations given by:

$$\nabla \times \vec{E}(\omega, r) = -i\omega\mu(r)\vec{H}(\omega, r)$$
$$\nabla \times \vec{H}(\omega, r) = \vec{J}(\omega, r) + i\omega\varepsilon(r)\vec{E}(\omega, r)$$
$$\nabla \times \vec{E}(\omega, r) = \frac{\rho_{free}}{\varepsilon(r)}$$
$$\nabla \times \vec{H}(\omega, r) = 0$$

(4.5)

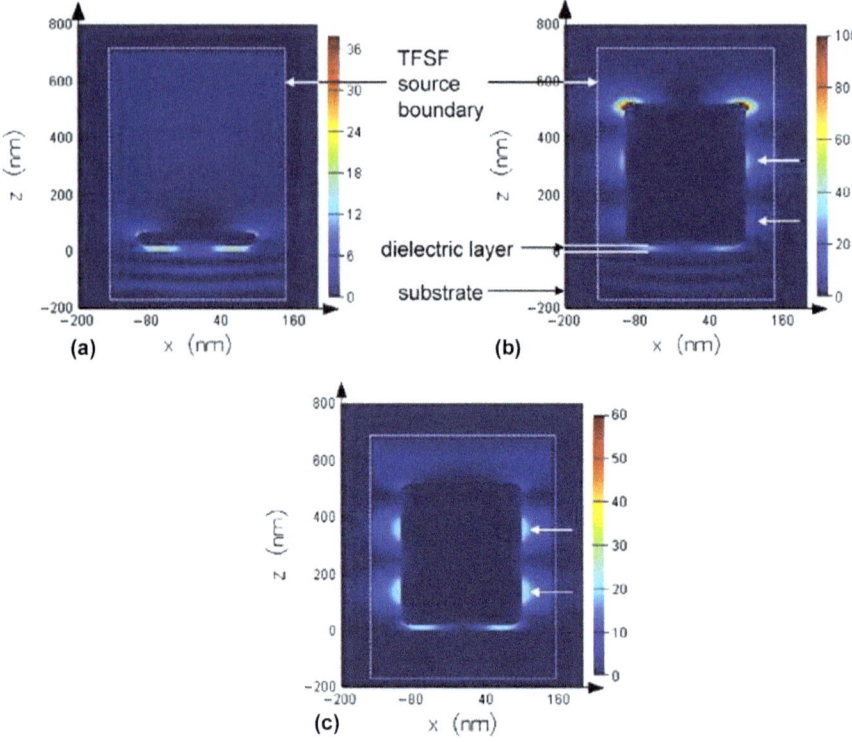

Figure 4.2 Electric field intensity for incident light at a wavelength of 500 nm, polarized in the *x* direction in the *x*–*z* plane through the axis of the nanoparticle. The nanoparticle has a base diameter of 200 nm and a height of a) 50 nm or b) 500 nm, c) 500 nm. The positions of the substrate and the dielectric layer are indicated in b). The light is incident on the nanoparticles from the substrate side for a) and b) and from the air side for c) (reprinted with permission from Mokkapati *et al.*,[37] Copyright 2011 IOP Science).

where ρ_{free} is the free electron density and ω is the oscillation frequency. After mathematical manipulation, reduced wave equations for $\vec{E}(\omega, r)$ and $\vec{H}(\omega, r)$ can be developed as follows:

$$\nabla \times \left(\frac{1}{\mu(r)} \nabla \times \vec{E}(\omega, r)\right) - \omega^2 \varepsilon(r) \vec{E}(\omega, r) = -i\omega \vec{J}(\omega, r)$$

$$\nabla \times \left(\frac{1}{\omega(r)} \nabla \times \vec{H}(\omega, r)\right) - \omega^2 \vec{H}(\omega, r) = \nabla \times \left(\frac{1}{\varepsilon(r)} \vec{J}(\omega, r)\right)$$

(4.6)

By expanding the fields in terms of finite elements it is possible to determine the spatial distribution of electromagnetic field magnitudes for an arbitrary choice of electric properties subject to an incident plane wave. Since FEM allows a flexible definition of the shape, size and connectivity for finite elements of an arbitrary geometry, the discretization resolution can be improved to any desired level close to interfaces to diminish numerical errors. As a result, FEM has been widely applied to study the optical response for abruptly rugged or angled objects, such as sharp tips and acute structures that usually demand a very small grid size in FDTD simulations to maintain numerical stability at the expense of central processing unit (CPU) cycles and memory allocations. However, FEM is a frequency domain approach that requires multiple simulations within the range of relevant incident wavelengths to render a full spectrum to be comparable with experiment.

In an FEM study of plasmon-enhanced light absorption, a periodic array of 10-nm diameter cylindrical silver nanowires was embedded in an photoactive layer consisting of poly(3-hexylthiophene):(6,6)-phenyl-C61-butyric-acid-methyl ester (P3HT : PCBM) with a 1 : 1 weight ratio (Figure 4.3).[38] It was found that when the thickness of the photoactive layer is increased to 200 nm, there is a plateau value of 60% for the overall absorption ratio (weighted by the AM 1.5G solar spectrum over the visible interval from 300–800 nm). In comparison to the case without the silver nanowires, a maximum absorption enhancement factor of ~ 1.45 was observed for an optimal spacing of 8 nm between the infinitely long cylinders (Figure 4.4). It is interesting to note than when the spacing is narrower than 4 nm, the overall absorption of light is significantly reduced by over 30%. This seemingly abnormal reduction can be ascribed to a decreased volume available to the photoactive components, in spite of the amplified electric fields in the narrow gap regions. When the silver nanowires are separated by over 20 nm, the light absorption efficiency starts to decline rapidly owing to decoupling of the surface plasmon modes, as clearly demonstrated by the spacing dependence of the electric fields.

For a spacing of 8 nm, the absorption spectrum exhibits two equally high peaks at 425 nm and 565 nm. The 425-nm peak is caused by the red shifted plasmon resonance because of coupling between the neighboring silver cylinders, while the 565-nm peak stems from absorption by the P3HT : PCBM

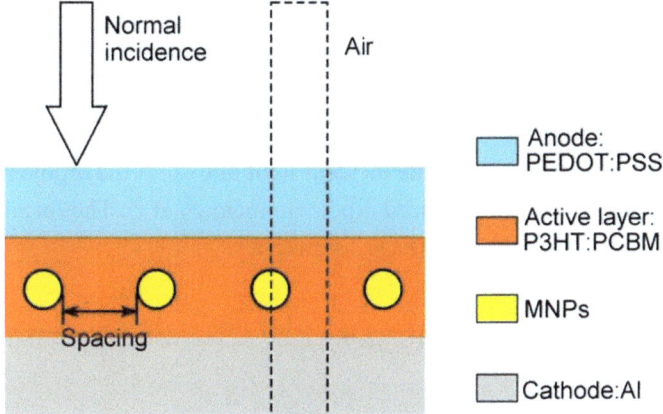

Figure 4.3   Periodic array of cylindrical silver nanowires embedded in P3HT:PCBM layer (reprinted with permission from Shen *et al.*,[38] Copyright 2009 American Institute of Physics).

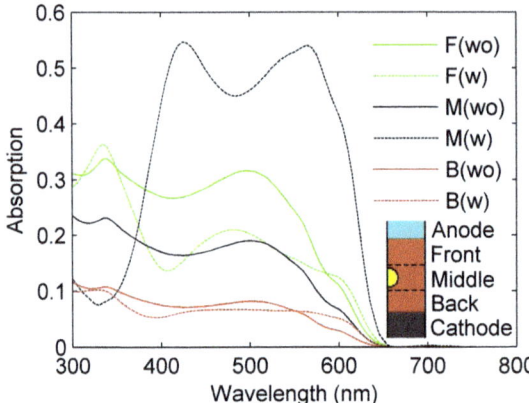

Figure 4.4   Absorption spectra with w) and without wo) silver nanowires in front f), middle m) and back b) layers with nanowires embedded in the middle layer with 8 nm spacing (reprinted with permission from Shen *et al.*,[38] Copyright 2009 American Institute of Physics).

mixture. Moreover, using a two-dimensional optimization, an optimal absorption enhancement factor of 1.56 was achieved at a spacing of 40 nm and a diameter of 24 nm for the cylinder arrays. With this nanopattern, the absorption spectrum of a 33-nm thick photoactive layer with silver cylinders is very similar to that of a 61-nm thick one without silver cylinders. Therefore, the presence of plasmonic nanoparticles can reduce the effective thickness of the photoactive layer by almost 50%, as indicated by the FEM simulations.[38]

4.2.3   Discrete Dipole Approximation (DDA)

If higher-order multipoles in an expansion of the electric field in terms of a discretizated grid of cubic elements are ignored, the local electric field, $\vec{E}_{loc}(\omega, r)$, has the following form:

$$\begin{aligned}\vec{E}_{loc}(\omega, r_i) &= \vec{E}_{inc}(\omega, r_i) + \vec{E}_{self}(\omega, r_i) \\ &= E_0 \exp(i\vec{K} \times \vec{r}_i) - \sum_{j \neq i} A_{ij} \times P_j\end{aligned} \quad (4.7)$$

where K is the wave vector of the incident light and A_{ij} is the dipole interaction matrix at $\vec{r}_i$ exerted by the induced dipole moment $\vec{P}_j$ at $\vec{r}_j$. The formula for the self-induced electric field, $\vec{E}_{self}(\omega, r)$, is given by:

$$A_{ij} \times P_j = \frac{\exp\left(i\omega c^{-1} r_{ij}\right)}{r_{ij}^3}\left\{\omega^2 c^{-2} \times (r_{ij} \times P_j) + \frac{\left(1 - i\omega c^{-1} r_{ij}\right)}{r_{ij}^2} \times \left[r_{ij}^2 P_j - 3r_{ij}(r_{ij} \times P_j)\right]\right\}$$

$$(4.8)$$

Under the approximation of linear response:

$$P_i = \alpha(\omega, r_i) \times E_{loc}(\omega, r_i) \quad (4.9)$$

where $\alpha(\omega, r_i)$ is the macroscopic polarizability. After inserting Equation (4.9) into Equation (4.7), it is easy to recognize that $\vec{E}_{loc}(\omega, r)$ can be determined in a self-consistent manner. Conveniently, the extinction, absorption and scattering cross sections are evaluated by the three following equations:

$$C_{ext} = \frac{4\pi\omega c^{-1}}{|E_0|^2} \sum_{j=1} \text{Im}(\vec{E}_{inc,j}^* \times \vec{P}_j)$$

$$C_{abs} = \frac{4\pi\omega c^{-1}}{|E_0|^2} \sum_{j=1} \left\{ \text{Im}\left[P_j \left(\alpha_j^{-1} \right)^* P_j^* \right] - \frac{2}{3}\omega^3 c^{-3} |P_j|^2 \right\} \qquad (4.10)$$

$$C_{sca} = C_{ext} - C_{abs}$$

To investigate the effect of size, shape, composition and aspect ratio of metal particles on the optical properties of a silicon solar cell, a DDA study was carried out on an infinite silicon plane upon which are deposited arrays of nanorods with different surface coverage ratios ranging from 4–16% (Figure 4.5).[39] When the nanorods are randomly fabricated, the extinction spectrum becomes very broad, in contrast to the case with a periodic structure,

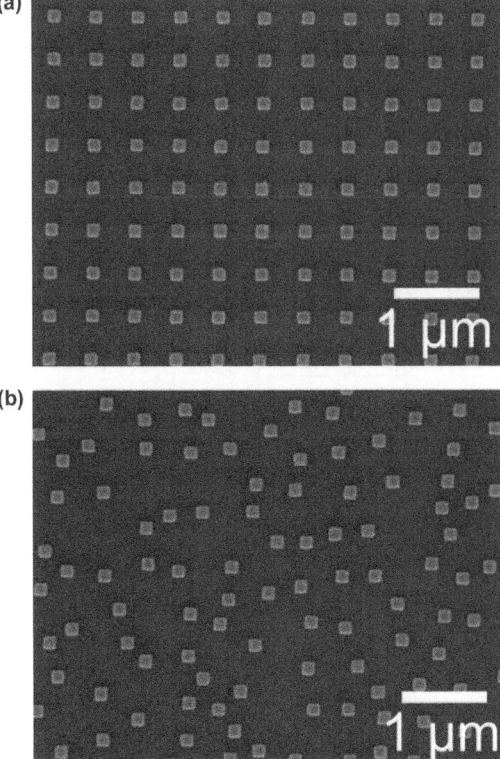

Figure 4.5   SEM images of a) periodic and b) random metal arrays of square particles (reprinted with permission from Temple *et al.*,[39] Copyright 2011 American Institute of Physics).

which exhibits a narrow and asymmetric peak. The difference can be ascribed to the broadened extent of diffraction in a randomly distributed pattern of the nanorods.

   In additional work in the same study, the extinction efficiency was studied for three cross-section shapes with a similar cross-sectional area of $10\,000\,nm^2$, namely a circle with a diameter of 115 nm, a square with a side length of 100 nm and a triangle with a side length of 150 nm (Figure 4.6). Interestingly, the square nanorods demonstrate the highest extinction peak for both gold and aluminum, while the circular nanorods are always the least efficient. Encouragingly, the ranking of extinction efficiency is in excellent agreement with experiment[39] and can be well explained by the high electric fields generated at the sharp tips of the square and triangular nanoparticles, leading to a strong near-field coupling. Although DDA does not impose boundary conditions as required by Maxwell equations and thus usually suffers from inaccurate electric fields at interfaces, it is nevertheless one of the most cost-effective computational electrodynamics methods and it can deliver qualitatively reliable and experimentally useful results.

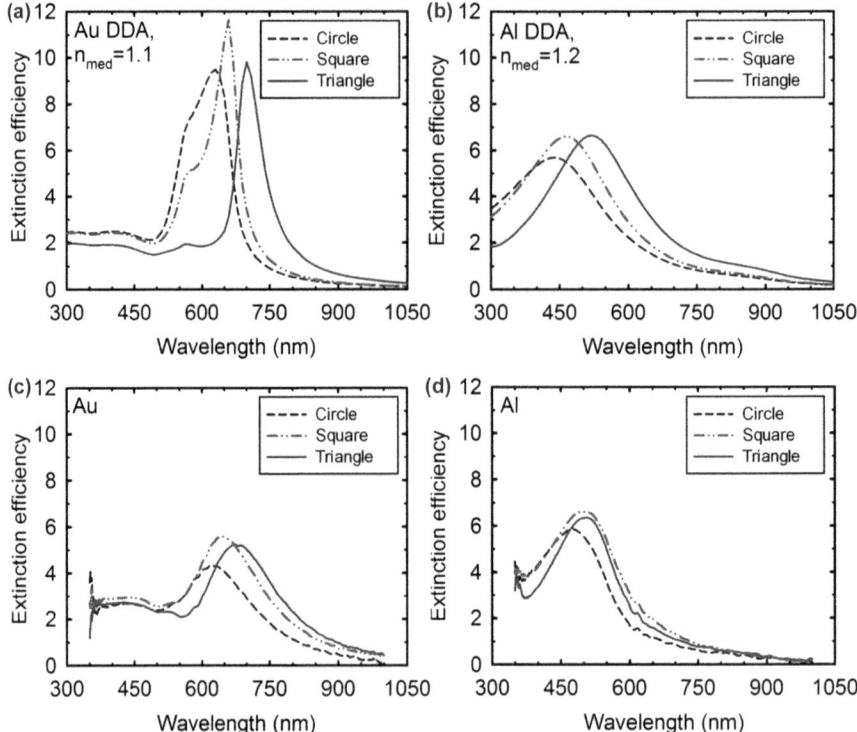

Figure 4.6   Simulated extinction spectra of a) Au and b) Al nanoparticles and experimental extinction spectra of c) Au and d) Al nanoparticles with different cross-section shapes (reprinted with permission from Temple *et al.*,[39] Copyright 2011 American Institute of Physics).

4.3 Many-body Theories

Although continuum models have been quite successful in assessing macroscopic optical response, they have intrinsic limitations for probing microscopic optical properties, such as molecular polarizability and photoconductivity. The limitations stem from the fact that continuum electrodynamics, as applied to metal nanostructures, are intended to describe the collective motions of the electrons and are thus not applicable to any physical phenomenon that occurs at small length scales (typically a few nanometers for typical condensed-phase systems). For small length scales, many-body theories need to be applied to account for the quantum characteristics of individual electronic transitions, for example, light absorption by an organic sensitizer and subsequent electron injection to semiconductor layer.

4.3.1 Linear Response Time-dependent Density Functional Theory (LR-TDDFT)

In a DSSC, the light absorption efficiency of an organic sensitizer can be calculated by LR-TDDFT because of its typically small size (less than 100 atoms). In LR-TDDFT,[48] the electronic excitation frequencies, ω, are determined by solving the non-Hermitian eigenvalue problem:

$$\begin{pmatrix} L & M \\ M^* & L^* \end{pmatrix}\begin{pmatrix} X \\ Y \end{pmatrix} = \omega \begin{pmatrix} -1 & 0 \\ 0 & 1 \end{pmatrix}\begin{pmatrix} X \\ Y \end{pmatrix} \qquad (4.11)$$

The response matrices, L and M, are constructed from the occupied-unoccupied Kohn–Sham orbital pairs using:

$$L_{hp,h'p'} = \delta_{pp'}\delta_{hh'}\left(E_p - E_h\right)$$

$$+ \int dr \int dr' \phi_h^*(r)\phi_p(r)\left(\frac{1}{|r - r'|} + \frac{\delta^2 E_{XC}}{\delta\rho(r)\delta\rho(r')}\right)\phi_{h'}^*(r')\phi_{p'}(r') \qquad (4.12)$$

$$M_{hp,h'p'} = \int dr \int dr' \phi_h^*(r)\phi_p(r)\left(\frac{1}{|r - r'|} + \frac{\delta^2 E_{XC}}{\delta\rho(r)\delta\rho(r')}\right)\phi_{h'}^*(r')\phi_{p'}(r')$$

where $\phi_h(r)$ and $\phi_p(r)$ are the occupied and unoccupied Kohn–Sham orbitals, E_h and E_p are the corresponding orbital energies, $\rho(r)$ is the electron density and E_{XC} is the exchange-correlation functional. In addition, the undetermined matrices, X and Y, consist of the expansion coefficients of the Kohn–Sham orbitals used to represent the vertically excited states.

N3 dye has four carboxyl groups capable of taking a proton, making its valence state charges range from -4 to 0. In a LR-TDDFT study[42] that accounts for aqueous solvation by means of a conductor-like polarizable continuum model (C-PCM),[49] the fully protonated form, $N3^0$, is found to be

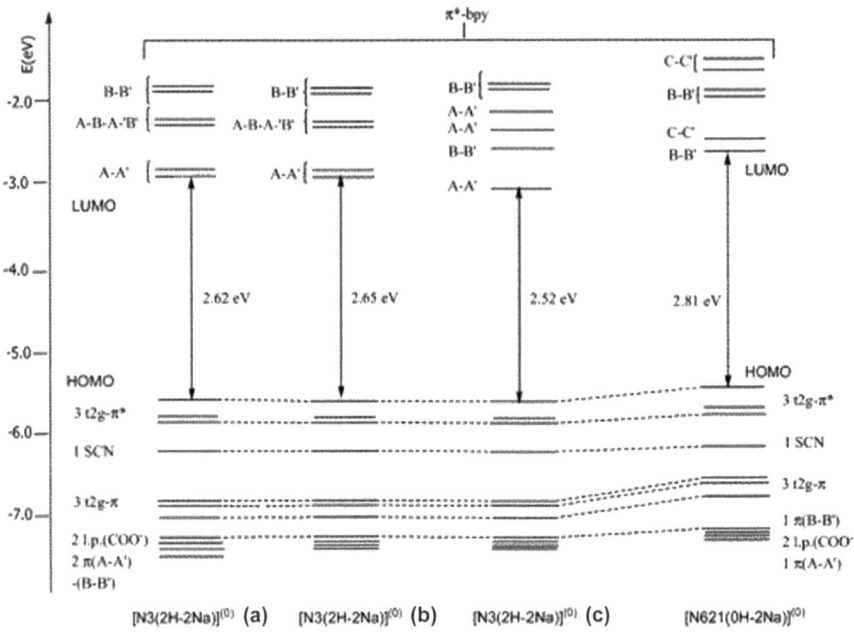

Figure 4.7 Energy diagram of the frontier orbitals of the three isomers of the fully protonated N3 dye, and the deprotonated N621 dye (reprinted with permission from Nazeeruddin et al.,[43] Copyright 2005 American Chemical Society).

most efficient for harvesting solar light, with a HOMO-LUMO (highest occupied molecular orbital-lowest unoccupied molecular orbital) gap of 2.59 eV (Figure 4.7), while the fully deprotonated form, N3^{-4} has the lowest visible light absorption efficiency with a HOMO-LUMO gap of 3.00 eV. By comparing the computed and experimental absorption spectra of the three isomers of N3^{0}, whose structures are known from X-ray scattering, three absorption peaks have been consistently assigned at 4.0 eV, 3.3 eV and 2.3 eV.

A further examination of the calculated molecular orbitals and electron densities reveals that the HOMO is primarily localized on the two thiocyanate (NCS) ligands, while the LUMO+4 is largely distributed over the bipyridine ligands. Therefore, the HOMO->LUMO+4 transition leads to a ligand-to-ligand charge transfer (LLCT) transition that can be stabilized by polar solvents. In fact, another theoretical study of the N3 dye showed that the primary absorption peak at ~4.0 eV due to the LLCT can be notably suppressed in the absence of solvent.[49] Another factor for the good photovoltaic performance of N3 dye on TiO$_2$ surfaces is the excellent match of the LUMO energy level of N3 to the conduction band (CB) of TiO$_2$, except for the neutral N3^{0}, whose LUMO is slightly below the CB's lower edge, all other protonation forms of N3 have a wide enough LUMO-CB energy gap to enable ultrafast photo-induced electron injection from the excited N3 dye to the TiO$_2$ layer (Figure 4.8).

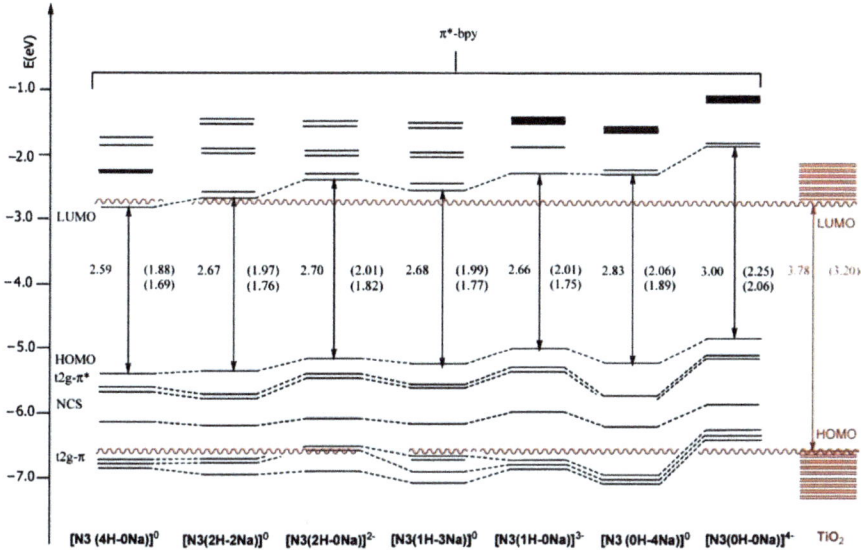

Figure 4.8   Energy levels and TDDFT excitation energies for N3 dye in water solution compared to the results for a model of a TiO$_2$ nanoparticle (reprinted with permission from Nazeeruddin *et al.*,[43] Copyright 2005 American Chemical Society).

4.3.2   Real-time Time-dependent Density Functional Theory (RT-TDDFT)

As implied by Equation (4.11) and Equation (4.12), the computational cost of LR-TDDFT grows rapidly with increasing system size, because the size of the response matrices is proportional to the product of the numbers of occupied and unoccupied molecular orbitals, that is, $N_{occ} \times N_{unocc}$. Moreover, the diffuse unoccupied orbitals usually require very large basis sets to satisfactorily converge the results. To circumvent these numerical challenges, a RT-TDDFT method was derived to track the response of an optical system upon external perturbation. In RT-TDDFT, an additional term, $\vec{E}(r) \times \vec{\mu}(r)$, is added to the system's quantum Hamiltonian, $\hat{H}_0$, to reflect light-matter interactions:

$$\hat{H} = \hat{H}_0 + \vec{E}(r, t) \times \vec{\mu}(r, t) \tag{4.13}$$

where $\vec{E}(r, t)$ is the external field and $\vec{\mu}(r, t)$ is the dipole moment operator. Subsequently, the system's wavefunction, $\varphi(r, t)$, is propagated according to the time-dependent Schrödinger equation:

$$i\hbar \frac{\partial \varphi(r, t)}{\partial t} = (\hat{H}_0 + \vec{E}(r, t) \times \vec{\mu}(r, t)) \varphi(r, t) \tag{4.14}$$

After a sufficiently long simulation time, the frequency-dependent molecular polarizability of the optical system, $\alpha(\omega)$, can be obtained by a Fourier transform on the time-dependent induced dipole moment, $P(t)$, normalized by the external perturbation, $E(t)$:

$$\alpha_{ij}(\omega) = \frac{\int dt e^{i\omega t} P_j(t)}{\int dt e^{i\omega t} E_j(t)} \tag{4.15}$$

Here one typically adds in an imaginary component to ω (*i.e.*, $\omega \to \omega + i\gamma$), to take into account the finite width of the electronic excited states due to quantum dephasing and vibronic relaxation. Under steady-state conditions, the absorption cross section, $\sigma(\omega)$, which is the ratio between the stimulated transition rate and the incident photo flux density, is determined by the imaginary part of $\sigma(\omega)$:

$$\sigma(\omega) = \frac{4\pi\omega}{c} \left\langle \overline{\alpha(\omega)} \right\rangle_{imag} \tag{4.16}$$

where $\overline{\alpha(\omega)}$ denotes the spatially averaged value.

Alizarin, an organic molecule, has received increasing attention as a harmless metal-free photosensitizer.[50] It can complete the electron injection into TiO_2 within 6 fs.[51] A theoretical study was carried out using RT-TDDFT to investigate the absorption spectra of alizarin bound to various TiO_2 nanoclusters of different sizes (Figure 4.9).[43] For the free neutral alizarin, the primary absorption peak appears at 4.5 eV and slightly shifts to the red upon increasing TiO_2 units (Figure 4.10), while the secondary peak is located at 2.67 eV in

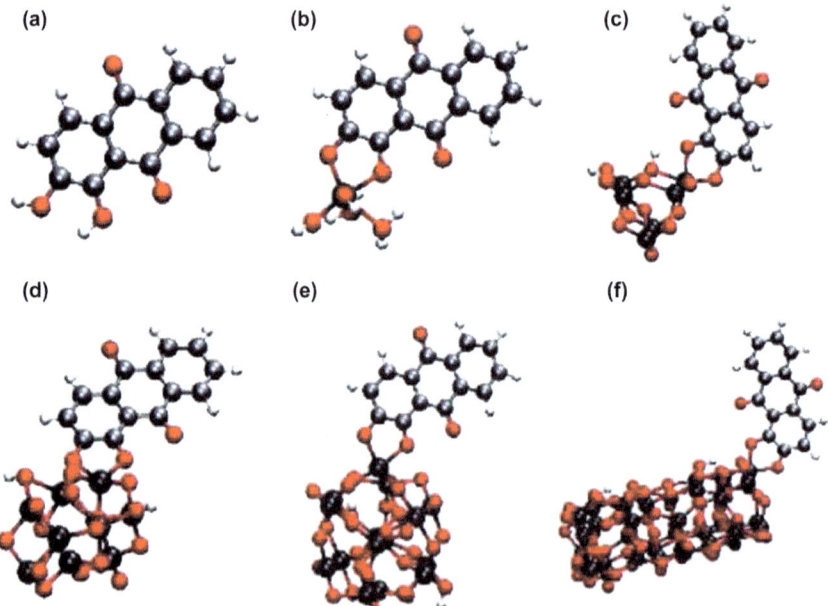

Figure 4.9 Free alizarin and absorbed alizarin on several TiO_2 clusters with different sizes (a) is the free alizarin while (b)–(f) are TiO_2 clusters of increasing complexity. (reprinted with permission from Sanchez-de-Armas *et al.*,[43] Copyright 2010 American Chemical Society).

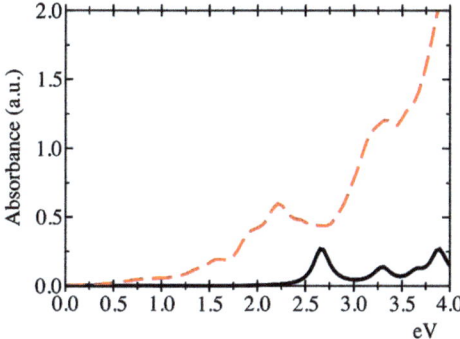

Figure 4.10 Calculated light absorbance of the free alizarin red) and alizarin bound to TiO$_2$)$_{38}$ cluster black) (reprinted with permission from Sanchez-de-Armas *et al.*,[43] Copyright 2010 American Chemical Society).

agreement with the experimental value of 2.88 eV. Interestingly, the red shift noted above is reversed for the deprotonated alizarin dianion, $Aliz^{2-}$ and its TiO$_2$ nanoclusters. In the absence of TiO$_2$, the primary absorption peak for $Aliz^{2-}$ is at 3.3 eV. Then, it shifts to 3.4 eV for $Aliz(TiO_2)_1^{2-}$ and even further to 3.7 eV for $Aliz(TiO_2)_6^{2-}$. The authors argue that the widened HOMO-LUMO gap can be ascribed to larger relative stabilization of the HOMO of $Aliz(TiO_2)_6^{2-}$ owing to its better negative charge delocalization over the larger TiO$_2$ clusters. On the other hand, both the HOMO and LUMO for the neutral alizarin are nearly equally stabilized by the TiO$_2$ clusters, making the red shift of the absorption peak barely observable. Eventually, these speculations gained support from a complementary LR-TDDFT simulation that determined the frontier molecular orbitals as a function of the TiO$_2$ cluster size. The combined RT-TDDFT and LR-TDDFT study suggests that the substrate binding the dye molecule should also be included in the explicit quantum mechanics treatment to ensure quantitative correctness, since the substrate could strongly influence the dye molecule's electronic transition through direct electronic couplings.

4.4 Hybrid Approaches

In light of the computational cost associated with many-body theories, it is not feasible to apply a quantum mechanics treatment to a full-scale PDSSC, which by definition requires a large number of conduction electrons to induce surface plasmon excitation. On the other hand, the continuum models have become the *de facto* means for describing the macroscopic optical response in these systems. Therefore, it is natural to divide a PDSSC conceptually into two parts, one for the photoactive component and the other for the plasmonic constituent. Here the photoactive component is usually composed of organic dye and/or in-organic metal complex molecules, the plasmon constituent is typically made of metallic or semiconductor nanoparticles and the coupling between them is commonly referred to as the "molecule–particle" interaction. Several optical theories have been developed to model this coupling and thereby to study the photovoltaic efficiency of PDSSC.

4.4.1 Multiscale Maxwell–Schrödinger Scheme (MMS)

In the MMS,[45] a molecule embedded in a dielectric medium is described as a two-level system under the random phase approximation (RPA), while the total electromagnetic fields, $\left(\vec{E}_{total}, \vec{H}_{total}, \vec{J}_{total} \right)$, at the location of the molecule, r_m, are decomposed into a contribution from itself, $\left(\vec{E}_m, \vec{H}_m, \vec{J}_m \right)$ and another one from everything else: $\left(\vec{E}_p, \vec{H}_p, \vec{J}_p \right)$:

$$
\begin{aligned}
\vec{E}_{total}(r_m, t) &= \vec{E}_m(r_m, t) + \vec{E}_p(r_m, t) \\
\vec{H}_{total}(r_m, t) &= \vec{H}_m(r_m, t) + \vec{H}_p(r_m, t) \\
\vec{J}_{total}(r_m, t) &= \vec{J}_m(r_m, t) + \vec{J}_p(r_m, t)
\end{aligned}
\tag{4.17}
$$

The time propagation of the molecule is described by the damped von Neumann equation:

$$
i\frac{\partial \rho(t)}{\partial t} = [H_0, \rho(t)] + [\vec{\mu}_m \times \vec{E}_p(r_m, t), \rho(t)] - i\gamma_m \rho(t)
\tag{4.18}
$$

where $\rho(t)$ is the density matrix, H_0 is the Hamiltonian in the absence of E_p, μ_m is the molecule's dipole moment and γ_m is the empirical damping factor. Accordingly, the electronic transition represented as the off-diagonal term of $\rho(t)$ is given by:

$$
i\frac{\partial \rho_{21}(t)}{\partial t} = \omega_{21} + \vec{\mu}_m \times \vec{E}_p(r_m, t) - i\gamma_m \rho_{21}(t)
\tag{4.19}
$$

where ω_{21} is the energy gap between the two molecular states. As seen above for the dielectric medium, the interior electromagnetic fields are described by the time-dependent Maxwell's equations, as shown in Equation (4.1), except for a change in the current density, $\vec{J}(r, t)$, which now includes a contribution from the molecule being excited, $\vec{J}_m(r_m, t)$:

$$
\vec{J}_m(r_m, t) = -2\omega_{21}\vec{\mu}_{21}\langle \rho_{21}(r_m, t) \rangle_{imag}
\tag{4.20}
$$

Thus, a dye molecule and its bound plasmonic nanoparticle can be interconnected by a mutual coupling *via* the molecule-induced current density.

This approach has been applied to a model system composed of a three gold spheres placed on a line that is taken to be the *X* axis, with radii of 2.5 nm and center-to-center spacing of 7.5 nm and with a pseudo-molecule placed midway between the second and third nanoparticles (Figure 4.11). The dipole transition moment of the pseudo-molecule is taken to be along the direction bisecting the *X* and *Y* axes, so the photo-induced current flow along the *X* axis between the

Array of spherical gold nanoparticles
with nearby molecule

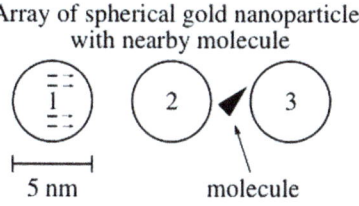

Figure 4.11   Pseudo-molecule placed between spherical gold nanoparticles (reprinted with permission from Lopata and Neuhauser,[45] Copyright 2009 American Institute of Physics).

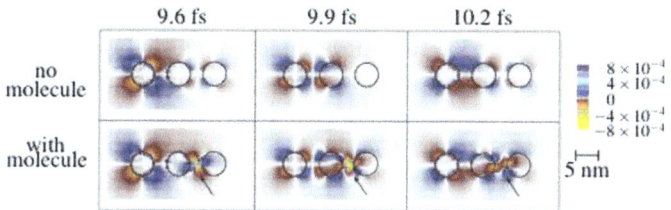

Figure 4.12   Three snapshots in time of the electric field with and without molecule (reprinted with permission from Lopata and Neuhauser;[45] Copyright 2009 American Institute of Physics).

nanoparticles is expected to be disturbed by absorption and scattering associated with the pseudo-molecule.

It is found that the energy transfer ratio between the nanoparticles is substantially influenced by the presence of the pseudo-molecule (Figure 4.12), particularly in the vicinity of its natural resonance frequency, which is distinct from the plasmon resonance frequency of the gold nanoparticles. By tracking the time-resolved electric field around the pseudo-molecule, a re-radiation of its absorbed light is revealed to be redirected along the Y axis. Thus, it can be imagined that an array of metallic particles is suitable to serve as a plasmonic switch that concentrates the incident photon energy into the desired dye molecules, depending on their orientation and location.

4.4.2   Hybrid RT-TDDFT/FDTD Approach (QM/ED)

Although MMS provides a notionally clear idea for dynamically incorporating the response of the dielectric environment into the molecular electronic states upon external electromagnetic perturbation, its practical application to realistic molecular systems is encumbered by the overly simplified description of the electronic degrees of freedom using the localized two-level RPA model. In fact, the two-level RPA model with a single resonance mode is bound to fail when describing electronic transitions in the photoactive components of solar cells since it is preferable to have a wide-band light harvester. Therefore, a theoretically more extensive electronic structure framework based on density functional theory is needed to incorporate many-body effects associated with the dye molecule into the hybrid theory. In addition, a computationally efficient approach such as FDTD is desired to describe the collective optical response of plasmonic particles that are near to the molecule.

In the recently developed hybrid QM/ED approach,[47] the ratio of scattered field to incident field was first evaluated by FDTD to obtain the so-called frequency-dependent scattering response function, $\lambda_{ij}(r,\omega)$, at the location of the dye molecule, r:

$$\lambda_{ij}(r,\omega) = \frac{E_{j,sca}(r,\omega)}{E_{i,inc}(r,\omega)} \qquad (4.21)$$

Then, the scattered field, $\vec{E}_{sca}(r,t)$ and the incident field, $\vec{E}_{inc}(r,t)$, are allowed to interact with the dye molecule in the time domain in a RT-TDDFT calculation. This is done using a two-dimensional Fourier transform *via* $\lambda_{ij}(r,\omega)$:

$$E_{i,sca}(r,t) = \frac{1}{2\pi} \sum_j \iint d\omega \, dt_1 e^{i\omega(t-t_1)} \lambda_{ij}(r,\omega) E_{j,inc}(r,t) \qquad (4.22)$$

The wavefunction is then propagated according to the time-dependent Schrödinger equation with both $\vec{E}_{inc}(r,t)$ and $\vec{E}_{sca}(r,t)$ included in the Hamiltonian:

$$i\frac{\partial\varphi(r,t)}{\partial t} = \left(\hat{H}_0 + \vec{E}_{inc}(r,t) \times \vec{\mu}(r,t) + \vec{E}_{sca}(r,t) \times \vec{\mu}(r,t)\right)\varphi(r,t) \qquad (4.23)$$

After this propagation, the molecular polarizability in the presence of the scattered field, $\alpha(\omega)$, can be obtained using Equation (4.15). The absorption cross section, $\sigma_k(\omega)$, for a particle-bound dye molecule irradiated by light with fixed propagation direction k is then given by:

$$\sigma_k(\omega) = \frac{4\mu\omega}{c} \left\langle \frac{1}{2}\left(\alpha_{ii} + \alpha_{jj}\right)\left(1 + \frac{1}{2}(\lambda_{ii}^* + \lambda_{jj}^*)\right)\right.$$
$$\left. + \frac{1}{8}\left((\alpha_{ii} - \alpha_{jj})(\lambda_{ii}^* - \lambda_{jj}^*) + (\alpha_{ij} + \alpha_{ji})(\lambda_{ij}^* + \lambda_{ji}^*)\right)\right\rangle$$

Calculations using this approach for a N3 dye molecule bound to a 20-nm diameter silver sphere (Figure 4.13), show that $\sigma_k(\omega)$ is substantially enhanced

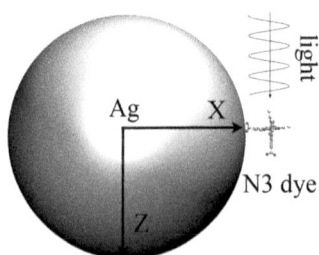

Figure 4.13   N3 dye molecule bound to a 20-nm diameter silver sphere.

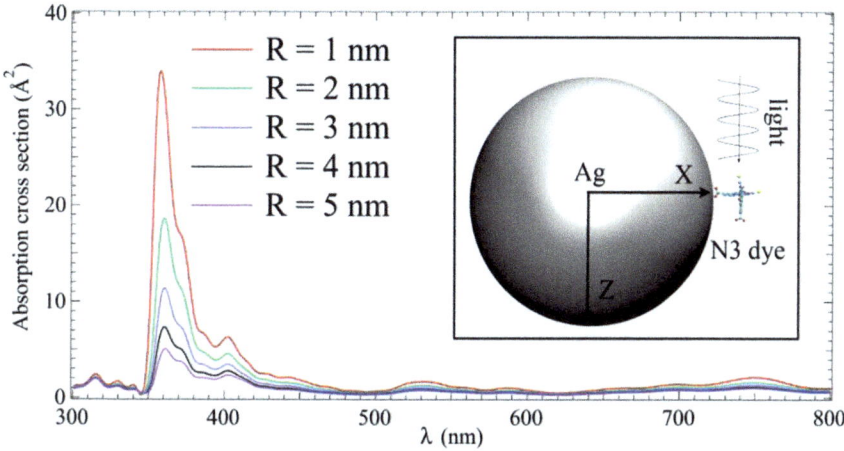

Figure 4.14 Calculated absorption profile of a silver-bound N3 dye with different values of the dye-particle separation, *R*.

at 357 nm, which is the plasmon resonance wavelength of the silver sphere in a vacuum (Figure 4.14).[47]

As expected, the on-resonance enhancement factor decreases rapidly from ~55 to ~8 when the dye–particle separation increases from 1–5 nm. It is very interesting to note that the two equally strong absorption peaks at 311 nm and 407 nm of the bare N3 dye are not equally amplified by the presence of the silver sphere, with a notably heightened peak at 407 nm and a nearly invariant peak at 311 nm. The apparent difference in the light absorption efficiency can be ascribed to the relative directions of the silver sphere's scattered electric field and the N3 dye's transition dipole moment for the two resonant modes, showing the importance of including the dye molecule explicitly in many-body electronic structure theory in determining plasmon-enhanced response.

An encouraging result from our recent study[52] of the N3 dye molecule bound to a TiO$_2$-coated oxidized silver sphere is that an IPCE enhancement factor of 9.9 was obtained for a TiO$_2$ thickness of 2.0 nm, in excellent agreement with the experimental value of 6.0. Furthermore, the observed red shift of the primary extinction peaks of the PDSSC relative to the plasmon maximum was also found in our simulations. By comparing the location of the experimental extinction peaks, we are able to estimate theoretically the thickness of an intermediate Ag$_2$O layer that is not experimentally measurable yet and is found to be essential in limiting the photovoltaic efficiency of the PDSSC.

4.5   Conclusions and Future Direction

The inclusion of sub-wavelength plasmonic particles and other plasmonic structures in solar cells has been shown to reduce notably the required thickness of the photoactive layer and to expand greatly the pool of its usable absorber materials. For most plasmon-enhanced solar cells, fabrication only involves small modifications of the existing architecture, adding to the technical feasibility of this approach. In general, there are two mechanisms for using plasmonic nanoparticles to improve solar cell photovoltaic efficiency, either (1) by increasing the effective optical path length by scattering in the front of the cell or by back reflection, or (2) by intensifying the near-field resonant coupling with the photoactive component through amplified local electric fields. As discussed in the present study, both enhancement mechanisms can be reliably modeled by classical electrodynamics, by quantum mechanics or by hybrid theories that combine quantum and classical approaches. However, it should be pointed out that although theoretical studies have often been brought to bear on classical electrodynamics of the plasmonic cells, it is rare to see studies of quantum properties such as the rate of plasmonically enhanced electron injection once a photon is absorbed. The combination of electrodynamics studies with our recently developed constrained real-time time-dependent density functional theory[53] (CRT-TDDFT) can thus play an important role, providing more accurate insight into working mechanisms that might result in design considerations for future plasmonic solar cells.

Acknowledgments

The research was supported by the ANSER Center, grant DE-SC0001059, funded by the US Department of Energy, Office of Science and Office of Basic Energy Sciences. The computational resources utilized in this research were provided by Shanghai Supercomputer Center.

References

1. U. S. Energy Information Administration, *Annual Energy Outlook 2011*, Office of Integrated and International Energy Analysis, U.S. Department of Energy, 2011.
2. P. Bozyk, *Globalization and the Transformation of Foreign Economic Policy Transition and Development*, Ashgate Pub Corporation, Warsaw, Poland, 2006.
3. K. A. Masarie and P. P. Tans, *J. Geophys. Res.*, 1995, **100**, 11593.
4. R. M. H. E. Van Eijs, F. M. M. Mulders, M. Nepveu, C. J. Kenter and B. C. Scheffers, *Eng. Geol.*, 2006, **84**, 99.
5. *BP Statistical Review of World Energy*, British Petroleum Global, 2009.
6. S. Dewitte, D. Crommelynck, S. Mekaoui and A. Joukoff, *Solar Phys.*, 2004, **224**, 209.
7. R. W. Miles, K. M. Hynes and I. Forbes, *Prog. Cryst. Growth Charact. Mater.*, 2005, **51**, 1.
8. D. M. Chapin, C. S. Fuller and G. L. Pearson, *J. Appl. Phys.*, 1954, **25**, 676.
9. B. O'Regan and M. Gratzel, *Nature*, 1991, **353**, 737.
10. M. A. Green, K. Emery, Y. Hishikawa and W. Warta, *Prog. Photovoltaics. Res. Appl.*, 2010, **18**, 346.

11. M. Grätzel, *J. Photochem. Photobiol. C. Photochem. Rev.*, 2003, **4**, 145.
12. N. S. Sariciftci, L. Smilowitz, A. J. Heeger and F. Wudl, *Science*, 1992, **258**, 1474.
13. C. J. Brabec, N. S. Sariciftci and J. C. Hummelen, *Adv. Funct. Mater.*, 2001, **11**, 15.
14. E. D. Gomez and Y.-L. Loo, *J. Mater. Chem.*, 2010, **20**, 6604.
15. D. L. Jeanmaire and R. P. Van Duyne, *J. Electroanal. Chem.*, 1977, **84**, 1.
16. M. Fleischmann, P. J. Hendra and A. J. McQuillan, *Chem. Phys. Lett.*, 1974, **26**, 163.
17. F. W. King and G. C. Schatz, *Chem. Phys.*, 1979, **38**, 245.
18. K. R. Catchpole and A. Polman, *Opt. Express*, 2008, **16**, 21793.
19. H. R. Stuart and D. G. Hall, *Appl. Phys. Lett.*, 1998, **73**, 3815.
20. S. Pillai, K. R. Catchpole, T. Trupke and M. A. Green, *J. Appl. Phys.*, 2007, **101**, 93105.
21. C. G. Granqvist and O. Hunderi, *Phys. Rev. B*, 1977, **16**, 3513.
22. F. J. Beck, A. Polman and K. R. Catchpole, *J. Appl. Phys.*, 2009, **105**, 114310.
23. D. Pacifici, H. J. Lezec and H. A. Atwater, *Nat. Photonics*, 2007, **1**, 402.
24. R. J. Walters, R. V. A. van Loon, I. Brunets, J. Schmitz and A. Polman, *Nat. Mater.*, 2010, **9**, 21.
25. Z. Lin, A. Franceschetti and M. T. Lusk, *ACS Nano*, 2011, **5**, 2503.
26. M. B. Smith and J. Michl, *Chem. Rev.*, 2010, **110**, 6891.
27. M. Ihara, K. Tanaka, K. Sakaki, I. Honma and K. Yamada, *J. Phys. Chem. B*, 1997, **101**, 5153.
28. S. D. Standridge, G. C. Schatz and J. T. Hupp, *Langmuir*, 2009, **25**, 2596.
29. I. Thomann, B. A. Pinaud, Z. Chen, B. M. Clemens, T. F. Jaramillo and M. L. Brongersma, *Nano Lett.*, 2011, 3440.
30. J. M. Rehm, G. L. McLendon, Y. Nagasawa, K. Yoshihara, J. Moser and M. Grätzel, *J. Phys. Chem.*, 1996, **100**, 9577.
31. S. D. Standridge, G. C. Schatz and J. T. Hupp, *J. Am. Chem. Soc.*, 2009, **131**, 8407.
32. C. Prasittichai and J. T. Hupp, *J. Phys. Chem. Lett.*, 2010, **1**, 1611.
33. T. Alsaka, M. Fujii and S. Hayashi, *Appl. Phys. Lett.*, 2008, **92**, 132105.
34. I. K. Ding, J. Zhu, W. Cai, S.-J. Moon, N. Cai, P. Wang, S. M. Zakeeruddin, M. Grätzel, M. L. Brongersma, Y. Cui and M. D. McGehee, *Adv. Energy Mater.*, 2011, **1**, 52.
35. P. Gay-Balmaz and O. J. F. Martin, *Appl. Opt.*, 2001, **40**, 4562.
36. F.-J. Haug, T. Soderstrom, O. Cubero, V. Terrazzoni-Daudrix and C. Ballif, *J. Appl. Phys.*, 2008, **104**, 064509.
37. S. Mokkapati, *J. Phys. D: Appl. Phys.*, 2011, **44**, 185101.
38. H. Shen, P. Bienstman and B. Maes, *J. Appl. Phys.*, 2009, **106**, 73109.
39. T. L. Temple and D. M. Bagnall, *J. Appl. Phys.*, 2011, **109**, 84343.
40. J. Jung, T. Søndergaard, T. G. Pedersen, K. Pedersen, A. N. Larsen and B. B. Nielsen, *Phys. Rev. B*, 2011, **83**, 085419.
41. X. Li, N. P. Hylton, V. Giannini, K.-H. Lee, N. J. Ekins-Daukes and S. A. Maier, *Opt. Express*, 2011, **19**, A888.

42. M. K. Nazeeruddin, F. De Angelis, S. Fantacci, A. Selloni, G. Viscardi, P. Liska, S. Ito, B. Takeru and M. Grätzel, *J. Am. Chem. Soc.*, 2005, **127**, 16835.
43. R. Sanchez-de-Armas, J. A. Oviedo Lopez, M. San-Miguel, J. F. Sanz, P. Ordejon and M. Pruneda, *J. Chem. Theory Comput.*, 2010, **6**, 2856.
44. T. B. Lynge and T. G. Pedersen, *Comput. Mater. Sci.*, 2004, **30**, 212.
45. K. Lopata and D. Neuhauser, *J. Chem. Phys.*, 2009, **130**, 104707.
46. D. J. Masiello and G. C. Schatz, *J. Chem. Phys.*, 2010, **132**, 064102.
47. H. Chen, J. M. McMahon, M. A. Ratner and G. C. Schatz, *J. Phys. Chem. C*, 2010, **114**, 14384.
48. E. Runge and E. K. U. Gross, *Phys. Rev. Lett.*, 1984, **52**, 997.
49. V. Barone and M. Cossi, *J. Phys. Chem. A*, 1998, **102**, 1995.
50. W. R. Duncan and O. V. Prezhdo, *Annu. Rev. Phys. Chem.*, 2007, **58**, 143.
51. R. Huber, J.-E. Moser, M. Grätzel and J. Wachtveitl, *J. Phys. Chem. B*, 2002, **106**, 6494.
52. H. Chen, M. G. Blaber, S. D. Standridge, E. J. DeMarco, J. T. Hupp, M. A. Ratner, G. C. Schatz, *J. Phys. Chem. C*, 2012, **116**, 10315.
53. H. Chen, M. A. Ratner and G. C. Schatz, *J. Phys. Chem. C*, 2011, **115**, 18810.

CHAPTER 5

Dynamics of Interfacial Electron Transfer in Solar Energy Conversion As Viewed By Ultrafast Spectroscopy

VILLY SUNDSTRÖM* AND ARKADY YARTSEV

Department of Chemical Physics, Lund University, Box 124, 22100 Lund, Sweden
*Email: villy.sundstrom@chemphys.lu.se

5.1 Dye Sensitized Nanostructured Metal Oxides for Grätzel Solar Cells

Sensitized wide band-gap semiconductors have long been considered as materials for photovoltaics. However, not until the discovery by Grätzel and co-workers,[1] that a material with high conversion efficiency could be produced by sensitizing a nanostructured thin film in a wide band-gap semiconductor, was real progress towards a competitive solar cell device made. A so-called dye-sensitized solar cell (DSC) typically consists of a thin ($\sim 1\,\mu m$) TiO_2 film of nanometre-sized ($\sim 10\,nm$) particles sintered together for electrical contact. This thin film is deposited on a conducting glass (ITO) electrode and brought into electrical contact with a counter electrode *via* an electrolyte. Light energy conversion in such a solar cell starts when light is absorbed by the sensitizing dye and electrons are injected from the excited state of the dye into the conduction band of the semiconductor. Electrons then migrate between

RSC Energy and Environment Series No. 8
Solar Energy Conversion: Dynamics of Interfacial Electron and Excitation Transfer
Edited by Piotr Piotrowiak

semiconductor particles until they reach the back contact and the external circuit, where they can perform work. Finally, electrons are lead back to a redox couple (often I^-/I_3^-), which regenerates the oxidized dye back to the neutral ground state and the dye can absorb another photon and start a new conversion cycle.

High efficiency of the electron injection step and a low yield of electron recombination between electrons in the semiconductor conduction band and oxidized dye are essential for an efficient material. The electron injection efficiency can be optimized by ensuring very fast injection, much faster than all competing processes deactivating the excited state of the sensitizer. The fraction of injected electrons that recombine with the oxidized dye may be influenced in several different ways. One possibility is to choose the sensitizer and redox couple in such a way that their electrochemical properties maximize the ratio of rates for oxidized dye reduction by the redox couple and electron recombination from the semiconductor. Another possibility is to decrease the rate of back-electron transfer from the semiconductor to the oxidized dye, by increasing the distance between the semiconductor and sensitizer (see *e.g.* Burfeindt *et al.*[2]).

As we will see below, design of sensitizers to control dye–semiconductor binding and distances of electron transfer may lead to counter intuitive results because of the multitude of factors controlling the binding. Still another approach that has been explored is to use a secondary electron donor attached to the sensitizer and hence move the positive hole on the sensitizer further away from the semiconductor surface (see *e.g.* Ghanem *et al.*[3]). The DSC field has been reviewed by different authors emphasizing various directions. An exhaustive and recent review of most aspects of the DSC can be found in Ardo and Meyer.[4] The most efficient DSC so far reported has an overall validated light-to-electricity conversion efficiency of $\sim 11\%$[1,5–8] and relies on TiO_2 anatase for the semiconductor and a transition metal complex (*e.g.* Ru(dcbpy)$_2$(NCS)$_2$ (dcbpy = 4,4'-dicarboxylate-2,2'-bipyridine, RuN3 for short) as sensitizer.

5.2 Electron Injection from Sensitizer to Semiconductor in Dye Sensitized Solar Cells

Since the discovery by Grätzel and co-workers of the DSC based on nanocrystalline metal oxide thin film electrodes,[1] transition metal complexes and in particular the RuN3 dye, have produced among the most efficient solar cells. This fact, in combination with the complex excited state structure (presence of both singlet and triplet states), was a strong motivation for the study of the excited state and electron transfer dynamics in these systems. More recently, progress in the development of other types of very efficient sensitizers, for example porphyrins, motivates similar studies of these molecules. We will present some results for RuN3 and for a family of Zn-porphyrin molecules as knowledge of the most fundamental electron transfer processes will contribute to a better understanding of DSC function and therefore optimization of cell performance.

The strong visible absorption of metal-polypyridyl molecules is due to singlet and triplet excited states. For the Ru-based molecules discussed here, the lowest excited state is a triplet metal to ligand charge transfer state (^{3}MLCT) and there is a ^{1}MLCT state at higher energy responsible for the main absorption band. Light absorption into this band therefore generates the excited ^{1}MLCT state, but within a very short time (<100 fs) the molecule has relaxed into the lowest ^{3}MLCT state, because of efficient intersystem crossing caused by the heavy metal atom. For efficient electron injection and energy conversion in the sensitized semiconductor system, the energy of the lowest ^{3}MLCT state has to be above the conduction band edge of the semiconductor. The resulting scenario is illustrated by Figure 5.1, showing the valence and conduction bands of the semiconductor, as well as the ground and excited states of the RuN3 sensitizer. Our work[9–11] shows that following light absorption to the ^{1}MLCT state, ~60% of the molecules inject electrons directly from this state into the semiconductor conduction band with a characteristic time constant of 50 fs. Upon excitation to higher-lying vibrational states of the ^{1}MLCT state, even faster injection occurs (<20 fs), in competition with vibrational energy relaxation and redistribution.[11] The residual ~40% of the excited sensitizers relax to the triplet state, from which they inject electrons much more slowly on the 1–100 ps time scale.[12] Below, we will show how this picture was obtained by using ultrafast spectroscopy to probe the various species involved in the light energy conversion process.

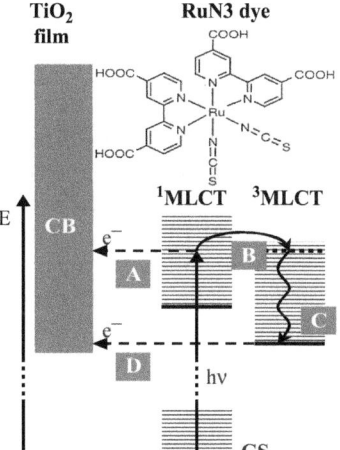

Figure 5.1 Schematic model of two-state electron injection and structure of RuN3. Following MLCT excitation (at 530 nm) of the RuN3-sensitized TiO$_2$ film, an electron is promoted from a mixed ruthenium NCS state to an excited π* state of the dcbpy-ligand and injected into the conduction band (CB) of the semiconductor. GS: ground state of RuN3. Channel A: electron injection from the non-thermalized, singlet^{1}MLCT excited state. Channel B and C: ISC followed by internal vibrational relaxation in the triplet ^{3}MLCT excited state. Channel D: electron injection from the thermalized, triplet ^{3}MLCT excited state.

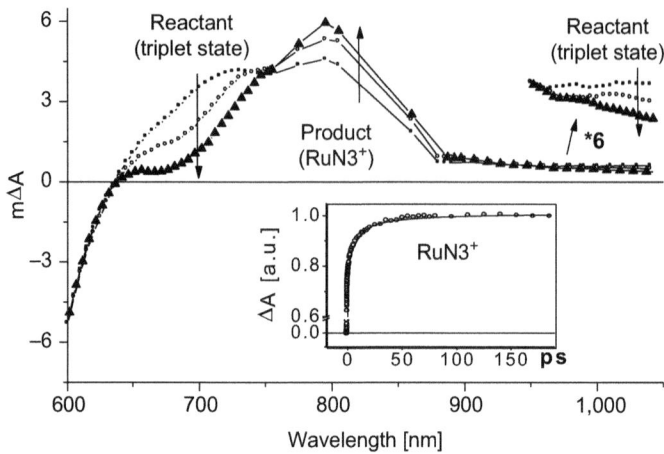

Figure 5.2   Visible and near-IR transient absorption spectra of RuN3-sensitized TiO$_2$ film. At time delays of 0.5 ps (—■—), 10 ps (—○—) and 150 ps (—▲—) between the pump and probe pulses we observe characteristic dynamics of the differential spectra with two isobestic points (the wavelength where $\Delta A_{reactant} = \Delta A_{product}$) at 760 and 940 nm, and one nearly time-independent $\Delta A = 0$ point at 630 nm. Inset, transient absorption kinetics of oxidized RuN3 cation-TiO$_2$ measured at 860 nm. Symbols are measured data, while the curve is a fit with the following time constants and amplitudes: rise within the laser pulse ($20 \pm 5\%$), 28 ± 3 fs (50%), 1 ± 0.1 ps (11%), 9.5 ± 1 ps (12%) and 50 ± 5 ps (7%).

The transient spectra of Figure 5.2 indicate the key spectral regions for probing these species – the appearance and decay of the triplet state is monitored around 700 nm and 1000 nm, while the formation of the oxidized dye is probed at ∼ 850 nm. The decay of the initially excited singlet state is monitored by stimulated emission and the kinetics measured at 600 nm are shown in Figure 5.3(a). The initial fast decay of the stimulated emission is characterized by a 30-fs time constant to a time-independent (on this time scale) signal level due to ground state bleaching. The stimulated emission decay of RuN3 in ethanol solution due to intersystem crossing is 70 fs,[11] showing that electron injection effectively competes with the very fast intersystem crossing. Ultrafast dynamics matching the decay of the singlet state can also be observed in other spectral regions. At 690 nm (Figure 5.3(b)), the instantaneous rise followed by the 30 ± 3 fs decay again reflects the evolution of the singlet state, while the 80 ± 5 fs rise corresponds to the formation and thermalization of the triplet state (channels B and C of Figure 5.1). It is intriguing that the decay of the singlet proceeds in ∼ 30 fs, but the rise of the triplet state occurs in ∼ 80 fs. However, the probe signal centered at 690 nm in the blue wing of the triplet absorption band (Figure 5.1) is sensitive to the spectral blue-shift caused by the triplet equilibration process. Thus, the ∼ 80 fs time constant encompasses both the intersystem crossing (ISC) process and the ensuing thermalization of the triplet state.

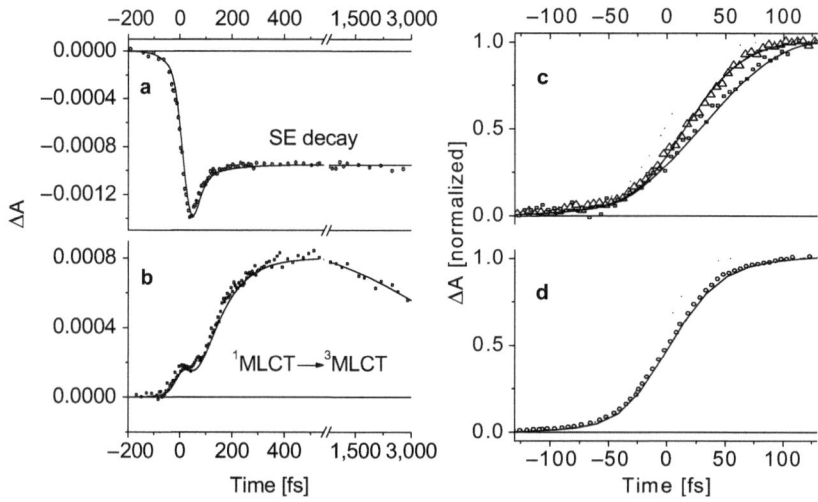

Figure 5.3 Transient absorption kinetics. Open symbols are measured data, curves are fits, and the instrument responses are represented by dotted curves. (a) SE decay probed at 600 nm. (b) excited state evolution (^{1}MLCT→^{3}MLCT) probed at 690 nm. (c) formation of the triplet state at 1050 nm; the data are well fitted with rise times of 30 ± 5 fs (for RuN3-TiO$_2$; triangles) and 70 ± 15 fs (for RuN3-EtOH; squares). (d) early-time transient absorption kinetics of oxidized RuN3 cation-TiO$_2$ measured at 860 nm.

The kinetics at 1050 nm (Figure 5.3(c)) provides one more possibility to monitor the formation of the triplet state. In this spectral region the absorption of the triplet is very broad and featureless and therefore not sensitive to spectral shifts that may be caused by the thermalization processes. Hence, a ~ 30-fs rise time is observed, corresponding directly to the decay of the singlet state monitored by stimulated emission. In Figure 5.3(c), the formation of the triplet state for RuN3 in solution is compared to that of RuN3-sensitized TiO$_2$ film and found to be considerably slower in solution (~ 70 fs). The faster formation of the triplet in RuN3-TiO$_2$ reflects the very fast and efficient electron injection from the singlet state prior to intramolecular vibrational relaxation (IVR), internal conversion (IC) and ISC. Finally, if the ~ 30 fs decay corresponds to electron injection from the RuN3 singlet state, then we also expect to observe this time constant for the formation of the oxidized RuN3. Indeed, the fit to the data at 860 nm (Figure 5.3(d)) confirms the presence of the ~ 30 fs time constant with high amplitude ($> 50\%$). Hence, we see that the optically excited singlet state decays with the same time constant of ~ 30 fs as the triplet state and oxidized dye products appear, thus coupling these species in a reactant–product relationship.

In a simplified reaction model, the rate constant related to the observed ~ 30-fs decay of the singlet state reflects the sum of the rates of electron injection from the ^{1}MLCT state (k_A) and ISC (k_B): $1/30 \ \mathrm{fs}^{-1} = k_A + k_B$.

The ratio between k_A and k_B of 1.5 is calculated from the amplitude ratio of femtosecond ($\sim 60\%$) and picosecond ($\sim 40\%$) electron injection contributions to the signal at 860 nm, yielding $k_A = 1/50$ fs^{-1} and $k_B = 1/75$ fs^{-1}. The value determined for $1/k_B$ is in good agreement with previous measurements of ISC for this class of molecules[13–15] and is in excellent agreement with the formation of the triplet state measured here for RuN3 in ethanol solution (Figure 5.3(c)). We conclude that after excitation of the sample at 530 nm, $\sim 60\%$ of the RuN3 molecules inject electrons from the singlet state into TiO$_2$ (channel A of Figure 5.1) and the rest undergoes ISC (channel B). After relaxation to the bottom of the triplet state (channel C), electrons are again injected into the semiconductor but now the reaction occurs on the picosecond time scale (channel D).

The highly non-exponential slow electron injection from the triplet state has been suggested to be a result of weak and highly inhomogeneous coupling between the sensitizer and semiconductor. However, with this explanation it is difficult to understand how triplet injection can be independent of semiconductor (the same ps injection rates were observed for RuN3 attached to TiO$_2$ and SnO$_2$),[16] but dependent on the solvent in contact with the dye–semiconductor film (a slower injection was observed in ethanol than in acetonitrile[9]). Also with the injection scheme of Figure 5.1, one would expect to observe excitation wavelength dependent (hot) injection from the triplet state, similar to what was observed for the singlet state. No evidence for hot triplet injection was however found.[9,11] These observations suggest that perhaps some intramolecular process within the sensitizer is the rate-limiting step of the triplet injection.

Various results in the literature have been interpreted to suggest that RuN3 is either attached by one bipyridyl ring [17–19] or both bipyridyl rings[20,21] to the semiconductor surface (the one-ring attachment is illustrated in Figure 5.1). With the one-ring attachment the two ligands become non-equivalent and it can be expected that injection from the attached and non-attached ligand is different. By comparing a number of spectroscopic properties, solar cell performance and electron injection characteristics for RuN3 and an analogue (Ru520DN) only capable of attachment to the semiconductor by one bipyridyl ring, we concluded that RuN3 is most likely to be attached to the semiconductor by only one bipyridyl ring. This results in a dramatically different picture of the mechanism of electron injection from the triplet state, illustrated by the schematic diagram in Figure 5.4. The resulting scenario is that the majority of the excited molecules inject electrons from the initially excited delocalized ^{1}MLCT state into the conduction band of the TiO$_2$ (Figure 5.4, pathway A$_1$) on the <100 fs timescale. In concert with this, the remaining population undergoes ISC and localizes in the ^{3}MLCT state (pathway A$_2$). After relaxation to the bottom of the triplet state (pathway C) and interligand electron transfer (ILET, pathway D), the electrons are injected from the ^{3}MLCT state of the attached ligand (pathway E). This pathway of electron injection is controlled by ILET, which proceeds with different picosecond time constants depending on the dye molecule–TiO$_2$ particle interaction and

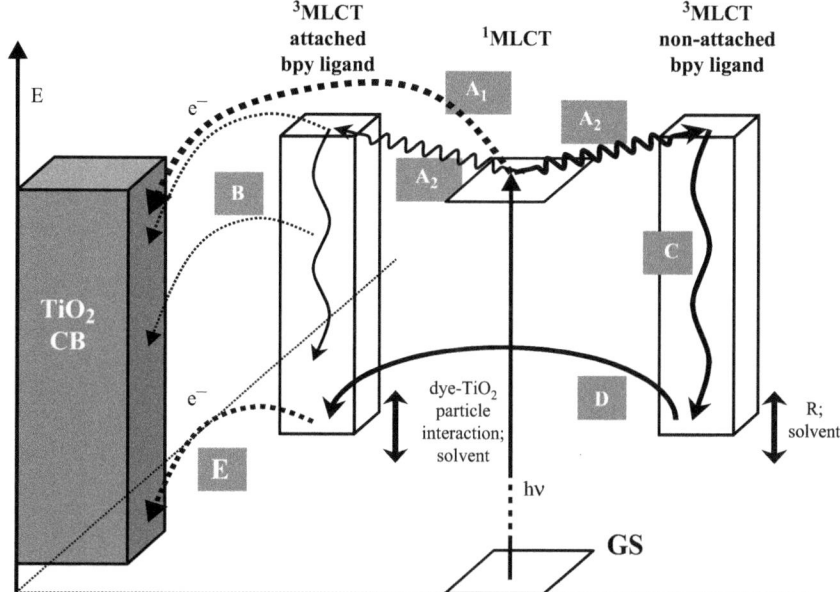

Figure 5.4 Schematic model of electron injection of RuN3. Following MLCT excitation of the RuN3-sensitized TiO$_2$ nanocrystalline film, electron injection
occurs from both excited states of the dye, ^{1}MLCT and ^{3}MLCT, into the
conduction band (CB) of the semiconductor. GS: ground state of RuN3.
Pathway A$_1$: electron injection from the initially excited delocalized
^{1}MLCT excited state. Pathway A$_2$: ISC and localization in the ^{3}MLCT
excited state. Pathway B: electron injection from the hot ^{3}MLCT excited
state of the attached bipyridine ligand (not observed in the present study).
Pathway C: internal vibrational relaxation in the ^{3}MLCT excited state of
the non-attached bipyridine ligand. Pathway D and E: ILET between the
bipyridine ligands and ensuing electron injection.

solvent environment. In total, up to $\sim 70\%$ of the electron transfer takes
place on the femtosecond timescale. This picture of the triplet injection
was obtained from a comparison of ILET dynamics of RuN3 in solution and
attached to the nanostructured TiO$_2$ film and correlation to the already discussed injection dynamics monitoring triplet state decay and oxidized dye
formation.[22]

The picture of singlet and triplet state electron injection given above holds
when the energies of the dye excited states are above the conduction band
edge of the semiconductor. For RuN3 the triplet state is expected to be located
close to the TiO$_2$ band edge; triplet injection could therefore be limited by
energetic mismatch, that is, the triplet state having energy lower than the
conduction band edge. The absorption spectrum of RuN3 has a wide red tail
owing to absorption to the ^{3}MLCT state, suggesting that there is a distribution
of triplet state energies. Measurement of triplet state electron injection
from RuN3 to TiO$_2$ with excitation into the red wing triplet absorption

revealed a decreasing quantum yield of electron injection, non-exponential and increasingly slower injection for progressively redder excitation wavelengths.[12] This was concluded to be due to energetic heterogeneity of the triplet states leading to excitation of sensitizer molecules having triplet state energies below the conduction band edge of TiO_2.[12] For injection to SnO_2, triplet injection efficiency is independent of excitation wavelength, the reason being that the triplet states lie significantly above the conduction band edge in this semiconductor. The energetics of triplet injection from RuN3 to TiO_2 and SnO_2 is illustrated in Figure 5.5.

Electron injection for other similar transition metal sensitizers is frequently observed to be ultrafast and not particularly sensitive to experimental conditions. Thus, this type of sensitizer offers a robust and very efficient vehicle for light absorption and electron injection in DSC and therefore electron injection is not generally a limiting factor for the efficiency of a solar cell. The reason for this we believe is a binding geometry providing relatively well-defined electron transfer pathways in combination with the metal-to-ligand nature of the electronic transition responsible for light absorption in these molecules. The latter will place the excited electron in the sensitizer close to the semiconductor surface, which gives rise to very fast electron transfer. The situation is quite different for the Zn-porphyrin sensitizers discussed here. The kinetics in Figure 5.6 illustrate the electron injection (rise of kinetics) and electron–cation recombination (decay) of a Zn-porphyrin sensitized TiO_2 film and we can see that by just changing the solvent used in the sensitization of the electrode surface, the electron injection time changes from a few-picosecond scale to ~ 100 ps. We attribute this to a change of sensitizer–semiconductor binding geometry induced by an interaction with the sensitization solvent molecules.[23] The effect is also reflected in a change of electron–cation recombination time (the decay of the kinetics in Figure 5.6) which we will discuss in more detail below in the context of electron–cation recombination.

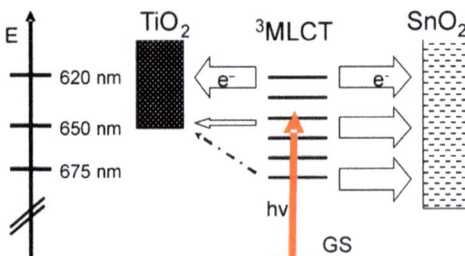

Figure 5.5   Triplet electron injection from RuN3 to TiO_2 and SnO_2 shows differences after red wing excitation of the visible absorption band of the dye. Low lying triplet states of RuN3 are situated below the conduction band edge of TiO_2 and this leads to activated slow electron injection. For SnO_2 the conduction band edge is sufficiently low for all triplets to inject efficiently.

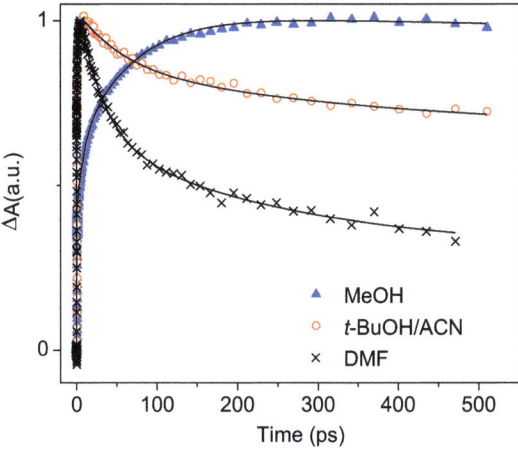

Figure 5.6 Transient absorption kinetics of 2,4,6-Me on TiO$_2$ sensitized in MeOH (closed, blue triangle), *t*-BuOH/ACN (open, red circle) and DMF (black cross) for 1 hour. Kinetics are normalized to their maxima and solid lines represent fits of the kinetics.

5.3 Electron–Cation Charge Recombination in Dye Sensitized Semiconductor Materials

For Ru-polypyridyl dyes (*e.g.* RuN3, the black dye) resulting in the most efficient solar cells, electron–cation recombination has been shown to be very slow (microsecond timescale)[24–26] and slower than regeneration of the oxidized sensitizer by the redox mediators,[27] and therefore not a limiting factor for the efficiency of a solar cell based on these dyes. This fact, established for some Ru-polypyridyl sensitizers often seems to have been extrapolated to suggest that this is also the case for other dyes,[28] leading to a picture where variations in solar cell efficiency have been directly correlated to the efficiency and rate of electron injection.[28] Below, we will show that electron–cation recombination can be a more important process for controlling solar cell efficiency.[23] This will also lead to a correlation of electron injection and recombination rates with the binding geometry of the sensitizer, and hence the steps towards the goal of designing nanostructured dye-sensitizer materials with predictable electron transfer properties. Many sensitizer molecules are developed with simple binding models guiding the design of the molecules. Detailed studies of the electron transfer dynamics (see below) show that considerably more advanced modelling of the dye-semiconductor interactions is required in order to predict the geometry correctly. We will use some of our recent results for Zn-porphyrin/TiO$_2$ electrodes to illustrate this. A combination of ultrafast spectroscopy, surface sensitive sum frequency generation (SFG) and measurements of solar cell power conversion efficiency leads to a picture where we can correlate binding geometry to electron transfer dynamics, which in its turn controls the conversion efficiency of the solar cell.

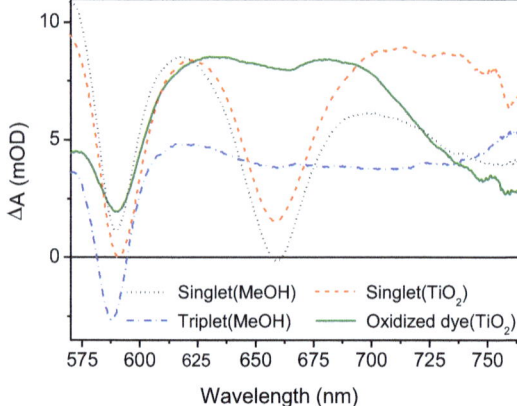

Figure 5.7 Spectral components extracted from measured transient absorption spectra. Dotted line represents singlet excited state spectrum (including stimulated emission) in methanol solution, while the dashed red line represents
the singlet excited state spectrum of TiO_2. The blue dash-dot line represents triplet spectrum in methanol solution and the solid green line
represents the oxidized dye of TiO_2.

The electron transfer dynamics were monitored by transient absorption
(TA) spectra and kinetics.[23] The time evolution of the TA spectrum of a
Zn-porphyrin/TiO_2 electrode can be described as a non-exponential process
involving only two species, the singlet excited state and the Zn-porphyrin
radical cation (their spectra are shown in Figure 5.7). Thus, it is seen that the
excited state of the sensitizer is formed within the time resolution (<100 fs) of
the experiment and then transforms with multi-exponential kinetics to the
oxidized sensitizer, which then decays back to the ground state by charge recombination with conduction band electrons. This recombination can be described for most dyes by two lifetimes on the tens and hundreds of picoseconds
timescale and a very slow >50 ns component. Finding that the electron in-
jection/recombination dynamics is dominated by two species, the singlet excited
state and dye radical cation, in addition to the ground state, implies that
kinetic measurements at a single wavelength can be used for precise characterization of the temporal evolution of the reactions. With the help of the
spectra in Figure 5.7 we can choose the most appropriate wavelength for this.
We see that stimulated emission of the sensitizer is at ~ 660 nm and the
oxidized dye absorbs strongly at the same wavelength. Thus, by measuring the
kinetics at 660 nm we can monitor the formation of the sensitizer excited state,
its decay and formation of the oxidized sensitizer, as well as the ensuing decay
of the sensitizer radical cation through recombination with conduction band
electrons.

By using the series of Zn-porphyrins depicted in Figure 5.8(a) we could vary
several molecular properties of importance for dye-semiconductor binding.[4,23]
From Marcus theory of electron transfer[29–31] (and its modifications for

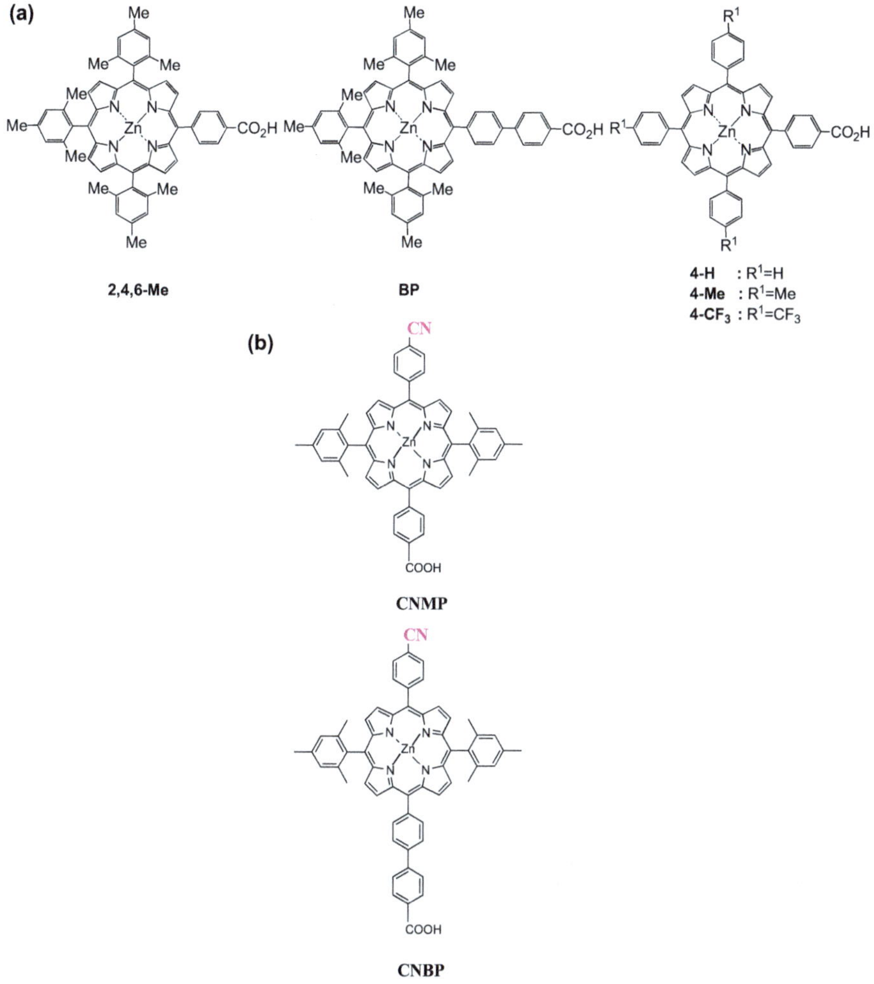

(a)

2,4,6-Me

BP

4-H   : R^1=H
4-Me   : R^1=Me
4-CF$_3$: R^1=CF$_3$

(b)

CNMP

CNBP

Figure 5.8   (a) Molecular structure of zinc porphyrins studied by Imahori *et al.*[23]
(b) CN-labelled Zn-porphyrins studied by Kathiravan *et al.*[34]

interfacial electron transfer)[32,33] it is expected that the electron transfer rate should have a strong (exponential) distance dependence. If electron transfer between the porphyrin core and the semiconductor occurs *via* the connecting spacer, as often envisaged, making this spacer longer should slow down the transfer. With the help of the sensitizers 2,4,6-Me and BP we can test this expectation; introduction of the extra phenyl moiety in the biphenyl spacer of TiO$_2$/BP relative to TiO$_2$/2,4,6-Me would result in approximately 1.5 times longer distance between the porphyrin core and the TiO$_2$ surface and therefore result in considerably smaller electronic coupling and much slower electron injection and recombination.

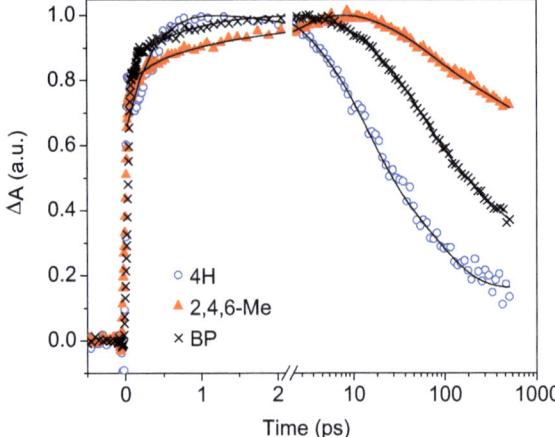

Figure 5.9 Transient absorption kinetics of 2,4,6-Me (full, red triangles), BP (black cross) and 4H (open, blue circles) sensitized TiO$_2$ in *t*-BuOH/ACN for 1 h. Kinetics are normalized to their maxima and black solid lines represent fits of the kinetics.

Figure 5.9 shows the transient absorption kinetics of the two molecules attached to the TiO$_2$ film over a time window of 500 ps. Already a superficial glance at the two kinetic curves shows that the charge recombination process (decay of the curves) does not meet this expectation – the BP sensitizer with the longer connecting spacer has a much higher amplitude of fast recombination components and a lower amplitude of long lived signal (32% *vs.* 64% for 2,4,6-Me) that accounts for long lived charges that can be harvested as photocurrent in a solar cell. Not only the amplitude of the >50 ns decay component is higher for 2,4,6,-Me/TiO$_2$ than for BP/TiO$_2$, but the lifetimes of the faster decays are shorter for BP/TiO$_2$. Also the electron injection (rise of the kinetics) is faster for BP/TiO$_2$. This shows that both electron injection and recombination are overall faster for the sensitizer with the longer connecting spacer. For recombination this appears in two ways – a lower amplitude of long lived (>50 ns) charges and faster recombination on the tens and hundreds of picoseconds timescale for the Zn-porphyrin with the longer connecting spacer (BP).

For the 4H molecule, lacking the methyl groups on the phenyl substituents on the porphyrin core, the effect is even more pronounced – there is almost no slow injection and most of the recombination is complete after 500 ps. Obviously, electron transfer does not occur as could be anticipated *via* the spacer connecting the porphyrin core to the TiO$_2$ surface. Instead, we suggest a picture where the single carboxyl anchoring group allows a flexible binding geometry; for some of the porphyrins, depending on length of the spacer group, bulkiness of the porphyrin core, and so on, a fraction of the molecules are bound at an angle to the semiconductor surface and electron transfer occurs through space rather than through the linker group connecting the porphyrin core to the anchoring COOH group (Figure 5.10).

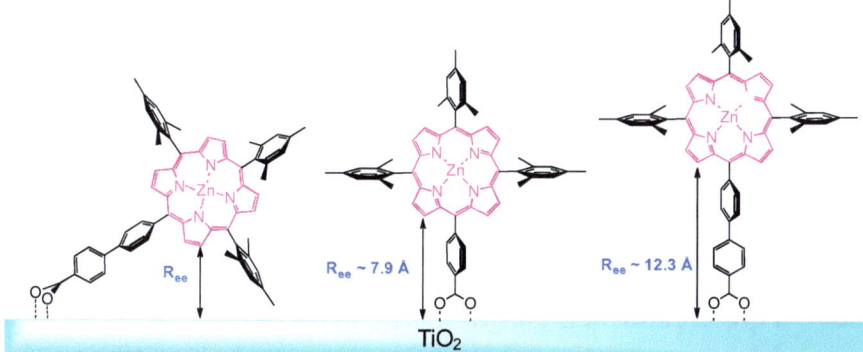

Figure 5.10 Binding model for Zn-porphyrins to TiO$_2$. The edge-to-edge distance (R_{ee}) is decreased upon tilting dye molecule.

When the tilt angle is changed as a result of a change of porphyrin molecule size or shape, the distance between the porphyrin core and semiconductor surface changes, which will lead to a change in the through-space electron transfer rate. Owing to the expected exponential distance dependence of electron transfer, only a modest change of distance (and thus angle) will have a dramatic impact on the transfer rate. As an example let us consider the situation when an upright (perpendicular) orientation of the porphyrin results in a distance of 12 Å between the porphyrin molecule and the TiO$_2$ surface and a recombination time of $\sim$10 ns (similar to the measured slow part of the recombination). A decrease of the porphyrin–TiO$_2$ distance to 8 Å, which corresponds to a tilt angle of 42° (relative to the surface normal), results in a dramatic shortening of the recombination time to $\sim$50 ps, assuming all other factors in the rate equation are constant. This simple picture is just to illustrate the idea, without consideration of the shape and size of the porphyrin core, which definitely will add complexity. Thus, we see that a modest variation in binding geometry can give rise to the very large variation in recombination rates that we observe for the studied Zn-porphyrins.

Electron transfer rates are, of course, not an unambiguous measure of binding geometry; for that an experimental method providing more direct structural information is required. To this end we have used SFG on Zn-porphyrin molecules (the same as in Figure 5.8(a)) labelled with a CN infrared-active chromophore (Figure 5.8(b)).[34] The IR transition dipole moment of the CN-group is along the symmetry axis of the Zn-porphyrin molecules; SFG will therefore give the orientation of the porphyrin relative to the semiconductor surface (the tilt angle). Measurement of the injection/recombination dynamics of the labelled molecules showed that both labelled and non-labelled molecules have identical or very similar electron transfer dynamics.[34] Since we can correlate electron transfer dynamics to solar cell efficiency (see below), this implies that we can also correlate solar cell efficiency to the binding geometry of the Zn-porphyrin sensitizers. The chain of logic for correlating power conversion efficiency (η) of solar cells based on non-labelled Zn-porphyrins to the binding geometry of the (labelled) molecules would then be the following:

$$^{\text{CN-labeled ZnP}}(\text{Tilt angle}) \leftrightarrow {}^{\text{CN-labeled ZnP}}(\text{CT dynamics})$$

$$= {}^{\text{ZnP}}(\text{CT dynamics}) \leftrightarrow {}^{\text{ZnP}}\eta \tag{5.1}$$

implying that the variation in solar cell efficiency that we observe for the different dyes and conditions can be correlated to the tilt angle of the molecules.

5.4 Dye Sensitized Solar Cell Performance in Relation Electron–Cation Recombination and Sensitizer Binding Geometry

Following the chain of logic defined above, we start by comparing electron transfer dynamics and solar cell power conversion efficiency. Thus, the results of the kinetic studies of the half cell electrode were compared to those of photovoltaic measurements on complete solar cells with the same sensitizers. The quantum efficiency (η) of a solar cell is generally defined as the external quantum efficiency (EQE = number of extracted electrons/number of absorbed photons). For a discussion of how key processes contribute to the overall quantum efficiency we can define the efficiency as:

$\eta \sim (efficiency\ of\ light\ harvesting) \times (yield\ of\ electron\ injection)$

$\times (yield\ of\ long\ lived\ conduction\ band\ electrons\ that\ can\ be\ extracted)$

$\times (1 - other\ losses)$

$$\tag{5.2}$$

Light harvesting is related to the surface coverage of the dye and how well it absorbs the wavelengths of the solar spectrum and it varies substantially between the different dyes and various experimental conditions. Variation in η as a result of variation in surface coverage was taken into account by normalizing cell conversion for surface coverage. The absorption spectra are very similar for all studied porphyrin dyes; variation in spectral coverage and its contribution to light harvesting efficiency can therefore be ignored. The rate of electron injection varies somewhat for the molecules/conditions studied, but it is still much faster than the intrinsic excited state lifetime of the sensitizer (~ 2 ns).[23] Therefore, in the porphyrin/TiO_2 combinations studied injection will not significantly influence the efficiency.

Recombination directly controls the number of long lived electrons in the conduction band contributing to the photocurrent. If there is fast electron–cation recombination, faster than the extraction of electrons or re-reduction of oxidized dye by the redox couple, this will decrease the quantum efficiency and power conversion of the solar cell. Thus, the number of long lived conduction band electrons, here monitored by the amplitude of long lived oxidized dye signal ($A_{>50\,ns}$) is expected to be correlated with the solar cell quantum efficiency and power conversion. From the relation above we see that taking into account variations in light harvesting (*i.e.* surface coverage) recombination should be the major determining factor for solar cell efficiency, if other losses are of minor importance. This is verified in Figure 5.11(a), for the Zn-porphyrin/TiO$_2$ electrodes studied and corresponding solar cells, by plotting solar cell power conversion efficiency normalized to surface coverage (η_{rel}) *versus* the amplitude of long lived (> 50 ns) charge recombination ($A_{>50\,ns}$).[23] Results for two sensitization times, 1 h and 12 h, are plotted separately and we can see that for 1 h sensitization, there is a linear relationship between η_{rel} and

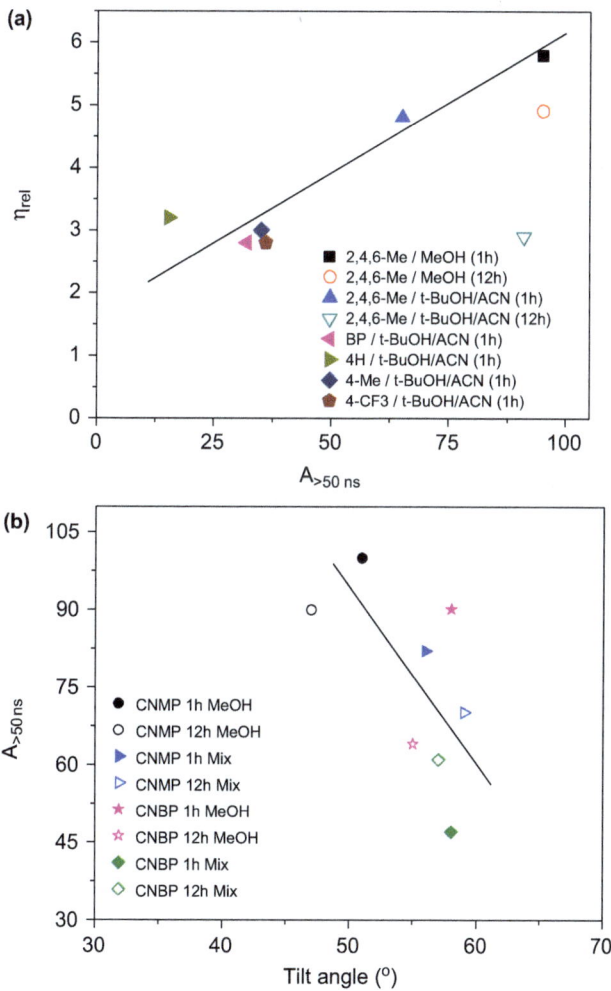

Figure 5.11 (a) Solar cell power conversion, for non-labeled porphyrins, normalized for surface coverage (η_{rel}) as a function of long lived recombination signal ($A_{>50\,ns}$). (b) $A_{>50\,ns}$ as a function of tilt angle for CN-labeled molecules.[34]

$A_{>50\,ns}$ (higher η_{rel} corresponds to higher $A_{>50\,ns}$). This result emphasizes the importance of considering electron–cation recombination as a key process for controlling DSC efficiency and it suggests a shift in focus from the injection process to the recombination.

The finding is perhaps what one would intuitively expect, but it also shows that for the series of Zn-porphyrins studied here, no other factors than those considered in the expression for solar cell efficiency have a significant impact on the variation of solar cell performance. The two η_{rel} vs. $A_{>50\,ns}$ points for 12 h sensitization do not follow the same trend as the 1 h plot, suggesting that for a particular value of the observed amplitude of slow recombination, the 12 h solar cell is less efficient than the cell that has an active electrode sensitized for 1 h. It was suggested that the reason for this is sensitizer aggregation caused by the long sensitization, which decreases the injection efficiency and therefore the cell efficiency.[23]

Now, we can take next step in the chain of logic above and relate measured electron–cation recombination to sensitizer tilt angle as measured by SFG on the CN-labelled molecules, Figure 5.11(b). Two observations can be made: first, for all molecules there is a considerable tilt angle between the symmetry axis of the molecule and the surface normal, thus verifying the initial proposal based on kinetic measurements only[23] and second, there is a clear correlation between tilt angle and amplitude of long lived dye–cation recombination signal, $A_{>50\,ns}$; a larger tilt angle leads to a lower amplitude of slow recombination. In other words, when the sensitizer molecule is more tilted and therefore closer to the semiconductor surface there is less very slow recombination and therefore fewer long lived conduction band electrons that can contribute to the photocurrent and the power conversion efficiency of the solar cell. For this class of porphyrin sensitizers the correlation obtained between binding geometry and solar cell efficiency suggests that one direction of future optimization work could be to find dye–semiconductor binding strategies that give a more upright and less flexible attachment of the sensitizer.

The results for the Zn-porphyrin/TiO_2 electrodes discussed above show that, at least for this class of molecules, we can use kinetic measurements of the electron transfer processes to provide information on the binding geometry of the sensitizer. This is an important result since ultrafast measurements on dye-sensitized nanostructured semiconductor films have been developed to a high sensitivity and a high degree of precision. A change in dye–semiconductor distance, by a change of tilt angle, should of course influence both the electron injection and recombination processes. If this is correct there should be a correlation between the rate of electron injection and the rate of recombination. Figure 5.12 shows that there is in fact such a correlation – the higher amplitude of instantaneous rise of the TA signal ($\ll 1$ ps electron injection) the lower amplitude of slow recombination. The very fast injection corresponds to a fraction of molecules making close contact with the TiO_2 surface and thus also very fast recombination (~ 100 ps timescale recombination). From this we can see that both the injection and recombination processes can be used to report on sensitizer binding geometry; trends in the rates of both processes when

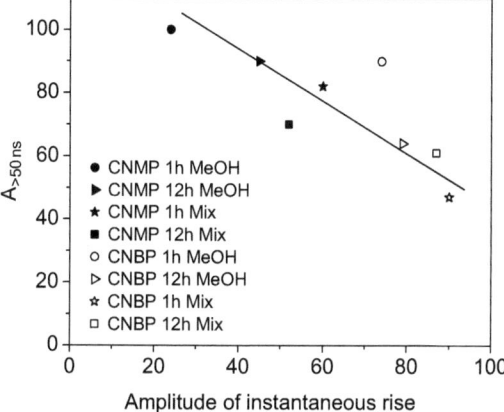

Figure 5.12 Correlation between instantaneous ($\ll 1$ ps) electron injection and long lived recombination signal ($A_{>50\,ns}$) of CN-labelled Zn-porphyrin molecules.[34] A high amplitude of ultrafast injection gives rise to fast recombination.

sensitizer attachment is modified can be converted into changes of electron transfer distance and therefore binding geometry.

5.5 Electron–Cation Interactions as a Source of Fast Recombination and Slow Charge Transport in Dye Sensitized Nanostructured Semiconductor Films

As described above, optical spectroscopy can provide detailed information about excited states and intermediates participating in the light-to-charge conversion process of DSC materials. As can be easily appreciated, electron transport within and between the semiconductor nanostructures is an important process for the function of a solar cell. Optical techniques unfortunately do not provide much information about charge transport. Better information is obtained from transient far-infrared conductivity spectra and kinetics measured by time-resolved terahertz (THz) spectroscopy (TRTS). This technique allows non-contact characterization of photoconductivity with sub-ps time resolution.[35,36] The amplitude of the photoconductivity is then a direct measure of the population of injected charge carriers. In addition, from the shape of the transient conductivity spectrum it is possible to infer the mechanisms of the charge transport or to distinguish between the response of free charge carriers and localized excitations.[37] The strong interaction of THz radiation with free charge carriers in semiconductors makes the TRTS an ideal tool for the investigation of charge carrier dynamics in semiconductors and DSC.

The generally accepted picture of DSC is that mobile electrons appear in concert with injection. Our results show that charge injection and formation of mobile charges are not necessarily directly connected and that charge transport in the active solar cell material can be different from that in non-sensitized semiconductors. This is related to strong electrostatic interaction between injected electrons and dye cations at the surface of the semiconductor nanoparticle. This interaction in addition may give rise to fast electron–cation recombination, decreasing the efficiency of a solar cell.

In this work, we take advantage of combining results of visible and THz pump–probe spectroscopy. As shown above, probing in the visible range typically provides information about the electronic state of the dye molecules, that is, the formation of oxidized dye molecules (cations) upon electron injection is monitored (see *e.g.* Figures. 5.2 and 5.7).

Probing in the THz spectral range is sensitive to the transport of mobile charge carriers (injected electrons). We investigated in detail the dynamics in ZnO and TiO$_2$ nanoparticle thin films sensitized by Zn(II)-5-(3,5-dicarboxyphenyl)phenyl-10,15,20-triphenylporphyrin (ZnTPP-Ipa) and RuN3 dyes.[38,39] Measurement of the THz kinetics after photoexcitation of the sensitizer reflects the population of mobile electrons injected into the semiconductor. For the Zn-porphyrin molecule, ZnTPP-Ipa, attached to a nanocrystalline ZnO film (Figure 5.13(a)), the rise in the population of injected mobile electrons is slow and occurs on the tens to hundreds ps timescale. After reaching its maximum value it does not decay for at least 10 ns which implies that recombination of injected electrons is slow. This behaviour is in sharp contrast to the transient absorption dynamics of oxidized dye (cation) (Figure 5.13(b)); the formation is essentially completed in 5 ps and the decay exactly matches the slow THz rise of mobile electrons in Figure 5.13(a).

This behaviour can be understood by the kinetic scheme depicted in Figure 5.14, involving an intermediate state between the excited state of the sensitizer and cation with mobile electrons in the conduction band of ZnO. Thus, photoexcitation of the sensitizer (1) leads to an electron–cation (EC) complex within ~ 5 ps (3) in which the electron is strongly bound to the cation and therefore does not contribute to the THz signal. The bound EC state can either recombine (5) or dissociate into a mobile electron in ZnO and cation (4).

Very similar results were obtained for RuN3/ZnO (Figure 5.13(a)), showing that this behaviour is a property of ZnO, rather than the sensitizer. Spectral features in the visible and near-IR observed for RuN3/ZnO in transient absorption measurements by Katoh *et al.*[40] were interpreted as an exciplex preceding electron injection. It appears very likely that the EC complex we observe in ZnTPP-Ipa/ZnO and RuN3/ZnO is a similar species. Since the observed intermediate state for ZnTPP-Ipa/ZnO has a clear cation signature from the transient absorption spectra[41] we have chosen the notation electron–cation (EC) complex. These results point to the key role of the EC complex in controlling charge recombination and therefore possibly solar cell efficiency.

The injection/recombination dynamics observed for sensitized nanocrystalline TiO$_2$ films is fundamentally different from that observed for sensitized ZnO. The THz and TA kinetics of both dyes (Figure 5.13(a) and (c); see also Figures. 5.2 and 5.3 above) show that both cations and mobile conduction band electrons appear on the same ultrafast timescale (100 fs – 10 ps). This suggests

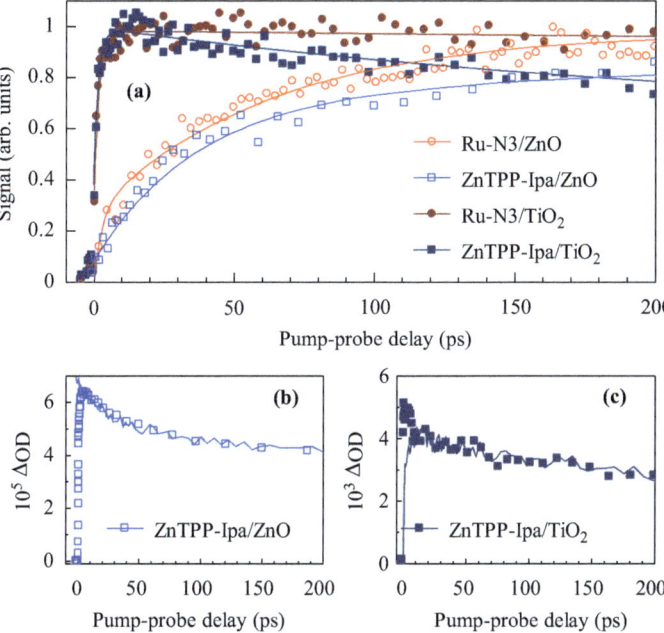

Figure 5.13 (a) Evolution of transient THz conductivity (normalized to unity). The lines serve only to guide the eye. (b) Transient absorption of ZnTPP-Ipa/ZnO probed at 655 nm (symbols). This wavelength was selected owing to the largest difference between the transient absorption of the initially excited state D* and of the excitation products. The line represents a vertically flipped and shifted transient THz conductivity of ZnTPP-Ipa/ZnO from (a). (c) Transient absorption of ZnTPP-Ipa/TiO$_2$ probed at 655 nm (symbols). The line represents a shifted and scaled transient THz conductivity of ZnTPP-Ipa/TiO$_2$ from (a) [note that the line is not flipped, unlike in panel (b)].

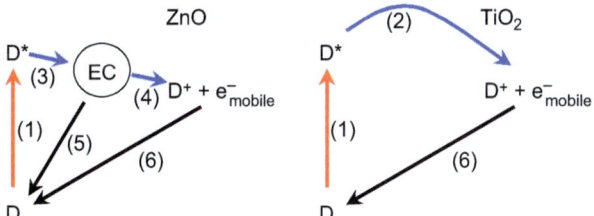

Figure 5.14 Scheme of the processes in dye-sensitized ZnO and TiO$_2$. (1) Dye excitation, (2) direct electron injection, (3) formation of an electron–cation (EC) complex, (4) dissociation of EC complex, (5) recombination of EC complex and (6) charge recombination. D = dye molecule.

that charge injection into TiO$_2$ is a direct process ((2) in Figure 5.14). Slow electron–cation recombination (6) occurring on a hundreds-ps timescale and longer is observed in ZnTPP-Ipa/TiO$_2$ as the decays in the THz and TA signals

(Figure 5.13(a) and (c)). The THz conductivity for RuN3/TiO$_2$ does not start to decay even on the nanosecond scale. Solar cells based on RuN3/TiO$_2$ can have a conversion efficiency of >10%. The somewhat faster electron–cation recombination in ZnTPP-Ipa/TiO$_2$ may contribute to a lower solar cell efficiency of 4–5% for similar porphyrin sensitizers.[23]

Transient far-infrared conductivity spectra (Figure 5.15) provide information on the local transport of carriers photogenerated in the semiconductor (*i.e.* generated by direct UV-excitation of the semiconductor), or injected into the semiconductor following excitation of a sensitizer. In order to describe the nanostructured semiconductor conductivity spectra we have used a microscopic picture of band-like transport in non-degenerate semiconductors.[38] This picture is in fact a Brownian motion of charge carriers, where most of the time the carriers move freely and only occasionally do they scatter, for example due to interaction with phonons. Inside semiconductor nanoparticles with sizes much larger than the charge carrier Bohr radius, it is reasonable to assume that the motion is the same as in the bulk material described by the mean scattering time τ_s. In the model, each scattering event results in a randomization of the

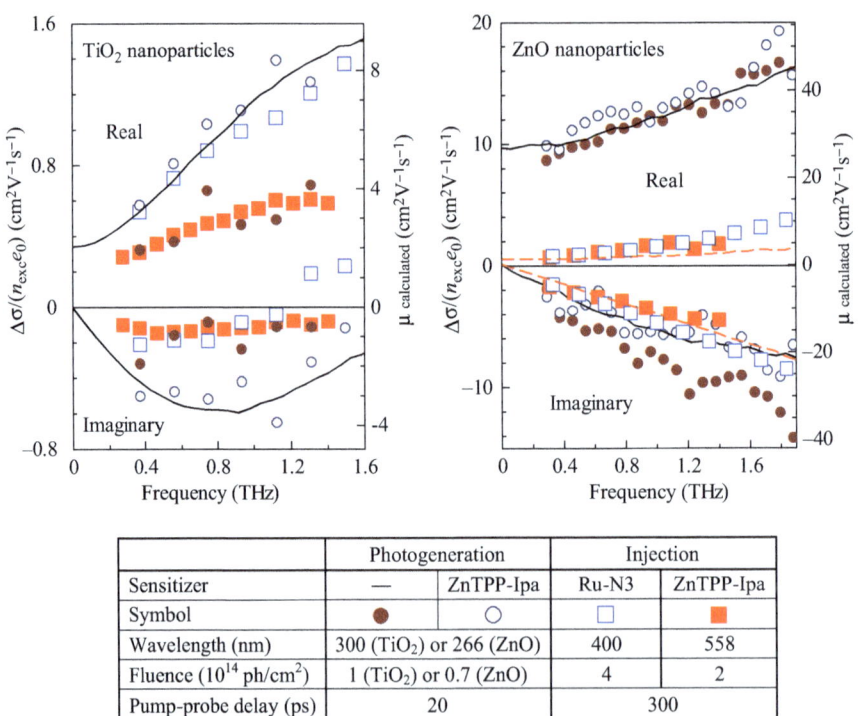

	Photogeneration		Injection	
Sensitizer	—	ZnTPP-Ipa	Ru-N3	ZnTPP-Ipa
Symbol	●	○	□	■
Wavelength (nm)	300 (TiO$_2$) or 266 (ZnO)		400	558
Fluence (10^{14} ph/cm^2)	1 (TiO$_2$) or 0.7 (ZnO)		4	2
Pump-probe delay (ps)	20		300	

Figure 5.15   Transient far-infrared conductivity spectra in bare and dye-sensitized ZnO and TiO$_2$ nanoparticles. Symbols (left axis): measured data; lines (right axis): calculated mobility of directly photogenerated electrons (solid) and injected electrons (dashed). These lines overlap in the left graph.

velocity vector $\boldsymbol{v}$ according to the Maxwell–Boltzmann distribution, which accounts for the carrier temperature. Unlike in bulk, carriers in semiconductor nanostructures can interact with the surface which can be described in terms of the following three probabilities ($p_t + p_s + p_r = 1$):[38]

- With a probability p_t, the carrier continues its motion without interacting with the nanoparticle boundary.
- The nanoparticle boundary isotropically scatters the carrier with probability p_s which means that there are equal probabilities that the carrier enters another nanoparticle or that the carrier remains in the original one.
- The carrier is reflected (scattered) back to the original nanoparticle with a probability p_r. Such a process arises owing to energy barriers between the nanoparticles and it is the only process leading to the carrier localization.

The probability that the carrier remains in the original nanoparticle (it is scattered back) is thus $p_b = p_r + p_s/2$ while the probability that it enters another one is $p_f = p_t + p_s/2$ and this probability can be understood as a permeability of the nanoparticle boundary.

A Monte Carlo method was used to simulate trajectories representing the stochastic motion of charge carriers. The mobility of carriers was then calculated using the Kubo formula:[42]

$$\mu_{ij}(\omega) = \frac{e}{k_B T} \int_0^\infty \langle v_i(0) v_j(t) \rangle \exp(i\omega t) dt \qquad (5.3)$$

where the averaging takes place over a canonical ensemble of carriers represented by a temperature T. It is possible to perform the simulation in almost any geometry. For the sake of simplicity, we mostly consider spherical or cubic nanoparticles. Such systems are macroscopically isotropic, therefore the mobility tensor is diagonal and the diagonal elements of the mobility tensor μ_{ii} are equal to each other. It turns out that the shape of the spectrum is determined only by the parameters p_t, p_r, p_s and $\alpha = d/l_{free}$ (d is the nanoparticle size and $l_{free} = v_{therm}/\tau_s$ is the carrier mean free path, where $v_{therm} = \sqrt{3k_B T/m}$ is the thermal velocity). The remaining parameters (τ_s, d, T) only determine the pertinent length scales, time, frequency and mobility amplitude.

The observed spectra of photogenerated carriers in bare semiconductors (Figure 5.15) can be explained by a band-like transport of electrons localized inside the semiconductor nanoparticles.[38] The assumptions of our model of transient conductivity closely resemble those of the flight phase of the "random flight model".[43] The Monte Carlo simulations revealed that the probability p_r that an electron is reflected by the surface is 80% for ZnO and 90% for TiO$_2$.[38] This means that long-range electron transport may exist; however, it is considerably reduced owing to the limited connectivity between the nanoparticles. The differences between the spectra of ZnO and TiO$_2$ are mainly related to the large difference in effective electron masses. The shape of the conductivity spectrum (*i.e.* the ratio between the real and imaginary part) is the same for carriers photogenerated in bare and sensitized nanoparticles. This means that attachment of neutral dye molecules does not influence the electron transport properties. Note that the magnitudes of THz conductivities both in bare and sensitized samples should be taken with care, since they are scaled by charge yields, which may be affected by inhomogeneities of the samples.

Apart from the amplitudes, the conductivity spectra of electrons *injected* into TiO_2 nanoparticles from either RuN3 or ZnTPP-Ipa are the same as the spectrum of *photogenerated* electrons in TiO_2 (Figure 5.15). We thus conclude that dye molecules either in their neutral, or oxidized state, do not influence electron transport in TiO_2. The small variance in the zero crossing of the imaginary part of conductivity can be attributed to slightly different nanoparticle diameters and/or probabilities p_r in individual samples. In contrast, the conductivity spectrum of electrons injected into ZnO differs from the spectrum of photogenerated electrons – the real part of conductivity is considerably lower. Our hypothesis is that cations restrict the motion of injected electrons even after they escape from the EC complex. The interaction between the cations and the injected electrons is necessarily electrostatic. This interaction is screened in TiO_2 nanoparticles owing to their high permittivity (80 compared to 8 in ZnO), so that the oxidized dye molecules do not influence electron transport in TiO_2, in agreement with our observations.

In order to confirm our hypothesis and to understand further the influence of oxidized dye molecules on the electron transport in ZnO, we investigated the response of electrons localized within the nanoparticles and moving in a model potential.[38] For the sake of simplicity, we consider that the semiconductor nanoparticles are nanocubes surrounded by a border; the border properties are the same as those of the bare nanoparticles (namely the electron reflection probability p_r). The attractive electrostatic force is approximated using a harmonic potential and its minimum is located at the nanoparticle surface (Figure 5.16). The advantage of this configuration is that the low permeability (high p_r) of the nanoparticle surface mimics the repulsive force which prevents recombination in the real system. Under these conditions, the calculations

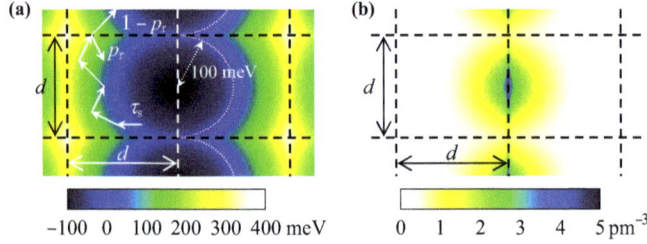

Figure 5.16   (a) Geometry employed in the Monte Carlo simulations of electrons interacting with dye cations. The cation is located in the centre of the picture. The dashed lines indicate nanoparticle boundaries. (b) Calculated electron density distribution within the configuration sketched in (a).

match the experimental data if a harmonic potential with an eigenfrequency of 10 THz is used. This model is, of course, very approximate, nevertheless it shows that the height of the potential barrier for the electron transport is about 100 meV. The degree of electron localization caused by the potential is illustrated by the electron distribution function plot in Figure 5.16.

Comparison of the THz and optical transient absorption dynamics obtained for sensitized and bare nanostructured ZnO and TiO_2 films results in the following picture. In ZnO, electron injection leads to the formation of a bound electron–cation complex, which breaks up into an electron and cation or recombines on the timescale of tens to hundreds of picoseconds. The mobility of electrons that escape from the complex is strongly impaired by attractive electrostatic interaction with the cation at the semiconductor surface. In TiO_2, injection results in instantaneous formation of mobile electrons and the charge recombination is very slow (for the dyes studied; of course recombination can be fast as illustrated above for unfavuorable dye–semiconductor geometries). We attribute the different charge transport and recombination in the two semiconductors to the screening of electrostatic interaction in TiO_2 owing to its high dielectric permittivity. Direct photoexcitation of the semiconductor results in instantaneous formation of mobile electrons for both ZnO and TiO_2. Electrons also jump between particles with low probability. ZnO is being explored as a promising material for many opto-electronics applications, including dye sensitized solar cells, owing to its ability to self assemble into various nanostructures and its high intrinsic electron mobility. The EC complex formation in dye sensitized materials, with its consequent fast charge recombination and drastically reduced mobility could limit the use of ZnO as a solar cell material. On the other hand, these drawbacks could possibly be remedied by improved technology in preparation, for example, by coating the ZnO nanoparticles with a thin layer of a high dielectric permittivity material.

5.6 Conclusions

The photovoltaic processes of nanostructured materials for solar energy conversion involve ultrafast key processes as initial steps in light to photocurrent conversion. Time resolved ultrafast spectroscopy is therefore a powerful tool for monitoring these processes. Excited states, charged species (ions) and electrons all have characteristic absorbance or emission spectra in the UV–vis–NIR spectral domain, facilitating their identification and mapping of dynamics. Far infrared (THz spectral domain) time resolved spectroscopy in addition provides information about the nature of excitations and charge carrier mobility. By combining time resolved spectroscopies ranging from the UV to THz, a comprehensive picture of the light induced excited state and carrier dynamics can be obtained.

Using these techniques for several different dye sensitized nanostructured semiconductor materials we have characterized injection dynamics in detail and shown that it can be very different for different dyes and strongly depends on dye–semiconductor binding geometry and the amount of excess excitation energy above the conduction band edge of the semiconductor. By combining time resolved methods and surface sensitive sum frequency generation measurements we could correlate electron transfer dynamics and solar cell performance to dye binding geometry and thus provide the basis for design of sensitizer–semiconductor interfaces with predictable electron transfer properties. A combined time resolved optical and THz approach gave detailed insight into the mechanism of formation of free mobile electrons and how electron–cation interaction may cause fast charge recombination and decreased electron mobility.

The results discussed in this chapter, as in most of other similar studies reported in the literature, are for thin sensitized films in contact with a solvent. Although results on carrier dynamics for devices are beginning to appear,[44,45] the picture is much less clear than for thin films. Future work should therefore increasingly focus on characterization of carrier dynamics, from femtoseconds to micro- and milliseconds, of functioning dye sensitized solar cells.

Acknowledgements

This work was supported by the Swedish Energy Administration, The Swedish Research Council, The Knut and Alice Wallenberg Foundation, and the Swedish Institute. We thank colleagues and collaborators at the Department of Chemical Physics, Lund University for their contributions and for many fruitful discussions.

References

1. B. Oregan and M. Gratzel, *Nature*, 1991, **353**, 737–740.
2. B. Burfeindt, C. Zimmermann, S. Ramakrishna, T. Hannappel, B. Meissner, W. Storck and F. Willig, *Zeitschrift Fur Physikalische Chemie-International Journal of Research in Physical Chemistry & Chemical Physics*, 1999, **212**, 67–75.
3. R. Ghanem, Y. H. Xu, J. Pan, T. Hoffmann, J. Andersson, T. Polivka, T. Pascher, S. Styring, L. C. Sun and V. Sundstrom, *Inorganic Chemistry*, 2002, **41**, 6258–6266.
4. S. Ardo and G. J. Meyer, *Chemical Society Reviews*, 2009, **38**, 115–164.
5. A. Hagfeldt and M. Gratzel, *Accounts of Chemical Research*, 2000, **33**, 269–277.
6. M. Gratzel, *Accounts of Chemical Research*, 2009, **42**, 1788–1798.
7. L. M. Peter, *Journal of Physical Chemistry Letters*, 2011, **2**, 1861–1867.
8. M. A. Green, K. Emery, Y. Hishikawa, W. Warta and E. D. Dunlop, *Progress in Photovoltaics*, 2011, **19**, 565–572.
9. J. Kallioinen, G. Benko, V. Sundstrom, J. E. I. Korppi-Tommola and A. P. Yartsev, *Journal of Physical Chemistry B*, 2002, **106**, 4396–4404.
10. G. Benko, P. Myllyperkio, J. Pan, A. P. Yartsev and V. Sundstrom, *Journal of the American Chemical Society*, 2003, **125**, 1118–1119.

11. G. Benko, J. Kallioinen, J. E. I. Korppi-Tommola, A. P. Yartsev and V. Sundstrom, *Journal of the American Chemical Society*, 2002, **124**, 489–493.
12. P. Myllyperkio, G. Benko, J. Korppi-Tommola, A. P. Yartsev and V. Sundstrom, *Physical Chemistry Chemical Physics*, 2008, **10**, 996–1002.
13. J. B. Asbury, R. J. Ellingson, H. N. Ghosh, S. Ferrere, A. J. Nozik and T. Q. Lian, *Journal of Physical Chemistry B*, 1999, **103**, 3110–3119.
14. N. H. Damrauer, G. Cerullo, A. Yeh, T. R. Boussie, C. V. Shank and J. K. McCusker, *Science*, 1997, **275**, 54–57.
15. A. T. Yeh, C. V. Shank and J. K. McCusker, *Science*, 2000, **289**, 935–938.
16. G. Benko, P. Myllyperkio, J. Pan, A. P. Yartsev and V. Sundstrom, *Journal of the American Chemical Society*, 2003, **125**, 1118–1119.
17. H. Rensmo, K. Westermark, S. Sodergren, O. Kohle, P. Persson, S. Lunell and H. Siegbahn, *Journal of Chemical Physics*, 1999, **111**, 2744–2750.
18. P. Persson and S. Lunell, *Solar Energy Materials and Solar Cells*, 2000, **63**, 139–148.
19. M. Haukka and P. Hirva, *Surface Science*, 2002, **511**, 373–378.
20. V. Shklover, Y. E. Ovchinnikov, L. S. Braginsky, S. M. Zakeeruddin and M. Gratzel, *Chemistry of Materials*, 1998, **10**, 2533–2541.
21. A. Fillinger, D. Soltz and B. A. Parkinson, *Journal of the Electrochemical Society*, 2002, **149**, A1146–A1156.
22. G. Benko, J. Kallioinen, P. Myllyperkio, F. Trif, J. E. I. Korppi-Tommola, A. P. Yartsev and V. Sundstrom, *Journal of Physical Chemistry B*, 2004, **108**, 2862–2867.
23. H. Imahori, S. Kang, H. Hayashi, M. Haruta, H. Kurata, S. Isoda, S. E. Canton, Y. Infahsaeng, A. Kathiravan, T. Pascher, P. Chabera, A. P. Yartsev and V. Sundstrom, *Journal of Physical Chemistry A*, 2011, **115**, 3679–3690.
24. S. A. Haque, S. Handa, K. Peter, E. Palomares, M. Thelakkat and J. R. Durrant, *Angewandte Chemie-International Edition*, 2005, **44**, 5740–5744.
25. J. E. Moser and M. Gratzel, *Chemical Physics*, 1993, **176**, 493–500.
26. B. Oregan, J. Moser, M. Anderson and M. Gratzel, *Journal of Physical Chemistry*, 1990, **94**, 8720–8726.
27. D. J. Fitzmaurice and H. Frei, *Langmuir*, 1991, **7**, 1129–1137.
28. H. Imahori, T. Umeyama and S. Ito, *Accounts of Chemical Research*, 2009, **42**, 1809–1818.
29. R. A. Marcus and N. Sutin, *Biochimica Et Biophysica Acta*, 1985, **811**, 265–322.
30. R. A. Marcus, *Journal of Chemical Physics*, 1956, **24**, 966–978.
31. R. A. Marcus, *Annual Review of Physical Chemistry*, 1964 **15**, 155–196.
32. H. Gerischer, *Photochemistry and Photobiology*, 1972 **16**, 243–260.
33. H. Gerischer, *Surface Science*, 1969, **18**, 97–122.
34. S. Ye, A. Kathiravan, H. Hayashi, Y. Tong, Y. Infahsaeng, P. Chabera, T. Pascher, A. P. Yartsev, S. Isoda, H. Imahori, V. Sundström. *Journal of Physical Chemistry C*, 2013, **117**, 6066–6080.

35. M. C. Beard, G. M. Turner and C. A. Schmuttenmaer, *Journal of Physical Chemistry B*, 2002, **106**, 7146–7159.
36. G. M. Turner, M. C. Beard and C. A. Schmuttenmaer, *Abstracts of Papers of the American Chemical Society*, 2002, **224**, 271-COLL.
37. H. Nemec, H. K. Nienhuys, F. Zhang, O. Inganas, A. Yartsev and V. Sundstrom, *Journal of Physical Chemistry C*, 2008, **112**, 6558–6563.
38. H. Nemec, P. Kuzel and V. Sundstrom, *Journal of Photochemistry and Photobiology A-Chemistry*, 2010, **215**, 123–139.
39. H. Nemec, J. Rochford, O. Taratula, E. Galoppini, P. Kuzel, T. Polivka, A. Yartsev and V. Sundstrom, *Physical Review Letters*, 2010, **104**.
40. R. Katoh, A. Furube, A. V. Barzykin, H. Arakawa and M. Tachiya, *Coordination Chemistry Reviews*, 2004, **248**, 1195–1213.
41. H. Nemec, J. Rochford, O. Taratula, E. Galoppini, P. Kuzel, A. Yartsev and V. Sundstrom, *Electron–Cation Electrostatic Interaction Controls Electron Mobility in Dye-Sensitized ZnO Nanocrystals*, 2009, unpublished work.
42. R. Kubo, *J.Phys.Soc.Jpn.*, 1957, **12**, 570–586.
43. R. Katoh, A. Furube, A. V. Barzykin, H. Arakawa and M. Tachiya, *Coordination Chemistry Reviews*, 2004, **248**, 1195–1213.
44. P. R. F. Barnes, A. Y. Anderson, J. R. Durrant and B. C. O'Regan, *Physical Chemistry Chemical Physics*, 2011, **13**, 5798–5816.
45. A. Listorti, B. O'Regan and J. R. Durrant, *Chemistry of Materials*, 2011, **23**, 3381–3399.

CHAPTER 6

Semiconductor Nanocrystals Studied by Two-Dimensional Photon Echo Spectroscopy

CATHY Y. WONG,[a] SHUN S. LO[b] AND
GREGORY D. SCHOLES*[c]

[a] Department of Chemistry, University of California, Berkeley, CA 94720,
USA; [b] Department of Chemistry and Biochemistry, University of Notre
Dame, Notre Dame, Indiana 46556-5670, USA; [c] Department of Chemistry,
Institute for Optical Sciences and Center for Quantum Information and
Quantum Control, University of Toronto, Toronto, Ontario, M5S 3H6,
Canada
*Email: gscholes@chem.utoronto.ca

6.1 Semiconductor Nanocrystals and Quantum Dot-Based Solar Cells

Semiconductor nanocrystals (NC) are structures whose size ranges between a
few to tens of nanometers. Because of their low dimensionality, quantum
confinement determines their optical and electronic properties,[1] allowing them
to be tuned by the size and shape of the nanoparticle. These nanostructures also
possess a high extinction coefficient and a wide spectral range for absorption.
All of these attributes make these nanostructures excellent materials for use in
the design and construction of solar cells.

RSC Energy and Environment Series No. 8
Solar Energy Conversion: Dynamics of Interfacial Electron and Excitation Transfer
Edited by Piotr Piotrowiak
© The Royal Society of Chemistry 2013
Published by the Royal Society of Chemistry, www.rsc.org

Semiconductor NCs have been incorporated into solar cells in different configurations, for example: (a) photoelectrodes composed of quantum dot arrays,[2] (b) metal-semiconductor photovoltaic cells,[2,3] (c) NC-polymer solar cells[3] and (d) quantum dot sensitized solar cells.[3–5] This field has been the focus of intense research in recent years because of the possibility that quantum dot-based solar cells can overcome the Shockley–Queisser photoconversion limit.[3–5] This possibility relies on two feasible processes: hot carrier extraction and multiple exciton generation (MEG).

The absorption of a photon with energies well above the band gap of the semiconductor produces free carriers or excitons with excess energy. These "hot carriers" or "hot excitons" usually relax to the lowest exciton state within hundreds of femtoseconds to picoseconds. The first strategy, hot carrier extraction, attempts to harness these hot carriers or excitons before they are able to relax to the lowest energy state. This scenario has been recently realized using PbSe NCs deposited on a single crystalline film of TiO_2[6] and observed in CdSe/ZnSe core/shell NCs.[7] The second strategy, MEG, involves the generation of multiple electron–hole pairs using this excess energy, which has been shown to occur when the energy of the incident photon is at least three times the band gap of the material, $E_{photon} \geq 3E_{gap}$. Initial studies argue that this process is very efficient in quantum confined system[8–10] while later reports have cast doubts over these claims.[11,12] Although there is still an ongoing debate about the efficiency of MEG in semiconductor NCs, [13,14] the fact that it is enhanced in NCs compared to bulk semiconductor is promising for photovoltaic applications.[2,14] Signatures of MEG have been reported in devices using PbS NCs[15] and polymer/PbSe,[16] while simultaneous extraction of multiple charges has been observed in CdSe NCs with adsorbed methylene blue molecules.[17] To improve upon these findings further, a better insight into multiexciton states and their dynamics is important.

In the past decade, developments in synthetic methods have opened up new and exciting possibilities for the fabrication of nanostructures that can be tailored to photovoltaic applications. Heterostructured NCs can be produced to improve the properties of existing NCs or others that exhibit entirely new properties. Some of the basic strategies are depicted in Figure 6.1. In Type-I, a NC is overcoated with a semiconductor material that has a larger band gap than the material in the core. These heterostructured NCs provide better passivation of the NC's surface, eliminating surface trap sites and improving the photoluminescence properties of the NC.[18–21] Furthermore, the shell also provides protection against photoinduced dissolution of the NC.[22] In Type-II heterostructured NCs, the band gap of the constituent materials are aligned in a staggered configuration.[23–29] With this configuration, it is energetically favourable for charge separation to occur after photoexcitation, making these materials very attractive for photovoltaic applications. Their recent incorporation into solar cell devices has been reported by McDaniel *et al.*[30] and Ning *et al.*[31]

Charge separation in Type-II heterostructures have been found to be ultrafast, in the hundreds of femtoseconds,[32–35] and the process may occur in the

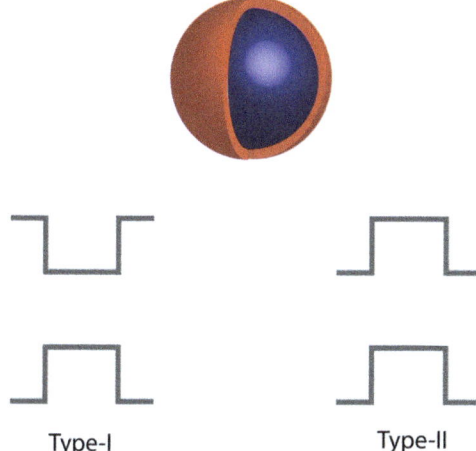

Type-I   Type-II

Figure 6.1   Band-alignment in heterostructured NCs. In Type-I heterostructures, a
larger band gap semiconductor material overcoats a core with a smaller
band gap. Meanwhile, the band alignment of the constituent materials in a
Type-II heterostructure is in a staggered configuration.

Marcus inverted region.[33,36] The spatial separation of charges decreases the
overlap between the electron and hole wavefunctions, which in turn modifies
the nature of the interactions between excitons. Interactions between excitons
can be studied by probing the spectral signatures of biexcitons. Biexcitons are
formed by sequential absorption of two photons. In core-only and Type-I
heterostructures, the exciton–exciton interactions are attractive and the energy
of the biexciton is smaller than twice the band gap, E_{gap}, by a binding energy,
$\Delta_{XX} \sim 15$–40 meV depending on NC size; meanwhile the corresponding
exciton–exciton interaction in Type-II heterostructures have been found to be
repulsive.[37,38] Furthermore, lifetimes of biexcitons in core-only and Type-I NCs
have been found to be on the picosecond timescale, with Auger recombination
being its main relaxation pathway; in the Type-II NCs, the reduced electron–
hole wavefunction overlap lengthens Auger decay times.[37–39]

The techniques that have been used to explore the different exciton and
multiexciton processes occurring in NCs includes femtosecond pump–probe,
fluorescence upconversion and time-correlated single photon counting, among
others. In the next sections, we describe the use of a different technique, two-
dimensional photon echo (2DPE), in order to study these processes and provide
case studies in NCs. In Section 6.2, a background explanation of this technique is
given, the advantages of this technique compared to those previously employed
are discussed, as is the type of information that can be obtained. In Section 6.3,
we provide experimental details of this technique and a walk-through of the data
processing required in order to obtain a two-dimensional spectra. Assignment of
the features observed in this spectra requires a simulation of the 2DPE signal;
this is discussed in detail in Section 6.4 for CdSe NCs. Finally, we present some
preliminary results on type-II CdTe/CdSe NCs in Section 6.5.

6.2 Two-Dimensional Photon Echo Spectroscopy: Background

Two-dimensional photon echo (2DPE) spectroscopy[40–44] is a fairly new technique which has the advantage of removing inhomogeneous broadening in the anti-diagonal direction, spreading out a crowded spectrum over two dimensions. This is of great benefit to the study of complex systems which produce exceedingly congested spectra, which can prove prohibitive for straightforward analysis. While 2D infrared spectroscopy has been broadly applied in the past decade to the study of such phenomena as solvent dynamics, vibrational coherences and protein folding, 2D spectroscopy was not brought into the visible wavelength region until the last decade since shorter wavelengths require more phase stability and delay stage accuracy. Diffractive optics-based schemes with passive phase stabilization were key to the development of experimental setups for 2DPE with adequate phase stability.

While 2DPE can no longer be considered a nascent technique, it still has yet to be applied to many problems in chemistry. It was originally conceived as a tool to study solvation dynamics, but it has proven to be a powerful spectroscopy, capable of providing insight into a variety of problems. This section will first explain the fundamentals of the 2DPE technique, then will present a brief discussion of a few of the phenomena it can be used to study.

6.2.1 Fundamentals of 2DPE Spectroscopy

2DPE is a four-wave mixing technique using a boxcars geometry, as shown in Figure 6.2. Also shown in this figure is the pulse sequence. Two "pump" pulses excite the sample and a "probe" pulse interrogates the sample, generating a signal field. The time delay between the arrival of the two pump pulses is termed the coherence time, τ. When k_1 arrives before k_2 and thus $\tau > 0$, "rephasing" pathways are observed. If k_1 arrives after k_2, $\tau < 0$ and "non-rephasing" pathways result. As will be discussed later, 2DPE spectra can be split into rephasing and non-rephasing portions for separate analysis.

The first pump pulse creates a coherence between the ground state and the excited state, either $|g\rangle \langle e|$ or $|e\rangle \langle g|$, depending on whether k_1 or k_2, respectively, arrives first. The second pump pulse creates either a population in the excited state, $|e\rangle \langle e|$, a population in the ground state, $|g\rangle \langle g|$, or a coherence, if the two pump pulses photoexcite a superposition of two states, $|e_x\rangle \langle e_y|$. After a population delay, T, a probe pulse arrives to examine the

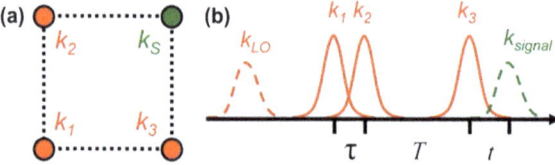

Figure 6.2   (a) Boxcars beam geometry and (b) pulse ordering for 2DPE experiments.

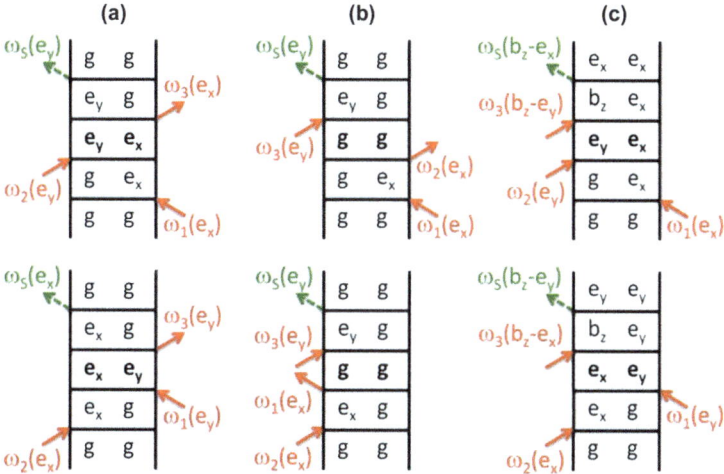

Figure 6.3 Feynman diagrams representing the possible Liouville pathways detected in the $k_S = -k_1 + k_2 + k_3$ direction, including (a) SE, (b) GSB and (c) ESA. The top row shows rephasing diagrams, where k_1 arrives before k_2. The bottom row shows non-rephasing diagrams.

sample. Three kinds of pathways can produce signal, stimulated emission (SE), ground state bleach (GSB) or excited state absorption (ESA), as shown in the Feynman diagrams of Figure 6.3. In the case of SE or GSB, the probe pulse produces a coherence between the ground state and excited state. In ESA, the probe will cause a coherence of the first excited state and a higher excited state. Signal is radiated into its phasematched direction to bring the system back to a population (ground state for SE or GSB, excited state for ESA).

The signal field is focused into a spectrograph attached to a charge coupled device detector (CCD), allowing the signal to be resolved spectrally. This provides one axis of our eventual 2DPE spectrum, $\hbar\omega_t$, the energy of the signal emitted to bring the system back to a population, that is, the energy of the coherence that exists during time t. This signal is collected as coherence time is scanned. A Fourier transform of the coherence time axis yields the second axis in the 2DPE spectrum, $\hbar\omega_\tau$, the energy of the coherence that exists during time τ. In this way, one axis can be thought of as the excitation energy and the other as the emission energy, analogous to plots of 2D fluorescence. While this is not strictly correct, since the energy of the second excitation pump pulse is not directly observable in the 2D spectrum, this analogy is still useful and aids in more intuitive interpretations of 2DPE data.

6.2.2 Solvation Dynamics

2DPE was originally developed for the study of solvation dynamics. The one dimensional version of this technique, photon echo (PE) spectroscopy, has long been a useful tool for this purpose.[45–47] Photon echo measurements provide information about the memory of a system. Note that in the above explanation of the pulse sequence there are two time periods where the system is in a coherence between the ground and excited state (we neglect ESA pathways for now). The first pump pulse and the probe pulse both send the system into a coherent superposition of the ground and excited states. During this time, the system evolves with an oscillatory component, $\exp(-i\omega_{eg}t)$, where $E_{eg} = \hbar\omega_{eg}$ is the energy difference between the two states. This gives the coherence of a certain phase. Over time, environmental fluctuations will change this transition energy and the frequency of this oscillation will change and thus the phase will change. If the first coherence and the second coherence are complex conjugates of each other, the two phases will be the opposite to one another. For example, if the first pump pulse produces the coherence $|g\rangle\langle e|$, its phase will evolve as $\exp(-i\omega_{eg}t)$. If the probe pulse produces the coherence $|e\rangle\langle g|$, the oscillatory component tof its phase will be $\exp(-i\omega_{ge}t)$; opposite to the first coherence. The evolved signal will be maximal when the system "rephases" to its original condition. Thus, if the system has perfect memory of its initial state, the coherence time, τ will equal the delay between the signal and the probe pulse, t. However, as each system in the ensemble loses memory of its initial transition energy and their phases are no longer the same, a signal maximum will no longer be observed at time $t = \tau$, since the individual systems will each rephase at their own rate. The two coherences will be complex conjugates if k_1 arrives before k_2, that is, in rephasing pathways, as is seen upon inspection of the SE and GSB diagrams of Figure 6.3.

The position of τ which produces the largest signal can be found by scanning coherence time. This value is termed the "peak shift" and the measurement of how peak shift changes with population time is called three-pulse photon echo peak shift (3PEPS). It has been shown[47–49] that at larger population times the 3PEPS signal directly follows the transition frequency correlation function:

$$M(t) = \frac{\langle\Delta\omega(0)\Delta\omega(t)\rangle}{\langle\Delta\omega^2\rangle} \tag{6.1}$$

where $\Delta\omega(t)$ is the difference between the ensemble's average transition frequency and the transition frequency at time t:

$$\Delta\omega(t) = \langle\omega_{eg}\rangle - \omega(t) \tag{6.2}$$

In the high temperature limit, $M(t)$ is identical to the Stokes shift function, $S(t)$.[48] In this way, PE (in particular, 3PEPS) is an excellent tool for the study of solvation dynamics; 3PEPS tracks a system's memory of its initial transition energy and this energy changes over time owing to the relaxation of the solvent molecules surrounding it. Even more information is now available through the extension of this spectroscopy into two dimensions. In 2DPE spectra, a signal along the diagonal is produced when $\omega_\tau = \omega_t$, that is, when the excitation and emission energies are identical, indicating that the transition energy has not changed. Thus, broadening along the diagonal is due to inhomogeneous broadening, while homogeneous broadening is observed in the direction perpendicular to the diagonal.

Experiments by Hybl *et al.*[50,51] and simulations by Kwac and Cho[52] have shown that solvation dynamics can be observed in both the real (absorptive) and imaginary (refractive) parts of 2DPE lineshapes. The peak shape in the real signal will lose its diagonal structure at larger population times, as the solvent configuration changes in response to the new charge distribution in the excited solute. In the imaginary signal, the slope of the nodal line between the positive and negative peaks of the refractive signal will change over population time and the correlation function can be directly obtained by this measure as well.[52] A number of other methods of extracting the correlation function from 2DPE spectra have been detailed in the work of Roberts *et al.*,[53] including the inhomogeneity index, ellipticity and dynamical linewidth.

6.2.3 Electronic Coherences

The previous section considered the utility of coherences between the ground and excited state which exist during the time delays τ and t. What of coherences during T? During the population time, coherent superpositions can be formed between intraband electronic states, that is, when the two pump pulses excite different electronic states. In the following, the example of an electronic coherence between two states, e_x and e_y, is examined and the four signals produced by this coherence are described. Figure 6.4 shows these four signals, their corresponding Feynman diagrams and the relationship of these four signals to each other. This type of analysis of coherent beating in rephasing and non-rephasing spectra was first proposed in a theoretical study by Cheng and Fleming.[54]

If k_1 photoexcites the sample to state e_x, followed by k_2 exciting e_y, the coherence $|e_y\rangle \langle e_x|$ exists during the population time. With the beam geometry of our setup, the detected signal is the relaxation of the ket side of the Feynman diagram, so this coherence would result in signal at $(\hbar\omega_\tau, \hbar\omega_t) = E_x, E_y$ which oscillates with a frequency $2\pi(E_y - E_x)/\hbar$. If, instead, k_1 excites e_y and K_2 excites e_x, the coherence $|e_x\rangle \langle e_y|$ would produce signal at $(\hbar\omega_\tau, \hbar\omega_t) = E_y, E_x$ and this signal would oscillate with a frequency $2\pi(E_x - E_y)/\hbar$; the signal would be in the opposite cross peak position (two points in the spectrum symmetric about the diagonal) and their oscillations would be exactly anti-correlated. Note that in both of these cases, k_1 arrives before k_2 and thus are rephasing pathways, shown in Figure 6.4(a).

Coherences can exist during the population time in non-rephasing pathways as well, Figure 6.4(b). If the two cases above are considered, except when k_2 arrives before k_1, the two resulting signals would be found at $(\hbar\omega_\tau, \hbar\omega_t) = E_y, E_y$, oscillating with a frequency $2\pi(E_y - E_x)/\hbar$ and $(\hbar\omega_\tau, \hbar\omega_t) = E_x, E_x$, oscillating with a frequency $2\pi(E_x - E_y)/\hbar$, respectively. The signals generated from these coherences appear on the diagonal but, as in the rephasing case, amplitude oscillation of the two signals should be anti-correlated. Note as well that if these electronic coherences exist, the signal at the upper cross peak position $(\hbar\omega_\tau, \hbar\omega_t) = E_x, E_y$ in the rephasing spectra, should be correlated with the signal at the upper diagonal position $(\hbar\omega_\tau, \hbar\omega_t) = E_y, E_y$ in the non-rephasing spectra. The two lower positions in the rephasing and non-rephasing spectra should be correlated as well.

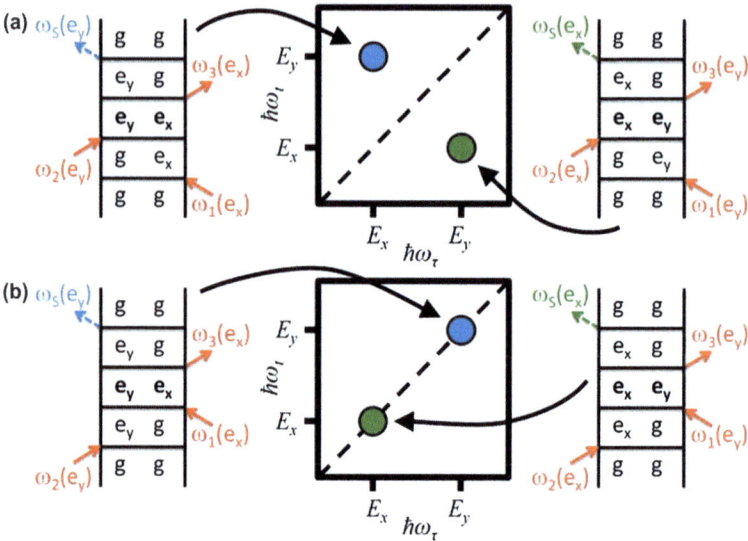

Figure 6.4   Feynman diagrams of four pathways which include a coherence between
state e_x and e_y during population time, and where these signals would
appear in (a) rephasing and (b) non-rephasing 2DPE spectra. Blue and
green circles indicate the position of signals that would oscillate with
frequency $2\pi(E_y - E_x)/\hbar$ and $2\pi(E_x - E_y)/\hbar$, respectively.

6.3   Two-Dimensional Photon Echo Spectroscopy: The Experiment

The experiments in this work were performed using a Ti:sapphire regeneratively
amplified laser system (Clark-MXR, CPA-2001) which produces 140 fs pulses at
775 nm with a repetition rate of 1 kHz at a power of 780 mW. The output was fed
into a non-collinear optical parametric amplifier (NOPA) that allowed the
wavelength of the pulses to be tuned in the range of 400–650 nm, with a duration
of ~ 30 fs after prism compression. A waveplate and polarizer were used to at-
tenuate the power to less than 5 nJ/pulse at the sample position to minimize
sample degradation. A beam splitter was used to divide the beam into pump and
probe beams and the probe beam was sent to a retroreflector mounted on a
stepper motor to produce the population time delay, T. The coherence time
delays, τ, were introduced by pairs of wedges, as shown in Figure 6.5. One wedge
from each pair was mounted on the stage of a small stepper motor with a range
of 10 cm. As one wedge moved in a plane perpendicular to the beam, the amount
of glass the beam must pass through changed, providing a small time delay. This
is discussed in detail in the next section.

In the 2DPE measurements described here, the signal was overlapped spatially with the local oscillator (LO), but the LO pulse arrived ~ 500 fs before the
probe pulse. This was achieved by inserting a glass window into the probe
beam, or equivalently, a pair of wedges. For measurements of CdSe nanocrystals, half-wave plates were inserted into each of the four beams. A chopper
was placed in the probe beam, set to ~ 3 Hz. The four beams were focused into

the sample in a boxcars geometry, overlapping the signal with the LO beam. The two pump beams and the probe beam were blocked after the sample to decrease contributions from scattering. The signal and LO were focused into a 0.63 m spectrograph with a 25 μm slit and recorded using a 16 bit, 400×1600 pixel, thermoelectrically cooled charge coupled device detector (CCD).

In 2DPE spectroscopy, coherence time is scanned using the wedge pairs, typically from -60 fs to 0 fs (where k_2 arrives before k_1, $\tau < 0$, the non-rephasing region), then from 0 fs to $+60$ fs (where k_1 arrives before k_2, $\tau > 0$, the rephasing region), in steps of 150 or 250 as. The population time, T, is the delay time between the second pump pulse (k_1 in the non-rephasing region, k_2 in the rephasing region) and the probe pulse. This delay is controlled by the position of a retroreflector mounted on a stepper motor. At each coherence time, an interferogram between the signal and the LO is collected with the CCD. After scanning through the entire range of coherence times, a 2D spectrum has been constructed, with one axis in time (coherence time) and the other axis in energy (determined by calibrating the CCD). The process is then repeated for the next population time. Linear absorption spectra are collected before and after the entire set of population scans, to ensure that the sample did not degrade during the measurement. All of the 2DPE measurements in this work are performed at ambient temperature.

6.3.1 Wedge Calibration

The delay between pulse 1 and pulse 2 is introduced using pairs of glass wedges. This delay is typically not very large (<200 fs), but the time steps are usually small (150–250 as). A typical delay stage, such as the stage used to control population time, is quite large and would not fit into our experimental setup and does not have the required precision. Instead, it has been found[40] that pairs of glass wedges mounted on smaller delay stages can readily accomplish the task of producing these small time steps. The wedges are aligned such that the inner faces

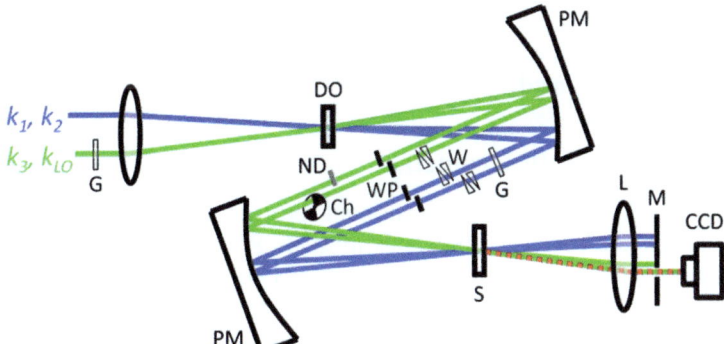

Figure 6.5 Experimental setup for polarization dependent 2DPE. CCD = spectrograph and CCD camera, Ch = chopper, DO = diffractive optic, G = glass window, L = lens, M = mask, ND = neutral density filter, PM = parabolic mirror, S = sample, W = wedge pairs, WP = half-wave plate. Blue lines originate from the pump beam, green lines originate from the probe beam. Dotted red line is the evolved signal, which overlaps with the LO beam.

are parallel to each other and the outer faces are normal to the beam. When aligned in this way, the two wedges act as a glass plate with a thickness that can be varied by lateral translation of one of the wedges in the plane of the inner faces of the wedges.

To calibrate these wedges and find time zero, a pinhole is placed at the sample position and aligned such that both pump beams can pass through. After the pinhole, the two incident beams are blocked and light diffracted by both beams into the signal direction is collected. This is simply for convenience; the key is that both beams are detected. Another option would be to place a second diffractive optic at the sample position which would combine k_1 and k_2 back together into one beam.[55] The pulses can be expressed as:

$$E_1(\omega, t) \cos(\omega(t - t_1 + \varphi_1)$$
$$E_2(\omega, t) \cos(\omega(t - t_2) + \varphi_2) \tag{6.3}$$

where E_i are pulse envelopes and $\cos(\omega(t - t_i) + \phi_i)$ are the carrier waves with frequency ω, a time delay of t_i and a phase shift of ϕ_i. Since the the two pump pulses are equal and phase locked, $E_1(\omega,t) = E_1(\omega,t)$ and $\phi_1 = \phi_2$, so the detector sees:

$$I(\omega, t) = |E\cos(\omega(t - t_1) + E\cos(\omega(t - t_2))|^2 \tag{6.4}$$

By setting $t_2 = 0$, this equation has only two variables: angular frequency, ω and t_1. In an experiment, we need to determine the position of the wedges in k_1 which results in an equal amount of glass in the k_1 and k_2 beams and the delay factor of the wedges, that is, how many fs of time delay are introduced per mm of glass, which depends on the angle of the wedge. Figure 6.6 shows plots of Equation (6.4) with different wedge delay factors.

The wedge position where both k_1 and k_2 traverse the same amount of glass is found by identifying the centre of the peak which shows no change in amplitude with angular frequency. The wedge delay factor can be found by taking a horizontal slice of the scan, determining the number of radians of oscillation per unit distance of wedge position, then dividing by the angular frequency at which the slice was made.

6.3.2 Spectral Interferometry

In order to obtain a final two-dimensional spectrum (where both axes are in an energy scale), the coherence time axis must be Fourier transformed into a frequency (or energy) axis. However, before this is done, the desired signal must be isolated and split into real and imaginary components. This section explains the spectral interferometry procedure used to isolate the signal, with the hope that it will aid other workers in trying to understand the complex algorithm used to process the experimental data. These procedures were developed by others. Particularly useful references which were used to develop the following section are: Brixner *et al.*[40] (background subtraction); Lepetit *et al.*[56] (spectral interferometry); and Hybl *et al.*[43] (phase wrapping). Other articles which include points of interest are cited in the text. This section shows the step-by-step process of

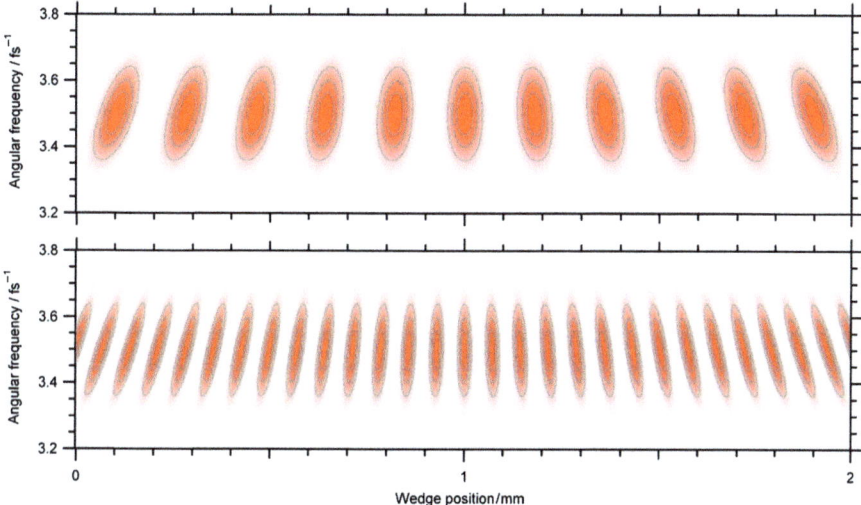

Figure 6.6 Plots of Equation (6.4) when the pulse is centered at an angular frequency of $3.5\,\text{fs}^{-1}$ with a full-width at half maximum of $0.2\,\text{fs}^{-1}$, t_1 is set at $0\,\text{fs}$ (an equal delay of k_1 and k_2) when wedge 1 is at a position of 1 mm, and the position of wedge 1 is varied. The wedge delay factor is set to be (a) $10\,\text{fs}\,\text{mm}^{-1}$ (b) $25\,\text{fs}\,\text{mm}^{-1}$.

extracting the complex signal field from the raw interferograms detected in the experiment. The data shown is collected from a CdSe quantum dot solution sample.

6.3.2.1 Collected Signal

Using the experimental setup described above with the CCD camera coupled to the signal in the k_S direction, the dominant fields which reach the detector are the signal field, E_S and the local oscillator field, E_{LO}, specifically:

$$I_{S,LO} = |E_S + E_{LO}|^2$$
$$= (E_S + E_{LO})(E_S^* + E_{LO}^*) \tag{6.5}$$
$$= E_S E_S^* + E_S E_{LO}^* + E_{LO} E_S^* + E_{LO} E_{LO}^*$$

where the notation for each field is:

$$E_\alpha = \sqrt{I_\alpha(\omega)}\exp(i\Phi_\alpha(\omega) + i\omega t_\alpha), \tag{6.6}$$

which contains information about the intensity, phase and average arrival time for each field pulse. Each of the terms in Equation (6.5) can be expressed as:

$$E_\alpha E_\beta^* = \sqrt{I_\alpha(\omega)I_\beta(\omega)}\exp(i\Phi_\alpha(\omega) - i\Phi_\beta(\omega))\exp(i\omega(t_\alpha - t_\beta)) \tag{6.7}$$

While the components of Equation (6.5) should dominate, scattered light from any of the other beams can also end up in the signal direction and be detected by the CCD. If we consider all of the possible contributions from actual signal and from scatter originating from each of the incident laser beams, the detected signal is:

$$
\begin{aligned}
I_{1,2,3,LO,S} = {} & |E_1 + E_2 + E_3 + E_{LO} + E_S|^2 \\
= {} & (E_1 + E_2 + E_3 + E_{LO} + E_S)(E_1^* + E_2^* + E_3^* + E_{LO}^* + E_S^*) \\
= {} & E_1 E_1^* + E_1 E_2^* + E_1 E_3^* + E_1 E_{LO}^* + E_1 E_S^* \\
& + E_2 E_1^* + E_2 E_2^* + E_2 E_3^* + E_2 E_{LO}^* + E_2 E_S^* \\
& + E_3 E_1^* + E_3 E_2^* + E_3 E_3^* + E_3 E_{LO}^* + E_3 E_S^* \\
& + E_{LO} E_1^* + E_{LO} E_2^* + E_{LO} E_3^* + E_{LO} E_{LO}^* + E_{LO} E_S^* \\
& + E_S E_1^* + E_S E_2^* + E_S E_3^* + E_S E_{LO}^* + E_S E_S^*
\end{aligned}
\tag{6.8}
$$

There are 25 possible contributions to the detected signal. The only contribution that is of interest in 2DPE spectroscopy is $E_S E_{LO}^*$. Some of the other contributions can be subtracted from the full signal using separate measurements, as described in Section 6.3.2.2. Others can be eliminated using time-ordering arguments, explained in Section 6.3.2.3.

6.3.2.2 *Background Subtraction*

If the k_3 beam path is blocked, the incident fields on the CCD would be:

$$
\begin{aligned}
I_{1,2,LO} = {} & |E_1 + E_2 + E_{LO}|^2 \\
= {} & (E_1 + E_2 + E_{LO})(E_1^* + E_2^* + E_{LO}^*) \\
= {} & E_1 E_1^* + E_1 E_2^* + E_1 E_{LO}^* \\
& + E_2 E_1^* + E_2 E_2^* + E_2 E_{LO}^* \\
& + E_{LO} E_1^* + E_{LO} E_2^* + E_{LO} E_{LO}^*
\end{aligned}
\tag{6.9}
$$

These terms include scattered homodyne contributions from the pump beams, as well as scattered pump probe contributions from each pump pulse and the LO pulse. The major contaminant here is the homodyne signal generated by the LO, since this signal would be directly coupled to the CCD. The component of interest, $E_S E_{LO}^*$, is not generated when the probe beam does not interact with the sample. Thus, by using a chopper in the k_3 beam path, these terms can be subtracted from the total signal. Since the time delays between these pulses are being varied during the measurements, the contribution from Equation (6.9) must be measured at each time step, which is why a chopper (or automated shutter) is ideal for this task.

If both of the pump beams are blocked, the CCD detects:

$$I_{3,LO} = |E_3 + E_{LO}|^2$$

$$= (E_3 + E_{LO})(E_3^* + E_{LO}^*) \quad (6.10)$$

$$= E_3 E_3^* + E_3 E_{LO}^* + E_{LO} E_3^* + E_{LO} E_{LO}^*.$$

which includes a scattered homodyne signal from the probe, scattered pump probe signals from the probe and LO and a directly coupled homodyne signal from the LO. Since the delay between k_3 and the local oscillator is fixed (it is created by a window of glass, so the delay will be constant even when the population time changes), the measurement of these contributions can be performed just once, by blocking the pump beams and recording the CCD image. In practice, the amplitude of this scatter is mostly dependent on the cuvette used to hold the sample. This scatter can be monitored on the CCD as a method of finding an optically clear spot on the cuvette for the measurement.

When $I_{1,2,LO}$ and $I_{3,LO}$ are removed from the collected signal, the resulting terms are:

$$I_{1,2,3,LO,S} - I_{1,2,LO} - I_{3,LO}$$

$$= |E_1 + E_2 + E_3 + E_{LO} + E_S|^2 - |E_1 + E_2 + E_{LO}|^2 - |E_3 + E_{LO}|^2$$

$$= \cancel{E_1 E_1^*} + \cancel{E_1 E_2^*} + E_1 E_3^* + \cancel{E_1 E_{LO}^*} + E_1 E_S^*$$

$$+ \cancel{E_2 E_1^*} + \cancel{E_2 E_2^*} + E_2 E_3^* + \cancel{E_2 E_{LO}^*} + E_2 E_S^* \quad (6.11)$$

$$+ E_3 E_1^* + E_3 E_2^* + \cancel{E_3 E_3^*} + \cancel{E_3 E_{LO}^*} + E_3 E_S^*$$

$$+ \cancel{E_{LO} E_1^*} + \cancel{E_{LO} E_2^*} + \cancel{E_{LO} E_3^*} + \cancel{E_{LO} E_{LO}^*} + E_{LO} E_S^*$$

$$+ E_S E_1^* + E_S E_2^* + E_S E_3^* + E_S E_{LO}^* + E_S E_S^*$$

and the resulting interferograms looks like what is shown in Figure 6.7. This figure shows a collection of interferograms, each one collected at a different coherence time, from -50 to 50 fs in steps of 0.25 fs.

While many of the undesired components can be removed by background subtraction, many scatter components remain which can interfere with the desired signal. The next step in the signal cleaning process can remove the rest of these terms, as described in the next section.

6.3.2.3 Time-Ordering Arguments

In the context of producing 2D spectra where both axes are in energy, the frequency axes is deemed the direct dimension, since each image acquired from the

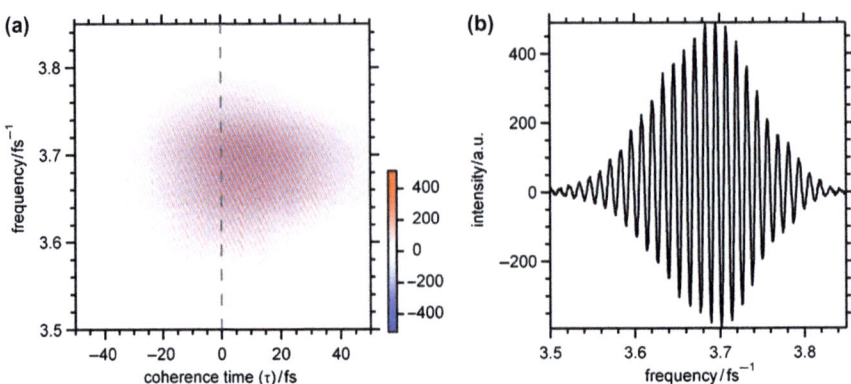

Figure 6.7 (a) Spectral interferograms collected using the 2DPE setup described in Section 6.3 and the background subtraction described in Section 6.3.2.2. (b) A slice from (a), shown by the grey dashed line, of a single spectral interferogram collected at a coherence time of $\tau = 0$ fs.

CCD camera contains interferograms already in a scale which can be converted to an energy scale without using a Fourier transform. It is in this dimension that much of the work of spectral interferometry must be done. We first perform an inverse Fourier transform of each interferogram from Figure 6.7 into the time domain. Figure 6.8(a) shows the absolute value of the complex results. After shifting the results so that the zero time component appears at the centre of the spectrum, we obtain the spectra shown in Figure 6.8(b), (c) and (d) for the absolute, real and imaginary components, respectively.

This signal contains the 13 uncrossed components from Equation (6.11). Recall that each of these components is of the form shown in Equation (6.7), which includes a term involving the difference in arrival time between the two pulses. This term allows components to be isolated by their particular arrival time differences. The signal of interest is the interference between the signal and the local oscillator, which has an arrival time difference of $t_S - t_{LO}$. Most of the amplitude in the inverse Fourier transformed signal shown in Figure 6.8 has a delay time of $\pm(t_S - t_{LO})$. If we apply a windowing function to isolate only these signals, we can eliminate all the terms that appear at other delay times, specifically, the terms shaded in grey:

$$
\begin{aligned}
W(t)\mathcal{F}^{-1}&(I_{1,2,3,LO,S} - I_{1,2,LO} - I_{3,LO}) \\
=& \; \cancel{E_1 E_1^*} + \cancel{E_1 E_2^*} + E_1 E_3^* + \cancel{E_1 E_{LO}^*} + E_1 E_S^* \\
&+ \cancel{E_2 E_1^*} + \cancel{E_2 E_2^*} + E_2 E_3^* + \cancel{E_1 E_{LO}^*} + E_2 E_S^* \\
&+ E_3 E_1^* + E_3 E_2^* + \cancel{E_3 E_3^*} + \cancel{E_1 E_{LO}^*} + E_3 E_S^* \\
&+ \cancel{E_{LO} E_1^*} + \cancel{E_{LO} E_2^*} + \cancel{E_{LO} E_3^*} + \cancel{E_{LO} E_{LO}^*} + E_{LO} E_S^* \\
&+ E_S E_1^* + E_S E_2^* + E_S E_3^* + E_S E_{LO}^* + E_S E_S^*
\end{aligned}
\tag{6.12}
$$

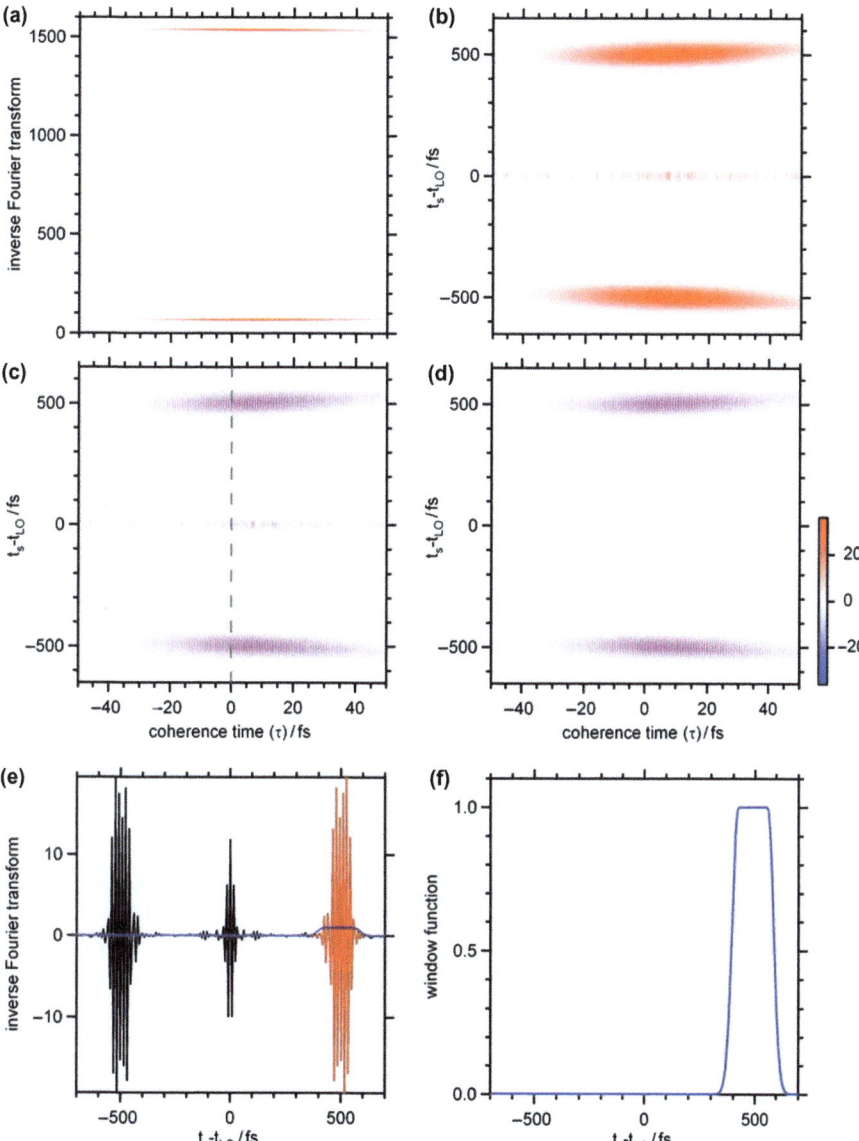

Figure 6.8 Isolating desired signal in the directly detected dimension. (a) Inverse Fourier transformed data before shuffling. (b), (c) and (d) are absolute, real and imaginary signal after shuffling. (e) is a slice of the real signal at $\tau = 0$, the dashed line in plot (c). The blue function in (e) and (f) is the windowing function with the Heaviside function, which selects the signal in red in (e).

The components removed by background subtraction include those which cannot be removed by time ordering arguments. In particular, for scans near $T = 0$, the delay between k_{LO} and k_S will be very close to the delays between k_{LO}

and the second pump beam to arrive (either k_1 or k_2). As well, since k_S is caused by the interaction of the probe beam, any components involving k_{LO} and k_3 also cannot be removed by time delay isolation. Fortunately, all of these components can be eliminated using the background subtraction described in Section 6.3.2.2

At this point, only two terms remain:

$$W(t)F^{-1}(I_{1,2,3,LO,S} - I_{1,2,LO} - I_{3,LO}) = E_S E_{LO}^* + E_{LO} E_S^* \qquad (6.13)$$

The inverse Fourier transform of this cosine function yields the two maxima, $t = t_S - t_{LO}$ from the $E_S E_{LO}^*$ component and $t = -(t_S - t_{LO})$ from $E_{LO} E_S^*$, shown in Figure 6.8(e). By the causality principle, the signal of interest cannot be generated until after the probe pulse arrives. By setting the local oscillator pulse to arrive much earlier than the probe pulse (*i.e.* many times the pulse duration), we ensure that the two terms in this inverse Fourier transformed spectrum do not overlap. In the experiments described here, $t_{LO} \approx -500$ fs, so only the signal at $t \approx 500$ fs satisfies causality. We can now apply a Heaviside function, $\Theta(t)$, to isolate the relevant term. The product of the windowing function, which in this case is a version of a Gaussian windowing function and the Heaviside function is shown in Figure 6.8(f). When this window is applied to the data, the result is as shown in red in Figure 6.8(e).

A forward Fourier transform back to the frequency domain yields the spectra shown in Figure 6.9, a complex electric field:

$$I(\omega) = \mathcal{F}[\Theta(t)W(t)\mathcal{F}^{-1}(I_{1,2,3,LO,S} - I_{1,2,LO} - I_{3,LO})]$$

$$= \sqrt{I_S(\omega)I_{LO}(\omega)} \exp(i(\phi_S - \phi_{LO} - \omega(t_S - t_{LO}))) \qquad (6.14)$$

As shown in Figures 6.8(e) and 6.9(c), the value of the real and imaginary signals alternates between being positive and negative. This is an artifact, owing to a technical aspect of the method used to perform the inverse Fourier transform. This can be overcome by multiplying each term in the result with alternating $+1$, -1, $+1$, and so on.

Also, note that the fast Fourier transform (FFT) algorithm requires equally spaced points, but the spectrum retrieved from the CCD is not linear in energy. The spectrograph attached to the CCD disperses the incident light, resulting in a spectrum which is (for most spectrometers) linear in wavelength, but not in frequency. If the non-linear correlation of wavelength and frequency is ignored and the FFT algorithm is used anyway, the signal will still show up at $\pm t_S - t_{LO}$ after the inverse Fourier transform described above, but as this time delay gets larger, the signal broadens in the time domain and the results are incorrect.[57] One might assume that the data must be interpolated into equally spaced points in frequency before the FFT algorithm is used. However, as shown in a paper by Dorrer *et al.*,[57] this is actually ineffective, as a significant amount of background noise is introduced upon inverse Fourier transform, affecting the resulting spectral phase, which will be discussed in Section 6.3.2.4. This is a consequence of error introduced by linearly interpolating the

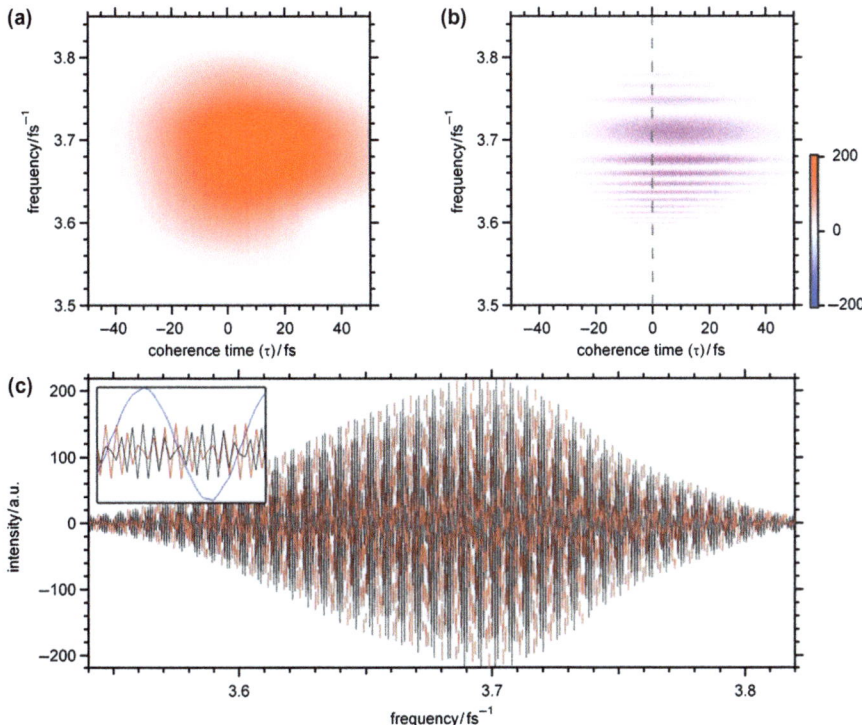

Figure 6.9 Absolute value (a) and real component (b) of the complex signal after application of the Heaviside function, windowing and Fourier transforming back to the frequency domain. (c) The real (black) and imaginary (red) components of a processed interferogram, a slice of the spectrum shown in (b) by the grey dashed line. The inset shows a small portion of these spectra overlaid with the original, purely real interferogram before processing. The imaginary signal is shifted by $\pi/2$ as expected.

experimental data. More advanced interpolation techniques are typically not feasible for these types of data, since there are not many data points taken per oscillation.

Instead of direct interpolation, Dorrer *et al.*[57] found that the most effective method was to use the FFT algorithm to perform the inverse Fourier transform, realizing that the resulting axis was not time, but spatial frequency. After windowing and a forward Fourier transform, again using the FFT algorithm, the axis is positioned in the spatial plane of the detector (*i.e.* linear in wavelength). At this point, phase can be extracted, without any artifacts introduced by interpolation. The non-linear calibration law for $\omega(x)$ can then be applied to yield the correct spectral phase. This is the procedure outlined above.

In order to extract the signal field from the expression of Equation (6.14), the data must be multiplied by $\exp\left(i(\phi_S - \phi_{LO} - \omega\,(t_S - t_{LO}))\right)$ and divided by $\sqrt{I_{LO}(\omega)}$. Experimentally, I_{LO}, the local oscillator spectrum, is recorded throughout the 2DPE measurement using shutters. One LO spectrum is

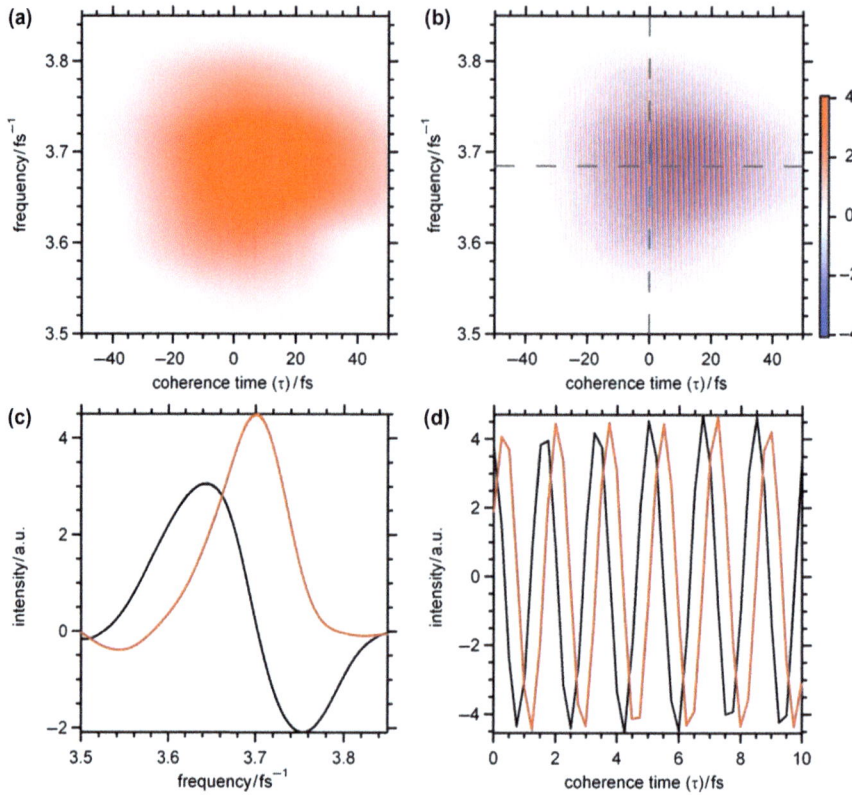

Figure 6.10 Absolute value (a) and real component (b) of the complex signal after isolating the desired signal. The real (black) and imaginary (red) components of a vertical (c) and horizontal (d) slice of the spectrum shown in (b) by the grey dashed line.

recorded for each population time, allowing confirmation that the pulse spectrum did not change during the scans. $\phi_S - \phi_{LO}$ is the phase difference between the signal and the local oscillator fields, which will be discussed in detail in the following section. To begin the phase extraction process, any phase value can be entered. When this procedure (and the sign rectification procedure described in the previous paragraph) is performed on the data of Figure 6.9, the results are as shown in Figure 6.10. Figure 6.10(a) is the absolute value of the signal field, which comprises real (Figure 6.10(b), with slices shown in black in 6.10(c) and 6.10(d)) and imaginary (slices shown in red) components. The separation of the detected signal into these two components is performed by "phasing" the spectra.

6.3.2.4   Phasing 2DPE Spectra

A complex number comprises real and imaginary components, which can also be defined by amplitude and phase. In the complex plane, if we defined phase as

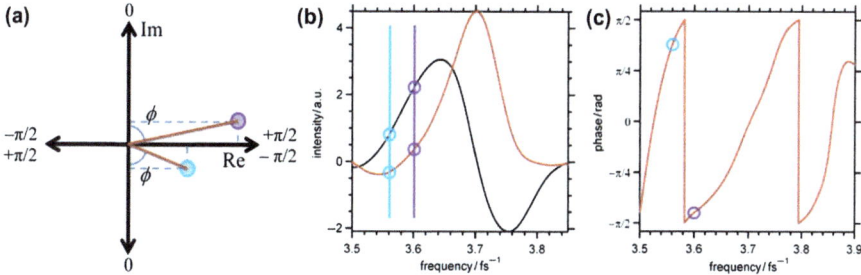

Figure 6.11 (a) Definition of ϕ in the complex plane. (b) Slice of the isolated signal, real (black) and imaginary (red), at $\tau = 0$. (c) Phase as a function of frequency, calculated by Equation (6.15). Blue and purple circles highlight two cases, showing a discontinuity in the phase caused by the limits of the arctan function.

the angle to the imaginary axis, as shown in Figure 6.11, phase could be solved for by:

$$\phi = -\arctan\left(\frac{\text{Re}}{\text{Im}}\right) \qquad (6.15)$$

The arctan (x) function can only have values between $-\pi/2$ and $+\pi/2$, so if the imaginary part of a continuous complex function passes through zero, it will cause a discontinuity in the phase. An example of this is shown in Figure 6.11. The curves in Figure 6.11(b) are slices of the real and imaginary signals (currently defined by an arbitrary value of $\phi_S - \phi_{LO}$, as mentioned in the previous section) at a particular coherence time, τ. For the real signal, this was shown by the vertical dashed line in Figure 6.10(b). By applying Equation (6.15) at each frequency point, the phase function at this coherence time can be found, shown in Figure 6.11(c). This function includes discontinuities in the phase whenever the imaginary part of the signal passes through zero, as shown by the circles. When Equation (6.15) is applied to every point in the real and imaginary spectra, the resulting phase is shown in Figure 6.12(a). The phase is not a smooth, continuous function, but instead includes many of these phase jumps.

To eliminate these jumps in the phase, a process called "phase wrapping" is performed, which can be visualized as wrapping the phase function of Figure 6.11(c) around a cylinder with a circumference of π. When the phase function is "unwrapped" from this cylinder, it forms a continuous function. As shown in Figure 6.12, after phase wrapping in the direct dimension at each coherence time, the phase function is smooth and continuous as a function of frequency. The phase surface has thus been smoothed in the direct dimension.

The phase must also be wrapped in the indirect dimension. Figure 6.13(b) shows a slice of the phase surface at a particular frequency value, revealing a large number of jumps in the phase as a function of coherence time. After phase wrapping in this dimension, the result, shown in Figure 6.13(c) and 6.13(d), is a smooth phase surface in both dimensions. This process has now yielded the

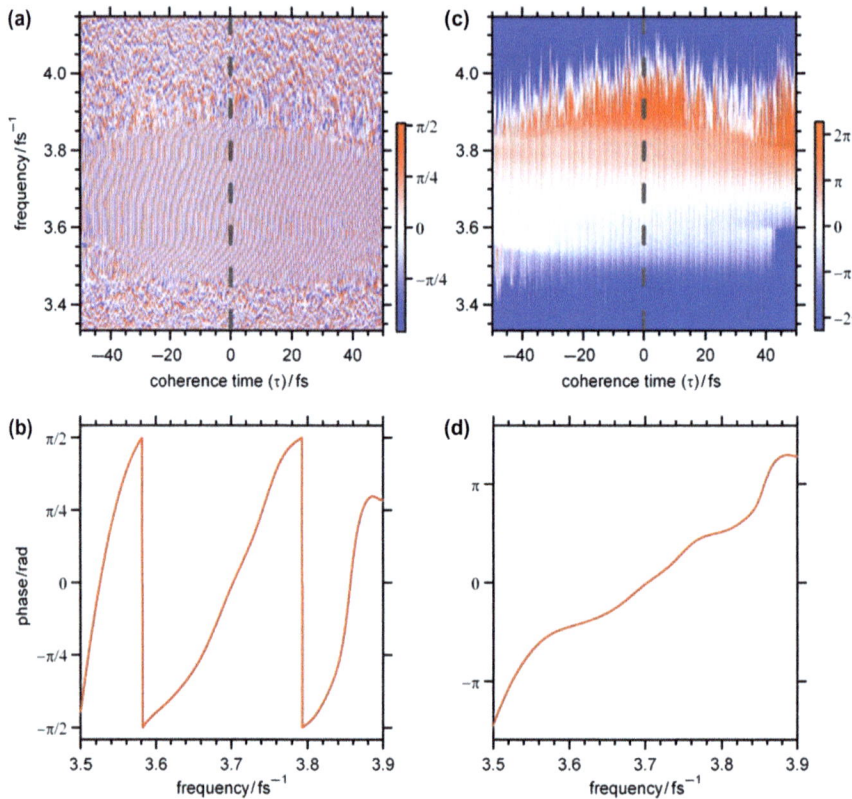

Figure 6.12 Phase wrapping in the directly detected dimension. (a) Original phase
surface, grey line shows (b) a slice of phase surface at $\tau = 0$. (c)
Phase surface after phase wrapping, grey line shows (d) a new slice of
phase surface at $\tau = 0$, after phase wrapping.

absolute value (Figure 6.10(a)) and phase (Figure 6.13(c)) of the isolated k_S
signal, which fully defines the complex signal field.

The complex signal field can now be Fourier transformed along the coherence time axis to yield 2D spectra in energy. Absolute, real and imaginary
2DPE spectra can be obtained in this way. By Fourier transforming only the
data at positive or negative coherence times, one can obtain separate rephasing
and non-rephasing 2D spectra.

The question we have not yet addressed is how to determine the value of
$\phi_S - \phi_{LO}$, the phase difference between the signal and local oscillator fields.
This relative phase factor will determine how the total signal is split into real
and imaginary components and currently an arbitrary number has been chosen.
To determine the relative phase, we use the projection slice theorem, which was
originally used in 2D nuclear magnetic resonance (NMR) spectroscopy for the
same purpose.[58] The theorem deals with a 2D spectrum in frequency, $S(\omega_1, \omega_2)$
which is a Fourier transform of a 2D spectrum in time, $S(t_1, t_2)$. It states that

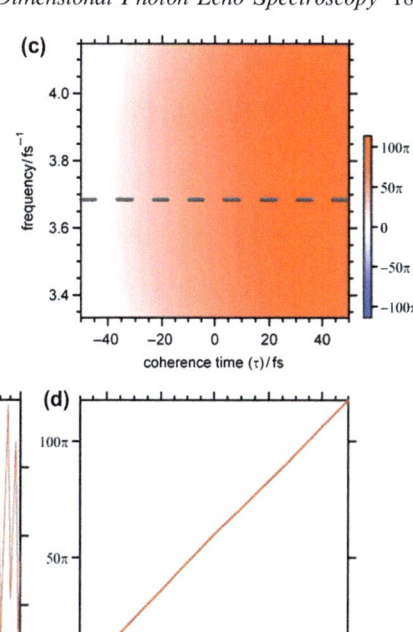

Figure 6.13 Phase wrapping in the indirectly detected dimension. (a) Phase surface after wrapping in the direct dimension, grey line shows (b) a slice of phase surface. (c) Phase surface after phase wrapping, grey line shows (d) a new slice of phase surface after phase wrapping.

the projection of the 2D spectrum $S(\omega_1, \omega_2)$ onto a line at angle ϕ through the origin will be the Fourier transform pair of a slice of the 2D spectrum $S(t_1, t_2)$ on the same line, at angle ϕ through the origin.[41]

For the 2DPE signal of interest here, when $\phi = 0$, the projection slice theorem relates the projection of the real 2DPE spectrum onto the ω_t axis and the spectrally resolved pump probe signal.[41] This pump probe signal was collected by blocking beams k_2 and k_3 and placing a chopper in the beam path of k_1. In this geometry, k_1 is the pump beam and k_{LO} is the probe beam. Since in the 2DPE setup the LO pulse arrives ~ 500 fs before the probe pulse, $T = 0$ for this pump probe measurement will occur when the population delay stage is at a positive value ~ 500 fs. The delay between the LO and probe pulses is determined by measuring an interferogram of just these two pulses, interpolating the interferogram to generate equally spaced points in frequency, then Fourier transforming to the time domain.

This theorem can be used to determine the correct relative phase factor. The entire spectral interferometry analysis described in this section is performed, changing the relative phase factor used until the projection of the real 2DPE spectrum in energy matches the spectrally resolved pump probe spectrum. In all of the work presented in this chapter, a non-frequency-dependent phase factor

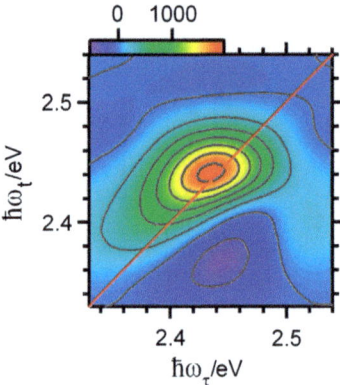

Figure 6.14   2DPE spectrum of CdSe NCs measured at $T = 60\,\mathrm{fs}$.

was chosen and the phase was assumed to be constant for all population scans. If the real projections of the entire set of population time spectra could not be adequately fitted to pump probe spectra by using a constant phase, the data set was discarded, with the interpretation that fluctuations in the environmental conditions in the laboratory led to unstable phase during the experiment.

After Fourier transform of the coherence time axis and determining the phase factor, the real component of the 2D spectrum for a sample of CdSe NCs is shown in Figure 6.14. The details of the sample used and the measurement conditions are given by Wong and Scholes.[59,60] The resulting spectra include positive and negative components, resulting from ground state bleaching and excited state absorption pathways, respectively. Simulations must be performed to analyze the origin of these signals and this will be discussed in Section 6.4.

6.3.3   Summary

To summarize, 2DPE spectra in energy are collected in our laboratory by the following process:

1 Obtaining $\sim 25\,\mathrm{fs}$, $\sim 25\,\mathrm{nm}$ bandwidth, $<5\,\mathrm{nJ}$ pulses which are resonant with the system to be studied, from a 1 kHz laser system using a NOPA to tune the central wavelength and bandwidth, prism pairs to compensate for dispersion and waveplate polarizer pairs to control the pulse energy.
2 Aligning the beams to have the wave vectors necessary to perform the 2DPE experiment.
3 Collecting interferograms while changing the coherence time, τ, for a series of population times, T.
4 Subtracting variable background signal by using a chopper in k_3 and constant background by collecting signal with both k_1 and k_2 blocked.
5 Inverse Fourier transforming each interferogram into the time domain to select only the signal which arrives near the time difference of $t_S - t_{LO}$, then forward Fourier transforming back to the frequency domain.

6 Dividing by the local oscillator spectrum, collected throughout the experiment and applying some relative phase factor $\phi_S - \phi_{LO}$ to obtain the absolute, real and imaginary signal fields.

7 Determining the phase surface by using the real and imaginary signal fields, then phase wrapping in both the direct and indirect dimensions to get a smooth phase surface.

8 Fourier transforming the coherence time axis to produce 2D spectra in energy.

9 Repeating the process (from step 6), changing the relative phase factor until the projection of the real 2D spectrum matches separately collected spectrally resolved pump probe spectra.

6.4 Simulation of the 2D Spectra: The Case of CdSe NCs

In order to assign the features observed in a 2D spectra, as in the case of Figure 6.14, a simulation of the 2DPE signals is performed. The procedure described in this section deals with the simplest case for NCs, a single material core-only situation, in particular, the case of CdSe NC is presented. This method can be extended for a type-II heterostructure, but owing to the length of the details, only an outline is provided in Section 6.5.1.

The first step in simulating the 2DPE signal is to determine the states that are accessible by the laser pulse; thus in the present case, it requires the calculation of the fine structure of the lowest exciton and the biexciton states for CdSe NCs. Once the relevant states are found, the optically allowed transitions and their relative strength are calculated. In this way, Feymann diagrams, similar to Figure 6.4, for all possible Liouville pathways are obtained and the 2D spectra is generated by considering all the pathways that give rise to a signal.

6.4.1 Fine Structure of the Lowest Exciton and Biexciton States in CdSe

Semiconductor NCs are composed of hundreds of unit cells and each cell has a HOMO (highest occupied molecular orbital) and a LUMO (lowest unoccupied molecular orbital). Taken together, these form valence and conduction bands, respectively. In the case of CdSe nanocrystals (NCs), an sp^3 hybridized semiconductor, $4p$ orbitals in Se form the valence band, while the conduction band is comprised of $5s$ orbitals in Cd. A single electron excitation from the highest occupied valence band to the lowest unoccupied conduction band creates an s-type excited electron and a p-type hole.[61]

Owing to this strong spin–orbit coupling in NCs, it is total angular momentum (instead of spin, as in molecules) which is a good quantum number. The states within the fine structure of NCs are labeled by their projected total angular momentum (F) on the unique crystalline c-axis of the NC. An excited electron can have $F_{electron} = \pm 1/2$ since it is in an s-type conduction band orbital, while a hole can be $F_{\text{hole}} = \pm 3/2$ or $\pm 1/2$ since it is in a p-type valence band orbital. Thus, an electron–hole pair can have $F = \pm 2$, ± 1, or 0. The lowest excitonic state of CdSe NCs comprises eight fine structure states, two with $F = \pm 2$, four with $F = \pm 1$ and two with $F = 0$. By considering all of the possible configurations for the spin of the electrons following an excitation of one electron from a valence band orbital (HOMO) to a conduction band orbital (LUMO), we find the configuration states. The wavefunctions of these configuration states are written in the electron representation. Electrons with spin α in the valence band can be represented by the following wavefunctions:

$$a = 2^{-1/2}|(X + iY)\alpha\rangle$$

$$b = |Z\alpha\rangle \tag{6.16}$$

$$c = 2^{-1/2}|(X - iY)\alpha\rangle$$

where:

$$\langle \theta, \phi | X \rangle = 2^{-1/2}(Y_{1,-1}(\theta, \phi) - Y_{1,+1}(\theta, \phi))$$

$$\langle \theta, \phi | Y \rangle = i2^{-1/2}(Y_{1,-1}(\theta, \phi) + Y_{1,+1}(\theta, \phi)) \tag{6.17}$$

$$\langle \theta, \phi | Z \rangle = Y_{1,0}(\theta, \phi)$$

and $Y_{l,m}$ are spherical harmonics. Similarly, electrons with spin β are represented by $\bar{a}, \bar{b}$ and $\bar{c}$. Conduction electrons are denoted by s and $\bar{s}$. The wavefunction corresponding to each exciton and biexciton configuration can be expressed by this notation in the following way: from left to right, electrons in the p-orbitals are labelled a, b or c with the corresponding "¯" if the spin is β; meanwhile, electrons in the s-orbital are labeled s or $\bar{s}$. Since electrons are indistinguishable, a permutator is applied to produce the final wavefunction for the each configuration, figure 6.15. The basis wavefunctions are finally obtained by taking the appropriate linear combinations of the configuration wavefunctions in order to construct spin-symmetry-adapted configuration state functions (CSF), which are eigenfunctions of the total spin operator $\hat{S}^2$.

The Hamiltonian of our system comprises the conduction and valence band energies, $\varepsilon_{c,v}$, the exchange interaction, $K_{i,j}$, the spin–orbit coupling, Δ and the crystal field splitting, Δ_{xf}. The matrix elements of this Hamiltonian in the basis of the Bloch wavefunctions a,b,c and s, Equations (6.16), whichcomprise the spin-symmetry-adapted configuration state functions, have been previously

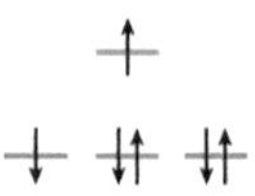

$$\begin{array}{c} \underline{\uparrow} \\[2pt] \underline{\uparrow}\ \underline{\uparrow\downarrow}\ \underline{\uparrow\downarrow} \end{array} \quad \equiv \quad |s\bar{a}b\bar{b}c\bar{c}|$$

Figure 6.15   Configuration assignment from pictorial representation – from left to right, electrons in the p-orbitals are labelled a, b, or c with the corresponding "¯" if the spin is β; meanwhile, electrons in the s-orbital are labeled s or $\bar{s}$.

determined using $\mathbf{k} \cdot \mathbf{p}$ theory.[62] The one-electron matrix elements, excluding exchange interactions, are given by:

$$H_{ss} = H_{\overline{ss}} = \varepsilon_c$$

$$H_{aa} = H_{\overline{cc}} = \varepsilon_v + \Delta/3$$

$$H_{cc} = H_{\overline{aa}} = \varepsilon_v - \Delta/3$$

$$H_{bb} = H_{\overline{bb}} = \varepsilon_v - \Delta_{xf}$$ (6.18)

$$H_{\overline{a}b} = H_{b\overline{a}} = \sqrt{2}(\Delta/3)$$

$$H_{\overline{b}c} = H_{c\overline{b}} = \sqrt{2}(\Delta/3)$$

With the above one-electron matrix elements, it is possible finally to construct the Hamiltonian matrix for our II-IV NCs (detailed Hamiltonian matrices can be found in Wong and Scholes[59] and Scholes *et al.*[63]). The eigenvalues and eigenfunctions for the lowest exciton and biexciton states, $\Psi_{X,BX}$, found after diagonalizing such Hamiltonian matrices, are summarized in Figure 6.16. Note that only the states whose energies overlap with the bandwidth of the laser pulse are presented.

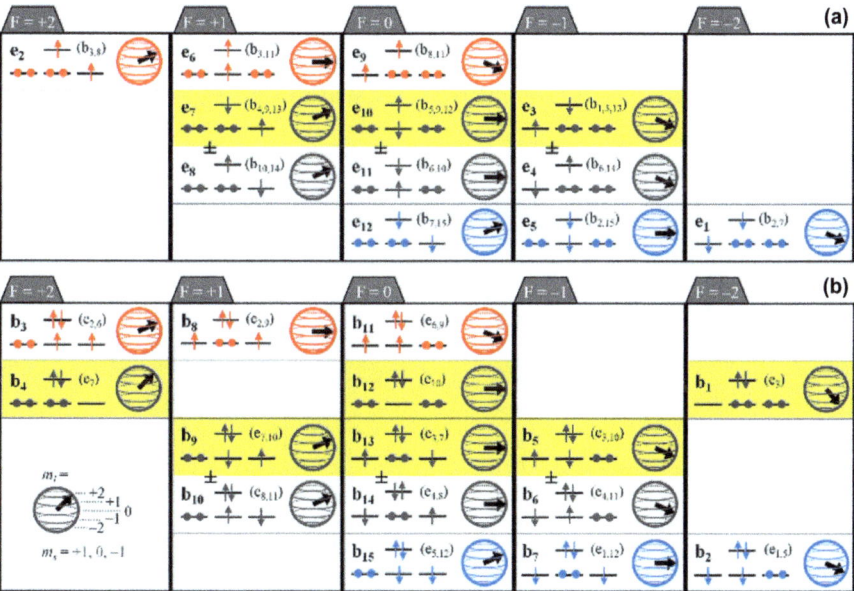

Figure 6.16 Configuration functions which comprise the ground state (a) exciton and (b) biexciton, organized horizontally by total angular momentum (F), and vertically by spin quantum number. Spin and orbital angular momenta are as shown in the inset legend. Bold face labels each configuration state, while smaller labels in brackets show the allowed transitions to single/biexcitonic configuration functions (reproduced from Wong and Scholes[59]).

6.4.2   Determination of Exciton–biexciton Transitions

The exciton and biexciton states, $\Psi_{X,BX}$, obtained from the diagonization of the Hamiltonian matrices, are linear combination of CSFs (labeled as e_i and b_i in Figure 6.16):

$$\Psi_X = \sum_k c_k^e e_k$$
$$\Psi_{BX} = \sum_k c_k^b b_k \tag{6.19}$$

The transition dipole moment between an exciton and a biexciton state is calculated using:

$$\mu_{X_\alpha \to BX_\beta} = \sum_{i,j} c_i^e \mu_{i,j} c_j^b, \tag{6.20}$$

where $\mu_{i,j} = 1$ when the transition between the exciton and the biexciton CSFs is allowed and zero when it is forbidden; c_i^e and c_j^b are the coefficients of the CSFs found in the exciton and biexciton wavefunctions, $\Psi_{X,BX}$, respectively, Equation (6.19).

Transitions between exciton and biexciton CSFs are allowed when all the configurations in the biexciton CSF are reachable from those in the exciton CSF. A biexciton configuration is reachable from an exciton configuration when there is no change in the total spin, that is, $\Delta S = 0$. A graphical representation, Figure 6.17, is most helpful to visualize this. For example, the first

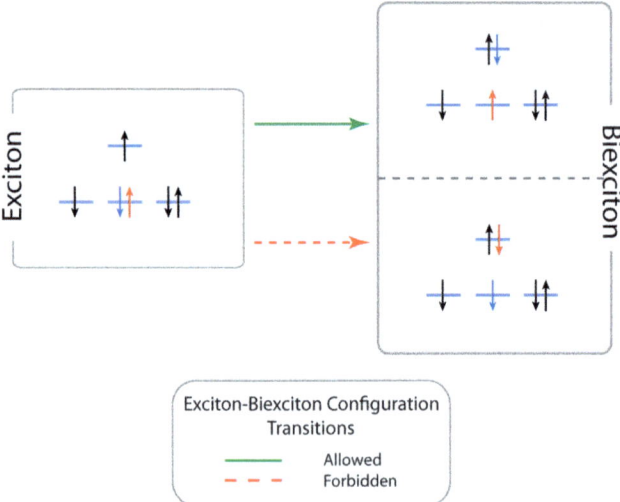

Figure 6.17   Assigning allowed and forbidden transitions between exciton and biexciton configurations – A biexciton configuration that can be obtained in a single excitation "step" from an exciton configuration is considered allowed (green arrows). Those that cannot be attained in this way are considered forbidden (red dashed arrows).

biexciton configuration (right panel, top) can be obtained from the exciton configuration (left panel) by taking the electron in the second valence-band level and placing it in the conduction band; therefore the transition among these configurations is allowed. On the other hand, the second biexciton configuration (right panel, bottom) requires two "steps", one to place the electron into the conduction band and then another to rotate its spin, in order to obtain the second biexciton configuration from the same exciton configuration, thus this transition is forbidden.

After determining the allowed transitions between exciton and biexciton configurations, those between CSFs can be obtained. For example, the transition $e_4 \rightarrow b_{12}$ is allowed because each exciton configuration in e_4 can reach a biexciton configuration in b_{12}, thus $\mu_{4,12} = 1$. On the other hand, the transition $e_4 \rightarrow b_{11}$ is forbidden because it requires two steps in order to produce the biexciton configuration from each of the exciton ones; therefore, $\mu_{4,11} = 0$. Transitions between exciton and biexciton states, $\Psi_X \rightarrow \Psi_{BX}$, are allowed when each e_i in Ψ_X has an allowed transition to a b_j in Ψ_{BX}. Figure 6.18 summarizes the allowed transitions between the lowest exciton and biexciton states in a wurtzite NC.

6.4.3 Simulating the 2D spectra

The 2D spectra are simulated using the formalism described by Abramavicius et al.,[64] specifically based on equations 42–48 of that paper. Careful book-keeping was done to count each of the possible SE, GSB and ESA Liouville

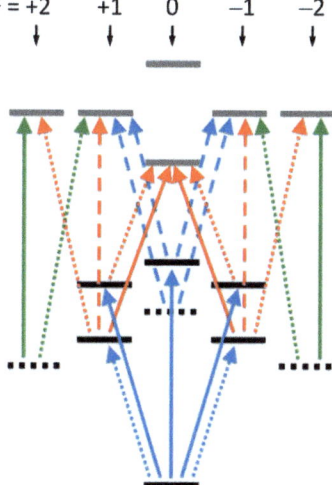

Figure 6.18 Optically allowed transitions from ground state to ground single exciton fine structure states to ground biexcitonic fine structure states. The horizontal position of state indicates F. Arrows in green, red, blue denote transitions originating from states with $F = \pm 2, \pm 1, 0$, respectively. Solid, dashed and dotted lines show strong, medium and weak transitions. Dashed levels are dark states (reproduced from Wong and Scholes[60]).

pathways between the ground state and the fine structure states of the exciton and biexciton. The signal intensity contributed by a particular pathway is determined by the product of the pulse envelopes and the relative transition dipole moments of each of the four transitions that comprise the pathway, as follows:

$$S_m(\omega, \omega_\tau, T) \propto [\mu_1^m g(\omega_0 - \omega_1^m)][\mu_2^m g(\omega_0 - \omega_2^m)][\mu_3^m g(\omega_0 - \omega_3^m)][\mu_S^m g(\omega_0 - \omega_S^m)]$$
$$\times [1 + LO(T)] N_i(T)$$

(6.21)

where ω_x^m is the frequency of transition x in pathway m, ω_0 is the carrier frequency of the laser pulse, $g(\omega_0 - \omega_x^m)$ is a Gaussian function and $N_i(T)$ is the population of exciton state i at population time T. Dynamics within the fine structure of the lowest exciton state are determined by the coupled differential equations:

$$\frac{d}{dt} N_i = \sum_j k_{ij} N_j$$

(6.22)

where the rates, k_{ij} are given by:

$$k_{ij} = \exp\left(-\frac{\Delta\varepsilon}{k_B T_{room}}\right)\left(\frac{1}{\tau_{ij}}\right)$$

(6.23)

Here $\Delta\varepsilon$ is the energy difference between state i and j and τ_{ij} is the transfer time constant between them. For transitions between exciton states $\pm 2 \leftrightarrow \mp 1L$, $\pm 1L \leftrightarrow \pm 1U$ and $0L \leftrightarrow 0U$, τ_{ij} is set to 500 fs. In the case of 0(L,U) $\leftrightarrow \pm 1(U,L)$, $\tau_{ij} = 1000$ fs. These are approximate values, which are in the same order of magnitude as those found in the literature for exciton dynamics within the fine structure.[63,65–69]

The LO phonon beating is simulated by a damped cosine function:

$$LO(T) = A_{LO} \exp\left(-\frac{T}{t_{LO}}\right) \cos\left(\frac{2\pi T}{T_{LO}} + \theta_{LO}\right)$$

(6.24)

where A_{LO} is the amplitude of the beating, t_{LO} is the damping time constant, T_{LO} is the period and θ_{LO} is a phase factor.

A Gaussian function was included for each pathway, centered at its signal position in the 2D plot, to simulate homogeneous broadening. In NCs, the dominant cause of inhomogeneous broadening is the size distribution of the particles,[70] observed experimentally as an elongation of the signal along the diagonal. In this simulation a Gaussian NC size distribution is assumed to result in a Gaussian distribution of fine structure energies. Shifting the energy of each state involved in a pathway by the same amount simulates the excitation of a NC of particular size and results in the elongated diagonal feature observed in measured 2DPE spectra. While this method of simulation does not consider the line shape functions in the response functions, it is an intuitive method which can be used to visualize quickly what can be expected in 2DPE spectroscopy.

6.5 CdTe/CdSe Core/Shell Nanocrystals Probed by 2DPE

In Type-II heterostructures, the possibility of populating a charge separated state (CS) after photoexcitation leads to qualitatively different exciton and multiexciton dynamics compared to core-only and type-I NCs. On one hand, CS states represent an additional pathway for the relaxation of excitons and maybe multiexcitons. On the other, interactions between excitons in the same NC are seen to be qualitatively different, resulting, for example, in repulsive exciton–exciton interactions and lengthening of Auger recombination decays.[37–39] In the first case, there are still open questions regarding the charge transfer (CT) process that populates CS states, for example, are the CT dynamics affected by the discrete nature of the phonon modes? What are the implications of adiabatic CT? What is the nature of the reaction coordinate?[71] In the second case, little is understood about the electronic fine structure of multiexciton states in type-II NCs. Recent studies using CdSe NCs show that knowledge of the biexciton fine structure is essential to understand the mechanisms controlling optical gain.[72,73] Similarly, it has been argued that multiple carrier generation efficiency may be determined by the density of biexciton states and their couplings.[74,75] Hence, elucidation of the biexciton fine structure in type-II heterostructures is of importance if these materials, with their intrinsic advantage (CT and longer Auger relaxation times) over core-only and type-I NCs, are to be used in such applications.

2DPE measurements were performed on CdTe/CdSe core/shell NCs using the experimental setup described in Section 6.3. The collected data were processed according to the procedure described in Section 6.3.2 to produce 2D spectra at different population times. For comparison, a sample of CdSe and type-I CdSe/CdS/ZnS was also measured.

The 2D spectra at population times $T = 140$, 180, 300 and 580 fs for CdTe/CdSe QDs are shown in the upper panels of Figure 6.19. Three main features are observed: a positive bleach on the diagonal, whose amplitude seemingly oscillates as a function of T; and two absorptive negative features diagonally below the bleach, labeled by a pink and a blue marker, whose amplitude also changes with T. The positive diagonal bleach feature reflects excitation into exciton states while the lower off-diagonal negative features indicate ESA. Since the laser spectrum was set to overlap with the red-edge of the absorption spectrum of the NC, we expect that contributions to the ESA mainly originate from transitions between the lowest exciton and lowest biexciton manifold of states and hopefully in the case of type-II, from CS states. Monitoring these features as a function of population time results in Figure 6.19(a) and 6.19(b). Fourier transforming the grey curve in Figure 6.19(a) yields a period of $\sim$200 fs ($\sim$20 meV) which is similar to the values previously reported for the LO phonon mode in CdTe NCs.[76,77]

In the case of the ESA features, at times there seems to be shuffling of the signal amplitude between the blue and the pink markers. A similar phenomenon was observed in 2DPE experiments on CdSe NCs, summarized in Wong

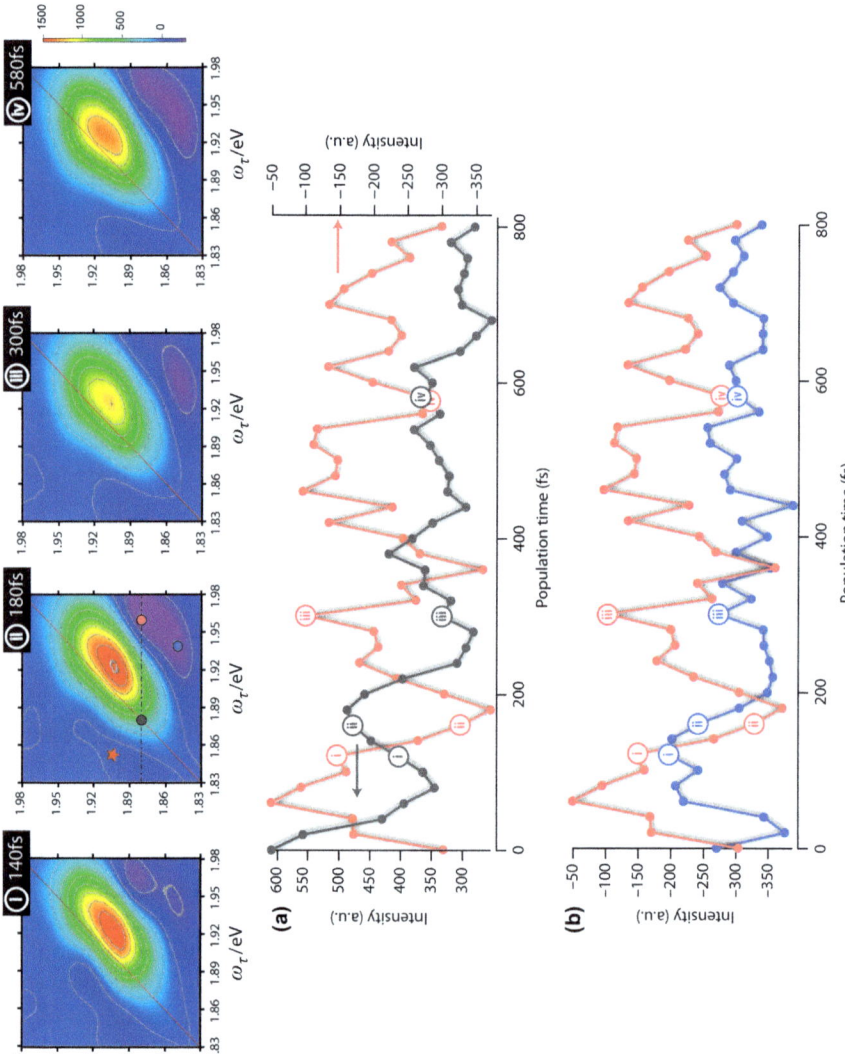

Figure 6.19   2D spectra of CdTe/CdSe (I) – Upper panels: 2D spectra at $T = 140, 180, 300$ and 580 fs. (a) Comparison of the evolution of an off-diagonal feature (pink marker) and a corresponding point in the diagonal feature (grey marker). (b) Same as (a) but for the two off-diagonal features indicated by pink and blue markers. Smaller features seems to be present in the region denoted by a red star.

and Scholes.[59,60] Using simulations of the 2D spectra, described in Section 6.4, it was shown that changes in these negative below-the-diagonal features correspond to different exciton–biexciton transitions. In that case, a higher energy peak is due to transition between the exciton state 0^u and the biexciton state $\pm 1^4$; meanwhile the lower energy feature was assigned to transitions between $\pm 1^u$ exciton and 0^4 biexciton state.

Other smaller features seem to be also present in the upper off-diagonal area of the 2D spectra (star in Figure 6.19), but they are obscured by the relatively large intensities of the bleach and the lower off-diagonal feature. In order to enhance them, a hyperbolic arcsine scaling is applied to the 2D spectra; this scaling is similar to a log-scale, but well-behaved for positive and negative features. The result of this scaling is presented in Figure 6.20 upper panels, where the upper off-diagonal features are now more visible. A negative feature, found at low energies, could be indicative of ESA into CS biexciton states, since CS states are lower in energy than locally excited (LE) states. Meanwhile, other features display a fair amount of change as a function of T. In particular, the feature indicated with a red marker shows an interesting behaviour; it appears as a positive feature at early T, then between 100 and 200 fs it suddenly turns negative, after which it becomes positive once again, remaining as such for rest of the T visited in our scans. The behavior of this particular feature was reproducible in all of the five different 2DPE scans performed on the sample. To understand better the information contained in the 2D spectra, it is therefore necessary to elucidate the origins of these features and a simulation needs to be performed.

6.5.1 Simulating the 2D Spectra of CdTe/CdSe Core/Shell Nanocrystals

To simulate the 2D spectra of CdTe/CdSe NCs we follow the procedure described in Section 6.4. The configurations for a type II system can be thought of as two sets of s and p-orbitals offset with respect to each other in a staggered alignment, as in Figure 6.21(a); here the blue lines represent the orbitals of the material with the smaller band-gap (CdTe), while the red lines, represent those of the larger band-gap material (CdSe). From this figure, it can be seen that in addition to the exciton and biexciton configurations found in Section 6.4.1, which will be denoted as LE from here on, we also find charge separated (CS) configurations, where the excited electron resides in the s-orbital of the second material.

Having CS configurations increases considerably the size of the exciton and biexciton configuration basis (from 12 to 24 configurations in the exciton case and from 15 to 90 configurations in the biexciton one), increasing the number of states accessible to the laser pulse. Once the wavefunctions of all the configurations are determined, we proceed to construct the CSF of the type-II model system as described in Section 6.4.1. Because there are biexciton configurations where there are four unpaired electrons, Figure 6.21(c), the branching diagram method, developed by Pauncz,[78] is employed to construct properly CSFs that are eigenfunction of $\hat{S}^2$; the details can be found in Lo.[79]

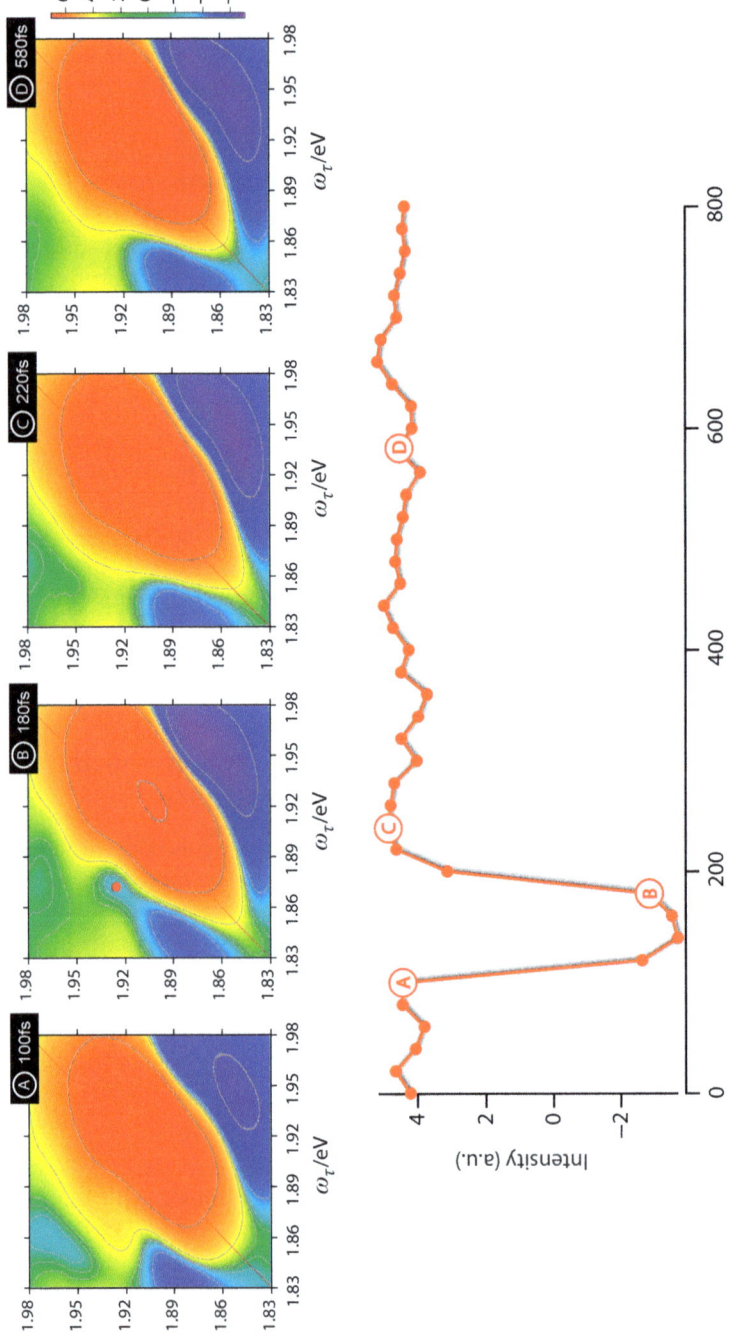

Figure 6.20 2D spectra of CdTe/CdSe (II) – Upper panels: hyperbolic arcsine of the 2D spectra shown in Figure 6.19. Smaller features, which were obscured in Figure 6.19, are now more apparent. Lower panel: evolution of an off-diagonal feature, red marker, as a function of population time.

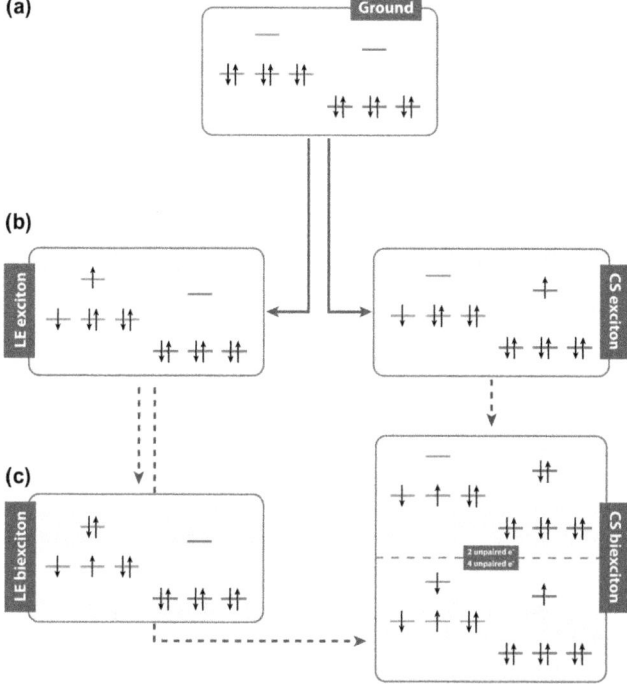

Figure 6.21   Pictorial representations of different configurations in a type-II system – The valance bands of each semiconductor comprising the type-II heterostructure are represented by three *p*-orbitals, and the conduction band by an *s*-orbital. Blue lines represent the conduction and valence band of the semiconductor with the smaller band-gap, while red lines represent those from the larger band gap material.

The Hamiltonian matrix for the lowest exciton and biexciton states in a type-II system is constructed using the one-electron matrix elements, Equations (6.18). By arranging the CSFs according to their total angular momentum F, the exciton and biexciton Hamiltonian is found to take a block-diagonal form, Figure 6.22 left panel. Diagonalization of each Hamiltonian produces the fine structure of the lowest exciton and biexciton state, Figure 6.22 center and right panels, respectively, of a system with a type-II band alignment. In this calculation, $E_{gap} = 2.29$ eV, $E_{CSgap} =$ and $k_{i,j} = 12$ meV; the spin–orbit coupling is chosen to be typical for CdTe, $\Delta = 927$ meV.[77,80,81] As a first approximation, we neglect exchange interactions when the electrons are in different subunits of the heterostructure.

The LE exciton states in the fine structure have an energy ordering (Figure 6.22 center panel, black lines) consistent with the literature.[82] We follow the conventional labelling for the fine structure states, that is, "F_x", where F is the total angular momnetum and x = u,l,csu,csl (upper, lower, charge separated upper and lower) to distinguish states with the same F values. The group of higher energy states contains the split-off band states, which are usually not shown because they are not accessible within the pulse-width of the

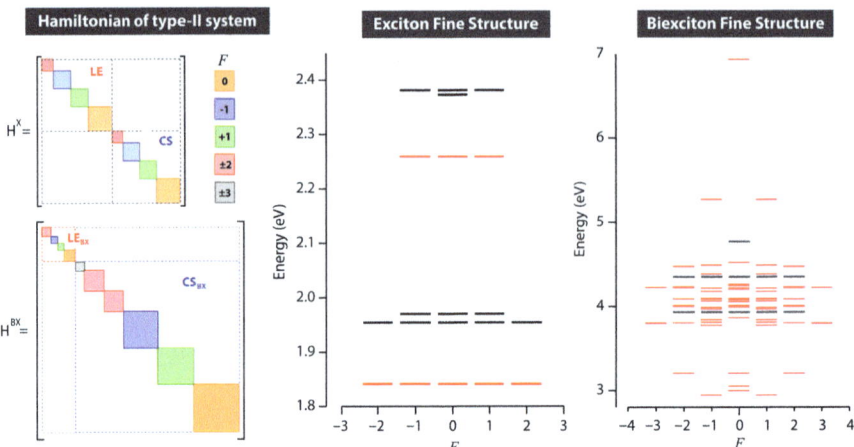

Figure 6.22   Density of states in a type-II system – Left panel: exciton and biexciton
Hamiltonian for a type-II system; center panel: Exciton fine strucuture
for a type-II system; and right panel: biexciton fine structure.

experiments. Inclusion of a second material in a staggered band alignment
configuration adds 12 extra (CS) exciton states (Figure 6.22 center panel, red
lines), with an ordering similar to that of the LE exciton states. Since we have
neglected exchange interactions between electrons located in different subunits
of the heterostructure, some degeneracy is recovered in the CS states compared
to the LE exciton states.

The LE biexciton states are shown in Figure 6.22 right panel, black lines. By
having CS configurations, there are an additional 75 biexciton states, making a
total of 90 biexciton states. In this case, we also follow the conventional
labeling of states, that is, "F_x", where F is the total angular momentum and
x = A,B,C,...; for CS biexciton states "cs" is added between F and x.

The procedure described in Section 6.4.2, is used to determine the exciton–
biexciton transitions. Equation 6.20 is modified to take into account the reduced transition dipole moment involving CS states:

$$\mu_{X_\alpha \to BX_\beta} = \sum_{i,j} c_i^{X_\alpha} \Theta \mu_{ij} c_j^{BX_\beta} \tag{6.25}$$

where $\mu_{ij} = 1$ when the transition between the exciton and the biexciton CSFs is
allowed and zero when it is forbidden. $\Theta \; \varepsilon \; [0,1]$ takes into account the reduced
transition dipole moment involving CS states. $c_i^{X_\alpha}$ and $c_j^{BX_\beta}$ are the coefficients
of CSF found in the wavefunction of exciton and biexciton states, respectively.
Figure 6.23 presents the allowed exciton–biexciton transitions within the
bandwidth of the laser. From this figure, it can be seen that many new pathways
are available to produce a signal compared to the single material case shown in
Figure 6.18. Finally, simulated 2D spectra for a type-II system are performed
following the prescription in Section 6.4.3.

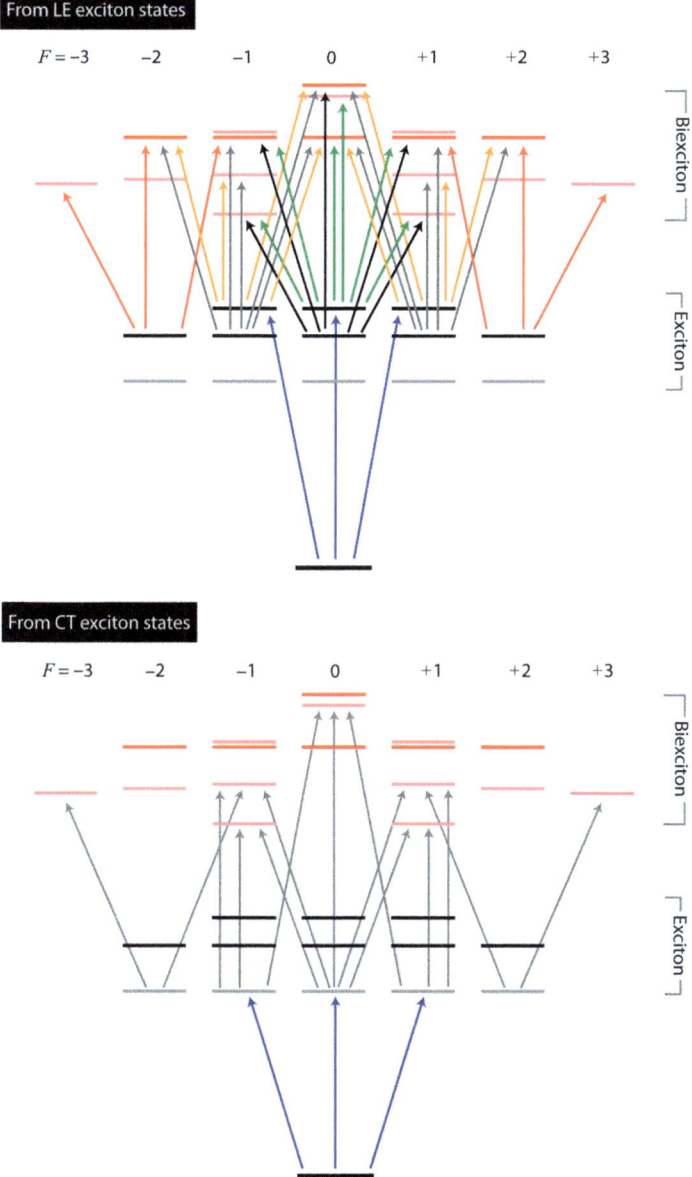

Figure 6.23 Exciton-biexciton transitions in a type-II system – Allowed transitions between: upper panel, LE exciton and biexciton states; lower panel, CS exciton and biexciton states.

6.5.2 Discussion

Figure 6.24(a) presents the simulated 2D spectra using the parameters described in the preceding subsections and plotted on an arcsinh intensity scale.

The inhomogeneous and homogeneous broadenings have been set artificially small to visualize better the different features originating from the various possible Liouville pathways. The GSB and ESA peaks along the diagonal region at ∼1.93 eV, result only from pathways involving LE states. Apart from these peaks, many other features are also seen in the simulated spectra; when

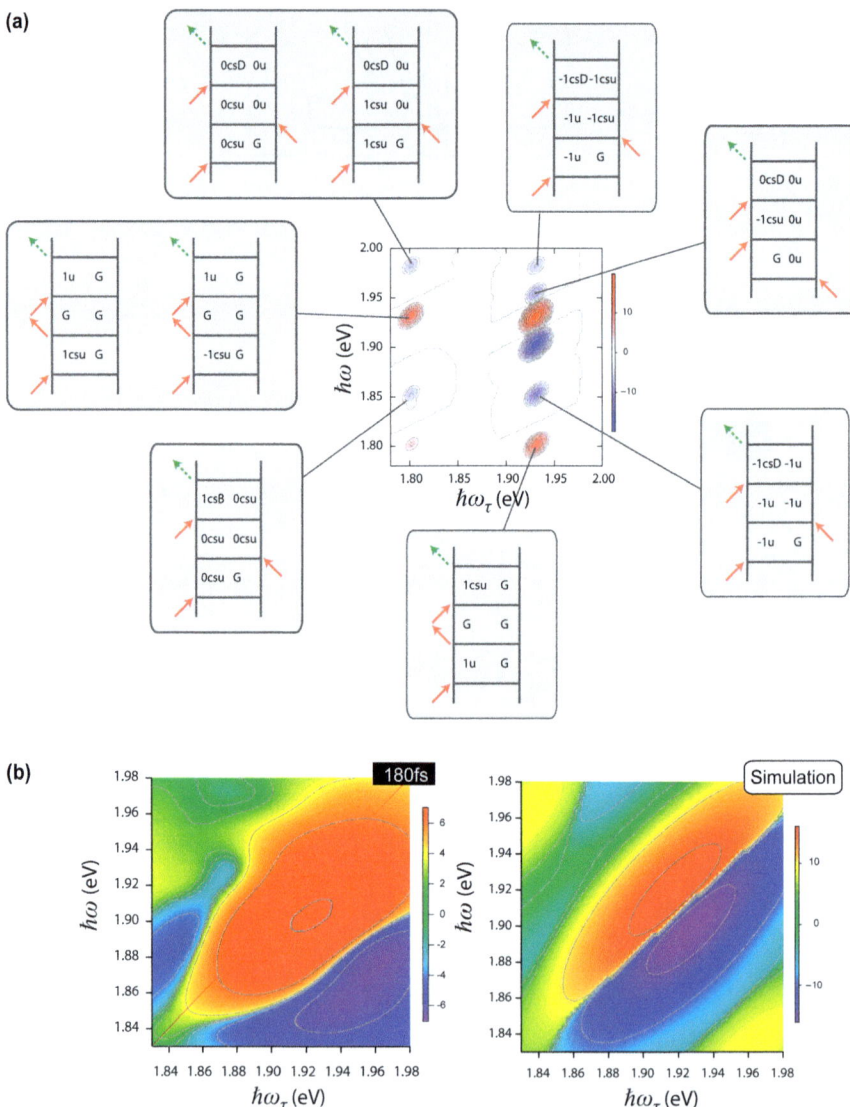

Figure 6.24   Simulated 2D spectra for a type-II system. (a) Simulated 2D with small broadening. Feymann diagrams depicting the Liouville pathway that gives rise to each signal are presented. (b) Comparison between experimental 2D spectra (left panel) and the simulated spectra (right panel) with more realistic broadening parameters.

the Liouville pathways that give rise to them are examined, it is found that CS exciton and biexciton states are involved. For example, the signal at $\hbar\omega_\tau \sim 1.80\,\mathrm{eV}$ and $\hbar\omega \sim 2.00\,\mathrm{eV}$ arises from the formation of a coherent superposition between the ground and the CS exciton state "0csu" (by interaction with the first pulse), followed by the emission from a coherent state between the CS biexciton "0csD" and the LE exciton "0u" (formed after interaction with the other two pulses). Some of the Liouville pathways leading to each of the other features are presented, showing that they would not appear in the 2D spectra if the CS exciton and biexciton states were absent.

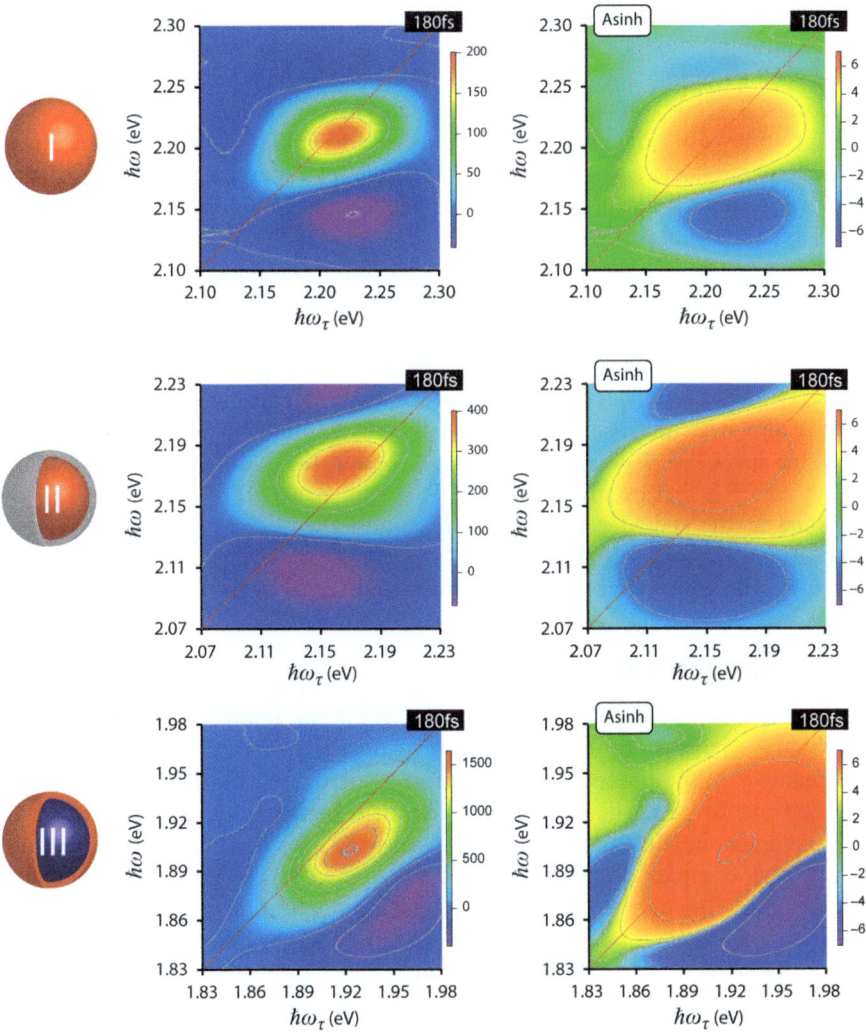

Figure 6.25   2D spectra. Representative 2D spectra at $T = 180$ fs. Sample I: CdSe, sample II: CdSe/CdS/ZnS and sample III CdTe/CdSe.

The simulated spectra, in a sinh scale, with more realistic broadening are presented in the right panel of Figure 6.24(b). Little agreement is observed in the off-diagonal features when compared to the experimentally obtained 2D spectra. This could be due to various factors: (1) Simulation of the signal might be too simplistic and a proper calculation of the lineshape function is required. (2) Exchange interactions between electrons located in different subunits may be important. (3) Strain effects, which have not been accounted in our model, may produced significant contributions. Manipulation of strain in hetero-structures has been shown significantly to modify the electronic structure of the nanoparticle, for example, Smith *et al.*[83] found that heterostructured NCs composed of CdTe and ZnSe, which has a type-I band alignment in the bulk, can be made to behave as a type-II heterostructure by varying lattice strain. The same authors suggest in a recent paper that large lattice mismatch between core and shell material can "bend" the band structure at the interface.[28] In another theoretical study on heterostructured nanowires (NW),[84] CdSe/CdTe were found to have a reduced band-gap compared to CdSe NW, with contributions from strain effects accounting for ~ 12% of this reduction. Therefore, proper incorporation of strain effects in our model may result in better agreement simulation and experiment.

Nonetheless, we find that features appearing in the 2D spectra of CdTe/CdSe heterostructures are qualitatively different from those found in CdSe core-only and CdSe/CdS/ ZnS type-I heterostructure, Figure 6.25. In the 2D spectra of CdSe and CdSe/CdS/ZnS samples, similar features are found; mainly an el-lipsoidal diagonal bleach signal and a strong ESA below-the-diagonal feature, consistent with the results presented in refs 59 and 60. Meanwhile, the 2D spectra for CdTe/CdSe shows an ESA beside the bleach and a low energy negative ESA feature above the bleach signal. Since all of these samples were probed at the red-edge of their corresponding absorption spectra, these ob-servations suggest that the resulting differences arises due to states present in type-II NCs that are absent in core-only and type-I NCs.

6.6 Conclusion

2D electronic spectroscopy extends the utility of femtosecond pump probe techniques because of it ability to correlate the transition energies of states populated by the pump pulse with those interrogated by the probe. This has turned out to be particularly useful for identifying exciton to biexciton tran-sitions in semiconductor nanocrystals. In principle, the method can help un-ravel the complex transient spectra associated with photoinduced charge separation. However, challenges include identification of spectral signatures of carriers or charge-separated states, which may have broad lineshapes or low intensity. An experimental issue is that a large spectral window needs to be probed to detect exciton bleach and documented polaron bands and that is difficult to achieve with a transform-limited laser pulse. Overall, though, 2D spectroscopy will enable more incisive probes of the interplay between excitons and charge separated states on ultrafast timescales.

Acknowledgements

The authors thank Roman Vaxenburg and Efrat Lifshitz for providing the CdTe/CdSe QD samples. The Natural Sciences and Engineering Research Council of Canada is gratefully acknowledged for support of this research.

References

1. A. Efros and M. Rosen, *Annu. Rev. Mater. Sci.*, 2000, **30**, 475.
2. A. J. Nozik, M. C. Beard, J. M. Luther, M. Law, R. J. Ellingson and J. C. Johnson, *Chem Rev.*, 2010, **110**, 6873.
3. P. V. Kamat, *J. Phys. Chem. C*, 2008, **112**, 18737.
4. I. Mora-Seró and J. Bisquert, *J. Phys. Chem. Lett.*, 2010, **1**, 3046.
5. S. Rühle, M. Shalom and A. Zaban, *ChemPhysChem.*, 2010, **11**, 2290.
6. W. A. Tisdale, K. J. Williams, B. A. Timp, D. J. Norris, E. S. Aydil and X.-Y. Zhu, *Science*, 2010, **328**, 1543.
7. A. Pandey and P. Guyot-Sionnest, *J. Phys. Chem. Lett.*, 2010, **1**, 45.
8. R. Schaller, M. Sykora, J. Pietryga and V. Klimov, *Nano Lett.*, 2006, **6**, 424.
9. V. I. Klimov, *Annu. Rev. Phys. Chem.*, 2007, **58**, 635.
10. V. I. Klimov, *J. Phys. Chem. B*, 2006, **110**, 16827.
11. G. Nair and M. G. Bawendi, *Phys. Rev. B*, 2007, **76**, 081304.
12. G. Nair, S. M. Geyer, L.-Y. Chang and M. G. Bawendi, *Phys. Rev. B*, 2008, **78**, 125325.
13. G. Nair, L.-Y. Chang, S. M. Geyer and M. G. Bawendi, *Nano Lett.*, 2011, **11**, 2145.
14. M. C. Beard, A. G. Midgett, M. C. Hanna, J. M. Luther, B. K. Hughes and A. J. Nozik, *Nano Lett.*, 2010, **10**, 3019.
15. V. Sukhovatkin, S. Hinds, L. Brzozowski and E. H. Sargent, *Science*, 2009, **324**, 1542.
16. S. J. Kim, W. J. Kim, A. N. Cartwright and P. N. Prasad, *Appl. Phys. Lett.*, 2008, **92**, 191107.
17. J. Huang, Z. Huang, Y. Yang, H. Zhu and T. Lian, *J. Am. Chem. Soc.*, 2010, **132**, 4858.
18. A. Kortan, R. Hull, R. Opila, M. Bawendi, M. Steigerwald, P. Carroll and L. Brus, *J. Am. Chem. Soc.*, 1990, **112**, 1327.
19. M. Hines and P. Guyot-Sionnest, *J. Phys. Chem.*, 1996, **100**, 468.
20. X. Peng, M. Schlamp, A. Kadavanich and A. Alivisatos, *J. Am. Chem. Soc.*, 1997, **119**, 7019.
21. B. Dabbousi, J. Rodriguez-Viejo, F. Mikulec, J. Heine, H. Mattoussi, R. Ober, K. Jensen and M. Bawendi, *J. Phys. Chem. B*, 1997, **101**, 9463.
22. S. F. Lee and M. A. Osborne, *Chem. Phys. Chem.*, 2009, **10**, 2174.
23. D. Milliron, S. Hughes, Y. Cui, L. Manna, J. Li, L. Wang and A. Alivisatos, *Nature*, 2004, **430**, 190.
24. J. E. Halpert, V. J. Porter, J. P. Zimmer and M. G. Bawendi, *J. Am. Chem. Soc.*, 2006, **128**, 12590.

25. R. Xie, U. Kolb and T. Basche, *Small*, 2006, **2**, 1454.
26. S. Kumar, M. Jones, S. S. Lo and G. D. Scholes, *Small*, 2007, **3**, 1633.
27. D. Dorfs, T. Franzl, R. Osovsky, M. Brumer, E. Lifshitz, T. A. Klar and A. Eychmueller, *Small*, 2008, **4**, 1148.
28. A. M. Smith and S. Nie, *Acc.Chem. Res.*, 2010, **43**, 190.
29. S. S. Lo, T. Mirkovic, C.-H. Chuang, C. Burda and G. D. Scholes, *Adv. Mater.*, 2011, **23**, 180.
30. H. McDaniel, P. E. Heil, C.-L. Tsai, K. K. Kim and M. Shim, *ACS Nano*, 2011, **5**, 7677.
31. Z. Ning, H. Tian, C. Yuan, Y. Fu, H. Qin, L. Sun and H. Ågren, *Chem Commun.*, 2011, **47**, 1536.
32. N. N. Hewa-Kasakarage, P. Z. El-Khoury, N. Schmall, M. Kirsanova, A. Nemchinov, A. N. Tarnovsky, A. Bezryadin and M. Zamkov, *Appl. Phys. Lett.*, 2009, **94**, 133113.
33. N. N. Hewa-Kasakarage, P. Z. El-Khoury, A. N. Tarnovsky, M. Kirsanova, I. Nemitz, A. Nemchinov and M. Zamkov, *ACS Nano*, 2010, **4**, 1837.
34. C.-H. Chuang, S. S. Lo, G. D. Scholes and C. Burda, *J. Phys. Chem. Lett.*, 2010, **1**, 2530.
35. C.-H. Chuang, T. L. Doane, S. S. Lo, G. D. Scholes and C. Burda, *ACS Nano*, 2011, **5**, 6016.
36. G. D. Scholes, M. Jones and S. Kumar, *J. Phys. Chem. C*, 2007, **111**, 13777.
37. D. Oron, M. Kazes and U. Banin, *Phys. Rev. B*, 2007, **75**, 035330.
38. Z. Deutsch, A. Avidan, I. Pinkas and D. Oron, *Phys. Chem. Chem. Phys.*, 2011, **13**, 3210.
39. R. Osovsky, D. Cheskis, V. Kloper, A. Sashchiuk, M. Kroner and E. Lifshitz, *Phys. Rev. Lett.*, 2009, **102**, 197401.
40. T. Brixner, T. Mancal, I. V. Stiopkin and G. R. Fleming, *J. Chem. Phys.*, 2004, **121**, 4221.
41. D. M. Jonas, *Annu. Rev. Phys. Chem.*, 2003, **54**, 425.
42. M. L. Cowan, J. P. Ogilvie and R. J. D. Miller, *Chem. Phys. Lett.*, 2004, **386**, 184.
43. J. D. Hybl, A. A. Ferro and D. M. Jonas, *J. Chem. Phys.*, 2001, **115**, 6606.
44. M. H. Cho, T. Brixner, I. Stiopkin, H. Vaswani and G. R. Fleming, *J. Chin. Chem. Soc.*, 2006, **53**, 15.
45. G. R. Fleming and M. H. Cho, *Annu. Rev. Phys. Chem.*, 1996, **47**, 109.
46. W. P. de Boeij, M. S. Pshenichnikov and D. A. Wiersma, *Annu. Rev. Phys. Chem.*, 1998, **49**, 99.
47. T. H. Joo, Y. W. Jia, J. Y. Yu, M. J. Lang and G. R. Fleming, *J. Chem. Phys.*, 1996, **104**, 6089.
48. Y. J. Yan and S. Mukamel, *Phys. Rev. A*, 1990, **41**, 6485.
49. W. P. deBoeij, M. S. Pshenichnikov and D. A. Wiersma, *J. Phys. Chem.*, 1996, **100**, 11806.
50. J. D. Hybl, S. M. G. Faeder, A. W. Albrecht, C. A. Tolbert, D. C. Green and D. M. Jonas, *J. Lumin.*, 2000, **87–9**, 126.
51. J. D. Hybl, A. Yu, D. A. Farrow and D. M. Jonas, *J. Phys. Chem. A*, 2002, **106**, 7651.

52. K. Kwac and M. Cho, *J. Phys. Chem. A*, 2003, **107**, 5903.
53. S. T. Roberts, J. J. Loparo and A. Tokmakoff, *J. Chem. Phys.*, 2006, **125**.
54. Y. C. Cheng and G. R. Fleming, *J. Phys. Chem. A*, 2008, **112**, 4254.
55. T. Brixner, J. Stenger, H. M. Vaswani, M. Cho, R. E. Blankenship and G. R. Fleming, *Nature*, 2005, **434**, 625.
56. L. Lepetit, G. Cheriaux and M. Joffre, *J. Opt. Soc. Am. B*, 1995, **12**, 2467.
57. C. Dorrer, N. Belabas, J. P. Likforman and M. Joffre, *J. Opt. Soc. Am. B*, 2000, **17**, 1795.
58. R. M. Mersereau and A. V. Oppenheim, *Proc. IEEE*, 1974, **62**, 1319.
59. C. Y. Wong and G. D. Scholes, *J. Lumin.*, 2011, **131**, 366.
60. C. Y. Wong and G. D. Scholes, *J Phys Chem A*, 2011, **115**, 3797.
61. M. Nirmal and L. Brus, *Acc. Chem. Res.*, 1999, **32**, 407.
62. P. K. Basu, *Theory of Optical Processes in Semiconductors*, Oxford University Press, 1997.
63. G. D. Scholes, J. Kim, C. Y. Wong, V. M. Huxter, P. S. Nair, K. P. Fritz and S. Kumar, *Nano Lett.*, 2006, **6**, 1765.
64. D. Abramavicius, B. Palmieri, D. V. Voronine, F. Sanda and S. Mukamel, *Chem. Rev.*, 2009, **109**, 2350.
65. V. Huxter, V. Kovalevskij and G. Scholes, *J. Phys. Chem. B*, 2005, **109**, 20060.
66. J. Kim, C. Y. Wong, P. S. Nair, K. P. Fritz, S. Kumar and G. D. Scholes, *J. Phys. Chem. B*, 2006, **110**, 25371.
67. J. Kim, C. Y. Wong and G. D. Scholes, *Acc. Chem. Res.*, 2009, **42**, 1037.
68. G. D. Scholes, J. Kim and C. Y. Wong, *Phys. Rev. B*, 2006, **73**, 195325.
69. C. Y. Wong, J. Kim, P. S. Nair, M. C. Nagy and G. D. Scholes, *J. Phys. Chem. C*, 2009, **113**, 795.
70. A. P. Alivisatos, A. L. Harris, N. J. Levinos, M. L. Steigerwald and L. E. Brus, *J. Chem. Phys.*, 1988, **89**, 4001.
71. G. D. Scholes, *ACS Nano*, 2008, **2**, 523.
72. R. R. Cooney, S. L. Sewall, D. M. Sagar and P. Kambhampati, *Phys. Rev. Lett.*, 2009, **102**, 127404.
73. S. L. Sewall, R. R. Cooney, K. E. H. Anderson, E. A. Dias, D. M. Sagar and P. Kambhampati, *J. Chem. Phys.*, 2008, **129**, 084701.
74. V. I. Rupasov and V. I. Klimov, *Phys. Rev. B*, 2007, **76**, 125321.
75. M. C. Beard, A. G. Midgett, M. C. Hanna, J. M. Luther, B. K. Hughes and A. J. Nozik, *Nano Lett.*, 2010, **10**, 3019.
76. V. S. Vinogradov, G. Karczewski, I. V. Kucherenko, N. N. Mel'nik and P. Fernandez, *Phys. Solid State*, 2008, **50**, 164.
77. H. Zhong, M. Nagy, M. Jones and G. D. Scholes, *J. Phys. Chem. C*, 2009, **113**, 10465.
78. R. Pauncz, *Spin Eigenfunctions*, Plenum Press, New York, 1979.
79. S. S. Lo, *Charge Transfer Processes in the Excited Dynamics of II-VI Semiconductor Nanocrystals*, PhD Thesis, University of Toronto, 2011.
80. S. Bloom and T. Bergstre, *Solid State Commun.*, 1968, **6**, 465.

81. Y. Al-Douri, R. Khenata, Z. Chelahi-Chikr, M. Driz and H. Aourag, *J. Appl. Phys.*, 2003, **94**, 4502.
82. A. Efros, M. Rosen, M. Kuno, M. Nirmal, D. Norris and M. Bawendi, *Phys. Rev. B*, 1996, **54**, 4843.
83. A. M. Smith, A. M. Mohs and S. Nie, *Nat. Nanotechnol.*, 2009, **4**, 56.
84. S. Yang, D. Prendergast and J. B. Neaton, *Nano Lett.*, 2010, **10**, 3156.

CHAPTER 7

Ultrafast Optical Imaging and Microspectroscopy

PIOTR PIOTROWIAK,*[a] LIBAI HUANG[b] AND
LARS GUNDLACH[c]

[a] Department of Chemistry, Rutgers University, Newark, New Jersey 07102, USA; [b] Radiation Laboratory, University of Notre Dame, Notre Dame, Indiana 46556, USA; [c] Department of Chemistry & Biochemistry, University of Delaware, Newark, Delaware 19716, USA
*Email: piotr@andromeda.rutgers.edu

7.1 Introduction

Ultrafast microscopy combines the capabilities of femtosecond pump probe spectroscopy with the ability to image and monitor individual objects, down to the single particle and single molecule level. This combination of temporal, spatial and spectral resolution is necessary when studying charge and energy transfer processes in heterogeneous and nanostructured materials, which are increasingly common in the fields of photovoltaics and photocatalysis. While tunable femtosecond lasers became ubiquitous excitation sources in multiphoton confocal microscopy, time resolved microscopy is still largely dominated by time correlated single photon counting (TCSPC) (Figure 7.1). Unfortunately, even the best $\sim 20\,ps$ (more typically $100\,ps$) temporal resolution of the increasingly popular TCSPC confocal microscopes is inadequate for the study of interfacial exciton and polaron dynamics in materials relevant to solar energy conversion which occur on a sub-picosecond timescale. Acquisition of images evolving on the femtosecond timescale, similar to the

RSC Energy and Environment Series No. 8
Solar Energy Conversion: Dynamics of Interfacial Electron and Excitation Transfer
Edited by Piotr Piotrowiak

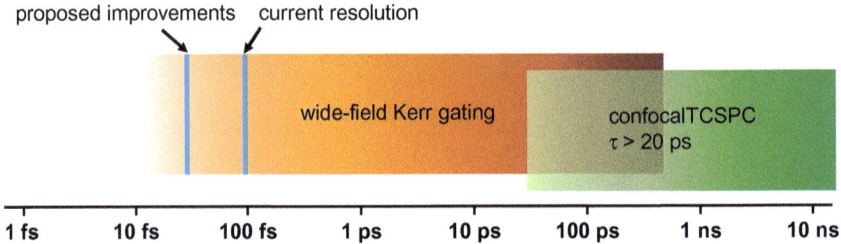

Figure 7.1   Time domains accessible to the wide field Kerr-gated fluorescence micro-
             scope and to confocal fluorescence microscopes relying on time-
             correlated-single photon-counting (TCSPC).

acquisition of femtosecond photoluminescence and absorption spectra, must
rely on optical gating techniques as there are no electronic devices capable of
operating in this time domain. While the literature on ultrafast microscopy is
still modest, several purely optical[1–5] and hybrid approaches[6–8] have been
successfully tested and implemented. The number of reports of microscopy
experiments in which the barrier of 1ps has been broken is steadily growing and
a few landmark demonstrations of sub-100 fs imaging already exist.[2,7]

   In this chapter we describe two all-optical realizations of femtosecond microscopy which are most general and straightforward to implement: wide-field
Kerr gated fluorescence microscopy (KGFM)[1,2] and scanning pump probe
transient absorption microscopy (PPTAM).[5] The purely optical approaches to
ultrafast microscopy can be classified on the basis of the location of the gating,
which can occur either in the sample itself or in an external medium and on the
wide-field *vs.* point-by-point scanning imaging mode, the latter including
confocal and near-field scanning optical microscopy (NSOM). The methods in
which the gating occurs in an external medium include Kerr gating,[2,9–11]
fluorescence up-conversion[1,12] and parametric light amplification.[13] Gating
the collected image away from the sample provides the mildest and arguably
most general means of probing ultrafast processes. In contrast, in pump probe
time resolved absorption,[5] second harmonic generation (SHG), sum frequency
generation (SFG) and in the *in situ* IR up-conversion,[4] the image gating and
hence the time resolution take place in the sample. These ultrafast imaging
methods can be very powerful and rich in chemical information, however, they
are inevitably most sample dependent and, except for the transient absorption,
least general. Their advantage, at least in most instances, is the simplicity of the
optical layout.

   Any optical gating scheme can be applied either in the scanning or wide-field
arrangement, although certain combinations are more advantageous and easier
to implement than others. Wide-field methods require an amplified laser as the
source of the gating pulse, which typically must be in the microjoule range.
Wide-field techniques allow simultaneous monitoring of the dynamics of
numerous objects in a typical $100 \times 100\,\text{m}\mu$ image area. In practice, this
multiplexing effect easily compensates for the lower per-photon efficiency of the

wide-field mode and allows collection of a number of trajectories in a single time scan. Most scanning ultrafast microscopy experiments can be performed by simpler to operate and less expensive femtosecond oscillators operating at typical 70–100 MHz repetition rates and offer higher spatial resolution (~ 200 *vs.* 500 nm) at the expense of being able to monitor only a single object or a single pixel at a time. As a result, in a typical experiment, most of the time is spent on scanning the background. The efficiency for each excitation photon is considerably higher for the confocal method but only when the emissive object has already been identified. It should be noted that in standard scanning confocal microscopy only the disappearance of the excited state species from the focal area is monitored. It is therefore not possible to distinguish population decay from energy migration away from the observation point. The wide-field approach allows monitoring of the spatial redistribution of the excited state species and in some instances the charge carriers within the view of the microscope, not only their decay to the ground state. The most powerful combination that has not yet been implemented in the ultrafast regime is selective confocal excitation coupled with wide-field image collection. The highest spatial resolution among all-optical methods is offered by near-field scanning fluorescence microscopy (NSOM), which unfortunately suffers from extremely low light throughput. Only isolated examples of time-resolved NSOM measurements with a relative low time resolution were reported,[3] attesting to the difficulty of this technique.

7.2 Kerr-Gated Femtosecond Fluorescence Microscopy and Micro-Spectroscopy (KGFM)

7.2.1 Background and Experimental Considerations

Kerr gating relies on transient birefringence induced by a light pulse in a nonlinear medium placed between crossed polarizers.[14] The collected luminescence can pass through the Kerr shutter only when the gating pulse is incident upon the Kerr medium. By delaying the gate pulse with respect to the excitation pulse in the same fashion as in pump probe experiments, the time evolution of the imaged object, $I(x,y;t)$, can be followed and the corresponding emission decays can be assembled. The gated wide-field image is recorded using a standard research-grade charge coupled device (CCD) camera. Using a thin silica plate as the Kerr medium we routinely reach a time resolution of 100 fs at excitation and detection wavelengths from 300–500 nm and 400–1000 nm, respectively. Importantly, a high-throughput spectrometer can be inserted into the path of the gated light allowing the collection of time-resolved emission spectra from the selected area of the sample (Figure 7.2). This methodology has significant advantages over the alternative approaches. It is a wide-field technique and as a result it allows collection of 'snapshots' of several of objects or of an area of the sample, without scanning. This feature enables us to acquire femtosecond emission 'movies' with the frame resolution of 100 fs (Figure 7.4).

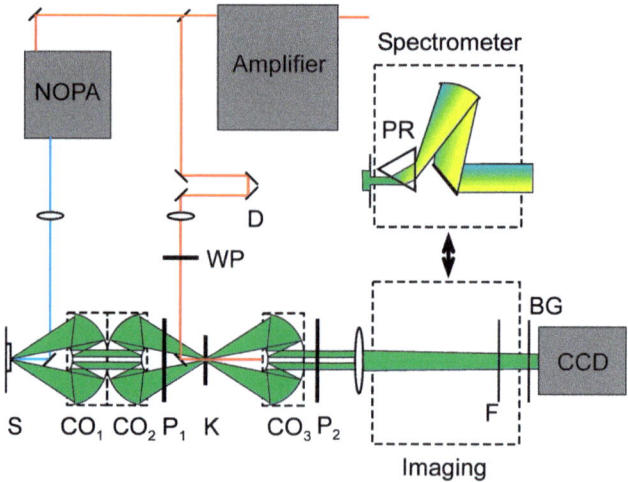

Figure 7.2   Optical block diagram of the wide field Kerr-gated microscope. Note the
position of the sample (S), the sequence of three matched Cassegrain
objectives (CO_1, CO_2 and CO_3), polarizers (P_1 and P_2), the Kerr medium
(K) and blocking filters (F). A prism spectrometer (PR) can be inserted
into the path of the gated light allowing monitoring of the collective
spectral dynamics of objects within the field of view.

The most salient technical aspects of Kerr-gated fluorescence microscopy are
briefly reviewed in the following paragraphs. The main challenge of the Kerr
shutter approach is the low throughput of the gate, which in the case of silica
reaches 5% before the onset of higher order nonlinear effects. Despite
numerous efforts,[15,16] no material combining silica's essentially instantaneous
response and high damage threshold with substantially higher third order
susceptibility $\chi^{(3)}$ has been found. The $\chi^{(3)}$ coefficient is correlated to the linear
refractive index of the material, n, which in turn depends on the band gap.
Heavy metal oxide glasses (Pb, Bs) have $\chi^{(3)}$ values on the order of 10^{-12} esu *vs.*
silica's 3×10^{-14}, however, their 2-photon absorption cross sections are large
and lead to strong background fluorescence. In an attempt to improve this
aspect of the Kerr-gated microscope, we tested the performance of several
crystalline media among which yttrium aluminum garnet (YAG) and gadolinium gallium garnet (GGG) emerged as possible substitutes for silica with a
3–6 higher gating efficiency albeit at the expense of a slightly slower response.[17]
It should be kept in mind that even with the current low light throughput, Kerr
gating exceeds the efficiency of NSOM by several orders of magnitude and
offers substantially better time resolution.

Since a molecule or a quantum dot can emit no more than one photon per
excitation, the optimum source for ultrafast fluorescence microscopy has a high
repetition rate (10^5–10^7 Hz) and a pulse energy sufficient to pump an excitation
NOPA (noncollinear optical parametric amplifier) and to drive the Kerr or up-
conversion gate. The current configuration of the Rutgers microscope relies on
a cryogenically cooled Ti:sapphire regenerative amplifier which produces 50 µJ

pulses of 50 fs duration at 100 kHz repetition rate. Approximately 5 μJ of the amplifier's output is used to pump the homebuilt NOPA and the remainder is directed to driving the Kerr gate. Our instrument utilizes exclusively Schwartzchild–Cassegrain reflective objectives because they are free of chromatic aberration and group velocity dispersion (GVD). The only optical elements contributing to the GVD are the first nanowire polarizer of the shutter and the 0.5 mm thick Kerr medium itself (Figure 7.3). The objectives are specifically designed for microscopy and allow the internal reflectors to be adjusted enabling us to reach the diffraction limited resolution of 0.4 μm (numerical aperture, NA = 0.5, λ = 500 nm).

The 'on-off' contrast of the Kerr shutter is a very important parameter which determines the range of emission lifetimes that can be monitored. The residual leakage through the closed gate is integrated by the CCD camera over the entire

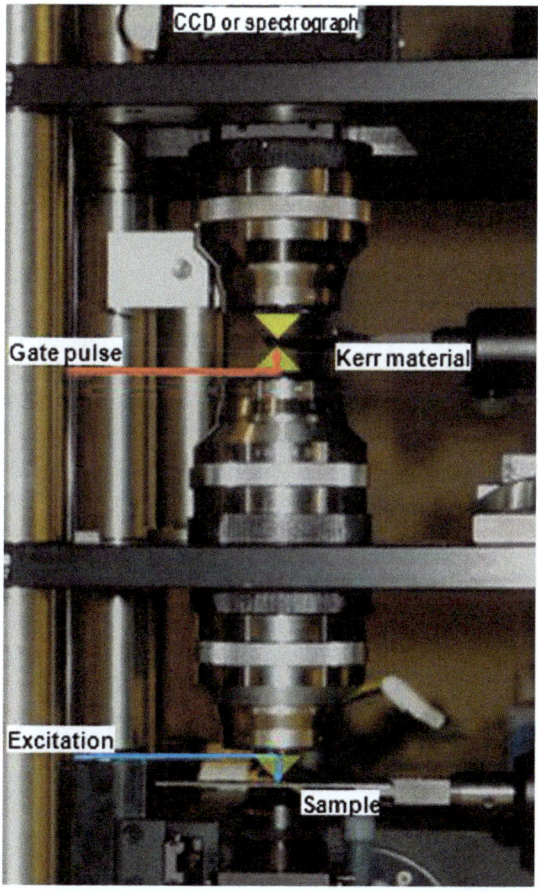

Figure 7.3 Light collecting objective and the Kerr-gate assembly of the second generation Rutgers microscope. The only non-reflective components are the nanowire polarizers and the 0.5. mm thick fused silica plate serving as the Kerr medium.

emission lifetime. If this leakage becomes too large, it can overwhelm the gated signal. The contrast depends on the extinction ratio of the crossed polarizers. With the extinction of 3000:1 (leakage of 0.033%) and a 100 fs gate (approximate parameters of our current system), measurements of emission lifetimes up to $\tau \approx 300$ ps are possible. Controlling the light leakage becomes more challenging when the sample exhibits a multi-exponential lifetime, a situation frequently encountered in inhomogeneous materials such as the dye-sensitized metal oxide particles. The longest lived components contribute the most to the time-integrated leakage and have to be suppressed. Nanowire polarizers of the newest generation are capable of extinction ratios in excess of 100 000:1, however, in the current Kerr-shutter layout the fluorescence incident on the polarizers is not collimated, leading to a substantially lower contrast and higher leakage. Further suppression of a long-lived background at kHz rates could be achieved by placing a Pockels cell in series with the Kerr shutter and post-gating the transmitted photoluminescence.

7.2.2 Excitation Fluence, Orientation, Shape and Size Dependence of Carrier Dynamics in CdS$_x$Se$_{1-x}$ Nanobelts

The Kerr-gated fluorescence microscope was applied to investigate the ultrafast carrier relaxation and carrier recombination pathways in individual semiconductor nanobelts.[10] On the basis of the single particle measurements we were able to disentangle the complex temporal evolution that was reported earlier for ensemble measurements on the same materials.[18] We found that the luminescence dynamics of highly excited CdS$_x$Se$_{1-x}$ nanobelts exhibits a non-linear dependence on the excited carrier density. We were able to reproduce the complex luminescence dynamics in a single particle using an amplified spontaneous emission (ASE) kinetic model containing three competing relaxation mechanisms. The high temporal resolution enabled us to determine the rates for relaxation pathways that are linear, quadratic and cubic in the carrier concentration.

Previous studies had to rely on ensemble averaging techniques with time resolution which was insufficient to resolve the very early dynamics. A group of CdS$_x$Se$_{1-x}$ particles is shown in Figure 7.4. Several clusters of small particles s well as one long nanobelt are present. The photoluminescence dynamics of this same ensemble was recorded in a Kerr-gated movie available online. Selected frames are shown in Figure 7.4. The entire movie spans 16 ps and consists of frames taken at 100 fs intervals which reveal the highly inhomogeneous excited state dynamics of the group of nanobelts. It can be seen that the fluorescence is not emitted homogeneously but is concentrated at certain areas and that each particle has its own characteristic dynamics.

Switching from the imaging to the spectrally resolved mode of the Kerr-gated microscope reveals the time dependent emission spectrum of the nanobelts (Figure 7.5). In this case, for the sake of simplicity we placed only two nanobelts in the view of the microscope. The luminescence intensity shown at the top

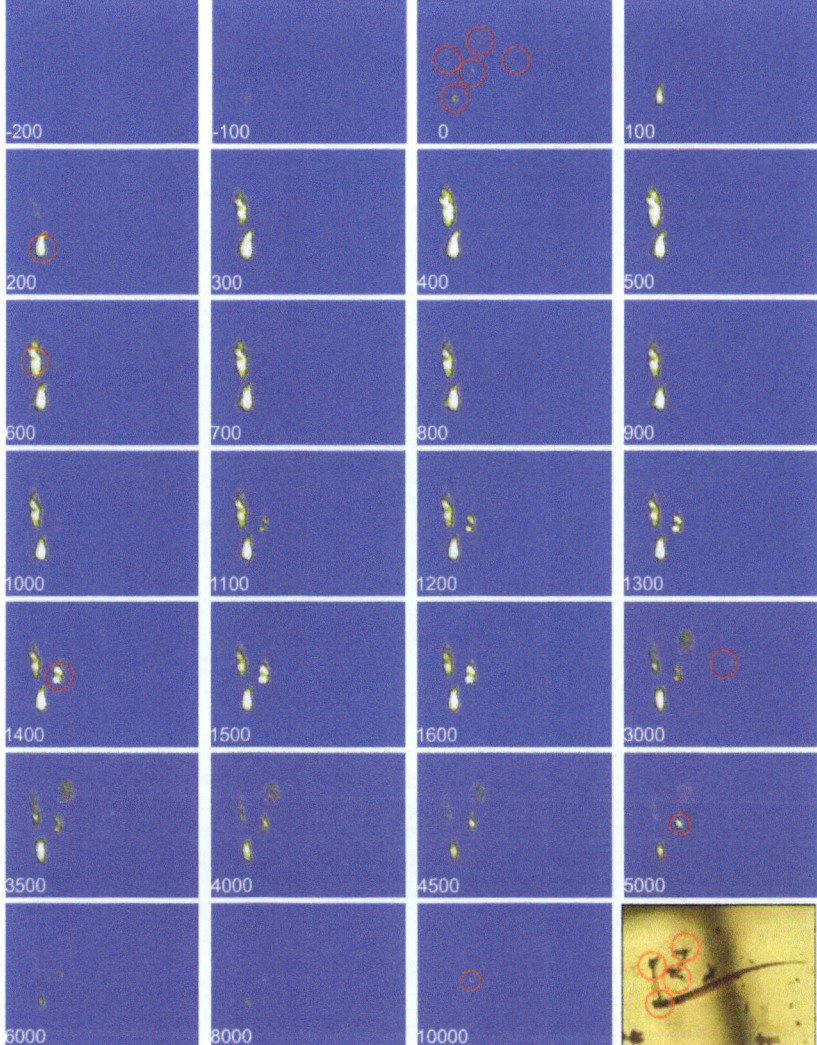

Figure 7.4 Sequence of 100 fs frames selected from a stroboscopic fluorescence movie of several simultaneously excited CdS_xSe_{1-x} nanobelts. The dark-field image of the same group is shown in the last frame (bottom right). The red circles highlight the distinct photoluminescence dynamics of individual nanobelts.

of Figure 7.5 exhibits unusual decay dynamics. We already know from the Kerr gated emission movie in Figure 7.4 that the time-dependent spectrum in Figure 7.5 is composed of contributions from individual particles which start emitting at different delay times. Since all the particles have very similar compositions and therefore very similar emission wavelengths, the contributions originating from spatially distinct particles cannot be disentangled in the time-resolved, ensemble averaged spectrum. The complexity of the

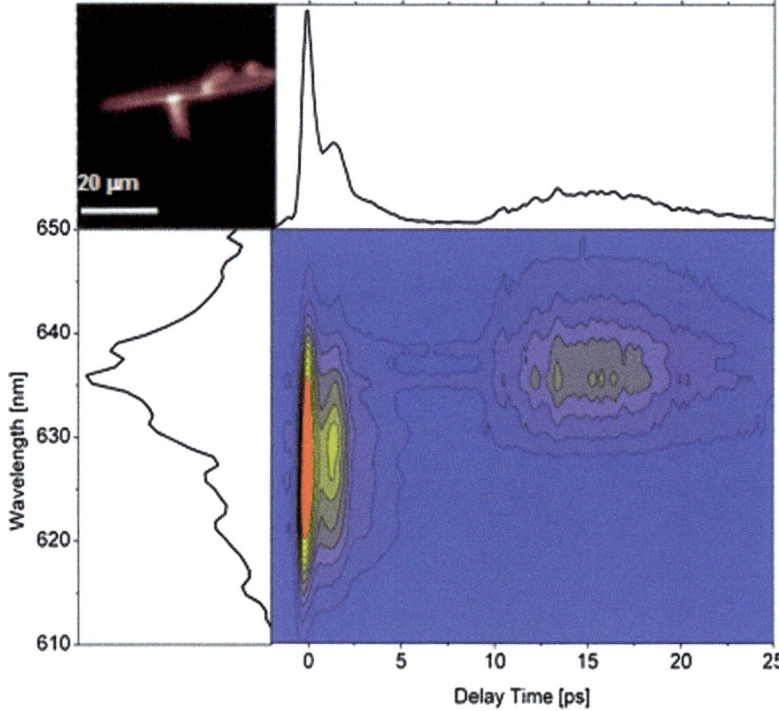

Figure 7.5   Ultrafast spectrally resolved dynamics of two CdS$_x$Se$_{1-x}$ nanobelts. The fluorescence image is shown in the upper left corner. Note that the broad, time-integrated photoluminescence spectrum shown on the left consists of two distinct features centered at 628 and 636 nm with vastly different dynamics that is clearly discernible in the 2D map.

temporal response of the ensemble and the superposition of the spectra underscore the need for single particle measurements.

The origin of the unusual and complex luminescence dynamics seen in Figures 7.4 and 7.5 becomes clear if one considers the strong excitation fluence-dependent shift of the emission onset. In Figure 7.4 different particles start emitting at different delay times owing to different densities of photogenerated charge carriers. The reason for the different initial exciton density is the strong absorbtion anisotropy of CdSSe. Since in our experiments the excitation pulse is linearly polarized and the different particles are randomly oriented, each nanobelt absorbs a different amount of light depending on its orientation. This leads to the generation of a different density of excitons and, as a consequence, charge carriers for every distinct orientation of a nanobelt. No trapping–detrapping processes invoked in other studies are involved on this short time-scale. In order to address this issue in detail, we studied the dependence of the photoluminescence dynamics on the initial exciton and carrier density in a single nanobelt which was excited by ≈ 30 fs pulses at 350 nm, leading to a hot electron distribution approximately 1 eV above the conduction band edge of bulk $CdS_{0.7}Se_{0.3}$. The excitation intensity was varied in order to simulate the orientation dependent absorption. The response of the emission intensity and dynamics to the excitation fluence was highly nonlinear and exhibited saturation effects, a behavior which is consistent with ASE and is ascribed to the onset of Auger recombination processes in semiconductor nanocrystals. We set up a kinetic model incorporating the Shockley–Reed, geminate and Auger recombination processes (Figure 7.6) and used it to model the nonlinear fluence dependent dynamics with the initial exciton concentration as the only adjustable parameter that was kept proportional to the intensity of the incident excitation beam. The success of this treatment underscores the importance of single particle ultrafast microscopy measurements as this level of analysis would not be possible on the basis of ensemble experiments.

Furthermore, the high spatial resolution of the microscope enabled us to map the luminescence intensity within a single nanobelt. The luminescence exhibited strong localization within the particle and originated mainly from the tip of the wedge-shaped crystal (Figure 7.7), indicating that confinement effects reported primarily for nanoparticles with radii of only a few nanometers, can be also important in larger objects where spatial confinement exists only in certain areas. There are two types of confinement effect which have important consequences for the charge carrier dynamics in semiconductor particles. The first is purely classical and leads to increased probability of charge carrier recombination in restricted areas of a particle. In smaller particles with nanometer dimensions the familiar quantum confinement effects appear. In this regime the band gap of the particle becomes strongly size dependent, affecting both the exciton and free carrier dynamics. Lastly, the nanobelt can behave as a low Q cavity leading to shape dependent local constructive and destructive interference effects. We do not yet fully understand the emission patter shown in Figure 7.7, however, it is clear that all above processes influence the performance of photovoltaic materials.

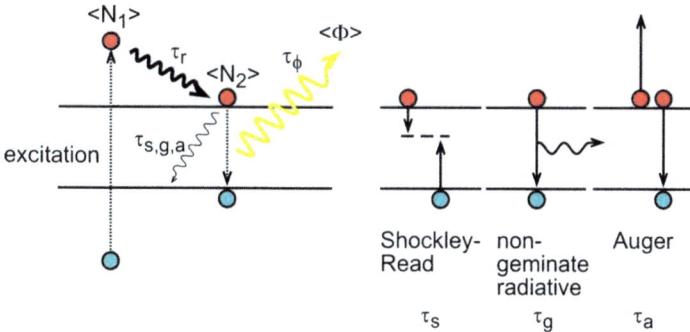

Figure 7.6 Schematic representation of three dominant decay channels responsible for the complex photoluminescence dynamics of the CdS_xSe_{1-x} nanobelts. The set of the corresponding rate reactions which were used to model the observed dynamics is shown below the diagram.

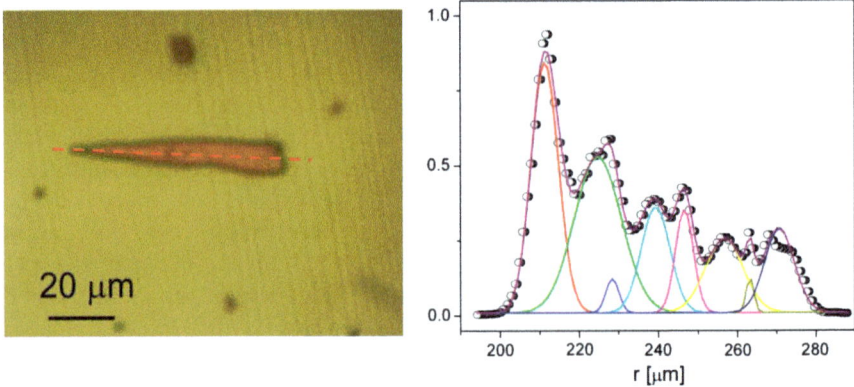

Figure 7.7 Photoluminescence profile of a 70 μm long wedge-shaped CdS$_x$Se$_{1-x}$ nanobelt. The integrated intensity has been projected onto the red line indicated in the dark-field transmission image on the left. Note that the emission is by far most intense at the sharp tip of the nanobelt.

7.2.3 Fluorescence Dynamics of Quantum Dots in Close Proximity to Metal Surfaces

Certain chromophore systems are intrinsically predisposed for ultrafast single molecule microscopy. Among these, emitters coupled to metal surfaces stand out as exceptionally well-suited subjects. Numerous observations of substantial radiative rate enhancement at the surface or in the vicinity of the surface of a metal were reported.[19,20] Radiative rate enhancements as large as 10^3 have been predicted for molecular fluorophores and for semiconductor quantum dots coupled to optimized nanoantennae.[21] Such accelerated emission rates put these systems well within the reach of the emerging femtosecond microscopy techniques. As a result, we decided to apply the Kerr-gated microscope to study of fluorescence dynamics of individual core-shell quantum dots in contact with smooth and nanostructured metal surfaces.

A homebuilt electrospray deposition system was used to disperse commercially available 2 nm diameter CdSe/ZnS core-shell quantum dots (Invitrogen Qdot 565 ETK) onto smooth and corrugated gold on the surface of a silicon wafer (Figure 7.8(a) and (b)). The sample was placed in a microscope cryostat and evacuated. Quantum dots (QDs) on smooth gold exhibited very weak fluorescence and we could not observe any excited state dynamics, most likely due to the combination of the long intrinsic excited state lifetime ($\sim$20 ns) and efficient quenching at the interface. We were only able to collect their static photoluminescence images with the Kerr shutter permanently open, that is, with polarizers P1 and P2 in parallel orientation (Figure 7.2). In contrast, the same quantum dots on corrugated gold produced much brighter images than could be detected in the gated mode (Figure 7.8(c)), despite the fact that the Kerr gate is only 100 fs long and has 5% transparency. On the rough metal surface, the sharp onset of emission is clearly visible despite the high noise level (Figure 7.8(d)).

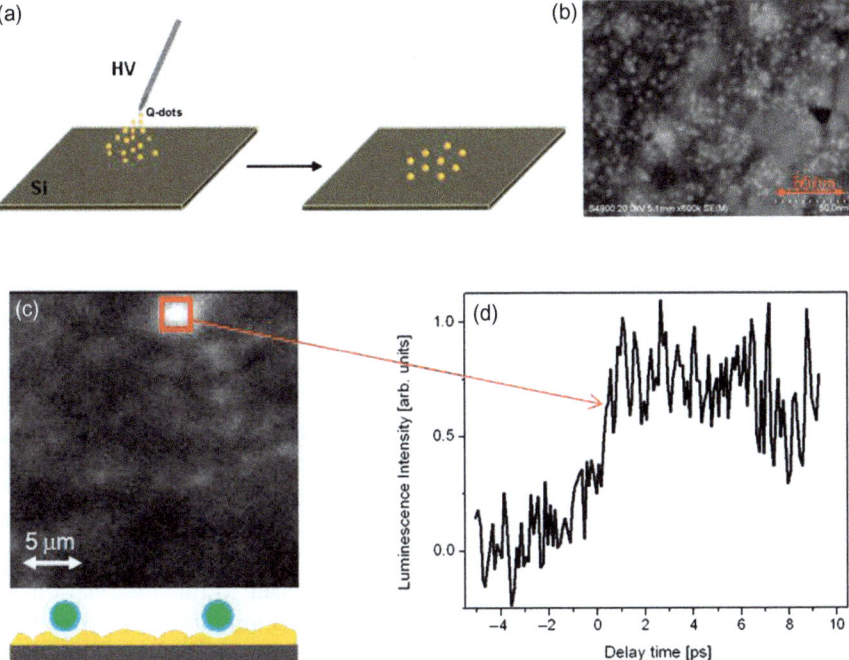

Figure 7.8 Fluorescence dynamics of core-shell quantum dots on the surface of corrugated gold. (a) Schematic representation of the electrospray deposition of quantum dots onto the metal surface; (b) SEM image of the quantum dots dispersed on the gold surface; (c) Kerr-gated photoluminescence image of quantum dots dispersed at a lower density on rough gold surface; (d) Fluorescence dynamics collected from the bright area indicated by the red square in the image shown in (c). The total length of the show scan is 15 ps. Each data point corresponds to a 50 s exposure, *i.e.* 5×10^6 excitations at a 100 kHz repetition rate. The FWHM duration of each Gaussian shaped gate is 100 fs.

These semi-quantitative observations suggest that at $t = 0$ the brightness of a quantum dot on a corrugated Au surface is 20 to 200 times higher than on smooth gold. The interplay between the ultrafast quenching at the metal surface and the emission enhancement is a fascinating fundamental problem with numerous practical applications and we intend to continue these studies in the future.

7.3 Femtosecond Pump Probe Transient Absorption Microscopy (PPTAM)

7.3.1 Background and Experimental Considerations

Pump probe spectroscopy transient absorption spectroscopy is the most widely applied method of probing excited state dynamics. In these experiments, a laser pulse (pump, ω_{pu}) places the nanostructure into the excited state; a second delayed pulse (probe, ω_{pr}) is used to monitor the population of a particular excited electronic state. Pump probe experiments measure pump induced differential changes of the probe, as given by:

$$\frac{\Delta I}{I} = \frac{I_{pump-on} - I_{pump-off}}{I_{pump-off}}$$

The pump probe signal is directly proportional to the population changes as given by $\frac{\Delta I}{I} \propto (\Delta n_f - \Delta n_i)$, where n_i and n_f are the populations of the initial and final states probed, respectively.

The implementation of single-particle pump probe microscopy requires a significant improvement in the signal-to-noise ratio (S/N) over traditional spectroscopy so that the response of a single nanostructure can be detected. There are two main sources of noise in these experiments: laser fluctuation and electronic noise of the detection system (for example the lock-in amplifier). Noise caused by laser intensity fluctuations can be to a great degree eliminated by using heterodyne lock-in detection with MHz modulation,[22] in which the intensity of the excitation beam or additional local oscillator is modulated by an acousto-optical modulator and the induced modulated signal can be then sensitively extracted by a lock-in amplifier referenced to this modulation frequency. Laser intensity fluctuations have the largest amplitude in the dc to 10 kHz range and correspond to the so-called *1/f* noise. When the modulation frequency *f* is in the MHz range, the laser intensity noise approaches the quantum shot noise limit, which is always present because of the Poissonian distribution of the photon counts at the detector. In this limit of the amplitude modulation and lock-in detection scheme, a $\Delta I/I$ on the order of 10^{-7} has been achieved, which is sufficient for single molecule detection.[5] When implemented in a scanning microscope, the pixel dwell time should be significantly longer than the modulation period, so that each pixel is reliably demodulated.

A typical experimental set up employed is shown in Figure 7.9. Pump and probe pulses can be derived either from a Ti:sapphire oscillator or from a

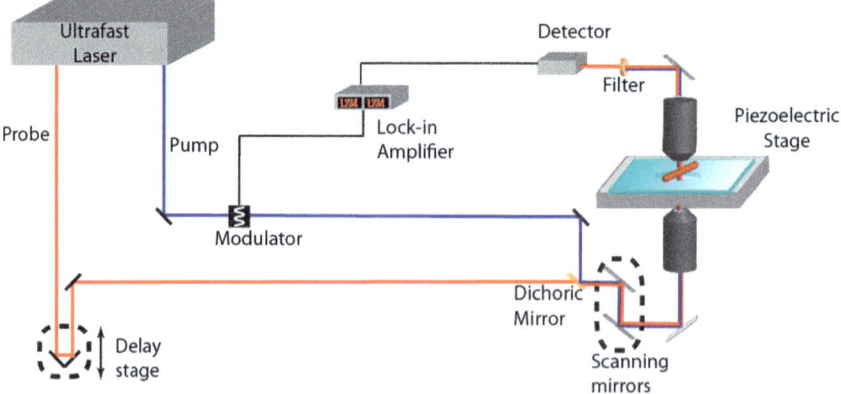

Figure 7.9   Block diagram of the femtosecond pump-probe transient absorption microscope.

synchronously pumped optical parametric oscillator (OPO) for wider wavelength tunability. Femtosecond oscillators with MHz repetition rates are preferred over kHz amplifiers because of the noise considerations discussed above. For samples with long excited state lifetimes pulse pickers can be employed in order to reduce the repetition rate, however, with the caveat of having to lower the modulation rate correspondingly.

Transient absorption traces are obtained by delaying the probe with respect to the pump with a high-precision mechanical translation stage. Images at a fixed pump probe delay can be acquired by either raster scanning the sample with a piezoelectric stage or by scanning the beams with the help of a scanning mirror. The galvanometer-driven beam scanning is generally preferred for high-speed imaging because of the much shorter pixel dwell times.

7.3.2 Examples of Applications of PPTAM to the Study of Exciton and Charge Carrier Dynamics in Nanostructures and Heterogeneous Materials

Single-particle pump probe microscopy provides attractive and complementary means for unraveling energy relaxation pathways in single nanostructures by offering an important advantage over photoluminescence based techniques: the measured signal is based on absorption, which means samples with low or even zero fluorescence quantum yield such as metallic nanoparticles, metallic single wall nanotubes (SWNTs) and graphene can be studied. As a result, at least in principle, transient absorption microscopy is a more general approach. Furthermore, often a distinct absorption signal is associated with the formation of excitons and free carriers, which is highly valuable in probing energy and charge transfer dynamics.

Pump probe transient absorption microscopy experiments on individual nanostructures were first performed on a single gold nanoparticle by van Dijk *et al.*[22] who used a birefringent crystal to generate perpendicularly polarized signal and reference beams in a common path interferometer, which provided shot noise-limited detection.[23] Later, a configuration based on a more straightforward optical pump probe set up was demonstrated and also achieved shot noise-limited detection by utilizing MHz modulation.[24–26] Since then, pump probe microscopy has been applied to study metallic nanostructures of various sizes and shapes, semiconductor nanowires, metal oxide nanostructures, carbon nanotubes, graphene and polymer blends.[27–38] Pump probe microscopy and its variants, for example stimulated emission microscopy and ground state depletion microscopy, have also been applied to biomedical imaging.[39]

Single nanostructures pump probe microscopy experiments revealed energy relaxation pathways that are obscured by ensemble averaging. For example, understanding of the intrinsic optical properties of single wall carbon nanotubes (SWCNTs) has been previously hindered primarily by the broad distribution of semiconducting and metallic nanotube types in as-synthesized samples. Even the best sorting or separation procedures fall far short of producing truly monodisperse batches of SWCNTs. Single chemical vapor deposition (CVD) grown SWCNTs and surfactant encapsulated SWCNTs were first imaged by transient absorption microscopy by Jung *et al.*[27] The sign of a transient absorption signal (ground state depletion and stimulated emission *versus* excited-state absorption) in combination with a resonance condition was utilized to differentiate metallic SWCNTs from semiconducting nanotubes.

More recently, Gao *et al.*[28] reported transient absorption microscopy and spectroscopy measurements on chirality-assigned individual SWCNTs. Transient absorption spectra of individual SWCNTs shown in Figure 7.10 were obtained by recording transient pump probe images at different probe wavelengths and reveal different origins of photo-induced absorption. Population

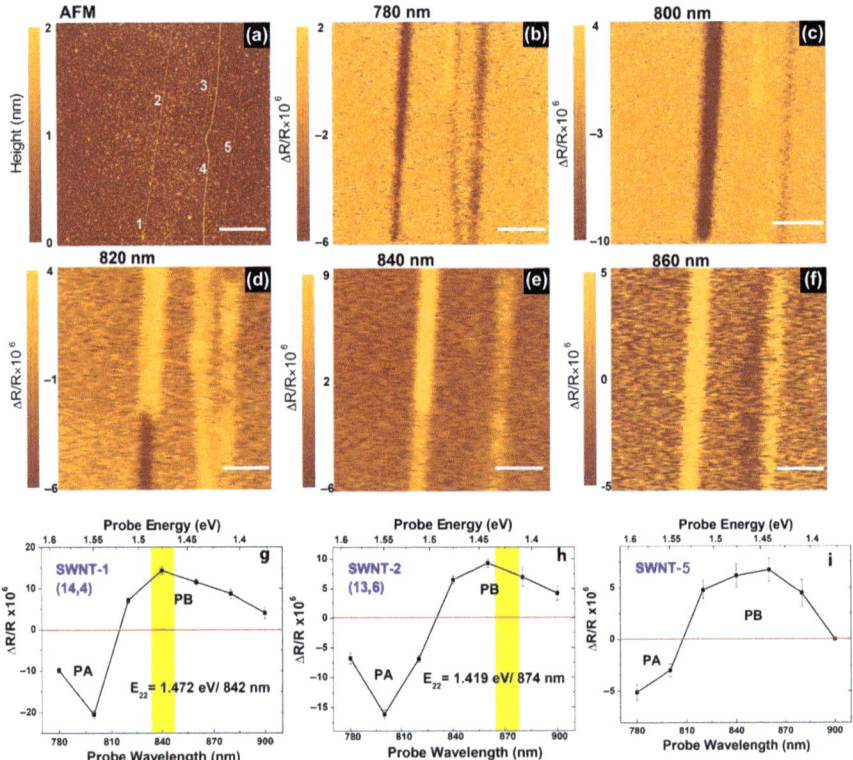

Figure 7.10   (a) Tapping mode atomic force microscopy (AFM) height image of the SWCNTs #1, #2, #3, #4 and #5. (b) to (f) transient absorption microscopy (TAM) images with pump/probe wavelength of 390/780 nm, 400/800 nm, 410/820 nm, 420/840 nm and 430/860 nm, respectively. All TAM images were collected at a pump probe delay of 0 ps. Note that the scale of the reflectivity change is different for each TAM image. Scale bars are 2 μm in all images. (h) to (i) Transient reflectivity changes ($\Delta R/R$) at 0 ps pump probe delay time *versus* probe wavelength for SWCNTs #1, #2 and #5.

dynamics at E_{11} and E_{22} transition frequencies were collected at the individual SWCNT level. The dynamics shows a fast $\sim 1\,ps$ decay for all the semi-conducting nanotubes studied. This is significantly faster than the previously reported decay times for SWCNT suspensions. The fast relaxation is attributed to the coupling between the excitons generated by the pump laser pulse and the underlying substrate.[28]

Variation in exciton dynamics at the single particle level was also resolved in semiconducting nanowires. Recently Lo *et al.*[29] reported transient absorption microscopy of single CdTe nanowires and found that the time constants vary for different wires owing to differences in the energetics and density of surface trap sites. Mehl *et al.*[33] used pump probe microscopy of individual needle-shaped ZnO rods to characterize spatial differences in their response to pho-toexcitation. Dramatically different recombination dynamics was observed in the narrow tips compared to the interior owing to different recombination mechanisms.

Pump probe mapping of spatially dependent carrier dynamics allows for detailed examination of the interaction between the nanostructures of interest and the environment. Huang *et al.*[5] and Gao *et al.*[36] used transient absorption microscopy as a novel tool to characterize the role of substrate in modulating carrier dynamics in epitaxial and CVD grown graphene. Significant spatial heterogeneity in the dynamics was observed owing to differences in coupling between graphene layers and the substrate for epitaxially grown graphene.[5] Later, monolayer CVD grown graphene suspended over micrometer-sized trenches, as well as graphene supported on SiO_2 were employed to investigate systematically the influence of substrate on the charge carrier cooling. As il-lustrated in Figure 7.11, at a given excitation energy, the dynamics for sus-pended graphene is much slower than for the substrate-supported graphene, owing to the presence of additional excess energy dissipation channels provided by the surface optical phonon of the substrate.[36] Ruijgrok *et al.*[35] combined single particle pump probe microscopy with optical trapping to study homo-geneous damping of the acoustic vibrations of single gold nanospheres and nanorods in water. Significant particle-to-particle variation in damping times was observed. The above experiments open up new opportunities for studying intrinsic and extrinsic energy dissipation mechanisms in nanomaterials in ways that ensemble measurements cannot.

Another important direction for pump probe microscopy is the interrogation of the morphology-dependent charge separation and recombination pathways in heterogeneous energy harvesting systems such as polymeric solar cells.[37] A proof-of-concept experiment in which morphology-dependent polaron dy-namics in thermally annealed P3HT-PCBM (poly(3-hexylthiophene-2,5-diyl and [6,6]-phenyl-C61-butyric acid methyl ester) blends was mapped with sim-ultaneously high spatial and temporal resolution by pump probe microscopy was performed by Wong *et al.*[38] These measurements revealed significant spatial heterogeneity in charge generation and recombination dynamics. By directly comparing the spatially resolved dynamics to ensemble dynamics (Figure 7.12), these results demonstrated that the apparent lifetimes recovered

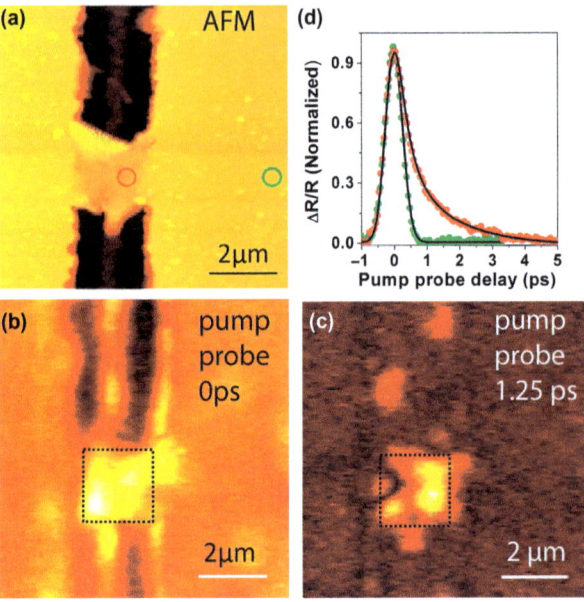

Figure 7.11 (a) AFM height image of a graphene sample. (b) Transient absorption image of the same area at 0 ps delay. (c) Transient absorption image at 1.25 ps delay. (d) Transient absorption traces collected at 800 nm for suspended graphene (red filled circles) and substrate-supported graphene (green filled circles). The beam positions at which the kinetics was collected are indicated by the red and green circles in (a). The pump fluence was $2\,\mathrm{mJ\,cm}^{-2}$ and the pump and robe wavelengths were 400 nm and 800 nm, respectively.

from ensemble measurements correspond to averaging over microscopically diverse areas and therefore are generally misleading.

7.4 Future Challenges: Single Molecule Detection, Higher Time Resolution and Spatial Super-Resolution in Femtosecond Microscopy

7.4.1 Single Molecule Femtosecond Microscopy

It is a natural goal to extend ultrafast microscopy to the single molecule, or more precisely single chromophore, or single emitter experiments. This is not merely a semantic distinction because these terms are not necessarily synonymous. A single conjugated polymer strand, a single non-metallic SWCNT, or single semiconductor nanowire, behave as a collection of many local chromophores defined by the exciton length and defects, rather than a single giant transition dipole. For example, depending on the conditions, a 1000 kDa strand of poly(p-phenylene vinylene) may contain as many as 1000 effective chromophores, making it a relatively easy potential target for single molecule

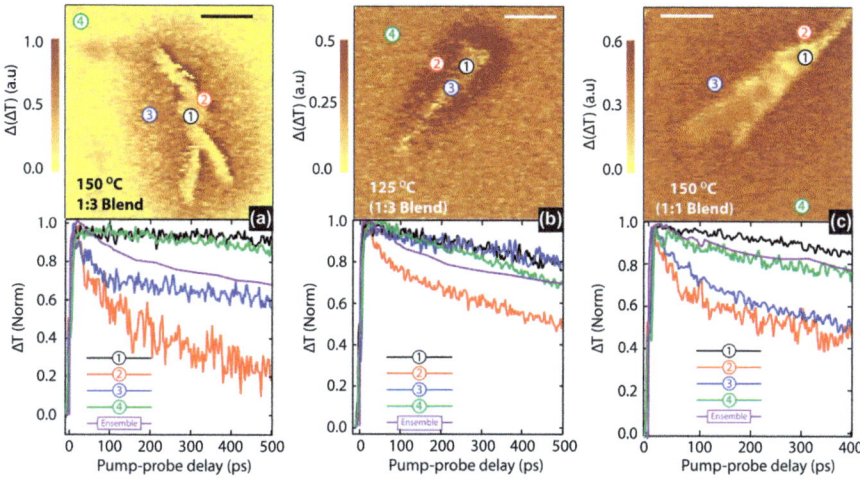

Figure 7.12 Upper graphs plot the relative change between TA images taken at 5 ps and 500 ps defined as $\Delta(\Delta T) = [\Delta T(5\ \text{ps}) - \Delta T(500\ \text{ps})]/\Delta T(5\ \text{ps})$. Lower graphs are the decay curves taken at different positions along with the ensemble kinetics at 800 nm. The positions where the decay curves are obtained are labeled in the upper graphs. Parts (a), (b) and (c) are for 1 : 3 blend with annealing temperature of 150 °C, 125 °C, and 1 : 1 blend annealed at 150 °C, respectively. Scale bars are 5 μm.

microscopy down to the highest time resolution. Indeed, it can be expected that sub-picosecond fluorescence and transient absorption experiments on single polymer strands and semiconductor nanowires similar to those reported by Huang *et al.*[5] will become reasonably common. However, ultrafast optical microscopy experiments on true single chromophore (*i.e.* single transition dipole) systems are far more challenging.

The ultimate single molecule detection limit is set by the probability with which a typical chromophore can emit a photon within a sub-picosecond time window. Even for the best chromophores with large transition dipoles and correspondingly high radiative rates this probability is low. The instantaneous brightness of a fluorophore is determined by the coefficient of spontaneous emission and hence by the extinction coefficient of the molecule. For an excellent molecular emitter with a radiative lifetime of 1 ns, such as for example the S_2 state of porphyrins ($\varepsilon_{max} = 600\,000$), the probability of emission within the initial 100 fs following excitation is only approximately 0.1%. Therefore, the molecule must be excited approximately 1000-times in order for one photon to be emitted with the specified 100 fs time window. Naturally, more than one photon must be collected in order to determine the dynamics of the system of interest and the collection efficiency of even the best microscope is far from 100%.

Köllner and Wolfrum performed a thoughtful and very useful error analysis of the minimum number of photons necessary in order to determine emission lifetime either in fluorescence imaging or spectroscopy.[40] In the absence of a background only 70 photons are needed to determine a single exponential lifetime with 25% accuracy. This, combined with the emission probability discussed above and a modest but realistic 1% detection efficiency, leads to a nearly 2 h acquisition time if a typical 1 kHz Ti:sapphire laser source is used. A cryogenically cooled 100 kHz source reduces this time to a much more manageable 70 s. In a spectrally resolved measurement, this number of photons must be collected in each wavelength channel and the length of the experiment increases accordingly. The optimum measurement window should be longer than five lifetimes, however, contrary to the common practice, little accuracy is gained by spreading the photons over more than 4–8 channels.

Samples displaying more complex decay dynamics are far more challenging. Here the photons must be distributed over more than 256 temporal channels in order to ensure that the lifetime components are adequately represented.[40] For example, resolving with 25% accuracy the components of a noiseless double exponential decay with $\tau_2/\tau_1 = 6$ and equal amplitudes $A_1 = A_2$ requires 900 photons and corresponds to a 15 minute long data acquisition with a 100 kHz excitation source. In the presence of noise and for a different ratio of amplitudes, these numbers can be considerably higher.

As mentioned earlier, certain single chromophore systems are intrinsically predisposed to ultrafast microscopy. Enhanced emission rates observed for molecules and quantum dots in the vicinity of a metal surface, as well as the theoretical predictions of the radiative rate acceleration factors in excess of 10^3 for optimized plasmonic nanoantennae, make these systems ideally suited to femtosecond imaging.[19–21,41–43] With the emission lifetime compressed (not quenched) from nanoseconds to picoseconds, the probability of emitting a photon within a 100 fs window increases dramatically and facilitates the measurements.

In conclusion, 'true single molecule' femtosecond fluorescence microscopy experiments are certainly possible, however, as a result of the constraints discussed above they will remain challenging even with optimum instrumentation. Methods which do not rely on spontaneous emission, for example SFG or stimulated emission, may hold an advantage at the single molecule level.

7.4.2 Sub 100 Femtosecond Time Resolution

The temporal resolution of the best ultrafast microscopy experiments that have been carried out so far is not always sufficient for monitoring the dynamics of processes which are closely tied to solar energy conversion. For example, interfacial electron transfer, the key redox reaction underlying hybrid photovoltaics, can occur within less than 20 fs.[44,45] Similarly, the dephasing of surface plasmons takes place on a 10–20 fs timescale. The maximum rate with which a chemical bond can be formed or broken is set by a half of the cycle of the respective vibrational mode. In the case of proton transfer reactions, which are central to light-driven water splitting and CO_2 reduction, this amounts to as little as 3 fs. While sub-10 fs light pulses can be routinely produced with the help of non-collinear parametric amplifiers (NOPAs) and broadband oscillators, performing imaging experiments with such high time resolution poses a major challenge.

The main technical obstacle is the pulse broadening in the intervening material. In microscopy the inherently thin samples contribute very little to the GVD, however, the complex optical components are major limiting factors. With full optimization of all optical elements, fluorescence microscopes based on either Kerr gating or parametric amplification, could lead to temporal resolution down to possibly 20 fs. Breaking this barrier will require further reduction of the GVD in the gating medium. The virtually unexplored Kerr-gating in the reflective mode (dielectric Kerr mirror) has the potential for negligibly low dispersion and a much higher time resolution that could be also extended into the UV domain.

The most daunting barrier to reaching higher time resolution in fluorescence microscopy is not of a technical nature, it is rather imposed by the photon bottleneck discussed above. The probability that an excellent fluorophore such as rhodamine 6G, will spontaneously emit a photon within a 10 fs window is smaller than 1×10^{-5}, leading to a dauntingly large number of necessary excitations. Pump probe transient absorption microscopy (PPTAM) is not limited by the radiative rate and with carefully optimized all-reflective imaging components should be able to reach 20 fs or better time resolution.

7.4.3 Spatial Super-Resolution

Combining femtosecond temporal resolution with higher than the diffraction limit spatial resolution is another very attractive direction for the development of ultrafast microscopy. Super-resolution in continuous wave (CW) fluorescence imaging was achieved using several wide-field and confocal schemes and has led to a true renaissance in optical microscopy.[46–48] The wide-field structured illumination approach introduced by Gustafsson[47] appears to be most suitable in the context of a time resolved imaging experiment, even though it offers only a modest, two-fold increase of resolution in the linear regime. Nonlinear methods which rely on the saturation of the entire wide-field image or selective suppression in the confocal mode are not suitable if the intrinsic excited state dynamics is the primary subject of interest. It is important to keep in mind that all super-resolution approaches require significantly more photons per image to be collected than in standard fluorescence microscopy. For example, the two-fold resolution increase achieved by structured illumination in the linear regime requires a six times higher number of collected photons. This puts more strenuous demands on the sample and extends the duration of the experiment. NSOM experiments, which currently show a time resolution of ~ 250 fs, have the weakest potential for further improvement, both in terms of spatial and temporal resolution. The difficulties of controlling dispersion of the excitation light propagating through the NSOM fiber and the tip as well as low photon throughput will become more severe if broadband 10–20 fs pulses are used.

A notable successful implementation of spatially super-resolved ultrafast imaging is the two-color IR fluorescence up-conversion microscopy developed by Fujii and co-workers.[4,49] In Fujii's method, a picosecond mid-IR OPA pulse prepares a vibrationally excited population of emitters and is followed by an up-conversion pulse in the visible range of the spectrum which promotes the molecules to the fluorescent electronic excited state. The final spatial resolution of the up-converted IR image is determined by the diffraction limit of the visible wavelength. As a result, temporally and spectrally resolved IR images can be collected from sub-micrometer areas of the sample.

7.5 Conclusions

As evidenced by the growing number of results, ultrafast microscopy is undergoing rapid development. Several approaches have crossed the line between a challenging demonstration and everyday application. Kerr-gated fluorescence microscopy experiments have reached a temporal resolution of 60 fs and diffraction limited spatial resolution.[2,10] Hybrid techniques complement all-optical methods and hold excellent promise for resolving femtosecond dynamics at the nanometer length scale. Some of the most successful hybrid ultrafast microscopy techniques, which merge femtosecond photoexcitation with non-optical detection modes, are described in other chapters of this monograph. Most notable in this category are the two-photon photoemission experiments of Petek *et al.*, who coupled phase sensitive interferometric time-resolved two-photon photoemission (ITR-2PP) with photoelectron emission microscopy (PEEM) detection and used it to image the plasmon dynamics in nanostructured silver with sub-femtosecond time resolution and 50 nm spatial resolution.[7,50] Methods combining femtosecond excitation with various modes of scanning tunneling microscopy (STM) were successfully tested by several groups. Nevertheless, it is likely that the all-optical schemes such as the Kerr-gated fluorescence microscopy (KGFM) and pump probe transient absorption microscopy (PPTAM) will emerge as the most general methods for ultrafast imaging thanks to their compatibility with a wide range of samples and environments and relative simplicity of implementation.

Acknowledgements

Research performed in the laboratories of P.P. at Rutgers University and L.H. at the Radiation Laboratory was supported by the Solar Photochemistry Program of the Office of Basic Energy Sciences of the US Department of Energy through grants #DE-FG02-06ER15828 to P.P. and #DE-FC02-04ER15533 to L.H. The key instrumentation components of the femtosecond Kerr-gated microscope constructed at Rutgers University were funded by the National Science Foundation Major Instrumentation (NSF-MRI) grants #1039828 and '0923345, as well as a Chemistry Research Instrumentation (CRIF) grant #0342432 to P.P. We acknowledge Dr. Zhihao Yu for his contribution to the development of the Kerr-gated microscope and thank Dr. Lynn Schneemeyer and Ms. Jianhua Bao for their assistance with the synthesis and characterization of the semiconductor nanobelts.

References

1. T. Fujino, T. Fujima and T. Tahara, *J. Phys. Chem. B*, 2003, **107**, 5120.
2. L. Gundlach and P. Piotrowiak, *Opt. Lett.*, 2008, **33**, 992.
3. V. Emiliani, T. Guenther, C. Lienau, R. Nötzel and K. H. Ploog, *J. Microsc.*, 2001, **202**, 229.
4. M. Sakai, T. Ohmori, M. Kinjo, N. Ohta and M. Fujii, *Chem. Lett.*, 2007, **36**, 1380.
5. L. Huang, G. V. Hartland, L. Q. Chu, R. M. Luxmi, C. Feenstra, K. Lian, Tahy and H. Xing, *Nano Lett.*, 2010, **10**, 1308.
6. D. Botkin, J. Glass, D. S. Chemla, D. F. Ogletree, M. Salmeron and S. Weiss, *Appl. Phys. Lett.*, 1996, **69**, 1321.
7. A. Kubo, K. Onda, H. Petek, Z. Sun, Y. S. Sung and H. K. Kim, *Nano Lett.*, 2005, **5**, 1123.
8. Y. Terada, S. Yoshida, O. Takeuchi and H. Shigekawa, *J. Phys.: Condens. Matter.*, 2010, **22**, 264008.
9. L. Gundlach and P. Piotrowiak, *Proc. SPIE, 6643*, 2007, 66430E1.
10. L. Gundlach and P. Piotrowiak, *J. Phys. Chem. C*, 2009, **113**, 12162.
11. T. Fujino, T. Fujima and T. Tahara, *Appl. Phys. Lett.*, 2005, **87**, 131105.
12. T. Fujino, T. Fujima and T. Tahara, *J. Phys. Chem. B*, 2005, **109**, 15327.
13. F. Devaux and E. Lantz, *Opt. Comm.*, 1995, **118**, 25.
14. S. Arzhantsev and M. Maroncelli, *Appl. Spectr.*, 2005, **59**, 206.
15. N. Sugimoto, H. Kanbara, S. Fujiwara, K. Tanaka, Y. Shimizugawa and K. Hirao, *J. Opt. Soc. Am. B*, 1999, **16**, 1904.
16. R. Nakamura and Y. Kanematsu, *Rev. Sci. Instrum.*, 2004, **75**, 636, 2004.
17. Z. Yu, L. Gundlach and P. Piotrowiak, *Opt. Lett.*, 2011, **36**, 2904.
18. A. Pan, H. Yang, R. Liu, R. Yu, B. Zou and Z. Wang, *J. Am. Chem. Soc.*, 2005, **127**, 15692.
19. Y. Ito, K. Matsuda and Y. Kanemitsu, *Phys. Rev. B*, 2007, **75**, 033309.
20. A. Ishida and T. Majima, *Analyst*, 2000, **125**, 535.
21. A. Kinkhabwala, Z. Yu, S. Fan, Y. Avlasevich, K. Müllen and W. E. Moerner, *Nat. Photonics*, 2009, **3**, 654.
22. M. A Van Dijk, M. Lippitz and M. Orrit, *Phys. Rev. Lett.*, 2005, **95**, 267406.
23. S. Savikhin, *Rev. Sci. Instrum.*, 1995, **66**, 4470–4474.
24. O. Muskens, N. Del Fatti and F. Vallée, *F. Nano Lett.*, 2006, **6**, 552.
25. H. Staleva and G. V. Hartland, *J. Phys. Chem C*, 2008, **112**, 7535.
26. H. Staleva and G. V. Hartland, *Adv. Funct. Mater.*, 2008, **18**, 3809.
27. Y. Jung, M. N. Slipchenko, C. H. Liu, A. E. Ribbe, Z. Zhong, C. Yang and J. X. Cheng, *Phys. Rev. Lett.*, 2010, **105**, 217401.
28. B. Gao, G. V. Hartland and L. Huang, L. *ACS Nano*, 2012, **6**, 5083.
29. S. S. Lo, T. A. Major, N. Petchsang, L. Huang, M. K. Kuno and G. V. Hartland, *ACS Nano*, 2012, **6**, 5274.
30. W. Min, S. Lu, S. Chong, R. Roy, G. R. Holtom and X. S. Xie, *Nature*, 2009, **461**, 1105.
31. G. V. Hartland, *Chem. Sci.*, 2010, **1**, 303.

32. T. Ye, D. Fu and W. S. Warren, *Photochem. Photobiol.*, 2009, **85**, 631.
33. B. P. Mehl, J. R. Kirschbrown, R. L. House and J. M. Papanikolas, *J. Phys. Chem. Lett.*, 2011, **2**, 1777.
34. L. Tong, Y. Liu, B. D. Dolash, Y. Jung, M. N. Slipchenko, D. E. Bergstrom and J. X. Cheng, *Nat. Nanotechnol.*, 2011, **7**, 56.
35. P. V. Ruijgrok, P. Zijlstra, A. L. Tchebotareva and M. Orrit, *Nano Lett.*, 2012, **12**, 1063.
36. B. Gao, G. V. Hartland, T. Fang, M. Kelly, D. Jena, H. G. Xing and L. Huang, *Nano Lett.*, 2011, **11**, 3184.
37. G. Grancini, D. Polli, D. Fazzi, J. Cabanilas-Gonzalez, G. Cerullo and G. Lanzani, *J. Phys. Chem. Lett.*, 2011, **2**, 1099.
38. C. T. O. Wong, S. S. Lo and L. Huang, *J. Phys. Chem. Lett.*, 2012, **3**, 879.
39. T. E. Matthews, J. W. Wilson, S. Degan, M. J. Simpson, J. Y. Jin, J. Y. Zhang and W. S. Warren, *Biomed. Opt. Express*, 2011, **2**, 1576.
40. M. Köllner and J. Wolfrum, *Chem. Phys. Lett.*, 1992, **200**, 199.
41. L. Brus, *Acc. Chem. Res.*, 2008, **41**, 1742.
42. K. T Shimizu, W. K. Woo, B. R. Fisher, H. J. Eisler and M. G. Bawendi, *Phys. Rev. Lett.*, 2002, **89**, 117401.
43. J. B. Khurgin and G. Sun, *J. Opt. Soc. Am. B*, 2009, **26**, B83.
44. L. Gundlach, R. Ernstorfer and F. Willig, *Appl. Phys. A*, 2007, **88**, 481.
45. G. Benkö, J. Kallioninen, J. E. I. Korppi-Tommola, A. P. Yartsev and V. Sundström, *J. Am. Chem. Soc.*, 2002, **124**, 481.
46. S. W. Hell and J. Wichmann, *Opt. Lett.*, 1994, **19**, 780.
47. M. G. L. Gustafsson, *PNAS*, 2005, **102**, 13081.
48. S. Bretschneider, C. Eggeling and S. W. Hell, *Phys. Rev. Lett.*, 2007, **98**, 218103.
49. M. Sakai, Y. Kawashima, A. Takeda, T. Ohmori and M. Fujii, *Chem. Phys. Lett.*, 2007, **439**, 171.
50. A. Kubo, Y. S. Jung, H. K. Kim and H. Petek, *J. Phys. B*, 2007, **40**, S259.

Ultrafast Multiphoton Photoemission Microscopy of Solid Surfaces in Real and Reciprocal Space

A. WINKELMANN,[a] C. TUSCHE,[a] A.A. ÜNAL,[a]
C.-T. CHIANG,[a] A. KUBO,[b] L. WANG[c] AND H. PETEK*[c]

[a] Max-Planck Institut für Mikrostrukturphysik, Weinberg 2, D-06120 Halle (Saale), Germany; [b] University of Tsukuba, Institute of Physics, Tsukuba, Ibaraki, 305-8571, Japan; [c] University of Pittsburgh, Department of Physics and Astronomy and Petersen Institute of NanoScience and Engineering, Pittsburgh, Pennsylvania, 15260, USA
*Email: petek@pitt.edu

8.1 Introduction

By combining ultrafast laser and photoelectron spectroscopy it is now possible to study electron dynamics in atoms, molecules and solid-state materials on the <100 as $(<10^{-16}$ s$)$ timescale.[1–5] In the case of semiconductors and metals the collective plasma response and interfacial charge transfer processes of strongly coupled adsorbates on solid surfaces occur on the sub-femtosecond timescale. Processes that previously could only be gleaned through frequency domain measurements can now be directly imaged in ultrafast pump–probe experiments utilizing multidimensional spectroscopic and microscopic analysis of photoemitted electrons. Through energy, momentum, space and spin analysis

RSC Energy and Environment Series No. 8
Solar Energy Conversion: Dynamics of Interfacial Electron and Excitation Transfer
Edited by Piotr Piotrowiak
© The Royal Society of Chemistry 2013
Published by the Royal Society of Chemistry, www.rsc.org

of photoelectrons we can gain differential energy, momentum, space and spin resolved information on electronically excited processes that is difficult to extract and interpret in integral measurements employing the scattering of optical fields.[8–10]

In this chapter we describe advances in the femtosecond time-resolved multiphoton photoemission spectroscopy (TR-MPP) as a method for probing electronic structure and ultrafast interfacial charge transfer dynamics of adsorbate-covered solid surfaces.[12–15] The focus is on surface science-based approaches that combine ultrafast optical pump–probe excitation to induce nonlinear multi-photon photoemission (MPP) from clean or adsorbate covered single crystal surfaces. The photoemitted electrons transmit spectroscopic and dynamical information, which is captured by their energy analysis in real or reciprocal space. We examine how photoelectron spectroscopy and microscopy yield information on the unoccupied molecular structure, electron transfer and relaxation processes, light induced chemical and physical transformations and the evolution of coherent single particle and collective excitations at solid surfaces.

Nonlinear optical spectroscopy has been highly successful in studies of ultrafast physical, chemical and biological phenomena of interest to solar energy conversion. In applications relating to electronic coherence, charge transfer and intermolecular interactions, optical spectroscopy offers several complementary approaches that may have significant advantages under different experimental conditions. Sophisticated multidimensional spectroscopic methods are able to follow coherent energy transport in light-harvesting systems or molecular interactions in liquids and more complex heterogeneous media.[16–19] Other methods, such as terahertz spectroscopy, can probe the free carriers, measuring the optical conductivity of a solid or nanoparticulate sample in response to optical excitation. Methods such as transient absorption, florescence upconversion, or nonlinear optical scattering can follow the decay of electronically excited adsorbates, the injection of carriers into a semiconductor or metal substrate and the recovery of adsorbate by back electron transfer or charge exchange with electrolytes. The main advantage of such techniques lies in their ability to probe dynamics in complex media, such as poorly defined buried interfaces that may exist in electrochemical environments or biological samples.

Optical techniques, however, have several intrinsic disadvantages relating to their integral nature and mismatch of dimensions between the optical dipoles of absorbers and emitters and wavelengths of optical fields. Fundamentally, optical techniques measure optical transitions, rather than electronic states. This may not be a drawback for a molecular system, in which optical absorption and emission can usually be understood from a simple Frank–Condon analysis but in the case of non-radiative processes, such as are necessarily involved in energy transduction, optical techniques struggle to follow energy and structural relaxation pathways. For example, intra and intermolecular charge transfer rates depend on energy differences between the donor and accepter states, which cannot be directly measured by optical methods unless transitions occur between them or to a common reference level. Furthermore, when applied to electronic properties of solids, optical absorption and emission spectra represent a response integrated over all momentum and energy states that are coupled by an optical transition moment. In principle, the range of states that can contribute to a linear optical spectrum is as wide as the photon energy and spans the momentum range of the entire Brillouin zone. This energy and momentum integration limits the spectroscopic information on the electronic band structure of solids to dominant contributions from critical points that have large joint density in the occupied and unoccupied states, or to free electrons that dominate the intraband response in metals and doped semiconductors. Moreover, light is absorbed and emitted by nanoscale objects with dimensions much smaller than the optical wavelengths. For this reason, optical microscopy struggles to image molecules and nanoparticles that serve as active media in solar energy transduction. Near field and super-resolution methods can probe optical chromophores with sub diffraction-limited resolution, but at the cost of perturbing the optical near-fields and therefore the measurement. Scanning a probe or a laser beam over the sample to generate an image is inefficient especially when time-resolved or multidimensional information is desired.

The past decade has witnessed a revolution in studies of surface electronic structure and charge carrier dynamics based on ultrafast laser-induced multiphoton photoemission. Some of the limitations of optical spectroscopy can be overcome by hybrid approaches which combine optical methods to induce multiphoton photoelectron emission with multichannel photoelectron spectroscopy and microscopy analysis and detection. Optical excitation by ultrafast pump and probe pulses provides insuperable time-domain resolution for studies of photoexcitation processes, electronic decoherence, electronic relaxation through electron elastic and inelastic scattering, interfacial charge transfer, electron solvation and surface femtochemistry. When combined with energy and momentum analysis of photoemitted electrons, time-resolved ultrafast multiphoton photoemission (TRUMP) resolves dynamical information on coherent polarization and carrier populations evolving within the occupied and unoccupied bands of solids, solid surfaces and solid surface–molecule interfaces. Momentum imaging of photoemission from localized states can be used to image electronic state probability distributions and through photoelectron diffraction, to extract element-specific atomic scale structural information. With electron spin analysis, the spin-dependent electronic structure and scattering dynamics in both magnetic and non-magnetic materials can be scrutinized. The ability to probe coherences and energy and momentum specified electronically excited populations directly make the TRUMP approach remarkably useful for studies of dynamics of interfacial electron and excitation transfer with relevance to solar energy conversion. For general reviews on applications to the electronic structure and charge transfer dynamics at molecule/solid interfaces, we refer the reader to excellent reviews by Zhu and co-workers[20–22] and Stähler and co-workers.[23,24] In this chapter, instead of covering every aspect of TRUMP methodology, we focus on some of the most advanced implementations that have been pioneered by the authors, including when a photoemission electron microscope is used for either the real-space <100 nm resolution imaging of photoelectron spatial distributions, or as an input electron optical column for a 2π-steradian energy-resolved imaging of photoelectron momentum distributions.

8.2 Frequency *vs.* Time Domain Measurements of Interfacial Electron Dynamics

Hybrid spectroscopic methods involving scattering of light to emit electrons (photoemission) or scattering of electrons to emit light (inverse photoemission) have been the primary tools for mapping, respectively, the occupied and un-occupied electronic bands of solids and solid surfaces. Angle-resolved photo-emission spectroscopy relates the energy and momentum resolved photoelectron distributions to the destiny of initial states (DOS) under the assumption that removal of one electron creates negligible deviation of the electronic system from its ground state. The spectral function of a quasiparticle in a solid (*e.g.* electron or a hole) is defined by its energy and interactions with the environment. Electrons and holes interacting with their environment have an associated complex self-energy, the real part corresponding to the energy renormalization by interactions with other carriers and the crystal lattice and the imaginary part describing the finite quasiparticle lifetime induced by the interactions. Specifically relating to photoinduced interfacial charge transfer, conventional photoemission spectroscopy measures the hole spectral function of adsorbates under the influence of strong electronic perturbation by the substrate. Simultaneously, photoemission spectra can also measure the energy and momentum resolved DOS of the substrate, which interact with the ad-sorbate and act as hole–acceptor states. The adsorbate and the substrate energy levels are measured with respect to a common reference energy, such as the Fermi (E_F) or the vacuum (E_{vac}) levels. Thus, photoemitted electrons carry information on the interacting adsorbate and substrate states and through coherent interference effects, their mutual interactions.

The momentum and energy relaxation timescales of electronically excited adsorbates (atoms and molecules) chemisorbed at metal or semiconductor surfaces extend from few to sub-femtosecond timescales. The dynamics of this strongly interacting regime used to be accessible only via frequency domain techniques that rely on spectral linewidth analyses of intrinsic or molecule-derived electronic states and bands on well-defined single crystal surfaces. The underlying assumption of such analyses is the energy–time Heisenberg un-certainty principle which relates the spectroscopic linewidths to electronic re-laxation rates. By now it is well established that linewidths include contributions from intrinsic and extrinsic processes, such as electron–phonon and defect scattering, as well as population relaxation processes, such as electron–electron scattering. Interfacial charge transfer, which can contribute dominantly to linewidths, is the decay of a localized electronic wave packet on adsorbate caused by interaction with the electronic continuum of the substrate. Inhomogeneous broadening from sample imperfections or from integration over dispersive momentum states can also contribute to spectroscopic line-widths. Therefore, only in exceptional cases has it been possible to extract quantitative dynamical information on specific scattering processes from line-width analyses. Even when the inhomogeneous contributions can be suppressed, interpretation of the frequency domain data has to rely on additional measurements and theoretical simulations to distinguish the elastic from inelastic scattering processes.

8.3 Time-Resolved Ultrafast Multiphoton Photoemission

The invention of pulsed lasers established the field of nonlinear optical spectroscopy of solids.[25] Multiphoton photoemission from metal and semiconductor surfaces was one of the earliest nonlinear optical processes to be studied. The spectroscopic applications of multiphoton photoemission were realized by implementing surface science techniques in the preparation of well-characterized and atomically ordered surfaces together, with the energy and momentum analysis of photoemitted electrons.[26–28] The field of ultrafast time-resolved photoemission, greatly benefitted from the invention of Ti:sapphire laser oscillators and amplifiers, which together with nonlinear optical frequency conversion techniques provide <100 fs pulses from IR to soft X-ray regions at kHz or higher repetition rates. High pulse repetition rates are essential for single photoelectron detection. Moreover the pulse durations available from Ti:sapphire laser-based systems, which now extend to <100 as regimes, are well-suited to studying ultrafast dynamical processes initiated by electron–hole pair excitation of solid state materials. Developments in nonlinear optical frequency conversion, ultrashort pulse generation and pulse energy and repetition rate continue to broaden rapidly the scope of research on the electronic structure and dynamics at solid–vacuum interfaces.

The process of nonlinear multiphoton absorption enables spectroscopic examination of unoccupied states of solids and solid surfaces. Like its linear counterpart, multiphoton photoemission probes the initial states below the Fermi level, E_F, within an energy range defined by $n\hbar\omega - \Phi = E_0$, where n is the order of the nonlinear process, Φ is the work function of the sample and E_0 is the positive binding energy of the initial state with respect to E_F.[29,30] The MPP process is enhanced by resonances with the intermediate and final states and therefore, MPP spectra also provide information on the unoccupied electronic structure of a sample. As in conventional photoemission spectroscopy, the parallel momentum, $k_{\parallel}$, is rigorously conserved, whereas relating the surface normal component of photoelectron momentum, $k_{\perp}$, to the band momentum requires knowledge of the potential discontinuity at the solid/vacuum interface. In the case of MPP, resonant interband transitions between initial and intermediate states below the vacuum level, E_{vac}, rigorously conserve energy and momentum within the constraints of Heisenberg uncertainty. Therefore, MPP can be used for the absolute band mapping of the occupied and unoccupied bands of solids. The resonance condition occurs at specific points in k-space, so that MPP with a tuneable light source can be used for absolute band mapping of occupied and unoccupied bands.[30–32]

Analysis of resonance features in MPP spectra with respect to the excitation photon energy, photoelectron energy and momentum, excitation laser pump–probe delay, as well as other features such as electron spin anisotropy, enable detailed investigations of the electronic structure, relaxation pathways and photoexcitation-induced nuclear dynamics. Of particular interest is the excitation and relaxation of molecule-induced surface states on metal and semiconductor surfaces. The spectroscopy of such states provides information on the energy alignment of molecular orbitals upon chemisorption with respect to reference levels of the substrate and on the interaction between adsorbate and substrate which can give rise to new electronic states that are specific to the interface. Furthermore, MPP studies provide information on the excitation process of adsorbed molecules, which can proceed by direct excitation within the molecular overlayer, the photoinduced charge transfer from substrate to adsorbate, or the putative hot-electron mechanism, whereby an electron excited within the substrate scatters into an adsorbate localized state. Time-resolved studies provide further information on the relaxation pathways of adsorbate localized states involving elastic or inelastic scattering to the substrate, stabilization of excited states through the structural rearrangement of the molecular overlayer, that is, solvation, or nuclear motion associated with bond breaking or correlated electron–nuclear motion, such as proton-coupled electron transfer.[23,24,33–36] Studies of molecular overlayers or interfaces between molecular materials provide further information on the molecular exciton dynamics, such as multiexciton generation, charge-transfer exciton formation and exciton fission, which are directly related to the charge and energy transport at molecular interfaces.[37,38] Thus, TR-MPP is an exceptionally capable method for multidimensional probing of ultrafast electron dynamics which constitutes the elementary steps in the solar energy conversion at molecule/solid interfaces.

In addition to the ability to probe single-particle dynamics of photoexcited electrons and holes in bulk and surface bands, TR-MPP can be a probe of the collective free-electron response of a metal. When an optical external field interacts with a flat metal surface at frequencies below the interband absorption threshold, it is screened on a sub- to few-femtosecond timescale by the collective plasma response of free electrons. The screening charge density oscillating with the opposite phase of the incident field causes nearly perfect reflection from a metal surface. For a surface with sub-wavelength asperities, which exist as defects or are engineered to mediate specific light-matter coupling, the light can be transiently captured in a localized or propagating surface plasmon mode. TR-MPP, being a nonlinear function of the surface electromagnetic field, responds to the local field enhancements associated with the excitation of the plasmonic modes, through the coupling of the collective response to single particle excitations. Therefore, when a textured metal surface is excited with light below the surface plasmon frequency, the localized and propagating plasmon modes are excited and their decay by a two- or multiple-plasmon process into single particle excitations can be detected as a photoemission current if photoexcited electrons exceed Φ. The resulting photoemission current can be detected with energy, momentum and spatial resolution. In this chapter, we will also describe time-resolved photoemission electron microscopy (TR-PEEM) studies of the localized and propagating surface plasmon dynamics in Ag films. Of particular interest is the role of plasmonic modes in local field enhancement in nanostructured metal films and in metal-semiconductor structures, where the excitation of plasmonic modes may enhance the light collection efficiency.

We structure our presentation according to the increasing instrumental complexity. First, we describe experiments that use a conventional PEEM to image MPP spatially from structured Ag surfaces. By combining equal pulse femtosecond laser pump–probe interferometry with PEEM imaging, we generate movies of ultrafast surface plasmon phenomena. Next we describe a similar PEEM microscope coupled with a double hemispherical electron analyser to enable energy resolved imaging of the real or reciprocal MPP distributions. Specifically, we focus on the momentum microscope application, where the momentum imaging of MPP maps the energy–momentum dispersions of the occupied and unoccupied bands of Ag(111) and Cu(111) surfaces and the probability distributions of unoccupied states of alkali atoms on a Ag(111) surface. We conclude with perspectives for applications of TR-MPP in fundamental studies of energy conversion at molecule–solid interfaces.

8.4   Time-Resolved Photoemission Electron Microscopy

Photoemission electron microscopy (PEEM) is a uniquely sensitive method for imaging surface electronic and magnetic structure of complex materials with nanometre spatial resolution. When combined with UV or an X-ray lamp, or synchrotron excitation sources, PEEM provides a particularly sensitive method for imaging spatially heterogeneous material properties such as elemental composition, magnetization, ferroelectric polarization, and so on.[39–42] Such measurements can be performed during a dynamical process, such as epitaxy, catalysis, while switching electric and magnetic fields. The ability of the PEEM approach to image specific aspects of a sample depends in part on the properties of photoelectron excitation source such as the photon energy, polarization and time-profile, as well as the electronic structure properties of materials that define the light-matter interactions leading to photoemission. The excitation source can be tailored to enhance imaging contrast with respect to chemical composition, electronic structure, electromagnetic mode, electronic state population, magnetization, and so on, with time resolution that only depends on the light source and therefore, can potentially be extended into the attosecond regime.[43]

Our focus here is on the TR-PEEM technique depicted in Figure 8.1(a), which combines the unsurpassed temporal resolution intrinsic to laser excitation sources, with the high spatial resolution of electron microscopy. This particular combination of ultrafast spectroscopy and electron microscopy techniques has the potential to image matter on the fundamental time and

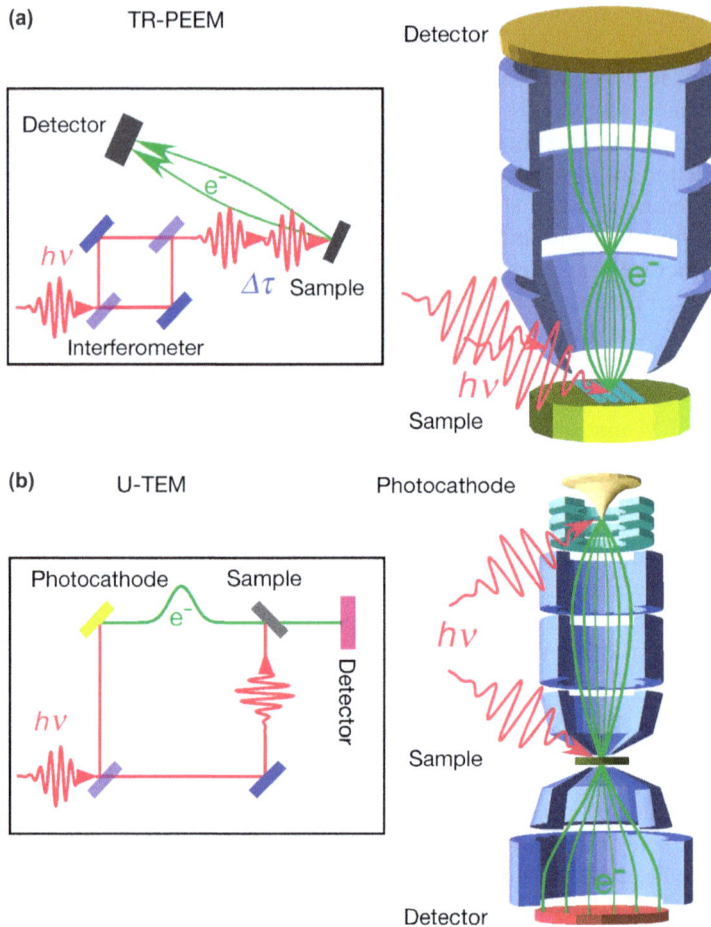

Figure 8.1   (a) Schematic representation of a TR-PEEM experiment using phase-correlated pump–probe excitation. The system time resolution is established by the laser pulse and the precision of the interferometer scanning. (b) Schematic representation of an ultrafast transmission electron microscope (U-TEM) experiment. The time resolution has multiple factors including the accuracy of laser and electron pulse synchronization and their respective pulse durations.

spatial scales of electron motion in atomic and molecular electron orbitals, although in practice such resolution is yet to be achieved. In practical TR-PEEM measurements, photoelectrons are excited stroboscopically by the joint action of pump and probe pulses with typical pulse durations of 10–100 fs. The pump pulse excites the sample within the field-of-view of the PEEM microscope, but usually lacks the photon energy to induce photoemission using a linear absorption process.

The delayed probe pulse excites the electronic system further to induce photoelectron emission from the transiently excited states prepared by the pump pulse. The time resolution is achieved optomechanically by scanning the optical path length of a delay line and therefore, is only limited by duration of the excitation pulses. Although the pump and the probe pulses may have sufficient intensity to induce nonlinear MPP individually, the primary interest is in the correlated signal involving the joint action where one photon is absorbed from each of the pump and probe pulses. The strength of the correlated signal depends on the dynamics within the electronic system of the sample and the pump–probe delay. The PEEM image records the spatial distribution of the excitation within the sample at a particular pump–probe delay. By acquiring PEEM images for a sequence of delays, one can generate a movie of the ultrafast photoinduced electron dynamics evolving spatially and temporally within the sample.

The spatial mapping of the nonlinear excitation in the sample is achieved by imaging the photoelectron emission with the electron optical system of the PEEM. By applying a high electric field between the sample and the microscope objective, a 2π solid angle distribution of photoelectrons is accelerated through the PEEM optical system and an image of either the real or reciprocal plane is projected onto a multichannel plate intensifier-based photoelectron detection system. As the de Broglie wavelength of 1 eV electrons (approximately 1 nm) is about a factor of thousand times smaller than the wavelength of comparably energetic photons, it enables the spatial resolution to exceed the diffraction limit of the excitation light substantially. In practice, however, the chromatic and spherical aberrations in electron optics limit the resolution of uncorrected PEEM instruments to ~50 nm. Advanced instruments with aberration correction achieve <4 nm spatial resolution for low energy (<10 eV) electrons.[44–46] These characteristics make TR-PEEM uniquely capable of imaging electronic excitations in solids and solid surfaces with femtosecond time resolution and sub-optical diffraction-limited spatial resolution.

TR-PEEM is one among several methods being developed for ultrafast microscopy applications. A common approach implemented within either a transmission electron microscope (TEM; see Figure 8.1(b)) or a reflection high-energy electron diffraction (RHEED) instrument uses femtosecond laser pulses to excite a sample as well as to generate electron pulses for time-correlated interrogation of the sample. Such methods, which use optical excitation and electron probing, are capable of atomic resolution in either the real or reciprocal space imaging modes, but the time resolution until now has been substantially lower than that of TR-PEEM because of the difficulty in generating and propagating femtosecond electron pulses at the point of measurement with minimum jitter and pulse broadening.

A particular advantage of the TR-PEEM approach, not shared by these more conventional embodiments of ultrafast microscopy, is the ability to image coherent phenomena in solids and surfaces. Because both the photoelectron excitation and imaging are sensitive to phases of both the excitation field and the electronic wavefunctions of the excited modes, a TR-PEEM experiment can uniquely probe coherent processes in solid-state materials.[6] Coupling light with the electronic system creates a coherent polarization field comprising collective and single particle excitations. The collective response of a sample, for example, can result in localized or propagating surface plasmon modes. The coherent polarization can also involve quantum mechanical superposition of single-particle electron–hole pair (*e–h*) excitations created by dipole coupling by the external field. Thus, TR-PEEM enables imaging of the transient coherent states that are involved in generating the photoemission signal. In addition to these coherent excitation channels, photoemission can proceed via incoherent pathways associated with the transient electronic occupation of the bulk and surface states.

In a coherent TR-PEEM experiment photoemission is generated by phase-related pump and probe excitation pulses with a duration that is comparable or shorter than the electronic dephasing times of the sample. These coherent fields can be generated in an interferometer by splitting a single pulse between two optical paths and interferometrically scanning the delay of one of the paths (Figure 8.1(a)),[12,47,48] or by employing more versatile methods involving phase and amplitude modulation of the excitation light.[49] Because photoemission can be imaged with single photoelectron counting sensitivity at MHz excitation rates, experiments can be performed in a quantum mechanical regime using single photoelectron imaging techniques.

Besides the coherence properties of light, TR-PEEM imaging depends on coherence in electron detection. Nonlinear photoexcitation propagates electrons from initially occupied states in the sample to free electron states in vacuum, which are imaged at the PEEM image plane. This coherent process can occur though multiple electron excitation and transmission channels resulting in quantum mechanical interference.[50] We will show how quantum state (energy and momentum) resolved imaging of the photoemission signal is particularly sensitive to such coherent effects.

8.5 Time-Resolved Photoemission Electron Microscopy (TR-PEEM) Imaging of Plasmonic Phenomena

The ultrafast imaging of plasmonic excitations at a metal/vacuum interface is particularly well suited for TR-PEEM studies.[6,14,51–55] In the following, we will describe the TR-PEEM imaging of localized and propagating plasmonic modes at Ag/vacuum interfaces. Although TR-PEEM is frequently applied in studies of plasmonic phenomena in Ag films, the method is broadly applicable to the imaging of spatiotemporal dynamics of coherent fields and electronically excited states in complex nanostructured materials.

Plasmonic excitations involve the coherent collective response of free electrons in metals or doped semiconductors to external particles or fields. The collective response of a solid-state plasma to external fields or particles occurs at a characteristic plasma frequency $\omega_p = \left(4\pi n e^2 / m^*\right)^{1/2}$, which is defined by the electron density, n and the effective mass, m^*.[56] Depending on the intrinsic or impurity doped material properties, plasma frequencies range from the infrared (IR) for semiconductors to vacuum ultraviolet (VUV) for metals reflecting their different free electron densities. Below the plasma frequency, the real part of the dielectric function of a solid-state plasma is large and negative and therefore external fields are efficiently reflected. Above the plasma frequency, however, fields propagate through plasmas as highly damped charge density oscillations. Metal–dielectric interfaces also support coherent charge density oscillations, but their properties are modified with respect to bulk plasmas by the reduced charge density in the selvage at the interface with the dielectric. In flat or nanostructured metal films, these collective surface modes (surface plasmons) are of particular interest because they can localize light on the nanometre spatial scale and thereby enhance the associated electromagnetic fields. This is useful for a variety of intriguing plasmonic phenomena, such as the surface-enhanced Raman effect, which enable single molecule sensing, spectroscopy and near-field microscopy applications.[57–59] The high field enhancement also facilitates nonlinear photoelectron emission.[60] The enhanced photoemission associated with the excitation of plasmonic modes makes TR-PEEM the ideal tool for studying the ultrafast nonlinear plasmonic phenomena. A TR-PEEM image of a metal film can usually be interpreted as a nonlinear map of the transient electromagnetic fields excited in the sample.

The collective electromagnetic modes of a metal/dielectric (vacuum) interface are determined by the dielectric properties of the component media and the geometry of the interface.[56,61] Here we are interested in two cases, where a silver film (i) has occasional nanometre scale asperities or (ii) is nearly atomically flat except for a sharp (compared to the wavelength of light) film discontinuity. The plasmonic modes of such films can be excited at frequencies that depend on the sample geometry and dielectric properties, as well as the wave vector of the excitation light. The resonance frequencies of surface plasmon (SP) modes can be calculated for metal particles with well-defined shapes and sizes, but for randomly structured metal films they are distributed broadly in frequency.[62] Excitation of localized SP modes in Ag, where the electromagnetic field is localized within a volume of dimensions that are small with respect to its wavelength, has been shown to enhance MPP in a broad range of frequencies in the visible and UV spectral regions up to the asymptotic SP resonance at 3.72 eV.[60] These localized modes, referred to as surface plasmon hot spots, have been shown to enhance the nonlinear photoelectron emission in numerous PEEM studies.[6,54,55]

In addition to the localized SP modes, metal/dielectric interfaces support surface plasmon polaritons (SPP), which propagate according to the dielectric properties of the interface at the local speed of light. The complex dispersion relationship for SPP propagation at a metal/dielectric interface is given by:

$$k = \frac{\omega}{c} \left(\frac{\varepsilon_1 \varepsilon_2}{\varepsilon_1 + \varepsilon_2} \right)^{\frac{1}{2}} \tag{8.1}$$

where ε_1 and ε_2 are the dielectric functions of each half-space forming the interface (Figure 8.2, inset).[63] The real propagation wave vector ($\mathrm{Re}(k)$ or k_{SPP}) of the Ag/vacuum interface, based on the experimental dielectric function of

Ag,[11,64] is shown in Figure 8.2. According to Figure 8.2, for small values of Re(k), the SPP field propagates at nearly the same group velocity as the external field, for which the dispersion is a function of the propagation k-vector at 65° from the surface normal (required by the PEEM geometry) and is indicated by the light line; in this case, SPP predominantly present an electromagnetic field in vacuum with a small component penetrating the skin depth of Ag. For large values of Re(k), the SPP is predominantly a charge density oscillation in Ag with an asymptotic surface plasmon (Ritchie plasmon) frequency of 3.72 eV,[65] for which the group velocity is zero. Because the SPP dispersion and light lines do not cross, the coupling of the external field incident from the vacuum into the SPP mode of a Ag/vacuum interface requires an additional source of momentum to compensate for the momentum mismatch, Δk, representing the difference between the Re(k) and the light lines at the excitation frequency. The momentum conservation for the mode coupling can be satisfied if Δk is supplied by the plasmonic medium or the dielectric with a sharp discontinuity or a periodic modulation, which provide, respectively, a broad spectrum and discrete comb k-vectors at integer multiples of the reciprocal grating wave vector.[7] In the following, we first illustrate the TR-PEEM imaging of localized SP modes and then SPP wave packet propagation at flat silver/vacuum interfaces.

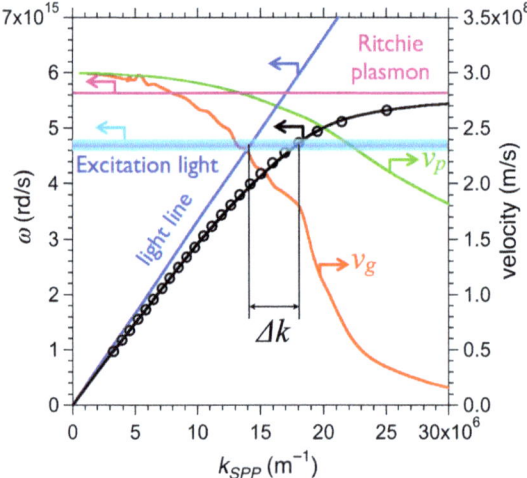

Figure 8.2   Real part of the SPP propagation wave vector, k_{SPP}, at a Ag/vacuum interface based on the optical constants of Ag. The light line is the 65° incidence angle. The blue line represents the spectrum of the excitation pulse. Δk is the momentum mismatch between the incident field and the SPP mode. v_g and v_p, expressed in units of the speed of light in vacuum c, represent the group and phase velocities implied by the dispersion function of Ag/vacuum interface.[6]
Reprinted (adapted) with permission from Kubo *et al.*[14] Copyright 2007 American Chemical Society.

8.5.1   TR-PEEM Imaging of Localized SP Modes

The type of plasmon mode supported by a metal film depends in part on its method of preparation. In Figure 8.3 we show a PEEM image of a silver grating structure with a period of 780 nm: the sample was fabricated by deposition of Ag onto a periodic array of mesa structures formed lithographically in a quartz substrate.[66] The Ag film is deposited on the mesa structures at an angle to minimize deposition of metal into the troughs between the mesas. These silver grating structures act as transmission filters, with optical characteristics that depend on the physical structure of the metal grating and its dielectric environment.[66] The PEEM image in Figure 8.3(a) is obtained with 4.9 eV photons from an Hg lamp, which have sufficient energy to induce single photon photoemission (1PP) from the Ag surface with a work function of 4.3 eV. The contrast of the 1PP image reflects both the spatial distribution of Ag, which gives bright contrast owing to its low work function and the shadowing effect of the oblique incidence of the Hg lamp light. The image of the same grating in the inset of Figure 8.3(a), obtained with higher resolution by scanning electron microscopy (SEM), shows surface roughness of the Ag film on the mesa structures with features on the 10–20 nm scale.[6] These random asperities, which result from the fabrication, are too small to image by the PEEM, which has ∼ 50 nm resolution.

When the same structure is imaged by 400 nm, ∼ 10 fs laser pulses incident at 65° from the surface normal (Figure 8.3(b)) the contrast is dramatically different. The TR-PEEM image is now dominated by emission from hot spots randomly distributed on the mesa structures. These hot spots correspond to localized SP modes, which are resonant or nearly resonant with the excitation light. The trapping of light in nanometre scale asperities enhances the local

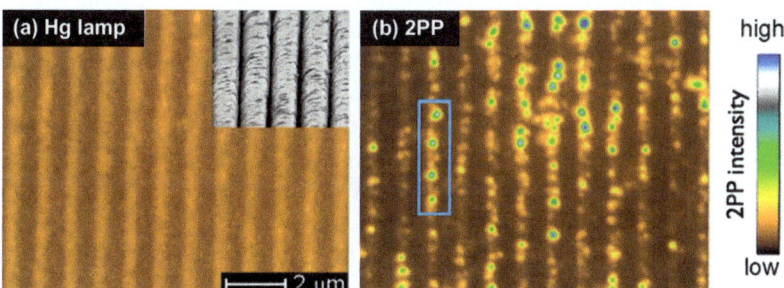

Figure 8.3   (a) PEEM image of the 780 nm period Ag grating structure excited with 4.9 eV photons from a Hg lamp. The contrast in this 1PP image is given by the material composition and shadowing effect of the excitation light. The inset shows a higher resolution SEM image, which shows the surface roughness. (b) 2PP-PEEM image of the same region of the sample excited by 400 nm, 10 fs laser pulses. The contrast is dominated by the distribution of the resonant SP modes (hot spots) at the sample surface.[6,11] The blue box shows the enlarged region of the sample that is the subject of Figure 8.4.
Reprinted (adapted and enlarged) with permission from Kubo *et al.*[6] Copyright 2005 American Chemical Society.

fields and the consequent nonlinear excitation leading to two-photon photo-emission (2PP). Because the excitation light has insufficient photon energy to overcome the work function of the sample, photoemission can only be excited by a nonlinear optical process and is therefore sensitive to nanoscale regions that support resonant SP modes, where the field enhancement is most effective. In the case of the 780 nm period grating, the incident light cannot couple into the propagating plasmon modes via the exchange of the grating momentum and, therefore, the 2PP signal is dominated by the randomly distributed local SP resonances. For resonant gratings we have observed uniform emission from similar mesa structures reflecting the efficient excitation of the grating modes.[14]

From the hot spot emission, we can conclude that the resonant SP modes enhance the local fields, but the mechanism for 2PP excitation requires further consideration. We will show that photoemission occurs by absorption of two photons from the field, but the identity of the single particle excitations and the excitation sequence is not unique. For example, the photons could be absorbed coherently or sequentially. Moreover, the coherence could involve the collective SP mode or single particle excitations. The measured photoelectron spectra for localized plasmon excitation in the Ag grating structure decays approximately exponentially with photoelectron energy.[6] Similar photoelectron distributions have been reported for Ag nanoparticles on graphite and metal oxides.[67,68] The lack of spectroscopic features from the bulk or surface electronic bands, such as are observed in 2PP spectra of atomically flat single crystal Ag surfaces,[69,70] indicates that the direct coupling of light to single particle excitations of Ag is unimportant. This is in part a consequence of the fact that resonant dipole e–h pair excitation channels are not available in bulk Ag below the onset of d- to sp-band absorption at 3.84 eV.[71,72] Therefore, light can only be absorbed through higher order processes, such as the intraband Drude absorption, where momentum is supplied by scattering of phonons or from a rough surface,[63] or by non-resonant two-photon absorption from the lower to the upper sp-band of Ag.[69,70] Photoelectrons can be generated by the sequential interband excitation hot electrons in the enhanced SP field, or by coherent excitation of a two-quantum (or higher) plasmon state followed by its decay into an e–h pair.[67] In either case, photoemission can occur only for electrons that have absorbed at least two photons from the incident field to acquire sufficient energy to over-come the work function of the surface. The contributions of sequential *versus* coherent absorption mechanisms can be discerned in TR-PEEM measurements from their dependence on the phase of the excitation light.

The dynamics of localized SP mode excitation and dephasing can be in-vestigated by recording TR-PEEM movies as a function of interferometrically scanned delay between identical, pump–probe pulse pairs. In order to illustrate typical SP mode dynamics, in Figure 8.4 we show excerpts from a TR-PEEM movie for a section of the image in Figure 8.3(b) containing four resonant SP modes. The excitation pulses for generating the movie frames are obtained by passing the 90 MHz repetition rate pulse train from the laser through a Mach–Zehnder interferometer (MZI). As indicated schematically in Figure 8.1(a), the MZI generates a phase correlated pulse pair by splitting each pulse into two equivalent replicas, passing them through two different paths and recombining them at the output, with a relative delay that depends on the difference in the transversed path length. The delay can be set precisely to a specific phase of the excitation field by monitoring the spectral interference between the pump and probe pulses in a grating spectrometer. Each frame of a TR-PEEM movie is obtained in a specific phase between the excitation pulses. The movie is recorded by advancing the path length of one arm of MZI by ~ 100 nm, or one-quarter of an optical cycle of 400 nm excitation light, which corresponds to advancing the delay by ~ 330 as and recording a PEEM image for each delay.

Figure 8.4 shows selected frames for $N - \frac{1}{4}$, $N + 0$, $N + \frac{1}{4}$ and $N + \frac{1}{2}$ cycle delays where $N = 0, 5, 10, \ldots, 30$. At $N = 0$ the four hot spots have maximum intensity, because the pump and probe pulses coincide in time and as the phase is advanced, their intensities oscillate at the driving frequency. As the delay is advanced and the pump pulse wanes, the coherent polarization no longer oscillates at the driving frequency but instead at the natural material and structure dependent resonant frequency of each SP mode. When N is advanced, the constructive interference between the SP mode and the probe pulse may not occur at $N + 0$ delay, but rather may fall behind or be advanced with respect to the oscillation of the excitation field. For example, at $N = 10$–20 cycle delays for spots labelled B and D photoemission, the intensity maximum remains at the $N + 0$ relative phase, whereas for A and C the maximum shifts, respectively, to a positive and negative delay with respect to $N + 0$.

This mode-specific phase dependence of 2PP intensity reflects the dispersion of SP mode frequencies that can be excited within the bandwidth of the 10 fs, 400 nm excitation pulses. The pump pulse drives free electrons in the Ag sample at the 750 THz excitation frequency. At a flat surface, the out-of-phase

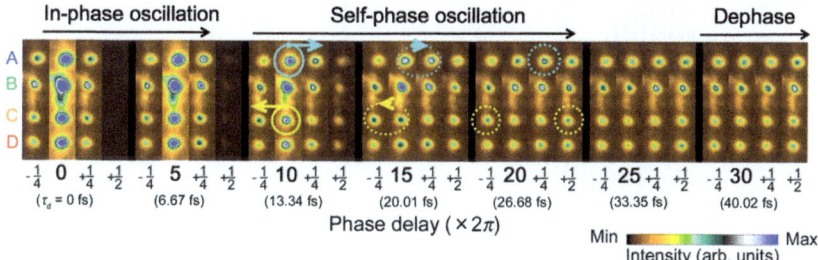

Figure 8.4 Selected frames of a TR-PEEM movie for $N - \frac{1}{4}$, $N + 0$, $N + \frac{1}{4}$ and $N + \frac{1}{2}$ cycle delays where $N = 0, 5, 10, \ldots, 30$, for the four hot-spots indicated in Figure 8.3. When the delay is short compared with the pulse duration, the spots labelled A–D oscillate in phase with the driving field. At intermediate delays ($N = 10$–20) the modes oscillate at their own frequency and the intensities reflect the interference between the driving and the resonance frequencies. For even longer delays, the sample has lost the memory of the phase of the driving field between the pump and probe interactions and therefore the intensity is independent of the excitation phase.[6] Reprinted (adapted) with permission from Kubo *et al.*[6] Copyright 2005 American Chemical Society.

oscillation of free electrons screens and reflects the incident field essentially instantaneously. Because 400 nm light does not excite resonant *e–h* pair transitions, the sample does not retain memory of the excitation field. At a rough surface, however, the incident light can excite and is transiently captured as a charge density oscillation of the resonant plasmon modes. The SP modes retain the memory of the phase of the excitation pulse, but they oscillate at their own resonant frequency, which may be different from the driving frequency, after the driving field has decayed. The SP modes dephase by a variety of processes including scattering of the excitation field or the decay into single particle excitations.[73] Only the latter process gives rise to the 2PP signal. In a complex sample the individual SP dipoles may be sufficiently close for two or more modes to be coupled leading to complex spatiotemporal dephasing dynamics of the coupled oscillators. In the case of the hot spots A–D, our measurements are consistent with simple, single-mode dynamics. Therefore, we attribute the phase advancement and retardation of A and C, respectively, to a higher and lower resonant frequency with respect to the driving field.

A more quantitative measurement of the coherent mode dynamics can be obtained by measuring the 2PP intensity of each mode while performing an interferometric scan of the pump–probe delay.[6] Such experiments, shown in Figure 8.5, give an SP mode specific interferometric two-pulse correlation (I-2PC) measurement of the induced polarization and hot-electron population dynamics. The signal in Figure 8.5 consists of an oscillatory component near

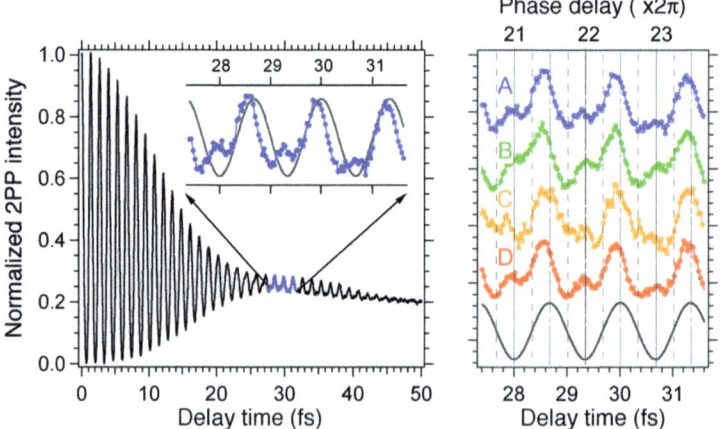

Figure 8.5 Interferometric two-pulse correlation measurements of the coherent response of spots A–D in Figure 8.4. Measurements of the 2PP signal intensity at each hot spot reflects the coherent SP polarization dynamics, near zero delay, as well as a contribution from the sequential 2PP, which is responsible for the slow decay of the signal after the coherent oscillations have died out. The enlarged image of the coherent oscillations shows the SP mode specific nonlinear polarization dynamics. The corresponding laser phase is also shown.[6]
Reprinted (adapted) with permission from Kubo *et al.*[6] Copyright 2005 American Chemical Society.

zero delay, which reflects the interference between the pump and probe induced coherent polarizations and an exponentially decaying signal on a longer timescale owing to a sequential (incoherent) 2PP process via intermediate hot-electron states.

The width of the oscillatory signal has contributions from the pulse duration and more significantly, the free-induction decay (FID) of the coherent polarization excited in the sample. Although the timescale of FID has contributions from both the intrinsic (homogeneous) and inhomogeneous dephasing processes,[74] in the case of single particle measurements such as in Figure 8.5, one can assume that only the intrinsic field scattering and hot-electron decay processes contribute. The FID times measured from I2PC scans for SP hot spots A–D range from 4.9–5.8 fs with the decay times correlating with the mode frequency, which is expected from the dielectric function of Ag.[6,52] Because the TR-PEEM measurements are integrated over a range of energies, the hot-electron decay on a longer timescale is dominated by the lowest energy hot electrons, which have the highest intensity in typical 2PP spectra via the SP excitation as well as the longest lifetimes.[6,75]

The example of random SP modes on rough Ag surfaces demonstrates how TR-PEEM can be used to monitor coherent spatiotemporal dynamics on the nanometre spatial and femtosecond temporal scales. These studies are not just confined to SP modes and can in principle be extended to more complex nanometre structures involving coherent excitations in variety of materials including metals, semiconductors and even biological materials. The only requirement is that different excitations be accessible to probing by photoemission techniques. In the examples provided, the SP modes are not strongly coupled to each other, so the analysis assumed single mode dynamics. It is more interesting, however, to be able to image the coupled mode dynamics that are known to be important in light harvesting systems.[76,77] Extensions of TR-PEEM techniques to multidimensional spatiotemporal coherent dynamics are likely to emerge through development of sophisticated coherent excitation techniques and through improvement in imaging resolution.[78]

8.5.2 TR-PEEM Imaging of Surface Plasmon Polariton (SPP) Dynamics

In this section we describe TR-PEEM applications to imaging of SPP dynamics in atomically smooth nanostructured metal films.[7,52] By creating a discontinuity in a smooth metal film, it is possible to couple SPP fields into a Ag/vacuum interface and to observe their coherent evolution, including propagation, diffraction, interference, and so on. Using nanofabrication techniques it is possible to manufacture sub-optical wavelength structures in Ag films that perform structure-specific coupling of SPP fields. The spatial distribution of such structures within the Ag film and the excitation geometry define the amplitude, phase and k-vector of SPP wave packets (WP), which in turn define the subsequent linear propagation of such WPs according to the complex dielectric function of the Ag/vacuum interface.[7,14,52,79]

Any discontinuity in the Ag film, which the Fourier spectrum of spatial frequencies translates into a spectrum of propagation wave vectors, can act as an SPP coupling structure. Such structures can be conveniently generated by focused ion beam (FIB) milling, or other lithographic techniques.[80] The structures can be formed either in the substrate for Ag film deposition, or directly in an Ag film. In experiments reported here, coupling structures with specified dimensions are formed by FIB in Si(001) wafer substrates. Subsequently, the SiO_2 layer is removed from the processed substrate by flashing its temperature to 1150 K under ultrahigh vacuum (UHV). After the bare Si surface is exposed, an Ag film of ~ 80 nm thickness is deposited at low temperature (100 K) from an e-beam evaporator source. Unless the Ag film thickness substantially exceeds the ~ 30 nm skin depth of Ag, the Si substrate can absorb the SPP field for photon energies that exceed its 1.1 eV band gap.[7] Deposition of Ag on lithographically structured Si substrates is useful for *in situ* preparation of clean and atomically smooth polycrystalline Ag films with well-defined coupling structures.[81] Alternatively, the coupling structures can be formed directly by FIB processing of Ag samples, but this may lead to contamination of the film surface from Ga ions or through transport of the sample between vacuum chambers. We have used both approaches and found no evidence that SPP WP dynamics as observed by TR-PEEM are sensitive to such contamination.

The complex interaction between an external field and a coupling structure that launches an SPP WP can be modelled by finite-difference time-domain (FDTD) methods.[82] We have performed such simulations recently in the case of a slit coupling structure for our experimental conditions, that is, 10 fs, 400 nm pulses incident at 65° from the surface normal (Figure 8.6).[7] The FDTD simulations identify three contributions, which add coherently to give the total SPP WP field. For an incident field propagating by a k_{light} wave vector from the left side and tilted with respect to the surface normal, the primary scattering in the SPP mode occurs at the right edge of the slit. Both experiment and theory indicate additional contributions that are caused by trapping the external field in a Fabry–Pérot resonance within the slit, which enhances the SPP field whenever the slit width is an integer multiple of $\lambda/2$, where λ is the vacuum wavelength of the incident light, as well as a cylindrical wave that is scattered by the left slit edge and propagates to the right edge, where it is coupled into the SPP mode. The contributions from these different sources can be discerned in experiments and FDTD simulations by determining how the SPP field amplitude varies as a function of the slit width.[7]

Excitation of an Ag film with a discontinuity acting as a coupling structure launches an SPP WP, which can be imaged as a spatial interference pattern by TR-PEEM, such as shown in Figure 8.6(c). For 400 nm wavelength laser pulses incident at 65° from the surface normal and propagating in a direction orthogonal to the length dimension of the coupling structure (Figure 8.6(a)), we observe an interference pattern forming fringes parallel to the coupling structure. The interference results from the coherent superposition of SPP WP launched from the coupling structure and propagating at the Ag/vacuum

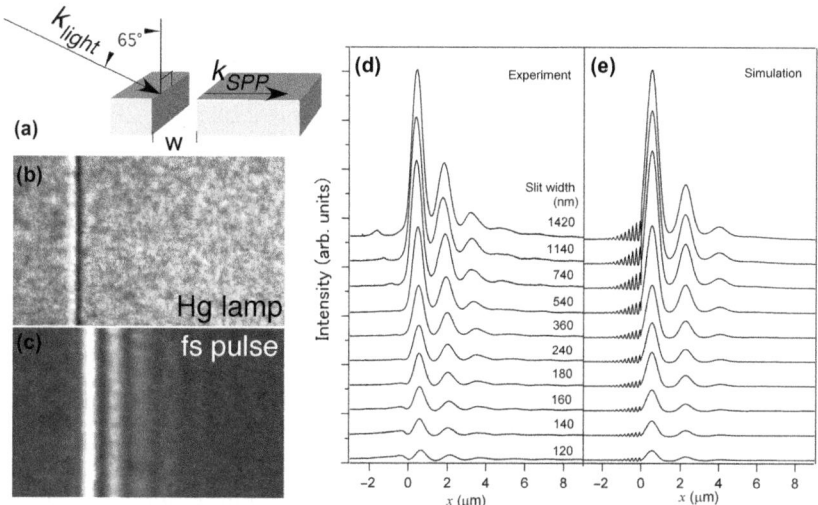

Figure 8.6   (a) Incident (k_{light}) and SPP (k_{SPP}) wave vectors involved in excitation of
an SPP WP at a slit with a width w fabricated in an Ag film. (b) the Hg
lamp and (c) 3.1 eV femtosecond laser excited 1PP and 2PP from an Ag
film with a 200 nm slit. The periodic pattern in (c) with a period of 1.65 μm
results from the interference between SPP WP and the excitation light. (d)
Measured and e) calculated interference patterns as a function of the slit
width. The rapidly oscillating interference in (e) caused by a backward
propagating SPP WP is not resolved in the PEEM measurements.[7] (b)–(e).
Reprinted (adapted) with permission from Zhang *et al.*[7] Copyright (2011)
by The American Physical Society.

interface with a k_{SPP} direction orthogonal to the coupling structure and its
magnitude given the complex propagation constants of an Ag/vacuum interface
and the field of the external excitation pulse, which irradiates the entire field-
of-view of the PEEM at its focal plane. The intensity of the beating pattern, but
not the wavelength, depends on the fabricated width w of the coupling structure, as can be seen in the intensity cross sections reported in Figure 8.6(d). The
surface projected wavelength of the incident light is $\lambda_x = 441$ nm, whereas the
SPP WP wavelength at the centre frequency is $\lambda_{SPP} = 348$ nm. The wavelength
of the interference between the two fields is $\lambda_{beat}^+ = \left(\lambda_{SPP}^{-1} - \lambda_x^{-1}\right)^{-1}$ or ~ 1.65 μm.
The period of the beating pattern can be immediately deduced from the mo-
mentum mismatch Δk between the light line and the SPP dispersion function in
Figure 8.2 from the relation $\lambda_{beat} = 2\pi/\Delta k$.[52]

In addition to the beating pattern of the forward propagating SPP field,
FDTD simulations in Figure 8.6(e) predict a backward propagating wave
for which the predicted beating pattern $\lambda_{beat}^- = \left(\lambda_{SPP}^{-1} + \lambda_x^{-1}\right)^{-1}$ is 190 nm. The
corresponding polarization gratings are weak and decay over a short distance
reflecting the spatial extent of the overlap of counter propagating fields. This
high-frequency grating is not resolved by TR-PEEM under the conditions used
for recording the forward propagating interference patterns.

The interference pattern in Figure 8.6 is produced by the total surface field a few nanometres from the Ag surface, which drives the 2PP process.[7,52] The visibility of the fringes depends on the relative strengths of the interacting fields and the relative contribution from coherent *versus* sequential 2PP processes. The high fringe visibility indicates that the internal SPP and external excitation field amplitudes are comparable and contributions from sequential processes are negligible. Therefore, the TR-PEEM image in Figure 8.6(c) gives a non-linear map of the SPP field excited at the Ag/vacuum interface. It represents the fourth power of the total electric field of the external and SPP pulses in the Ag film integrated over time. From knowledge of the external field, it is possible to determine the dispersive and dissipative SPP WP propagation as long as the interacting fields overlap in time and space.[7,52]

The spatial extent of the interference pattern from the coupling structure depends on (i) the SPP decay into single particle excitations; (ii) the dispersion of the SPP WP propagation (according to Figure 8.2 the SPP phase velocity is twice the group velocity); and (iii) the overlap of the interfering fields. Because the external field propagation is faster than the SPP WP, in the case of slow SPP dephasing at long wavelengths, the overlap of the two fields could diminish before the SPP dephasing has occurred. In such a case it is still possible to study SPP wave packet dynamics by using phase correlated pump–probe pulse pairs. With two-pulse excitation, the interference pattern observed by TR-PEEM consists of self-interference of each pulse with its own SPP WP. This signal does not depend on the pump–probe delay. In addition, when the delay is short or comparable to the SPP dephasing time, a component corresponding to the SPP WP excited by the pump pulse interferes with the external field of the probe pulse. In TR-PEEM movies this component can be seen to propagate away from the coupling structure as the pump–probe delay is advanced.[52]

8.5.3 Nanoplasmonic Optics

Coupling into the SPP mode can be accomplished with arbitrarily shaped coupling structures. When the incident field vector is not parallel to a coupling structure, the SPP WP along the slit is generated by a position-dependent phase, which causes k_{SPP} to deviate from the direction of the projection of k_{light} onto the surface plane. In this case, it turns out that the interference fringes remain parallel to the coupler even though both k_{SPP} and the projection of k_{light} are not orthogonal to the slit length direction. Coupling structures of different shapes can be used as nanometre scale optical elements for shaping SPP fields.

As an example of TR-PEEM imaging of SPP WP dynamics imposed by a plasmonic optic, we show two images from a TR-PEEM movie for 17.0 and 17.5 optical cycle delay for a "V" coupling structure where two slits are fabricated in a Ag film such that they meet at an apex with a mutual angle of 60° (Figure 8.7(a)).[83] The external field propagates from the right exciting the "V" coupling structure starting at its apex. Each side of the coupling structure launches an SPP wave packet, which propagates with a tilted wave front with respect to the excitation field. As the field propagates from right to left, it

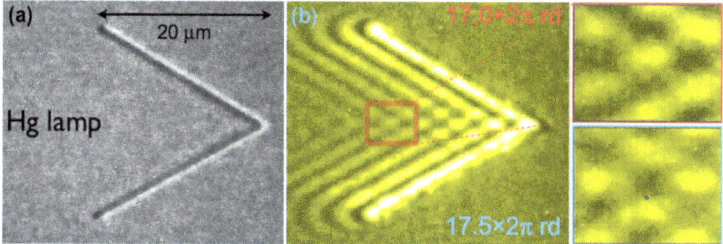

Figure 8.7 (a) PEEM image of a 60° 'V'-shaped coupling structure upon Hg lamp excitation. (b) A TR-PEEM image of the same structure with 10 fs, 400 nm laser pulse excitation using a phase-locked pulse pair with 17.0 optical cycles delay (22.6 fs). The interference pattern between the external excitation and SPP WP fields produces a polarization grating with phase fronts parallel to the coupling structure. The expanded region of (b) shows the complementary contrast in the interference pattern when the phase is advanced by 0.5 cycles.

encounters the coupling structure at different points along the propagation direction. Each point of the coupling structure acts as a source of circular SPP waves. The waves emanating from each point add coherently to form a wave front. The wave front is tilted with respect to the incident wave because of the slower propagation of the SPP WP with respect to the exciting field. The interference between the external and SPP fields gives a polarization grating with lines of maximum and minimum intensity parallel to the two sides of the coupling structure. These interference wave fronts represent the total field rather than the propagation of SPP WP, for which the k_{SPP} points in a direction between that of the external field and the propagation of the interference pattern.

In addition to the excitation of SPP WP by tilted slits, the images in Figure 8.7 show two additional phenomena associated with plasmonic optic structures. Of interest are the diffraction patterns of SPP waves at the open ends of the "V" structure. The diffraction behaviour is expected for SPP WP, just like any other electromagnetic wave. It suggests that lithographically formed coupling structures could be used to tailor complex surface electromagnetic fields, just as has been demonstrated by Nelson and co-workers on longer spatial and temporal scales in the case of femtosecond laser excitation of surface phonon polaritons in ionic crystals.[84,85] Furthermore, such structures could be excited by spatially shaped pulses to generate optimal SPP fields for controlling electromagnetic phenomena on the nanometre scale.[86]

As an example of nanoscale manipulation of electromagnetic fields we consider the interference patterns in Figure 8.7(b) that are generated by the "V" coupling structure. Each segment of the coupling structure generates an SPP WP that propagates with a component in the direction of the other segment. Where the SPP WPs overlap their interference ensues. Thus, the "V" structure acts as a simple SPP interferometer. The effect of advancing the phase of the interferometrically locked pulse pair excitation field from 17.0 to 17.5 optical cycles is shown in expanded sections of Figure 8.7(b). We can see that the total field at a designated location integrated over the interaction time can be minimized or maximized by the choice of the phase (pulse delay) of the excitation waveform. Therefore, the simple "V" coupler acts as an SPP interferometer when excited by a plane wave; its function can be made more versatile by addressing each side of the interferometer with a separate pulse. We emphasize that the method of SPP detection is a second-order nonlinear process and therefore the SPP wave packets are sufficiently intense to drive coherent nonlinear excitation processes in Ag. Therefore, similar coherent interactions using separately addressable coupling structures could be used for nonlinear coherent control and optical switching in plasmonic nanostructures[87] on the nanometre spatial scale and femtosecond timescale. TR-PEEM is an ideal instrument for studying such coherent nonlinear phenomena in complex nanostructured materials.

8.6 Photoelectron Momentum Mapping

Having described real space imaging with a PEEM, we now turn to energy-filtered momentum imaging of the occupied and unoccupied electronic bands and states of single crystal surfaces. We emphasize that both spatial and momentum imaging is possible within the same instrument,[88] making PEEM a highly versatile microscopic and spectroscopic instrument for imaging ultrafast nanoscale electronic phenomena.

8.6.1 Momentum Mapping of Delocalized Bands

In its most basic version, imaging in the photoelectron emission microscope does not involve explicit energy filtering but instead relies on localized changes in the absolute electron yield to reveal the real-space structures of interest. This can be achieved, for example, by the use of photon energies near the photoemission threshold in order to image work function changes, or by exploiting element-specific absorption edges using tuneable synchrotron radiation. In contrast to these indirect approaches, the addition of an electron energy filter to the PEEM allows direct element-specific photoelectron imaging based on the chemical information provided, for example, by core level photoelectrons. Moreover, in addition to the energy-resolved real-space mapping of surfaces and nanostructures, the information contained in the angular distributions of the optically excited photoelectrons is also of significant interest, for example in a spatially resolved investigation of the valence band electronic structure. Based on very general principles of electron-optical imaging, the combined energetic and directional information that is contained in the photoelectron momentum vector can be accessed in parallel using a PEEM-based momentum microscope.[89,90]

The momentum microscope[89] is a combination of a conventional PEEM column, which serves as an input to a double-hemispherical imaging electron energy analyser, shown schematically in Figure 8.8. The addition of the energy

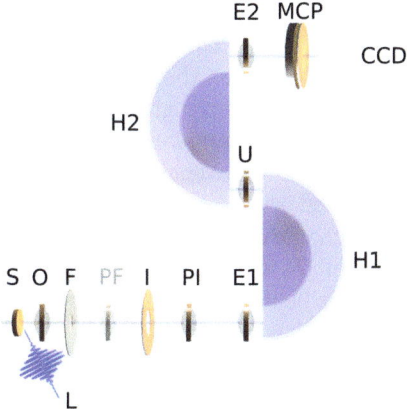

Figure 8.8 Simplified working principle of the momentum microscope (for explanations of the symbols see text).

analysers on the output of the PEEM column serves the dual purpose of energy filtering the photoelectrons and simultaneously preserving the spatial information delivered by the PEEM stage. Here we are primarily concerned with laser-based experiments, which excite photoelectrons.

Laser pulses (L) from a Ti:Sapphire oscillator are frequency doubled ($hv = 3.1$ eV) for 2PP or quadrupled ($hv = 6.0$ eV) for 1PP. The PEEM objective electron lens (O) forms a real-space image of the sample (S) at the image plane (I), as well as a Fourier transformed image at the focal plane (F). As is well known from the properties of lenses in conventional light optics, at F the spatial intensity distribution is arranged according to the surface-parallel momenta (k_x, k_y) of the emitted photoelectrons contributing to the image in real (x, y) space at I. The aperture in the focal plane F allows control of the image quality by restricting the electron k-vectors contributing to the real-space image and the aperture in the image plane I allows selection of a real-space region for spatially resolved k-space imaging of microstructures. Imaging of the focal plane or the image plane is selected by the electron optical system shown schematically by lenses PF and PI, respectively. The respective image selected by one or the other lens from the PEEM column is focused by lens E1 to the entrance slit of the hemispherical analyzer H1. The first analyser H1 works in tandem with the second analyser H2 to provide energy filtered and aberration corrected imaging. Aberrations introduced by energy filtering with H1 are corrected by passing the energy-filtered electrons through the inversion lens U then effectively reversing their trajectories through H2. The aberration correction is achieved by the principle of closed Kepler-ellipses for the trajectory of a particle in a central potential, which can be established between an inner and an outer spherical electrode. The lens U then separates the start and the end point of a closed trajectory in the central potential by the use of two separate hemispherical analysers.[90] After passing H2, the electrons are finally projected (E2) onto the multi-channel plate with a phosphor screen (MCP) and the distribution is imaged by a charge coupled device (CCD) camera.

The momentum microscope is especially well suited for studying the full-hemispherical momentum emission distributions of photoelectrons, as the extraction voltage of several kV between the objective lens and the sample leads to the collection of practically all emitted electrons. The momentum distributions in the x and y directions are recorded for selected photoelectron energies to obtain a three-dimensional energy-momentum dispersion function from microscopically defined samples. The full-hemispherical patterns are collected without any mechanical movement of the sample in a single run, which is particularly useful for measurements of symmetry related intensity changes, for example, under helicity reversal of circularly polarized light (circular dichroism in angular distributions, CDAD[91]) or under magnetization reversal of magnetic samples. For example, magnetic dichroism[40] can be studied directly without the influence of any external mechanical adjustments in the complete accessible k-space. Moreover, full band mapping measurements can be performed on micron-sized samples, such as microcrystallites in a polycrystalline sample.[89,90]

The momentum microscope directly delivers constant energy maps of the parallel photoelectron momenta (a collection two dimensional (2D) momentum distribution curves). As instructive examples, we will discuss the formation of photoelectron emission patterns from quasi-free electron systems. In the bulk of a three dimensional (3D) crystal, the constant energy surfaces in momentum space are spheres with radii k_E given by the quasi-free electron dispersion $(k_E)^2 \sim 2m_{eff} E$ with the effective electron mass m_{eff}. Similarly, for 2D surface state systems, the relevant constant energy surfaces are cylinders in three-dimensional k-space. For direct optical transitions in these systems, one has to consider the restrictions of energy and momentum conservation. With a free-electron final state characterized by a spherical k-surface at the final state energy E_F, we can see in Figure 8.9(a) for the 3D case that k conservation after photoexcitation is only possible under the action of a reciprocal lattice vector G when the photon energy is so low as to provide insignificant momentum (this is generally the case in the UV and VUV regime). After translation by G, the initial state surface (blue) and the final state surface (yellow) can intersect. At the intersection points, the final k vector is equal to the initial k vector (the photon gives no momentum to the initial state) and the energy difference corresponds to the photon energy. For an isotropic quasi-free 3D dispersion, the intersection is a circle and the photoelectrons of a fixed energy are excited along a cone of directions inside the sample.[92] For the 2D systems, the initial k vector is defined only parallel to the surface and this momentum is conserved in an optical transition to the free electron final state. This means we must consider the intersection of a cylindrical constant energy surface of the 2D state with the final state sphere, again leading to circle and conical emission direction distributions, as shown in Figure 8.9(b).

As an example of photoelectron momentum imaging, we discuss 2PP measurements of Ag(111) and Cu(111) surfaces. With 3.1 eV photon excitation, 2PP can be excited either from the sp-bands of Ag and Cu, or the Shockley surface (SS) states that exist on these surfaces. The optical transitions between

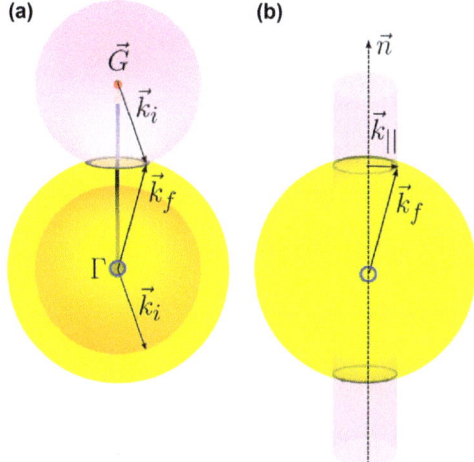

Figure 8.9 Direct optical transitions for nearly-free electron (a) bulk states and (b) surface states. Both types of transitions lead to nearly circular features in the momentum distributions of the photoemitted final states. n is the surface normal. The final state sphere is shown in yellow, the initial states isosurfaces are light blue.

the occupied sp-like electronic states of Cu(111) and Ag(111) to final states at energies near 6 eV can be treated according to the quasi-free 3D model explained above.[70] The additional presence of the prototypical 2D SS states makes these surfaces ideal objects for a 2PP momentum microscope study using 3 eV photon energy excitation. In Figure 8.10, we show the momentum microscope images of photoemission from near the E_F of Ag(111) (left) and Cu(111) (right). In both cases, the photoemission intensity is limited to k-vectors near 0.6 Å^{-1} prescribed by the maximum kinetic energy of photoelectrons moving nearly parallel to the sample. The value of the maximum parallel momentum is determined by the work functions of both samples and the photon energy.

The Ag(111) surface in Figure 8.10(a) shows a central disk-like intensity due to the SS state (binding energy ~ 60 meV, Fermi vector $\sim 0.08 \text{ Å}^{-1}$), which is not fully resolved into an open circle owing to the momentum (0.04 Å^{-1}) and energy (<200 meV) resolution of the momentum microscope. One can still distinguish, however, the sp-band L-gap by the very low intensity in the immediate vicinity of the Shockley surface state emission as seen by a white halo around the black disk. The SS state on Cu(111), because of its larger binding energy (~ 400 meV) and larger Fermi vector ($\sim 0.2 \text{ Å}^{-1}$), is clearly resolved as a circle in Figure 8.10(b). The Cu(111) L-gap cannot be clearly distinguished because it is already very near to the surface state at E_F. The approach of the SS state to the Cu bulk sp-band moreover leads to a departure from the quasi-free parabolic dispersion, as has been shown by three-photon photoemission (3PP) and scanning tunnelling spectroscopy for the unoccupied region of the SS state band.[93,94]

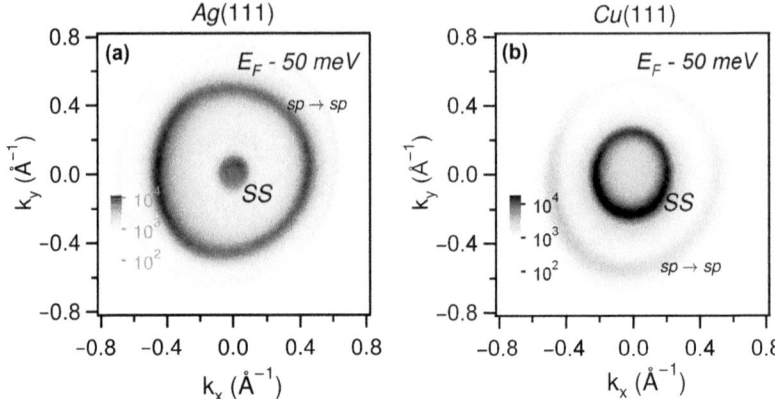

Figure 8.10   Two-photon photoemission momentum distributions from initial states
50 meV below E_F for (a) Ag(111) and (b) Cu(111) surfaces as recorded by
the momentum microscope.
Reprinted (adapted) with permission from Petek.[109] Copyright 2013
American Institute of Physics.

Surrounding the central surface state feature, both surfaces show an additional ring-like intensity originating from bulk sp–sp transitions. This is a 3D
transition shown in Figure 8.9(a), involving the reciprocal lattice vector
$G = <111>$. As we see by the different intensities, this type of transition is
clearly more pronounced on the Ag(111) surface relative to the surface state
emission compared to the Cu(111) surface. A quantitative theoretical explanation for this difference is lacking. As the bulk band structures of both materials
are very similar, this might point to possible interference of surface photoemission with the bulk emission in Ag(111),[95] which should be different because
of a different surface electronic response of both systems, or different electronic
screening at optical frequencies, which is related to the different binding of the
d-bands of the two metals.

From the point of symmetry, we can see on both surfaces that the sp–sp
transition has a three-fold deformation as compared to the circular distribution
expected from a perfect nearly-free electron system. This indicates that the
measurement is sensitive to the bulk band structure surrounding the [111] bulk
direction, which also depends on the direction of the vector k, not only on its
magnitude as for the quasi-free electron dispersion. Because the initial states
relevant in the measurement are near the Fermi level near the [111] direction, the
observed three-fold deformation reflects the distortion of the Ag Fermi surface
near its [111] neck-regions, as has also been seen in Cu(111).[96] This is a general
manifestation of bulk electronic states near the sp-band gap of noble metals.

8.6.2   Momentum Mapping of Localized Electronic States

Adsorption of atoms and molecules on metal surfaces introduces new electronic
states. These states can be associated with the formation of new adsorbate–

substrate chemical bonds, or adsorbate states strongly perturbed by the Coulomb and Pauli exclusion interactions with the substrate. In some cases, adsorbate–adsorbate interactions are sufficiently strong for the formation of adsorbate localized electronic bands. Therefore, in order to understand the interfacial electronic structure it is useful to image photoelectron energy and momentum distributions from the adsorbate electronic states.

The chemisorption of alkali atoms on noble metal surfaces [*e.g.*, Ag(111) and Cu(111)] is particularly interesting because of the relatively simple electronic structure and photoexcitation induced femtochemistry.[34,97–100] Upon chemisorption, owing to the repulsive image–charge interaction, alkali atoms at low coverage chemisorb predominantly in an ionic state losing their valence *s* electron to the substrate. 2PP and inverse photoemission spectra of alkali covered metal surfaces have identified an unoccupied state <3 eV above E_F, which is predominantly of the *s* valence state origin.[101,102] This unoccupied state, hereafter referred to as the σ-resonance, can be detected as a resonant intermediate state in the 2PP process excited by 400 nm light.[100]

Study of the unoccupied states of alkali atoms systems with the momentum microscope provides an extended view of their electronic structure in *k*-space. The unoccupied states of chemisorbed alkali atoms have been measured by angle resolved 2PP spectroscopy, but the mechanical rotation of the sample, which changes the surface projection of the incident electric field, precluded a quantitative analysis of the data.[103] As an example of photoelectron angular distribution imaging from an adsorbate localized unoccupied state, in Figure 8.11 we present results for Cs adsorption on Ag(111) measured by 2PP with the momentum microscope. Figure 8.11(a) shows the 2D (k_x and k_y) momentum dependent distribution of photoelectrons on the intermediate state energy $E_{Cs} \sim 2.6$ eV above E_F corresponding to the energy of the Cs σ-resonance. In addition to the bulk sp–sp transition of Ag(111), which was explained in Figure 8.10, Figure 8.11(a) includes contributions from the Cs σ-resonance, which at the coverage of this measurement serves as an intermediate state in the 2PP process. The high intensity near normal emission in the central region ($k_{\parallel} = 0$) is consistent with the *s* character of the parent orbital. The Cs σ-resonance intensity falls off gradually with an approximately Gaussian distribution as k_x and k_y increase, reflecting the localized nature of excitation on Cs atoms, in other words the Cs overlayer coverage is too sparse for the 6*s* orbitals to form a delocalized and therefore, dispersive band. The outer ring in the 2D momentum map corresponds to the *unoccupied* Ag(111) SS state.

These assignments are more evident from the $E(k_x)$ dispersion map shown in Figure 8.11(b), which is assembled from a 3D data set of momentum maps, such as in Figure 8.11(a), for photoelectron energies from 3.4 to 6.2 eV. The 2D $E(k_x)$ map corresponds to a cut through the 3D data set along $k_y = 0$. The k_x and k_y range of the 2PP signal is limited by a parabolic curve corresponding to the maximum surface-parallel momentum of photoelectrons at a specific photoelectron energy. The largest observable *k*-vectors correspond to electrons moving parallel to the surface at each energy. The range of *k*-vectors for particular a photoelectron energy is wider for the Cs-covered surface than for

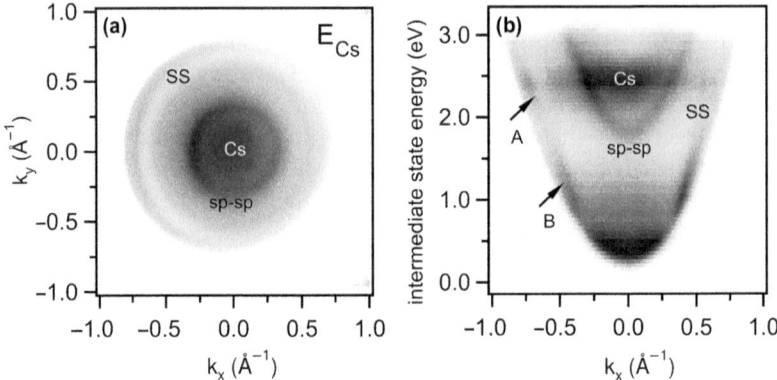

Figure 8.11 Momentum microscope imaging in 2PP of adsorption of Cs on Ag(111). Dark contrast corresponds to high 2PP intensity. (a) Photoelectron momentum map corresponding to intermediate states at the energy E_{Cs} of the Cs σ-resonance (for the Cs coverage $E_{Cs} = E_F + 2.6$ eV was used). The high intensity near normal emission (central region) is consistent with the σ character of the Cs state, which extends over the whole $k_\parallel$ spectrum owing to its localized nature in real space. The outer ring corresponds to the unoccupied Ag(111) Shockley surface state (SS) and the inner ring to the sp–sp transition of bulk Ag(111), as can be seen by comparison with the dispersion map shown in (b). b) The dispersion map is obtained by recording momentum maps for different photoelectron energies and taking a cut through the 3D data at $k_y = 0$. The Cs σ-resonance shows up as a non-dispersing (horizontal) feature around $E_F + 2.6$eV. The anomalous inversion of contrast between regions B and A in the dispersion map reveals evidence of interference between the SS band and the degenerate Cs σ-resonance. By contrast, the sp–sp (initial to final state) transition with mainly bulk character is not noticeably influenced by the degeneracy with the intermediate Cs state. Reprinted (adapted) with permission from Petek.[109] Copyright 2013 American Institute of Physics.

the clean Ag(111) surface, because Cs adsorption reduces the surface work function. By lowering the work function, large $k_\parallel$-momentum states of the clean surface shown in Figure 8.10, which would have undergone total internal reflection at the surface, now have sufficient energy simultaneously to conserve $k_\parallel$ and to overcome the surface work function upon photoemission.[29]

In the dispersion map, the Cs σ-resonance is distinguished by its non-dispersing ($k_\parallel$-independent energy) distribution around E_{Cs}, as already noted in Figure 8.11(a). The dispersion map also shows the unoccupied part of the SS state of Ag(111), which becomes visible in 2PP owing to the extended $k_\parallel$ range of the final states. By contrast, for the clean Ag(111) surface, only the occupied part of the SS state can be observed by 2PP for initial states with $k \leq 0.08$ Å^{-1} for 2PP from near E_F.[70] Mapping the dispersion for the SS state, the bulk sp–sp transition over an extended $k_\parallel$ range for both Ag(111) and Cu(111) makes it evident that both the surface and bulk states deviate from the paradigmatic nearly-free electron parabolic shape of the effective-mass picture.[93]

From the dispersion map in Figure 8.11(a), we can also see that at the energy levels of the unoccupied Cs σ-resonance and the SS state are degenerate for $|k_\parallel|$ values near 0.7 Å^{-1}, as indicated by the arrow A. This means that the Cs σ-resonance and the SS state could in principle be excited simultaneously from the same initial state belonging to the bulk sp-band and to the same final state that is detected by the momentum microscope. In addition, the Cs σ-resonance is known to interact with the resonant SS state and sp-band via resonant charge transfer, which is the elastic contribution to the Cs σ-resonance decay.[34,97,104]

The existence of coherent excitation pathways via two-photon resonant transitions and resonant interactions between intermediate states creates several scenarios for how quantum interference can affect the adsorbate photoexcitation and 2PP processes. That interference affects the 2PP process via the degenerate Cs σ-resonance and the SS state is evident from the momentum and dispersion map data in Figure 8.11. Specifically, the SS state intensity is modified as it crosses the Cs σ-resonance in energy and momentum space in a manner that can only be explained by quantum interference. As indicated by the arrows in Figure 8.11(b), the SS state band at energies below E_{Cs} shows higher intensity (dark contrast) than the background (arrow B), as can be expected if it serves as a resonant intermediate state in the 2PP process. At higher energies close to the Cs σ-resonance (arrow A) the contrast becomes inverted, with SS appearing in Figure 8.11(b) as a light band crossing the large $k_\parallel$ wing of the Cs σ-resonance. This contrast inversion is also clearly seen in the two-dimensional momentum map in Fig. 11(a) where SS appears as the light ring. A close examination of Figure 8.11(a) finds that the contrast is actually modulated by parallel momentum in the $-k_x$ direction first dipping below the uniform Cs σ-resonance background for >-0.7 Å^{-1} and then increasing above the background for <-0.7 Å^{-1}. Significantly, the contrast modulation is much more modest in the $+k_x$ direction. The contrast modulation is also seen in the k_y direction, but is dependent only on the magnitude of k_y and not its sign. Therefore, the quantum interference is clearly $k_\parallel$ dependent. The interference depends on the $k_\parallel$-vector of the photoelectron, because in a coherent 2PP process it is conserved.

Next we will examine possible origins of quantum interference in 2PP via the Cs σ-resonance on the Ag(111) surface. Quantum interference occurs when a transition from the same initial and final states can occur coherently *via* multiple excitation pathways. When taking the coherent sum over all pathways, phase differences cause constructive and destructive interferences between different amplitudes. Accounting only for the resonant interactions, the photoemitted intensity of the 2PP process depends on one-photon transition amplitudes promoting an electron first from an initial state $\left|\psi_i(\vec{k})\right\rangle$ to two possible, intermediate states $|\psi_{Cs}\rangle + |\psi_{SS}\rangle$ and then through another photon transition to the final state $\left|\psi_f(\vec{k})\right\rangle$. The relevant one-photon matrix elements are of the type:

$$M = \{\langle\psi_{Cs}| + \langle\psi_{SS}|\}V_I\left|\psi_{i/f}(\vec{k})\right\rangle \qquad (8.2)$$

where V_I is the operator of the light-matter interaction to be specified later. Thus, the matrix elements M for each one-photon transition are the sum of two coherent contributions corresponding to the Cs state $|\psi_{Cs}\rangle$ and the Shockley surface state $|\psi_{SS}\rangle$. The 2PP intensity is ultimately determined by terms of MM* for each one-photon transition, where interference can result owing to the phase difference between the excitation via $|\psi_{Cs}\rangle$ or $|\psi_{SS}\rangle$.

Before proceeding further we note that $|\psi_{Cs}\rangle$ and $|\psi_{SS}\rangle$ are not eigenstates of the system, although we have treated them as such. We know that $|\psi_{Cs}\rangle$ decays into $|\psi_{SS}\rangle$ and the bulk sp-band continuum through resonant charge transfer. This interaction can lead to a Fano interference,[105] because each intermediate state can be excited directly by an optical transition or via the other state through their residual interaction.[104] Although Fano interference is usually considered to occur between a discrete state and a continuum, in the case of Cs/Ag(111) surface the interaction is between one state, which is discrete in energy and continuous in $k_\parallel$ (Cs σ-resonance) and another, which is continuous in energy and discrete in $k_\parallel$ (SS state). Thus, the Fano resonance is expected in the $E(k)$ space.

In addition to the already described interference mechanisms, we also consider the special properties of light matter interactions at a metal surface. The transition moment V_I can be expressed as a function of the vector potential operator $\mathbf{A}$ of the excitation field and the momentum operator $\mathbf{p} = -i\hbar\vec{\nabla}$. It has been shown that the dipole approximation has significant deficiencies in a quantitative description of the optical excitation of adsorbates, which require explicit accounting of the variation in the vector potential at the surface in the interaction operator.[106,107] This leads to an interaction of the form:

$$V_I \sim (\mathbf{A} \cdot \mathbf{p} + \mathbf{p} \cdot \mathbf{A}) = 2\mathbf{A} \cdot \nabla + div\mathbf{A} \qquad (8.3)$$

consisting of "dipole" and "surface" contributions. The $\mathbf{A} \cdot \nabla$ term with a spatially constant $\mathbf{A}$ is the usual dipole approximation for photoexcitation in the bulk of a solid, where $div\mathbf{A}$ can be set to zero. The $div\mathbf{A}$ part, however, contributes at the surface where the surface normal gradient of the vector potential has a non-zero value. The "dipole" and "surface" contributions in V_I provide an additional possibility of interference through the cross terms in evaluating MM* using the forms of M and V_I in Equations (8.2) and (8.3). In comparison to the interference between two coherently excited states $(|\psi_{Cs}\rangle + |\psi_{SS}\rangle)$ and the Fano interaction, this can be described as the interference between two coherent excitation mechanisms $(V_I^{dipole} + V_I^{surface})$. Such interference, for example, is evident in 1PP spectra of the bare Ag(111) surface with ~ 6 eV photons,[50] where for initial states in the sp band, the k-conserving direct optical bulk sp–sp transition ($\mathbf{A} \cdot \nabla$ term) interferes with the "surface photoemission" ($div\mathbf{A}$ term). As such, this mechanism is relevant for the excitation of bulk states near the surface, intrinsic quasi-two dimensional surface states and adsorbate states alike and *per se* does not require coherent excitation of different states.

Although the relative importance of the various interference mechanisms that are simultaneously at work in the case of the Cs σ-resonance and the SS state has not been assigned so far, qualitatively the influence of the various coherent and $k_\parallel$-dependent contributions to the photoemission signal can be analysed by a consideration of the symmetry of the experimental setup. Applied to our problem, the symmetry of the $k_\parallel$-dependent intensity, as measured in Figure 8.11(a), is determined by the relative orientation of the $k_\parallel$-vector of the photoelectron and the direction of the vector potential **A**. Because **A** is fixed by the geometry of the experiment to a 65° off-normal incidence in the *x,z*-plane, the interference contrast is allowed to vary with $k_\parallel$ on going from positive to negative values in the *x,z*-plane.

For comparison, considering the symmetry with respect to the *y,z* plane, the transition strength should be independent of the sign of k_y, as is confirmed by the symmetry with respect to $k_y = 0$ seen in Figure 8.11(a). The characteristically different change of interference contrast upon inverting the surface parallel components of the photoelectron in k_x on the one hand or k_y on the other allows the influence of a specific dipole contribution to be pinpointed by the following argument. In the case of the symmetric interference in the k_y direction, the symmetry is in principle related to the *z*-component of the vector potential **A** and we expect that a circular interference feature would be observed for **A** exactly normal to the surface. This case would be consistent with a two-state interference via the SS or the Cs states, which owing to their orbital symmetry are excited most effectively by the A_z component via the dipole and surface photoemission terms. Now, in the case that we break the symmetry by tilting **A** and thereby introduce a finite A_x component, there would be a contribution from A_x only in the dipole part of Equation (8.3), as the surface photoemission term is governed by A_z alone. The relative orientation of A_x and k_x reverses when switching from $+k_x$ to $-k_x$ and thus this dipole term contributes constructively and destructively to the symmetric interference features, which would be formed by the A_z contribution acting alone. Overall, this is consistent with the symmetry breaking in the interference features along k_x and the symmetric interference distributions along k_y. Irrespective of the quantitative theoretical description still to be achieved, the interference effects observed with the highly sensitive momentum mapping of the momentum microscope provide unequivocal proof that the excitation of the Cs σ-resonance is a coherent process (rather than involving hot-electron mechanism).

Whereas we see in Figure 8.11(a) and 11(b) significant interference effects of the Cs σ-resonance in the momentum region of the SS state, the bulk sp–sp transition seems to be relatively unaffected. Compared with the intensity modifications of the outer surface state ring, the intensity of the sp–sp transition over the $k_\parallel$ cone in Figure 8.11(b) does not show obvious anomalies at E_{Cs}. The reduced sensitivity of the sp–sp transition to interference effects with Cs can be expected from wave function overlap arguments. Any sp–sp transition possibly involving an intermediate Cs state must be located close to the surface. Those bulk sp–sp transitions, however, which are excited below the immediate surface but within the inelastic mean free path can still produce a dominating number of photoelectrons in direct bulk two-photon transitions without a contribution from an intermediate state of a surface Cs atom. If these bulk electrons dominate, the possible interference effects from the surface will be obscured.

8.7   Summary

In this chapter we have described ultrafast laser-excited nonlinear MPP experiments using PEEM imaging of photoelectron energy, space and momentum distributions. The photoemitted electrons carry nearly complete information on the eigenspace of electrons in a solid. In a coherent MPP process this eigenspace is mapped onto that of free electrons in vacuum. By energy, momentum and spatial mapping of photoelectrons with a PEEM or momentum microscope it is then possible to reveal the eigenspace within the uncertainty imposed by non-conservation of the perpendicular momentum in photoemission. In a coherent MPP process, the information gained can be used to reveal the coherent electromagnetic fields excited in a sample or the properties of localized and delocalized surface and bulk wave functions. Full multidimensional mapping of the eigenspace can reveal interactions such as between the alkali atom resonances and the bulk and surface bands of the substrate. Indeed, multidimensional maps provide far more complete characterization of interactions, which occur in one-dimensional energy and 3D momentum spaces, than is possible by conventional methods of angle resolved photoemission. In addition to the studies of prototypical systems we have introduced here, further applications of the momentum microscope will include time-resolved studies of adsorbate systems revealing the coupling of electronic and nuclear motions, such as occur in surface femtochemistry,[34,98,99] as well as studies of magnetic and other spin-dependent systems with explicit spin-analysis of the photoelectrons.[108]

Acknowledgements

The research in Pittsburgh was supported by the Division of Chemical Sciences, Geosciences and Biosciences, Office of Basic Energy Sciences of the US Department of Energy through Grant DE-FG02-09ER16056, NSF grants ECS-0403865 and CHE-0507147 and PRESTO JST. We thank Jürgen Kirschner for numerous helpful discussions and for his continuing support of the momentum microscope experiments performed in Halle.

References

1. G. Sansone, E. Benedetti, F. Calegari, C. Vozzi, L. Avaldi, R. Flammini, L. Poletto, P. Villoresi, C. Altucci, R. Velotta, S. Stagira, S. De Silvestri and M. Nisoli, *Science*, 2006, **314**, 443.
2. A. L. Cavalieri, N. Muller, T. Uphues, V. S. Yakovlev, A. Baltuska, B. Horvath, B. Schmidt, L. Blumel, R. Holzwarth, S. Hendel, M. Drescher, U. Kleineberg, P. M. Echenique, R. Kienberger, F. Krausz and U. Heinzmann, *Nature*, 2007, **449**, 1029.

3. P. B. Corkum and F. Krausz, *Nat Phys*, 2007, **3**, 381.
4. M. F. Kling and M. J. J. Vrakking, *Annu. Rev. Phys. Chem.*, 2008, **59**, 463.
5. F. Krausz and M. Ivanov, *Rev. Mod. Phys.*, 2009, **81**, 163.
6. A. Kubo, K. Onda, H. Petek, Z. Sun, Y. S. Jung and H. K. Kim, *Nano Lett.*, 2005, **5**, 1123.
7. L. Zhang, A. Kubo, L. Wang, H. Petek and T. Seideman, *Phys. Rev. B*, 2011, **84**, 245442.
8. E. Beaurepaire, J. C. Merle, A. Daunois and J. Y. Bigot, *Phys. Rev. Lett.*, 1996, **76**, 4250.
9. N. Del Fatti, C. Voisin, M. Achermann, S. Tzortzakis, D. Christofilos and F. Vallée, *Phys. Rev. B*, 2000, **61**, 16956.
10. M. Bonn, D. N. Denzler, S. Funk, M. Wolf, S. S. Wellershoff and J. Hohlfeld, *Phys. Rev. B*, 2000, **61**, 1101.
11. P. B. Johnson and R. W. Christy, *Phys. Rev. B*, 1972, **6**, 4370.
12. H. Petek and S. Ogawa, *Prog. Surf. Sci.*, 1997, **56**, 239.
13. M. Weinelt, *J. Phys.: Condens. Matter*, 2002, **14**, R1099.
14. A. Kubo, Y. S. Jung, H. K. Kim and H. Petek, *J. Phys. B*, 2007, **40**, S259.
15. U. Bovensiepen, H. Petek and M. Wolf, eds., *Dynamics at Solid State Surfaces and Interfaces, Volume 1*, 1st edn., Wiley-VCH, Weinheim, 2010.
16. S. Mukamel, *Annu. Revi. Phys. Chem.*, 2000, **51**, 691.
17. S. T. Roberts, K. Ramasesha and A. Tokmakoff, *Acc. Chem. Res.*, 2009, **42**, 1239.
18. J. C. Wright, *Annu. Rev. Phys. Chem.*, 2011, **62**, 209.
19. S. Garrett-Roe, F. Perakis, F. Rao and P. Hamm, *J. Phys. Chem. B*, 2011, **115**, 6976.
20. X. Y. Zhu, *Surf. Sci. Rep.*, 2004, **56**, 1.
21. X.-Y. Zhu, *J. Phys. Chem. B*, 2004, **108**, 8778.
22. C. D. Lindstrom and X. Y. Zhu, *Chem. Rev.*, 2006, **106**, 4281.
23. J. Stähler, U. Bovensiepen, M. Meyer and M. Wolf, *Chem. Soc. Rev.*, 2008, **37**, 2180.
24. J. Stähler, M. Meyer, U. Bovensiepen and M. Wolf, *Chem. Sci.*, 2011, **2**, 907.
25. J. A. Armstrong, N. Bloembergen, J. Ducuing and P. S. Pershan, *Phys. Rev.*, 1962, **127**, 1918.
26. K. Giesen, F. Hage, F. J. Himpsel, H. J. Riess and W. Steinmann, *Phys. Rev. Lett.*, 1985, **55**, 300.
27. J. Bokor, *Science*, 1989, **246**, 1130.
28. R. Haight, *Surf. Sci. Rep.*, 1995, **21**, 275.
29. F. Bisio, M. Nývit, J. Franta, H. Petek and J. Kirschner, *Phys. Rev. Lett.*, 2006, **96**, 087601.
30. A. Winkelmann, W.-C. Lin, C.-T. Chiang, F. Bisio, H. Petek and J. Kirschner, *Phys. Rev. B*, 2009, **80**, 155128.
31. W. Schattke, E. E. Krasovskii, R. D. Muiño and P. M. Echenique, *Phys. Rev. B*, 2008, **78**, 155314.

32. Z. Hao, J. I. Dadap, K. R. Knox, M. B. Yilmaz, N. Zaki, P. D. Johnson and R. M. Osgood, *Phys. Rev. Lett.*, 2010, **105**, 017602.
33. N.-H. Ge, C. M. Wong, R. L. Lingle, J. D. McNeill, K. J. Gaffney and C. B. Harris, *Science*, 1998, **279**, 202.
34. H. Petek, M. J. Weida, H. Nagano and S. Ogawa, *Science*, 2000, **288**, 1402.
35. B. Li, J. Zhao, K. Onda, K. D. Jordan, J. Yang and H. Petek, *Science*, 2006, **311**, 1436.
36. R. D. Muiño, D. Sánchez-Portal, V. M. Silkin, E. V. Chulkov and P. M. Echenique, *Proc. Nat. Acad. Sci.*, 2011, **108**, 971.
37. W. A. Tisdale, K. J. Williams, B. A. Timp, D. J. Norris, E. S. Aydil and X.-Y. Zhu, *Science*, 2010, **328**, 1543.
38. W.-L. Chan, M. Ligges, A. Jailaubekov, L. Kaake, L. Miaja-Avila and X.-Y. Zhu, *Science*, 2011, **334**, 1541.
39. H. H. Rotermund, *Surf. Sci. Rep.*, 1997, **29**, 265.
40. C. M. Schneider and G. Schönhense, *Rep. Prog. Phys.*, 2002, **65**, 1785.
41. G. Schönhense, H. J. Elmers, S. A. Nepijko and C. M. Schneider, *Adv. Imaging Electron Phys.*, 2006, **142**, 159.
42. A. Locatelli and E. Bauer, *J. Phys.: Condens. Matter*, 2008, **20**, 093002.
43. M. I. Stockman, M. F. Kling, U. Kleineberg and F. Krausz, *Nat Photonics*, 2007, **1**, 539.
44. R. Wichtendahl, R. Fink, H. Kuhlenbeck, D. Preikszas, H. Rose, R. Spehr, P. Hartel, W. Engel, R. Schlögl, H.-J. Freund, A. M. Bradshaw, G. Lilienkamp, T. Schmidt, E. Bauer, G. Benner and E. Umbach, *Surf. Rev. Lett.*, 1998, **5**, 1249.
45. T. Schmidt, H. Marchetto, P. L. Lévesque, U. Groh, F. Maier, D. Preikszas, P. Hartel, R. Spehr, G. Lilienkamp, W. Engel, R. Fink, E. Bauer, H. Rose, E. Umbach and H. J. Freund, *Ultramicroscopy*, 2010, **110**, 1358.
46. R. M. Tromp, J. B. Hannon, A. W. Ellis, W. Wan, A. Berghaus and O. Schaff, *Ultramicroscopy*, 2010, **110**, 852.
47. H. Petek, A. P. Heberle, W. Nessler, H. Nagano, S. Kubota, S. Matsunami, N. Moriya and S. Ogawa, *Phys. Rev. Lett.*, 1997, **79**, 4649.
48. S. Ogawa, H. Nagano, H. Petek and A. P. Heberle, *Phys. Rev. Lett.*, 1997, **78**, 1339.
49. M. M. Wefers and K. A. Nelson, *Opt. Lett.*, 1993, **18**, 2032.
50. T. Miller, W. E. McMahon and T. C. Chiang, *Phys. Rev. Lett.*, 1996, **77**, 1167.
51. M. Cinchetti, A. Gloskovskii, S. A. Nepjiko, G. Schönhense, H. Rochholz and M. Kreiter, *Phys. Rev. Lett.*, 2005, **95**, 047601.
52. A. Kubo, N. Pontius and H. Petek, *Nano Lett.*, 2007, **7**, 470.
53. M. Bauer, C. Wiemann, J. Lange, D. Bayer, M. Rohmer and M. Aeschlimann, *Appl. Phys. A*, 2007, **88**, 473.
54. L. Douillard and F. Charra, *J. Phys. D: Appl. Phys.*, 2011, **44**, 464002.
55. S. J. Peppernick, A. G. Joly, K. M. Beck and W. P. Hess, *J. Chem. Phys.*, 2011, **134**, 034507.

56. J. M. Pitarke, V. M. Silkin, E. V. Chulkov and P. M. Echenique, *Rep. Prog. Phys.*, 2007, **70**, 1.
57. K. A. Willets and R. P. Van Duyne, *Annu. Rev. Phys. Chem.*, 2007, **58**, 267.
58. C. C. Neacsu, S. Berweger, R. L. Olmon, L. V. Saraf, C. Ropers and M. B. Raschke, *Nano Lett.*, 2010, **10**, 592.
59. M. I. Stockman, *Opt. Express*, 2011, **19**, 22029.
60. P. Monchicourt, M. Raynaud, H. Saringar and J. Kupersztych, *J. Phys.: Condens. Matter*, 1997, **9**, 5765.
61. F. J. García de Abajo, *Rev. Mod. Phys.*, 2010, **82**, 209.
62. M. I. Stockman, *Phys. Rev. B*, 2000, **62**, 10494.
63. U. Bovensiepen, H. Petek and M. Wolf, eds., *Dynamics at Solid State Surfaces and Interfaces, Voume. 2*, 1st edn., Wiley-VCH, Weinheim, 2012.
64. E. Louis, A. E. Yakshin, T. Tsarfati and F. Bijkerk, *Prog. Surf. Sci.*, 2011, **86**, 255.
65. R. H. Ritchie, *J. Phys.: Condens. Matter*, 1993, **5**, A17.
66. Z. Sun, Y. S. Jung and H. K. Kim, *Appl. Phys. Lett.*, 2003, **83**, 3021.
67. J. Lehmann, M. Merschdorf, W. Pfeiffer, A. Thon, S. Voll and G. Gerber, *Phys. Rev. Lett.*, 2000, **85**, 2921.
68. F. Evers, C. Rakete, K. Watanabe, D. Menzel and H.-J. Freund, *Surf. Sci.*, 2005, **593**, 43.
69. N. Pontius, V. Sametoglu and H. Petek, *Phys. Rev. B*, 2005, **72**, 115105.
70. A. Winkelmann, V. Sametoglu, J. Zhao, A. Kubo and H. Petek, *Phys. Rev. B*, 2007, **76**, 195428.
71. M. A. Cazalilla, J. S. Dolado, A. Rubio and P. M. Echenique, *Phys. Rev. B*, 2000, **61**, 8033.
72. A. Marini, R. Del Sole and G. Onida, *Phys. Rev. B*, 2002, **66**, 115101.
73. L. Cao, N. C. Panoiu, R. D. R. Bhat and J. R. M. Osgood, *Phys. Rev. B*, 2009, **79**, 235416.
74. B. Lamprecht, A. Leitner and R. R. Aussenegg, *Appl. Phys. B*, 1999, **68**, 419.
75. S. Ogawa, H. Nagano and H. Petek, *Phys. Rev. B*, 1997, **55**, 10869.
76. N. S. Ginsberg, Y.-C. Cheng and G. R. Fleming, *Acc. Chem. Res.*, 2009, **42**, 1352.
77. A. Ishizaki, T. R. Calhoun, G. S. Schlau-Cohen and G. R. Fleming, *Phys. Chem. Chem. Phys.*, 2010, **12**, 7319.
78. M. Aeschlimann, T. Brixner, A. Fischer, C. Kramer, P. Melchior, W. Pfeiffer, C. Schneider, C. Strüber, P. Tuchscherer and D. V. Voronine, *Science*, 2011, **333**, 1723.
79. A. Kubo and H. Petek, in *Ultrafast Phenomena XVI*, eds. P. Corkum, S. di Silvestri, K. A. Nelson, E. Riedle and R. W. Schoenlein, Springer-Verlag, Berlin Stresa, Italy, 2008, vol. 92, p. 687.
80. P. Nagpal, N. C. Lindquist, S.-H. Oh and D. J. Norris, *Science*, 2009, **325**, 594.
81. I. Matsuda and H. W. Yeom, *J. Electron Spectrosc. Relat. Phenom.*, 2002, **126**, 101.

82. L. Zhang and T. Seideman, *Phys. Rev. B*, 2010, **82**, 155117.
83. A. Kubo and H. Petek, unpublished results.
84. T. Feurer, N. S. Stoyanov, D. W. Ward and K. A. Nelson, *Phys. Rev. Lett.*, 2002, **88**, 257402.
85. T. Feurer, N. S. Stoyanov, D. W. Ward, J. C. Vaughan, E. R. Statz and K. A. Nelson, *Annu. Rev. Mater. Res.*, 2007, **37**, 317.
86. M. Sukharev and T. Seideman, *J. Phys. B*, 2007, S283.
87. V. V. Temnov, K. Nelson, G. Armelles, A. Cebollada, T. Thomay, A. Leitenstorfer and R. Bratschitsch, *Opt. Express*, 2009, **17**, 8423.
88. Y. Fujikawa, T. Sakurai and R. M. Tromp, *Phys. Rev. Lett.*, 2008, **100**, 126803.
89. B. Krömker, M. Escher, D. Funnemann, D. Hartung, H. Engelhard and J. Kirschner, *Rev. Sci. Instrum.*, 2008, **79**, 053702.
90. M. Escher, N. Weber, M. Merkel, C. Ziethen, P. Bernhard, G. Schönhense, S. Schmidt, F. Forster, F. Reinert, B. Krömker and D. Funnemann, *J. Phys.: Condens. Matter*, 2005, **17**, S1329.
91. G. Schönhense, *Phys. Scr.*, 1990, **1990**, 255.
92. G. D. Mahan, *Phys. Rev. B*, 1970, **2**, 4334.
93. A. A. Ünal, C. Tusche, S. Ouazi, S. Wedekind, C.-T. Chiang, A. Winkelmann, D. Sander, J. Henk and J. Kirschner, *Phys. Rev. B*, 2011, **84**, 073107.
94. L. Bürgi, L. Petersen, H. Brune and K. Kern, *Surf. Sci.*, 2000, **447**, L157.
95. T. Miller, E. D. Hansen, W. E. McMahon and T. C. Chiang, *Surf. Sci.*, 1997, **376**, 32.
96. M. Hengsberger, F. Baumberger, H. J. Neff, T. Greber and J. Osterwalder, *Phys. Rev. B*, 2008, **77**, 085425.
97. S. Ogawa, H. Nagano and H. Petek, *Phys. Rev. Lett.*, 1999, **82**, 1931.
98. H. Petek, H. Nagano, M. J. Weida and S. Ogawa, *J. Phys. Chem. B*, 2001, **105**, 6767.
99. H. Petek and S. Ogawa, *Annu. Rev. Phys. Chem.*, 2002, **53**, 507.
100. J. Zhao, N. Pontius, A. Winkelmann, V. Sametoglu, A. Kubo, A. G. Borisov, D. Sanchez-Portal, V. M. Silkin, E. V. Chulkov, P. M. Echenique and H. Petek, *Phys. Rev. B*, 2008, **78**, 085419.
101. M. Bauer, S. Pawlik and M. Aeschlimann, *Phys. Rev. B*, 1997, **55**, 10040.
102. D. A. Arena, F. G. Curti and R. A. Bartynski, *Phys. Rev. B*, 1997, **56**, 15404.
103. A. G. Borisov, V. Sametoglu, A. Winkelmann, A. Kubo, N. Pontius, J. Zhao, V. M. Silkin, J. P. Gauyacq, E. V. Chulkov, P. M. Echenique and H. Petek, *Phys. Rev. Lett.*, 2008, **101**, 266801.
104. A. G. Borisov, J. P. Gauyacq, E. V. Chulkov, V. M. Silkin and P. M. Echenique, *Phys. Rev. B*, 2002, **65**, 235434.
105. U. Fano, *Phys. Rev.*, 1961, **124**, 1866.
106. D. Bejan and G. Raşeev, *Surf. Sci.*, 2003, **528**, 163.
107. G. Raşeev and D. Bejan, *Surf. Sci.*, 2003, **528**, 196.
108. C. Tusche, M. Ellguth, A. A. Ünal, C.-T. Chiang, A. Winkelmann, A. Krasyuk, M. Hahn, G. Schönhense and J. Kirschner, *Appl. Phys. Lett.*, 2011, **99**, 032505.
109. H. Petek, *J. Chem. Phys.*, 2012, **137**, 091704.

CHAPTER 9

Light at the Tip: Hybrid Scanning Tunneling/Optical Spectroscopy Microscopy

JAO VAN DE LAGEMAAT*[a,b] AND MANUEL J. ROMERO[a,b]

[a] Chemical and Materials Sciences Center, National Renewable Energy Laboratory, 1617 Cole Boulevard, Golden, Colorado, 80401, USA;
[b] National Center for Photovoltaics, National Renewable Energy Laboratory, 1617 Cole Boulevard, Golden, Colorado, 80401, USA
*Email: Jao.vandelagemaat@nrel.gov

9.1 Introduction

With its introduction in 1981, the scanning tunneling microscope (STM) has sparked a revolution in nanoscale characterization techniques and enabled a detailed view of nanoscale phenomena that was heretofore not possible or could only be inferred by more macroscopic, external observations.[1] The STM technique has been progressively improved, recently even allowing imaging of the precise wavefunctions of a single pentacene molecule.[2] In this case, a single absorbed CO molecule at the apex of the tip provided chemically specific phase information on the local wavefunction sign and allowed unprecedented insight into local chemical and electronic properties. New modes of feedback that do not rely on current made it possible to explore non-conducting samples and today a large range of nanoscale interactions are explored using chemically specific tips and forces in a field more generically called scanning probe microscopy (SPM).

RSC Energy and Environment Series No. 8
Solar Energy Conversion: Dynamics of Interfacial Electron and Excitation Transfer
Edited by Piotr Piotrowiak
© The Royal Society of Chemistry 2013
Published by the Royal Society of Chemistry, www.rsc.org

One major field of research combines optical access with SPM methods. An interesting example of this is based upon light emission that is induced by current flow between tip and sample. There is a considerable history of observations of this luminescence. Not long after the introduction of the STM, light emission from an STM tip was observed and reported on metallic samples.[3] By detecting this scanning tunneling luminescence (STL) in the far field and correlating with the tunneling current and topography, it proved possible to obtain atomic-scale resolution using this technique and to detect emission from individual atoms on metallic surfaces[4] as well as to detect light emission from molecules[5,6] or only parts of molecules.[7] The extraordinary spatial resolution in these optical measurements, especially in comparison with more traditional optical far field techniques, as well as near-field optical microscopies, is owed to the ability of an STM tip to inject electrons very locally, whereas the emitted photons are collected in the far field. This limits the resolution to that of the tunneling process instead of to an intrinsic physical property of the interaction with light and allows the study of excited state dynamics on nanoscale length scales. The chemical specificity of the emission process and wavelength allows new imaging techniques that combine high spatial resolution with high energetic resolution as well as chemical information. Currently, researchers are pushing the limits of this even further by adding time-resolved capabilities to these experiments so that not just energetics of the nanoscale light generation can be studied but also the (ultrafast) dynamics.[8,9]

This review will focus on light emission from nanoscale structures. It will first discuss the mechanism of light emission, the spatial resolution possible and the addition of time-resolved detection. Subsequently, it will discuss several examples from the literature that illustrate the power of the technique.

9.2   Light Emission from the STM

9.2.1   Mechanism

There are four mechanisms that are generally considered to explain light emission induced by current injection from a STM tip. These are illustrated in Figure 9.1. In the first, illustrated in Figure 9.1(a), injection of minority carriers into a semiconductor from the tip leads to minority carrier diffusion and subsequent recombination. The length scale over which the carrier travels is related to its diffusion length in the semiconductor material. The efficiency of this process is generally in the order of 10^{-4} photons/electron.

Second Figure 9.1(b), is a mechanism that occurs on metal samples and highly doped semiconductors, carriers are injected "hot" from the tip (or from the sample if the polarity is reversed) and excite tip-induced surface plasmons (TISP).[10] These states can decay radiatively in the tip/sample plasmonically active gap. The efficiency of this process is generally around 10^{-4} photons/electron. Third, a direct transition between surface states on the tip and band edge states on a semiconductor substrate can be radiative with an efficiency of

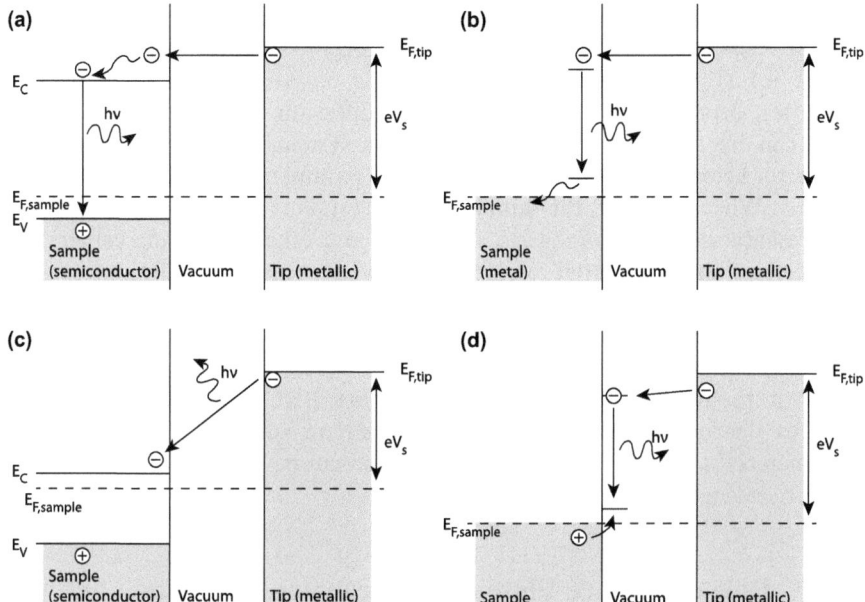

Figure 9.1   Four major mechanisms for light emission in a scanning tunneling microscope. (a) Injection of minority carriers from the tip into an unoccupied
band of a doped semiconductor leading to carrier diffusion and radiative
recombination. (b) Surface plasmon mediated emission. Hot carrier injection into a metal leads to radiative decay of a tip-induced surface
plasmon. (c) Direct dipole radiation. The transition from tip to substrate
results in a direct dipole that can emit light. (d) An absorbed molecule or
nanostructure accepts opposite carriers into its states leading to radiative
recombination on the molecule.

10^{-6} photons/electron or less. Lastly, a molecular or nanoscale species that is
present on a metal surface can undergo simultaneous (but separate) hole and
electron transfer from the substrate and the tip, respectively, leading to an
excited molecular species that can recombine radiatively. The efficiency of this
process depends on the radiative efficiency (internal conversion) of the molecular or nanoscale species and on the quenching efficiency owing to the
nearby metal. The metallic species can also specifically enhance the photon
emission from the absorbed molecule by surface plasmon enhancement. Efficiencies are generally similar to those in Figure 9.1(a) and (b).

A fifth mechanism that is not drawn in Figure 9.1 is a combination of
Figures 9.1(b) and (d). In this mechanism, a surface plasmon is created by the
tunneling event, which then subsequently transfers energy to a fluorescent
species on the surface.[11] This surface species in turn emits light. This interaction
might also occur by the creation of a hybrid surface-plasmon/exciton species[12–15] or be enhanced by the strong optical field enhancement afforded by
plasmonically active structures such as formed by the tip/substrate
geometry.[6,16–21]

As one would expect, the spatial resolution of light emission from STM is strongly dependent on the mechanism responsible for the light emission. In Figure 9.1(a) and when researching a single crystalline semiconductor, the resolution depends on the minority carrier diffusion length,[22] which in some cases can be on the micrometer-scale. For systems such as organic semiconductor heterojunction blends, the relevant parameter is the exciton diffusion length.[17] When however, the sample is polycrystalline or even nanocrystalline, the presence of grain boundaries, impurities and other quenching centers will start determining the spatial resolution of the technique.[22] Resolutions as good as single atom scale have been demonstrated in the literature[4,23–28] and it has even been demonstrated that light emission can be induced from individual parts of fluorescent molecules.[7] In most cases where surface plasmons are involved in the light emission mechanism (Figure 9.1(b)), the length scale is related to the length scale of the surface plasmon which ranges from a few nanometers[29] to hundreds of nanometers, but can be selected to be at shorter and larger length scales.[30]

9.2.2 Role of Surface Plasmons in Scanning Tunneling Luminescence

It is well known that surface plasmons play a major role in tip-enhanced Raman spectroscopy as well as in surface-enhanced luminescence.[16,31] However, the strong role of surface plasmons in light emission induced by tunneling electrons on light emission from surface species has not always been recognized even though it was quickly identified by the IBM group that first demonstrated STL emission.[10,18] As mentioned already above, one of the major mechanisms (Figure 9.1(b)) is creation of surface plasmons that decay radiatively. The specific geometry of the tip and substrate is known to create new surface plasmon states dubbed tip-induced surface plasmons (TISPs), which, owing to the specific geometry, can be radiative. This decay in turn excites absorbed nanoscale species such as molecules or quantum confined semiconductors by energy transfer which in turn emit light. One can expect this energy transfer to be enhanced if the excitonic states on the absorbed species are strongly coupled to the plasmonic states.

In Figure 9.1(b) the relevant length scale that is created is that of the surface plasmon. In general, two different modes can be created in the tip/sample cavity. The first is a localized surface plasmon (LSP) which resonates between the tip and the sample. The second is the more commonly discussed propagating surface plasmon (PSP) which spreads out on the sample surface from the location of the tip like the wave pattern generated by a rock breaking a water surface (see Figure 9.2(a)). Romero *et al.*[30] were able to image the local prevalence of such modes on a flame-annealed gold surface that had single crystalline terraces (where PSPs are more likely to occur) interspersed with rough boundaries where LSPs are more likely. Their results are shown in Figure 9.2. Such selectivity of the length scale of the plasmon allows for differentiation between local excitation and wide area excitation. Last, but not

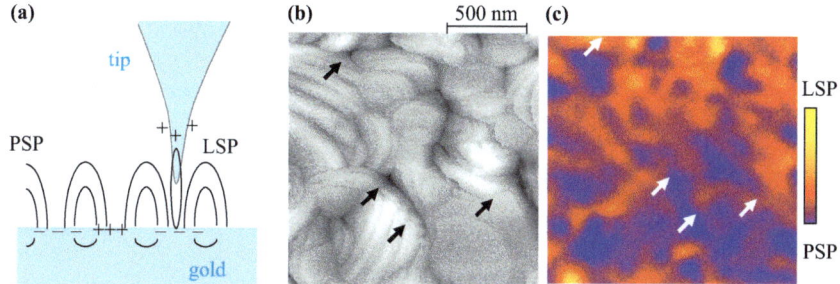

Figure 9.2   (a) Schematics of excitation of localized and propagating modes of the plasmon (LSP and PSP modes, respectively) during STM observations. (b) STM image of a flame-annealed gold substrate at $V = 700$ mV, $I_t = 1$ nA. (c) Corresponding map of the tunneling luminescence reflecting the relative contribution of the LSPs and PSPs to the plasmon excitation; $V = 3$ V, $I_t = 10$ nA.

least, the presence of TISPs, strongly enhances the rate of fluorescence locally. Calculations have shown that enhancements of 4 to 5 orders of magnitude are easily possible even when accounting for losses in the metal structures.[32]

9.2.3   Experimental Considerations

It is an obvious requirement of scanning tunneling luminescence that the experimental setup is highly efficient in collecting light emitted from the tip-substrate area. To a first approximation one can assume that light is emitted more or less in a hemisphere (this is not true for light emitted by TISPs – see below), so ideally one has a collection geometry that collects at as large a solid angle as possible. When performing measurements in air, this can be achieved by simply placing a detector (*e.g.* a photomultiplier tube) or an optical fiber close to the tunneling gap.[33–35]

In ultra high vacuum (UHV) setups in general, space is more restricted and systems of lenses, reflectors, or bundles of optical fibers are employed.[18] Figure 9.3 illustrates the detection geometries that are generally used. Figure 9.3(a) shows a lens collecting light emitted from the tunneling gap which is then directed through an optical viewport and imaged on a detector or monochromator,[36,37] Figure 9.3(b) shows a bundle of fibers all collecting light emitted in a different direction.[38] Figure 9.3(c) shows an ellipsoidal mirror that directs the emitted light upwards. Figure 9.3(d) shows a parabolic mirror that directs the emitted light into a lightpipe.[39]

All of these setups have been used to enable highly spatially selective and sensitive detection of STL and as such each setup simply presents a tradeoff between possible optical access and light detection efficiency and the capabilities of the scanning probe microscope. It is clearly possible to obtain high numerical aperture (*e.g.* 0.5 steradians or more) detection with most STM designs with some clever engineering. This is further helped by the fact that in most geometries, a strong angular dependence of the light emission is expected.

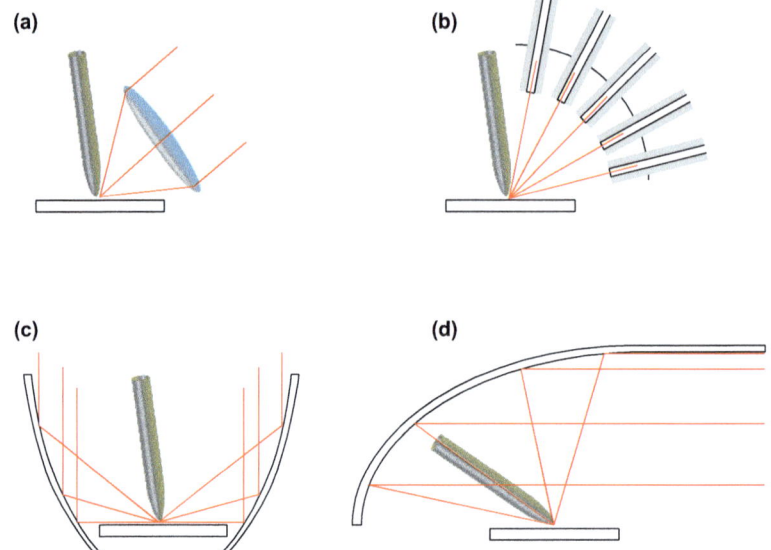

Figure 9.3   Different geometries used to detect light emission in high vacuum STM setups. (a) Lens system projects the light out through an optical viewport. (b) Bundle of optical fibers is used to collect light. (c) Ellipsoidal mirror is used to direct light out. (d) Parabolic mirror directs light into a light guide.

Rendell and Scalapino calculated that for light emission by TISPs in the geometry of a metallic tip near a surface, emission angles near 55° are most prevalent with a fairly narrow angular distribution.[40] This realization can significantly simplify the experimental design and allow use of only a few fiber collectors while still retaining large collection efficiency.

9.3   Light Emission from Nanostructures

9.3.1   Individual Atoms, Molecules and Surface States

Soon after the discovery of light emission from metallic substrates in STM, the technique was applied to different metallic substrates and contrast was found on the nanometer[41] and even subnanometer scale, indicating near atomic resolution of light emission.[42] Later, the technique was applied to surface reconstruction of Au(110) and the (1×2) reconstruction was resolved.[23] Figure 9.4 shows an example of observed atomic resolution of the photon emission intensity from such surfaces. In an impressive recent experiment, Chen *et al.* imaged the light emission of chains of small numbers of Ag atoms on Ni/ Al alloy that were prearranged using the STM itself.[26] In this experiment, particles in a box-like states were observed in the emission maxima along the chain, directly visualizing Fermi's golden rule that states that the rate of a transition correlates with the overlap of the initial and final states. The authors even observed a subtlety in this rule when applied to electron tunneling where it

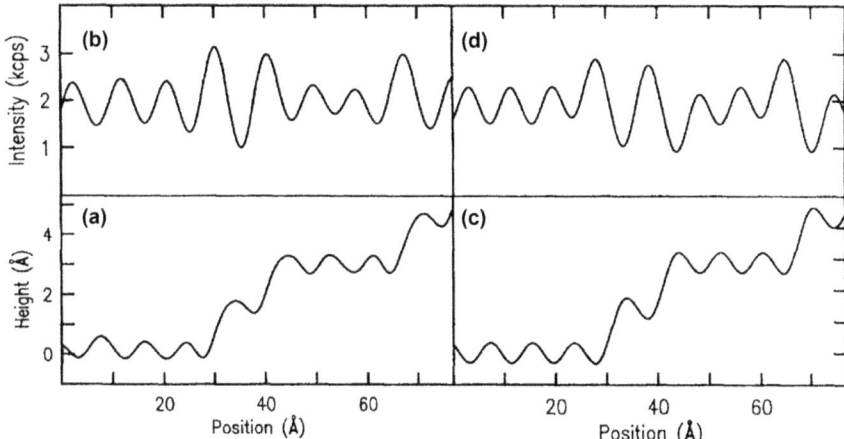

Figure 9.4 Cross sections through a stepped area of a Au surface and results of a model calculation. (a) Section of a STM topograph, (b) corresponding section of the photon map, (c) surface corrugation in the simulation and (d) calculated photon intensity. Section (b) has been low-pass filtered to decrease the photon shot noise.
Reproduced with permission from Berndt *et al.*[23]

turns out that the maximum emission should be observed at locations where the *derivative* of the wavefunction of the states on the atom chain is largest, resulting in a shift of the maxima with respect to the atom position.

Researchers also applied these methods to look at individual absorbed surface species. An early example is that of C60 molecules on Au(110) substrates where light emission appears to come from single C60 molecules even when they are arranged in dense layers.[43,44] This study set a record for the spatial resolution for detecting light emission of 4 Å. Later studies ensured electrical separation of the molecules from the underlying substrate using insulating tunneling barriers. Figure 9.5 shows an example of such a study[6] in which NaCl layers were used between the Au(111) substrate and the fullerene. This allowed clear observation of molecular fluorescence from the fullerene that appears to be excited by a hot electron injection in the fullerene, followed by molecular fluorescence. Even phosphorescence could be observed from the fullerenes in these experiments indicating that excited state quenching by the metal must be surprisingly small (Figure 9.5, right panel).

Recently, researchers were able to excite parts of a molecule separately in the STM and to observe the luminescence from individual lobes of a porphyrin.[7] Differences were observed in these spectra owing to conformational differences associated with different binding modes and differences in the height of the lobes above the substrate surface which was an ultrathin alumina film grown on NiAl(110). The spectra show vibrational patterns, indicating that a fluorescence mechanism is involved in the light emission (Figure 9.6).[7] These results stand in stark contrast to measurements of porphyrins directly on metal surfaces or using a thin self-assembled monolayer, where the molecular luminescence is

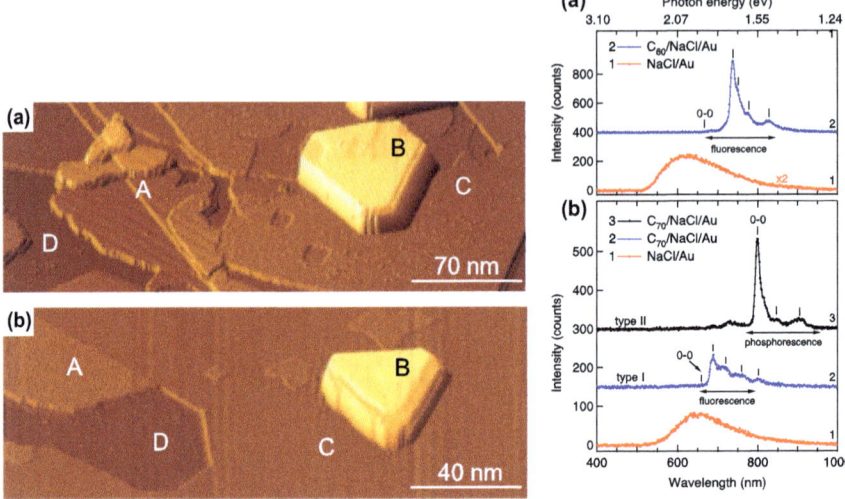

Figure 9.5 Left: STM topographic image of C60 (top) and C70 (bottom) molecules on thin NaCl (C) grown on Au(111) (D). In regions (A) individual C60 are visible on the gold surface, in (B) C60 nanocrystals are observed. Right: STM-light emission spectra obtained over C60 (top) and C70 (bottom) nanocrystals on NaCl/Au substrates.
Reproduced with permission from Rossel *et al.*[6]

completely quenched and the presence of the porphyrin increases the surface plasmon emission from the substrate by modulating the tip/substrate distance.[37] In an interesting recent development, while not specifically a single molecule, in STL of multilayers of porphyrins on gold, emission from higher-lying states than the normal S_1 lowest vibrational state and hot electroluminescence from higher lying vibrational states were observed owing to strong coupling to the tip-induced surface plasmons in the gap and because the plasmons behave as a strong coherent source of energy that can actively control the radiative emission channel of molecular species.[45] A similar conclusion was reached by Wiederrecht *et al.* for *J*-aggregates that were strongly coupled to surface plasmons generated on a prism geometry[46] and by Johnson *et al.* for pentacene layers on nanohole arrays in silver.[47]

9.3.2 Thin Film and Organic Semiconductors

STL has been widely applied to inorganic semiconductor surfaces and thin-film semiconductors especially those with grain boundaries and nanoscale impurities that lead to excitation quenching. It has also been applied to thin films of organics where properties such as the exciton diffusion length can be studied.[17,36,45,48]

In Section 9.2.1, we provided a description of the processes involved in luminescence excited by hot tunneling electrons in inorganic semiconductors. The first mechanism (the tunneling of minority carriers from the tip to the

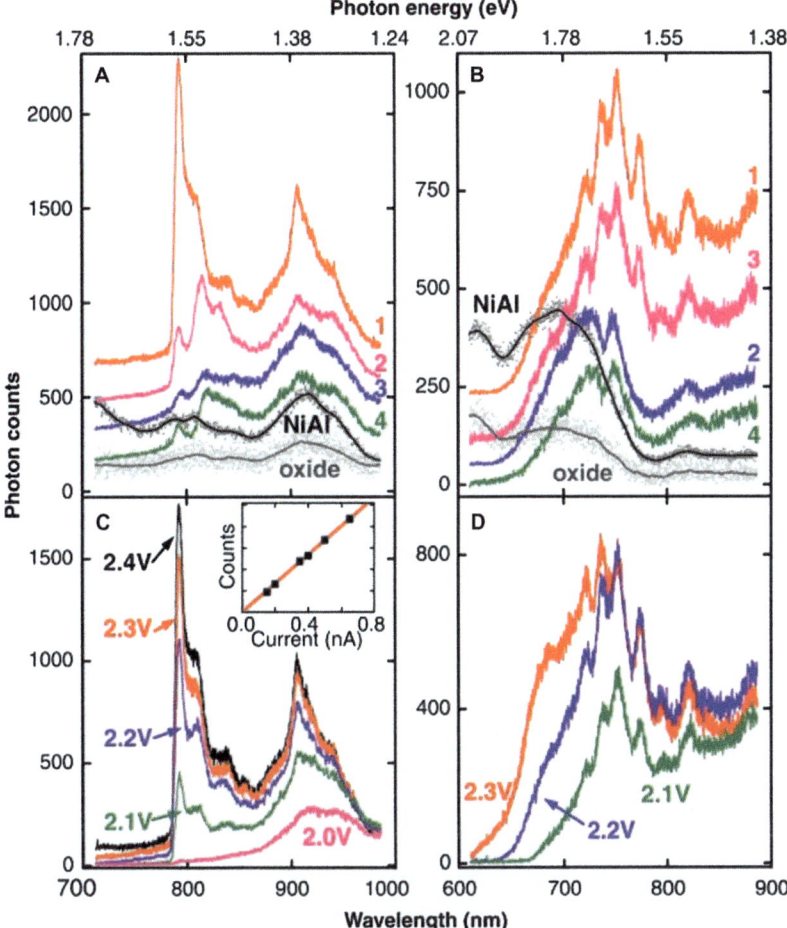

Figure 9.6 Light-emission spectra acquired on the different lobes of two Zn-porphyrin molecules on $Al_2O_3/NiAl(110)$ surfaces together with spectra acquired on bare NiAl and $Al_2O_3/NiAl(110)$ surfaces. The numbers 1–4 correspond to different lobes of the porphyrin. Reproduced with permission from Qiu *et al.*[7]

semiconductor and subsequent recombination with the available majority carriers) should be universal to all semiconductors with a direct band gap and, therefore, commonly observed during STM measurements. However, the efficiency of this process is extremely low and nanoscale imaging of semiconductors is generally only achieved at cryogenic temperatures. The efficiency of the luminescence improves dramatically in semiconductors with a depletion (or inversion) region near the surface. This is observed in chalcopyrite thin films, where spontaneous deviations in the stoichiometry of the surface cause a depletion (or even a type of inversion) region that reinforces the homojunction.[49] In this case, nanoscale electroluminescence can be excited when the STM tip locally drives the surface homojunction into forward bias. Because the attenuation length of the hot tunneling electrons is estimated to be in the tens of nanometers, this STM-based luminescence can be used to investigate the surface electronics in semiconductors. For illustration purposes, we present results for chalcopyrite thin films. Figure 9.7(a) and (b) show the STM image and the corresponding photon energy map for tunneling luminescence (color-coded so that red and blue shifts in the emission spectrum are intuitive). Figure 9.7(c) and (d) show the scanning electron microscopy image (over the same area) and the corresponding photon energy map for the luminescence excited by the electron beam, which originates in the bulk of the film. We clearly see how variations in the transition energy within the grains are largely mitigated at the surface.

Another approach to improving the efficiency of the tunneling luminescence is by intentionally introducing a junction near the surface of the semiconductor. As is the case when a surface band bending is present, electroluminescence can be observed when the tip drives the shallow junction into forward bias.[50,51]

Other mechanisms responsible for the STM-based luminescence are the excitation of surface plasmons[29] or, in some cases, *bremsstrahlung* radiation[52,53] resulting from the acceleration of hot tunneling electrons into the semiconductor and their subsequent deceleration accompanied by the emission of

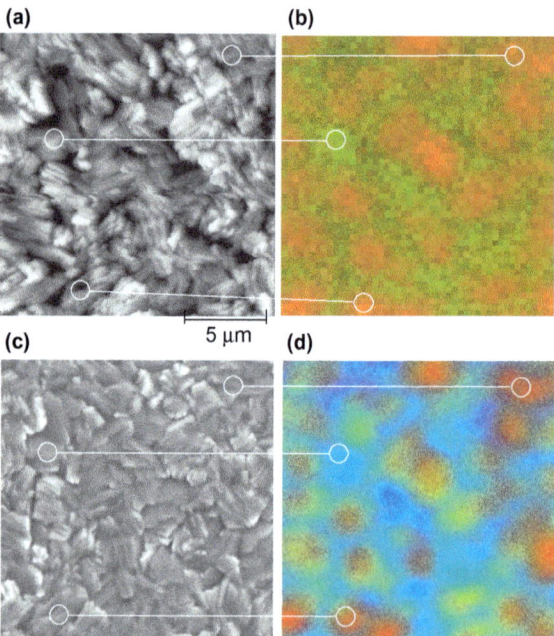

Figure 9.7   (a) STM image of a chalcopyrite thin film and (b) corresponding photon energy map of the tunneling luminescence ($V_t = -5$ V, $I_t = 50$ nA). (c) SEM image matching the previous STM and (d) corresponding photon energy map of the cathode luminescence ($E_b = 5$ keV, $I_b = 750$ pA). The red-green-blue (*RGB*) scale in (b) and (d) corresponds to *R*: 1.17 eV, *G*: 1.20 eV, *B*: 1.22 eV.

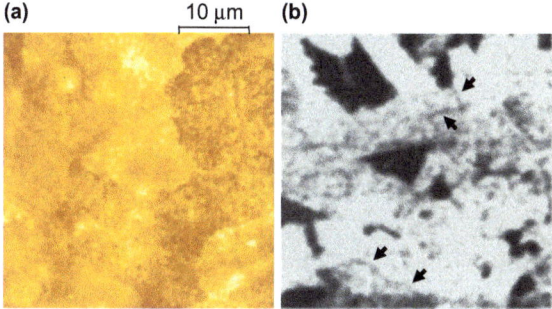

Figure 9.8   (a) Constant current STM image of a silicon thin film at $V = +1.7$ V and $I_t = 1$ nA. (b) Corresponding image of the photon intensity obtained in a second scan at $V = +2.7$ V and $I_t = 20$ nA. Time/pixel $= 125$ ms.

photons – a sort of STM-induced local breakdown. Using this *bremsstrahlung* radiation, we identified regions of low conductivity near the surface of silicon thin films obtained by epitaxy as well as high recombination associated with grain boundaries (see Figure 9.8).

In contrast to inorganic semiconductors, the mechanism responsible for the tunneling luminescence observed in organic semiconductors is molecular (or excitonic) fluorescence. The efficiency of this process can be enhanced by coupling with TISP, which can be generated by laser illumination of a metallic tip[31] or hot tunneling electrons.[17] Using the enhancement effect of the TISP, we have demonstrated nanoscale luminescence imaging of organic semiconductors consisting of the conjugated polymer poly(3-hexylthiophene) (P3HT, electron donor) blended with the C60 derivative 1-(3-methoxycarbonyl)-propyl-1-phenyl-(6,6)C61 (PCBM, electron acceptor). The excitonic luminescence is significantly enhanced when the conjugated polymer is coupled to the plasmon excitation at the tip (tip-enhanced luminescence). This effect allows dramatic improvement in the detection efficiency of the excitonic luminescence and, consequently, resolution of individual domains of the conjugated polymer in which the exciton will recombine before dissociation at the P3HT/PCBM interface. To illustrate this, Figure 9.9 shows the topography and corresponding images of the excitonic luminescence for P3HT:PCBM films deposited on gold. Annealing the films promotes the segregation of P3HT and PCBM and, therefore, the exciton generated within a domain of P3HT can recombine before reaching the P3HT:PCBM interface. This effect explains the increase in the luminescence intensity observed with annealing post-treatment and can be used to investigate the exciton transport and recombination in multiphase organic semiconductors.

9.3.3   Nanocrystals

The STL technique has by now been extensively applied to generate light emission from individual semiconductor and metal quantum dots. These nanoparticles are generally smaller than 100 nm and larger than a few atoms. In the case of semiconductors, quantum confinement causes the energy level structure

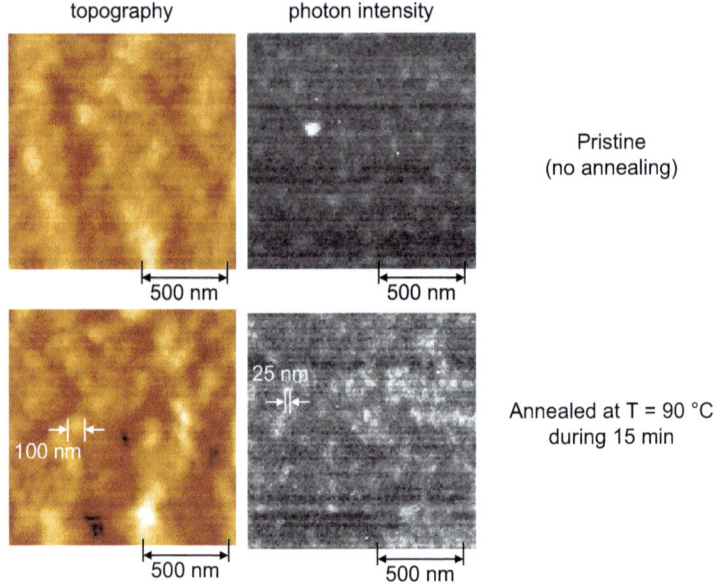

Figure 9.9   Topography and corresponding photon intensity images of the molecular luminescence of P3HT:PCBM films deposited on gold and annealed at different temperatures. $V = +3\,V$, $I_t \sim 100\,nA$. The observed morphologies are consistent with previous results indicating phase separation of the order of 50–100 nm. The photon intensity images can resolve features of the order of 20–30 nm.

to become more like an atom in that there are discrete energy states instead of a continuous band as in macroscopic semiconductors.[54] It also causes new and unique effects such as multiple exciton generation that are potentially useful in solar energy conversion as well as light emitting diodes.[55] Metal nanoparticles, especially those consisting of the coinage metals such as Au, Ag, Pt, and so on, can become strongly surface-plasmon active and are useful in surface-enhanced Raman spectroscopy as well as to enhance solar energy conversion in solar cells.[56,57] Initial STL experiments on nanoparticles involved individual metallic nanoparticles as well as arrays of metallic nanoparticles.[58–62] Single quantum dots and arrays of quantum dots were studied as embedded dots inside hetero-structures[63–72] as well as colloidal quantum dots on surfaces.[11,39,73]

Figure 9.10 shows an example of a STL measurement on a self-organized hexagonal array of silver nanoparticles on a Au (111) surface.[58] At the bias applied in Figure 9.10, the plasmon mode that is excited is the lowest energy coupled mode for which it is predicted that the emission maxima occur between particles exactly as observed. At higher applied biases, higher energy modes are excited and as can be seen in Silly *et al.*[58] the emission maxima occur on top of the particles. In other work, self-organized arrays of Ag nanoparticles were compared with disordered arrays and from this, evidence for coupled plasmon modes was obtained.[61]

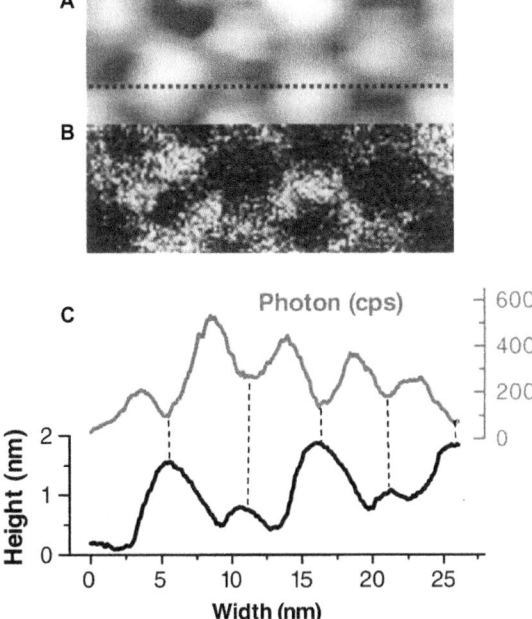

Figure 9.10   Simultaneously recorded STM topography (A) and photon map (B) of hexagonally ordered silver nanoparticles on a Au substrate. Scan area 26×10 nm. Graph (C) shows a topography cross-section along the dotted line in (A). The emission intensity shows maxima when the tip is between the particles indicating that the emission is caused by coupled plasmon modes between the particles.
Reproduced with permission from Silly *et al.*[58]

The tunneling luminescence spectroscopy of embedded semiconductor dots formed by epitaxy (*i.e.*, InP, InAs, InGaAs, *etc.*) has been investigated by several research groups.[63–72] Most applications of these self-assembled structures (lasers, solar cells, *etc.*) demand a very strict control and uniformity on the nanoscale (distribution, size, geometry of the QDs) and STM-based measurements are well suited to resolving the properties of single dots. This is demonstrated in Figure 9.11 for self-assembled InGaAs QDs grown by metalorganic vapor phase epitaxy (MOVPE) on (311)B *n*-type GaAs substrates.

Figure 9.11(a) shows the STM image of the dots obtained at $V = 4$ V and $I_t = 60$ pA. At positive bias, holes are transferred directly to QDs, recombining with available electrons (*n*-type) and individual dots can be resolved without the interference from the wetting and barrier layers. Figure 9.11(b) shows the STL spectrum obtained when the tip is positioned on the A and B QDs selected from the STM image and the tunneling current is increased to 10 nA. The unresolved spectrum obtained by cathodoluminescence (CL) is shown for comparison. CL is excited from the InGaAs QDs and the GaAs barriers. In contrast, no emission from the GaAs is seen on the tunneling luminescence spectrum, corroborating the idea that tunneling holes recombine within the

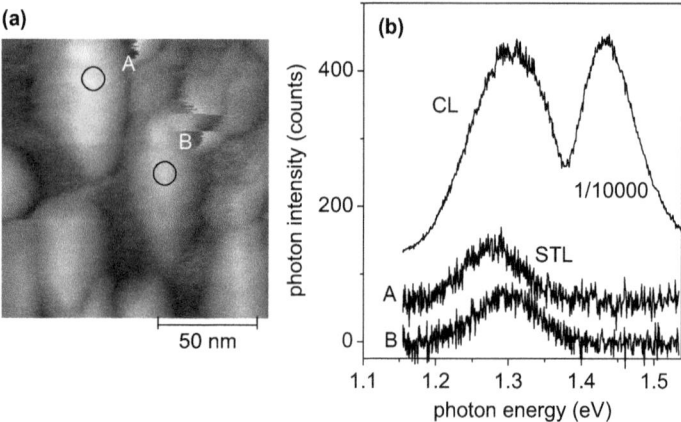

Figure 9.11   (a) A constant current mode STM image of InGaAs QDs ($V = 4$ V, $I_t = 60$ pA, $T = 300$ K, Z range = 5.5 nm) and (b) corresponding STL emission spectra obtained from QDs A and B on the STM image. Acquisition time = 100 s at $I_t = 10$ nA. The CL spectrum is shown for comparison.

dots. We investigated these structures to assess the degree of uniformity on both the micro- and nanometer scales and found substantial fluctuations in the photon energy associated with QD transitions at different spatial scales. This is a very interesting aspect of self-assembled structures that should be addressed to improve QD uniformity for optoelectronic applications.

In contrast to embedded dots inside heterostructures formed by hetero-epitaxy, colloidal quantum dots have not been extensively studied using STL techniques. However, single semiconductor quantum dots[74] as well as multi-layers of quantum dots[75] have been studied by standard scanning tunneling spectroscopy techniques extensively. We have studied STL of samples of isolated CdSe/ZnS core/shell quantum dots bound to Au surfaces using hex-anedithiol self-assembled monolayers (SAMs). Two main effects are observed. One effect is strong quenching of the TISP emission when the voltage applied to the tip is in resonance with the band gap emission of the quantum dot.[39] Under these conditions, simultaneous tunneling of an electron and hole from the tip and the substrate respectively (or *vice versa*) prevents the formation of an emissive TISP. This phenomenon was shown to be useable to detect the presence and the band gap of quantum dots absorbed to rough gold substrates where topography cannot be used to detect their presence (Figure 9.12). In this plasmon resonance imaging technique, the tip–substrate voltage is scanned at each pixel and a resonance can be detected by a dip in the STL intensity as a function of applied bias. In this way chemical information (the band gap of the nanoscale species in this case) can be detected at small length scales. We demonstrated 10-nm resolution, limited only by the radius of the tip.[39]

The second main effect that can be observed is emission from the quantum dot. Similar to the observations of light emission from single molecules

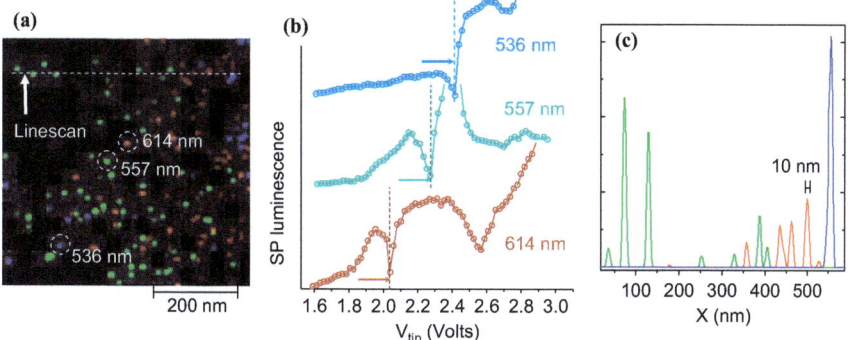

Figure 9.12 Image of the energy of the plasmon resonance for a multidot sub monolayer consisting of three different types of CdSe/ZnS core/shell quantum dots with emission (in solution) at 536 (blue), 557 (green) and 614 (red) nm. (b) Tunneling luminescence spectra showing the resonances in the extinction of the free plasmon for the three different types of dot. (c) A line scan over the image (a) demonstrates a resolution better than 10 nm.

discussed above, because the bias onset of this emission coincides with the point where the TISP emits at a short enough wavelength to excite the quantum dots, this emission is due to energy transfer from the TISP to excitons on the quantum dots, followed by light emission.[11] It was also shown that the excitonic states on the quantum dots can quantum mechanically hybridize with the surface plasmon at Rabi splittings of around 100–200 meV, similar to what has been observed in literature,[15] with surface plasmons being excited in a prism geometry hybridizing with similar CdSe quantum dots as those employed in the STL study.[11] In contrast to the above observations, a study of STL on CdSe quantum rods that were tightly coupled to the metal substrate because of the absence of a SAM in between showed direct luminescence from electron–hole recombination on the rods without exciting TISPs.[76]

The STL technique has also been applied to multilayers of CdSe quantum dots.[73] Interestingly, the opposite behavior is found in results on single dots on Au surfaces[39] in that the presence of the quantum dots leads to an increase in the plasmon emission because of a more conducive dielectric environment in the multilayer geometry. It was also observed that the onset of STL in these cases occurred at the band gap of the quantum dots divided by a factor of two, indicating that tunneling to the dots is an important step in the light emission process. It also indicates that better modeling of the potential distribution inside the quantum dot layers is necessary.[73]

9.4 Time-Resolved Studies of Photon Emission in the STM

Within the field of STL there is a drive towards adding time resolution to the technique. As shown above there is very good spatial resolution as well as energetic resolution possible. The addition of dynamic measurements would allow for the elucidation of the dynamics of charge and energy transport in individual nanostructured systems. Currently this field is still very much in its infancy but promising results are starting to be obtained. In one example, the voltage at the tip is pulsed with a pulse width of 100 ns and STL emission traces based on the statistics of when the first photons are detected after the pulse are obtained as a result. The technique was used to study single crystalline samples of epitaxially grown multilayer quantum confined structures in GaAs.[77] In a similar fashion, one can use a tunneling current from a tapping-mode atomic force microscope (AFM) tip to perform short-lived excitation of the tunneling current and detect electroluminescence as a result. This technique was applied to polycrystalline layers of $CuGaSe_2$ to show how the presence of grain boundaries affects current flow and correspondingly the locality of electro-luminescence. To date this technique has only been applied to study the intensity and wavelength of the emission, but it also seems amenable to detecting the electroluminescence as a function of time after the tapping event at time resolutions on the order of 100 ns to microseconds and longer.

Much faster time resolutions can be obtained by exciting the tip/substrate gap using ultrafast laser pulses at high repetition rates. At these high repetition rates and short pulses, the STM feedback loop cannot keep up and the gap is maintained at a constant distance, although current pulses can be ultrashort as was determined by Uehara *et al.* by detecting the light emission as a consequence of the current injection pulses.[8] In this study, the current pulses were shown to be of the same duration as the laser pulses, opening up the ultrafast time domain to these techniques. The same study also showed that the laser light can directly generate surface plasmons at the tip/substrate gap that can radiate out, providing an alternative mechanism for light emission. The observation that the laser pulse duration determines the duration of the current pulse opens up an even faster than picosecond time domain for study using STL emission.

While not specifically using time-resolved excitation, Silly and Charra applied time-autocorrelated two-photon counting measurements to STL emission of bare gold surfaces.[78] They found that because of the strong local nature of the inelastic electron tunneling process that is at the basis of STL, photons emitted show strong bunching behavior on the nanosecond timescale. Such measurements should also be elucidative when applied to quantum nanoemitters such as quantum dots.

9.5   Conclusions and Outlook

STL is a technique that provides extraordinary spatial resolution as well as optical resolutions limited by the resolution of the tunneling process. Several different kinds of luminescence mechanisms have been observed with the most prevalent ones involving the creation of (radiative) surface plasmons induced by the tip. The technique is suitable for the study of charge and energy transport in nanoscale systems, even allowing for the study of the emission characteristics of the individual parts of small molecule chromophores, as they are influenced by binding geometry to a substrate. STL and related techniques will increasingly be used to address charge and exciton transport in nanoscale systems, especially those relevant to energy conversion in solar energy conversion structures, as well as light emitting diodes based on nanoscale emitters such as semiconductor quantum dots and others. The addition of time resolution to the STL technique will allow the study of the dynamics of charge and energy transport in interesting nanoscale systems.

Acknowledgment

The authors acknowledge funding from the Solar Photochemistry Program of the Division of Chemical Sciences, Geosciences, and Biosciences, Office of Basic Energy Sciences of the DOE through contract DE-AC36-08G028308.

References

1. G. Binnig and H. Rohrer, *Rev. Mod. Phys.*, 1987, **59**, 615.
2. L. Gross, N. Moll, F. Mohn, A. Curioni, G. Meyer, F. Hanke and M. Persson, *Phys. Rev. Lett.*, 2011, **107**, 086101.
3. J. Coombs, J. Gimzewski, B. Reihl, J. Sass and R. Schlittler, *J Microsc-Oxford*, 1988, **152**, 325–336.
4. C. Thirstrup, M. Sakurai, K. Stokbro and M. Aono, *Phys. Rev. Lett.*, 1999, **82**, 1241–1244.
5. R. Berndt, R. Gaisch, W. Schneider, J. Gimzewski, B. Reihl, R. Schlittler and M. Tschudy, *Surf. Sci.*, 1994, **307**, 1033–1037.
6. F. Rossel, M. Pivetta, F. Patthey and W.-D. Schneider, *Opt Express*, 2009, **17**, 2714–2721.
7. X. Qiu, G. Nazin and W. Ho, *Science*, 2003, **299**, 542–546.
8. Y. Uehara, A. Yagami, K. Ito and S. Ushioda, *Appl. Phys. Lett.*, 2000, **76**, 2487–2489.
9. A. Anderson, K. S. Deryckx, X. J. G. Xu, G. Steinmeyer and M. B. Raschke, *Nano Lett.*, 2010, **10**, 2519–2524.
10. R. Berndt and J. Gimzewski, *Surf. Sci.*, 1992, **269**, 556–559.
11. M. J. Romero and J. van de Lagemaat, *Phys. Rev. B*, 2009, **80**, 115432.
12. C. Symonds, J. Bellessa, J.-C. Plenet, A. Brehier, R. Parashkov, J. S. Lauret and E. Deleporte, *Appl. Phys. Lett.*, 2007, **90**, 091107.
13. N. I. Cade, T. Ritman-Meer and D. Richards, *Phys. Rev. B*, 2009, **79**, 1–4.
14. G. P. Wiederrecht, G. Wurtz and J. Hranisavljevic, *Nano Lett.*, 2004, **4**, 2121–2125.
15. D. E. Gómez, A. K. C. Vernon, P. Mulvaney and T. J. Davis, *Appl. Phys. Lett.*, 2010, **96**, 073108.
16. R. M. Roth, N. C. Panoiu, M. M. Adams, R. M. Osgood, C. C. Neacsu and M. B. Raschke, *Opt. Express*, 2006, **14**, 2921–2931.
17. M. J. Romero, A. J. Morfa, T. H. Reilly, J. van de Lagemaat and M. M. Al-Jassim, *Nano Lett.*, 2009, **9**, 3904–3908.

18. F. Rossel, M. Pivetta and W.-D. Schneider, *Surf. Sci. Rep.*, 2010, **65**, 129–144.
19. I. Gryczynski, J. Malicka, W. Jiang, H. Fischer, W. Chan, Z. Gryczynski, W. Grudzinski and J. R. Lakowicz, *J. Phys. Chem. B*, 2005, **109**, 1088–1093.
20. F. Tam, G. P. Goodrich, B. R. Johnson and N. J. Halas, *Nano Lett.*, 2007, **7**, 496–501.
21. P. Anger, P. Bharadwaj and L. Novotny, *Phys. Rev. Lett.*, 2006, **96**, 113002.
22. I. Chizhov, G. Lee, R. Willis, D. Lubyshev and D. Miller, *J. Vac. Sci. Technol., A*, 1997, **15**, 1432–1437.
23. R. Berndt, R. Gaisch, W. Schneider, J. Gimzewski, B. Reihl, R. Schlittler and M. Tschudy, *Phys. Rev. Lett.*, 1995, **74**, 102–105.
24. T. Yokoyama and Y. Takiguchi, *Surf. Sci.*, 2001, **482**, 1163–1168.
25. M. Sakurai, C. Thirstrup and M. Aono, *Surf. Sci.*, 2003, **526**, L123–L126.
26. C. Chen, C. A. Bobisch and W. Ho, *Science*, 2009, **325**, 981–985.
27. H. Imada, M. Ohta and N. Yamamoto, *Appl Phys Express*, 2010, **3**, 045701.
28. M. Reinhardt, G. Schull, P. Ebert and R. Berndt, *Appl. Phys. Lett.*, 2010, **96**, 152107.
29. A. Downes and M. Welland, *Phys. Rev. Lett.*, 1998, **81**, 1857–1860.
30. M. J. Romero, J. van de Lagemaat, G. Rumbles and M. M. Al-Jassim, *Appl. Phys. Lett.*, 2007, **90**, 193109.
31. A. Hartschuh, H. Qian, A. Meixner, N. Anderson and L. Novotny, *Nano Lett.*, 2005, **5**, 2310–2313.
32. P. Johansson, *Phys. Rev. B*, 1998, **58**, 10823–10834.
33. N. Venkateswaran, K. Sattler, J. Xhie and M. Ge, *Surf. Sci.*, 1992, **274**, 199–204.
34. V. Sivel, R. Coratger, F. Ajustron and J. Beauvillain, *Phys. Rev. B*, 1992, **45**, 8634–8637.
35. S. Manson-Smith, C. Trager-Cowan and K. O'Donnell, *Phys Status Solidi B*, 2001, **228**, 445–448.
36. S. Alvarado, W. Riess, P. Seidler and P. Strohriegl, *Phys. Rev. B*, 1997, **56**, 1269–1278.
37. X. Guo, Z. Dong, A. Trifonov, S. Yokoyama, S. Mashiko and T. Okamoto, *Jpn. J. Appl. Phys., Part 1*, 2003, **42**, 6937–6940.
38. R. Arafune, K. Sakamoto, K. Meguro, M. Satoh, A. Arai and S. Ushioda, *Jpn. J. Appl. Phys., Part 1*, 2001, **40**, 5450–5453.
39. M. J. Romero, J. van de Lagemaat, I. Mora-Sero, G. Rumbles and M. M. Al-Jassim, *Nano Lett.*, 2006, **6**, 2833–2837.
40. R. W. Rendell and D. J. Scalapino, *Phys. Rev. B*, 1981, **24**, 3276–3294.
41. A. Downes, P. Guaino and P. Dumas, *Appl. Phys. Lett.*, 2002, **80**, 380–382.
42. R. Berndt, J. Gimzewski and R. Schlittler, *Ultramicroscopy*, 1992, **42**, 528–532.
43. R. Berndt, R. Gaisch, J. Gimzewski, B. Reihl, R. Schlittler, W. Schneider and M. Tschudy, *Science*, 1993, **262**, 1425–1427.

44. R. Berndt, R. Gaisch, W. Schneider, J. Gimzewski, B. Reihl, R. Schlittler and M. Tschudy, *Appl Phys A*, 1993, **57**, 513–516.
45. Z. C. Dong, X. L. Zhang, H. Y. Gao, Y. Luo, C. Zhang, L. G. Chen, R. Zhang, X. Tao, Y. Zhang, J. L. Yang and J. G. Hou, *Nat Photonics*, 2010, **4**, 50–54.
46. G. P. Wiederrecht, J. E. Hall and A. Bouhelier, *Phys. Rev. Lett.*, 2007, **98**, 083001.
47. J. C. Johnson, T. H. Reilly, A. C. Kanarr and J. van de Lagemaat, *J. Phys. Chem. C*, 2009, **113**, 6871–6877.
48. S. Alvarado, L. Libioulle and P. Seidler, *Synth. Met.*, 1997, **91**, 69–72.
49. M. J. Romero, C. S. Jiang, R. Noufi and M. Al-Jassim, *Appl. Phys. Lett.*, 2005, **86**.
50. M. J. Romero, C.-S. Jiang, J. Abushama, H. R. Moutinho, M. M. Al-Jassim and R. Noufi, *Appl. Phys. Lett.*, 2006, **89**, 143120.
51. M. J. Romero, M. A. Contreras, I. Repins, C.-S. Jiang and M. M. Al-Jassim, *Mater. Res. Soc. Symp. Proc.*, 2009, **M09-11**, 1165.
52. A. L. Lacaita, F. Zappa, S. Bigliardi and M. Manfredi, *IEEE Trans. Electron Devices*, 1993, **40**, 577–582.
53. N. Akil, S. E. Kerns, D. V. Kerns, A. Hoffmann and J. P. Charles, *Appl. Phys. Lett.*, 1998, **73**, 871–872.
54. L. E. Brus, *J. Chem. Phys.*, 1984, **80**, 4403–4409.
55. A. J. Nozik, *Chem. Phys. Lett.*, 2008, **457**, 3–11.
56. A. J. Morfa, K. L. Rowlen, T. H. Reilly, M. J. Romero and J. van de lagemaat, *Appl. Phys. Lett.*, 2008, **92**, 013504.
57. E. T. Yu and J. Van De Lagemaat, *MRS Bull.*, 2011, **36**, 424–428.
58. F. Silly, A. Gusev, A. Taleb, F. Charra and M. Pileni, *Phys. Rev. Lett.*, 2000, **84**, 5840–5843.
59. A. Taleb, A. Gusev, F. Silly, F. Charra and M. Pileni, *Appl Surf. Sci.*, 2000, **162**, 553–558.
60. N. Nilius, N. Ernst and H. Freund, *Phys. Rev. B*, 2002, **65**, 115421.
61. N. Nilius, H. Benia, C. Salzemann, G. Rupprechter, H. Freund, A. Brioude and M. Pileni, *Chem Phys Lett*, 2005, **413**, 10–15.
62. S. Katano, K. Toma, M. Toma, K. Tamada and Y. Uehara, *Phys Chem Chem Phys*, 2010, **12**, 14749–14753.
63. J. Lindahl, M. Pistol, L. Montelius and L. Samuelson, *Appl. Phys. Lett.*, 1996, **68**, 60–62.
64. K. Yamanaka, K. Suzuki, S. Ishida and Y. Arakawa, *Appl. Phys. Lett.*, 1998, **73**, 1460–1462.
65. A. Zrenner, F. Findeis, E. Beham, M. Markmann, G. Bohm and G. Abstreiter, *Journal of Luminescence*, 2000, **87–9**, 35–39.
66. U. Hakanson, M. K.-J. Johansson, M. Holm, C. Pryor, L. Samuelson, W. Seifert and M. Pistol, *Appl. Phys. Lett.*, 2002, **81**, 4443–4445.
67. U. Hakanson, M. K.-J. Johansson, J. Persson, J. Johansson, M. Pistol, L. Montelius and L. Samuelson, *Appl. Phys. Lett.*, 2002, **80**, 494–496.
68. U. Hakanson, H. Hakanson, M. K.-J. Johansson, L. Samuelson and M. Pistol, *J. Vac. Sci. Technol., B*, 2003, **21**, 2344–2347.

69. U. Hakanson, V. Zwiller, M. K.-J. Johansson, T. Sass and L. Samuelson, *Appl. Phys. Lett.*, 2003, **82**, 627–629.
70. S. Jacobs, M. Kemerink, P. Koenraad, M. Hopkinson, H. Salemink and J. Wolter, *Appl. Phys. Lett.*, 2003, **83**, 290–292.
71. T. Tsuruoka, Y. Ohizumi and S. Ushioda, *Appl. Phys. Lett.*, 2003, **82**, 3257–3259.
72. T. Tsuruoka, Y. Ohizumi and S. Ushioda, *J. Appl. Phys.*, 2004, **95**, 1064–1073.
73. A. J. Maekinen, E. E. Foos, J. Wilkinson and J. P. Long, *J. Phys. Chem. C*, 2007, **111**, 8188–8194.
74. L. Jdira, K. Overgaag, R. Stiufiuc, B. Grandidier, C. Delerue, S. Speller and D. Vanmaekelbergh, *Phys. Rev. B*, 2008, **77**, 205308.
75. L. Jdira, K. Overgaag, J. Gerritsen, D. Vanmaekelbergh, P. Liljeroth and S. Speller, *Nano Lett.*, 2008, **8**, 4014–4019.
76. T. Lutz, A. Kabakchiev, T. Dufaux, C. Wolpert, Z. Wang, M. Burghard, K. Kuhnke and K. Kern, *Small*, 2011, **7**, 2396–2400.
77. M. Stehle, M. Bischoff, H. Pagnia, J. Horn, N. Marx, B. L. Weiss and H. L. Hartnagel, *J. Vac. Sci. Technol., B*, 1995, **13**, 305–307.
78. F. Silly and F. Charra, *Appl. Phys. Lett.*, 2000, **77**, 3648–3650.

CHAPTER 10

Time Resolved Infrared Spectroscopy of Metal Oxides and Interfaces

AKIHIRO FURUBE

National Institute of Advanced Industrial Science and Technology (AIST), Tsukuba Central 5, 1-1-1 Higashi, Tsukuba, Ibaraki 305-8565, Japan
Email: akihiro-furube@aist.go.jp

10.1 Introduction

Metal oxides are utilized as an essential part of solar energy conversion.[1–3] When suitable sensitizing molecules, semiconductor nanoparticle, or metal nanoparticles are adsorbed on the surface of metal oxide, they can inject electrons after being photo-excited by the visible light into the conduction band of the metal oxide. These electrons can be taken out as photocurrent from the metal oxide if a suitable electrode is formed using the metal oxide combined with a counter electrode and a charge transporter between the two electrodes. The photo-voltage is usually obtained as the difference between the conduction band energy level and the redox potential of the charge transporter.

One successful application of a solar cell using metal oxide as an electron acceptor is the dye-sensitized solar cell (DSSC).[2–4] In DSSCs, organic or metal complex dyes are usually chemically adsorbed on the surface of a wide band gap metal oxide semiconductor nanocrystalline electrode by an anchor group(s) (for example the carboxyl group). Nanocrystalline electrodes are made by sintering nanoparticles, so that a large surface area is obtained on the porous

RSC Energy and Environment Series No. 8
Solar Energy Conversion: Dynamics of Interfacial Electron and Excitation Transfer
Edited by Piotr Piotrowiak
© The Royal Society of Chemistry 2013
Published by the Royal Society of Chemistry, www.rsc.org

electrode. The working principle of the initial photon-to-current conversion process in DSSC is based on the following photophysical and photochemical processes: upon photoexcitation of the dye by visible light, electrons are excited from the ground state of the dye to its excited state and then electron injection occurs from the excited dye into the conduction band of the metal oxide semiconductor film. The injected electron relaxes to the bottom of the conduction band and then the electrons are trapped in a defect site below the conduction band, followed by diffusion of electrons in the electrode toward the transparent conductive electrode. The electrons travel to the external electric circuit and then to the counter electrode, usually platinum. Finally, the electron returns to the oxidized sensitizer dye via redox mediators, such as the I^-/I_3^- pair. The initial electron injection process on metal oxide must be certain (with nearly 100% efficiency) and faster than the other competing processes, such as fluorescence and internal conversion, to achieve a high injection yield. Electron injection is thus a very important primary process in dye-sensitized solar cells.

Among wide-gap metal oxide semiconductor nanoparticles, TiO_2 is the most widely studied material in the field of DSSC because of its favorable physical/chemical properties, low cost, ease of availability and high stability. TiO_2 has a band-gap energy of about 3.2 eV for the commonly used anatase phase and the conduction band edge is located at about –0.5 V *vs.* NHE at pH 0.[5] This band edge energy level must be energetically lower (more positive) than the lowest unoccupied molecular orbital (LUMO) of sensitizer dye to accept an electron from the photoexcited dye. Other metal oxides are also used in DSSC. For example, ZnO is widely studied as an alternative material to TiO_2,[6–12] because it has similar band-gap energy levels of the conduction and valence bands to those of TiO_2 and it has faster electron mobility in conduction, although there still is a problem of chemical instability. Since SnO_2 has a lower conduction band edge than TiO_2 by ~ 0.4 eV, it is energetically favorable as an electron acceptor and is used for the DSSC electrode employed with dye sensitizers that absorb in the red to near-infrared region of sunlight.[8,9,13–16] This is because such dye molecules have usually lower LUMO levels than those absorbing only visible light. In_2O_3 has similar properties to SnO_2 in terms of energy levels.[8,9,17]

The nature of the electron accepting conduction band of these metal oxide semiconductors can be characterized by the density of states (DOS). The DOS, $g(E)$, is related to the effective mass, m^*, by the following equation:

$$g(E) = \frac{\sqrt{2}}{\pi^2 \hbar^3} m^{*3/2} \sqrt{E - E_{CB}} \qquad (10.1)$$

where E_{CB} is the energy of the conduction band edge.[9,18] Roughly speaking, in metal oxide materials, the transfer integrals for the d-orbitals of the neighboring metal atoms are smaller than the integrals for s-orbitals. Therefore, the effective mass of the conductive electron is expected to be larger for d-orbital materials, to which TiO_2 is categorized. When the effective electron mass is large, the DOS near the conduction band edge becomes large. This large DOS for TiO_2 provides a large number of electron acceptor states for electron injection. On the other hand, ZnO, SnO_2 and In_2O_3 are categorized as having *s*-orbital materials in their conduction bands and they have small DOS near the conduction band edge. The difference in the DOS should be sensitive to the rate of electron injection from sensitizer dye molecules. Not only the conduction band but also surface defect states below the conduction band can work as electron acceptors. Such defect states provide trapping levels of electrons and their DOS is practically decaying exponentially towards low energy.[19]

The energetics of electron injection from a photo-excited sensitizer dye molecule into a conduction band of a metal oxide semiconductor are illustrated in Figure 10.1. The electron in the highest occupied molecular orbital (HOMO) is excited to LUMO by absorbing a photon with energy *hν*. Electron injection to the metal oxide occurs, competing with deactivation processes back to the ground state, where both the conduction band and the surface states work as electron acceptors. The rate of electron injection becomes sensitive to the acceptor DOS. A larger DOS results in a larger rate of injection qualitatively. Since the yield of electron injection, Φ_{inj} is expressed as:

$$\Phi_{inj} = k_{inj}/(k_{inj} + k_{deact}) \tag{10.2}$$

in the simplest form, k_{inj} (electron injection rate) must be much larger, or faster in time, than the rate of all other deactivation process (k_{deact}) to have Φ_{inj} close to unity.

The electron injection rate is a very important parameter for characterizing and understanding the interfacial electron injection reaction and it can be measured by using time resolved spectroscopy. In 1996, Tachibana and co-workers successfully measured the timescales of electron injection from an

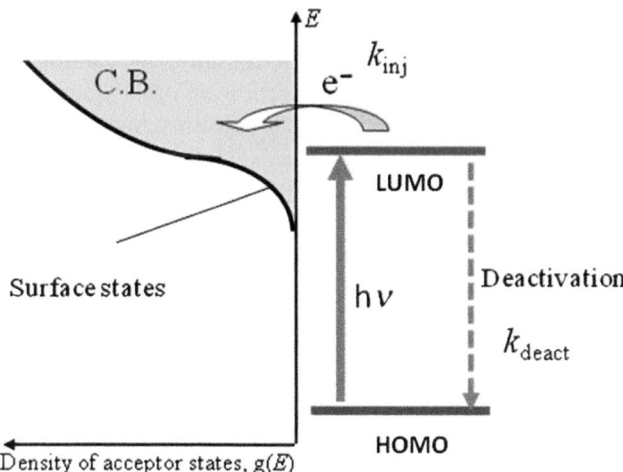

Figure 10.1 Illustration of the energetics of electron injection from a photo-excited sensitizer dye molecule into a conduction band of a metal oxide semiconductor.

efficient Ru-complex sensitizer, (Ru(dcbpy)$_2$(NCS)$_2$, generally called N3) into a TiO$_2$ nanocrystalline film using femtosecond visible-pump/visible-probe transient absorption spectroscopy.[20] After exciting the metal-to-ligand charge transfer (MLCT) band at 605 nm, generation of oxidized N3 at around 800 nm was clearly observed. Interestingly, the reaction was biphasic: the first part occurred within the time resolution (<150 fs) and the second one occurred in the picosecond timescale. This biphasic nature is ascribed to very fast intersystem crossing in N3 dye, which competes with electron injection in <100 fs timescale. After intersystem crossing, the generated lowest triplet state injects the electron slowly near or below the conduction edge. This study has been followed by numerous studies of similar or different systems of many kinds of dyes and several metal oxides in order to understand the primary process in DSSC from the viewpoint of surface physical chemistry.[5,12,14,15,17,18,21–54]

Probably, the most common technique is transient absorption observed in the visible and near-IR region. An amplified Ti:sapphire laser system is typically utilized as the light source. The fundamental femtosecond light at around 800 nm wavelength is split into two beams. Then, one beam is converted to visible light to excite dye as pump light and the other is converted to visible and near-IR light as probe light. As shown in Figure 10.2, these pump and probe pulses are introduced to the sample with an adjusted delay time, τ. The change in intensity of probe light induced by the excitation of the sample is recorded as a function of delay time following the rise and decay kinetics of the reaction intermediate which absorbs the probe light. Visible and near-IR probes are technically rather easy to use because of the ready availability of light detectors such as silicon and InGaAs pin-photodiodes, which are sensitive in the 400–1000 nm region and ~600–1600 nm region, respectively, Also optical parametric amplification and white-light continuum generation techniques easily generate tunable or broad-band probe light. Time-resolved fluorescence spectroscopy is technically easier than transient absorption but only emissive excited states of sensitizers can be detected. Generally, one cannot know whether fluorescence decay is due to electron injection or other quenching processes. Quantitative application of this technique for electron injection yields is usually very difficult.

Many laboratories have built transient absorption spectroscopic systems to investigate the electron injection process in dye-sensitized metal oxide semiconductors. However, because of the often observed complicated reaction

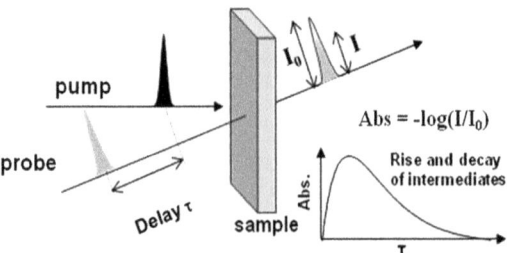

Figure 10.2 Principle of pump-probe transient absorption spectroscopy.

kinetics caused by the inhomogeneous nature of the surface and overlapping transient absorption bands of intermediate states of the dye (in the excited states and the oxidized state), transient absorption probing in the visible and near-IR region suffers from difficulty of spectral assignment of the intermediates. In this case, transient absorption in the IR region gives very useful information. The excited dye changes into an oxidized dye and the injected electron: $dye^* \rightarrow dye^+ + e^-$. An electron injected into the conduction band of the metal oxide, in addition to a shallowly trapped electron just below the conduction band, is known to give strong absorption in the IR region.[18,55,56] This absorption shows steeply increasing intensity with increasing wavelength, which is characteristic of the conducting electron in semiconductors, that is, optical transition from free electrons in the bottom of a conduction band (or shallow trapped electrons) to the upper levels of the conduction band. According to theory, for an ideal crystal, the shape of the spectrum can be reproduced by the power law λ^n. This tendency to depend on wavelength is based on the fact that phonons need to couple in order to cause optical transition from the bottom of the potential, expressed as a parabolic shape in k-space for the dispersion relation of the free electron (see Figure 10.3). In the mid-IR wavelength region, this optical absorption is strong. Simply by introducing an IR generation option for the optical parametric amplifier (OPA), IR detectors and some optical components, one can extend the visible-to-near-IR transient absorption setup to a mid-IR one. The technical details will be presented in the following section.

There are some advantages of IR probe transient absorption for investigating dye-sensitized metal oxide semiconductors. Since the absorption of injected electrons does not usually overlap other absorption bands of the dye molecule, it is easy to analyze the electron injection process. It is easy to compare electron injection processes in different kinds of dye molecule, because the reaction product is always the same, the conductive electron, unless one changes the metal oxide nanocrystalline film. Then not only the reaction rates but also the reaction yields can be compared by measuring the absorption intensities in the timescale just after the reactions are completed. This information is important especially when discussing the relation between the electron injection process and solar cell performance such as the photocurrent generation yield. It is hard to evaluate and compare electron injection yields for

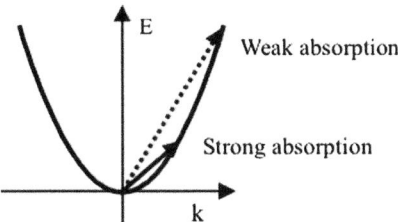

Figure 10.3 Intraband optical transition of an electron at the bottom of the conduction band shown in E–k space.

different dye sensitizers by only observing absorption of oxidized dyes, because different dyes give different absorption band positions and coefficients. Also, by choosing an appropriate IR wavelength region, one can measure the vibrational bands of the dye in the excited and oxidized forms.[18,38,44,49,57] These bands overlap with the electron absorption, although it can be easily subtracted because of its broad and featureless shape compared with the structured vibrational bands. The change in the structure of the dye upon electron transfer can then be discussed.

10.2 Experimental Techniques

Here, as an example of visible-pump/IR-probe femtosecond transient absorption spectrometers, the setup of our group is shown in detail.[9,58–70] As shown in Figure 10.4, the light source for transient absorption spectroscopy is a femtosecond titanium sapphire laser with a regenerative amplifier (Hurricane, Spectra Physics, 800 nm central wavelength, 130 fs FWHM pulse duration, 1 mJ/pulse intensity, 1 kHz repetition). The fundamental output of the laser is divided into two beams. One of the beams is used for the pump pulse at a visible wavelength and another is used for the probe pulse at an IR wavelength between around 2500 and 5000 nm. To obtain the visible pump light, an 800-nm beam is introduced into an optical parametric amplifier (OPA; TOPAS, Quantronix) with a mixing crystal. The IR probe beam is obtained by introducing the 800-nm beam into another OPA (TOPAS, Quantronix) with a differential frequency generation crystal. The pump light enters the decay stage to change the pulse distance between the pump and probe beams, thus achieving a time delay. We use a mechanical chopper at the half repetition (500 Hz) of the pulsed laser to control pump-on and -off. The probe beam is divided into two beams: one beam passes through the sample and another beam acts as a reference to compensate for the intensity fluctuation of each pulse. The light intensity of each laser beam is adjusted by a neutral density filter (NDF). The pump beam radius on the film surface is about 0.2–0.4 mm. Signals from

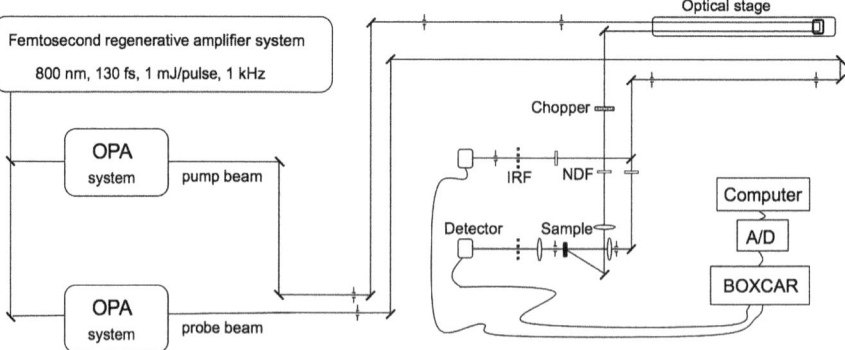

Figure 10.4   Schematic diagram of a visible-pump/IR-probe femtosecond transient absorption spectrometer.

mercury–cadmium–telluride (MCT, Hamamatsu, P3257-10) photodetectors are gated and acquired using boxcar integrators (Stanford Research, SR250) and an A/D board for a shot-to-shot basis. To select only the IR light, an IR band pass filter (IRF) is placed before each detector. The transient absorption intensity was calculated from the fluctuation-corrected pulse intensity of the probe with and without excitation, typically using thousands of pulses. Samples can be scanned to minimize the effect of possible sample degradation during the experiments. Transient absorption measurements are performed at a room temperature of 295 K.

Since a 150 fs laser pulse has an energy bandwidth of $\sim 200 \text{ cm}^{-1}$, it is possible to measure the vibrational transient absorption spectrum of sensitizer dyes by spectrally dispersing the IR pulse with an imaging spectrograph and observing its intensities with IR array detector elements. Using a 32-element array detection scheme, Lian's group has successfully measured transient absorption bands of C=O or O–H stretching and followed electron injection dynamics as time-dependent spectral shapes changing from the electronically excited state into the oxidized state.[18,38,44,49,57] Since electron absorption overlapping is structureless, it is easy to extract only the vibrational absorption bands of the sensitizer dye.

These spectroscopic data give complementary information about electron absorption. Thinking of the electron injection reaction in its simplest from; $dye^* \rightarrow dye^+ + e^-$, the reaction rate can be obtained just by observing one of the three species; dye^*, dye^+, or e^-. In actual cases, however, the inhomogeneous surface property of metal oxide nanoparticles results in complicated reaction processes. For example, in the case of coumarin-derivative sensitized ZnO nanocrystalline film, we found three reaction paths using visible-to-IR probe transient absorption spectroscopy: The first path was direct electron injection and the second and third paths were indirect injection through exciplex intermediates.[63,65] We reported that such exciplexes were formed by interaction between a dye molecule and a surface state on ZnO. Similar interface-bound charge-separated pairs were reported recently by Lian's group.[71] In these studies, the rise in electron absorption was delayed compared with that in oxidized dye. Although this unconventional kinetics was followed by combination of injected electron signals in the IR region and the oxidized dye signals in visible/NIR region, IR vibrational bands would also be useful in studying this kind of new reaction process at the surface of metal oxides.

10.3 Mechanism of Interfacial Electron Injection into Metal Oxides

In this session, some representative results of interfacial electron injection studies at metal oxide surface using time-revolved IR spectroscopy will be presented. The effects of metal oxide materials, sensitizer dyes, solvents and additive ions on dye-sensitized metal oxides will be discussed. Also, a recent result regarding electron injection from a photoexcited gold nanoparticle into a TiO_2 nanoparticle will be shown and its characteristics will be discussed.

10.3.1   Metal Oxide Semiconductor Dependence

In the introduction, it was mentioned that the electron transfer rate should increase when the density of acceptor states for the metal oxide becomes larger. Experimentally, this has been well proved using IR probe transient absorption spectroscopy. As mentioned earlier, IR transient absorption for dye-sensitized metal oxides reflects signals from injected electrons. So it is very easy to evaluate the rates of electron injection from photo-excited sensitizer dye molecules. Figure 10.5(a) indicates such rise kinetics of IR transient absorption.[18] Here, as a sensitizer, Ru-complex dye molecules called N3 were adsorbed on several kinds of metal oxide (TiO_2, SnO_2, ZnO) nanocrystalline films. The MLCT band of the dye was excited and the electron injection reaction that followed was monitored at a wavelength of around 2 µm. A similar study is also reported for SnO_2, In_2O_3 and ZnO, as shown in Figure 10.5(b).9 Here the same dye, N3, was excited at 540 nm (within the same MLCT band) and a near-IR or IR wavelength was used as a probe (see the detail in the caption). TiO_2 shows the fastest injection, being easily judged from the figures. Actually there are many studies of the electron injection process from N3 (or similar N719, where the protons on two of the four carboxylic groups are replaced by tetra-butylammonium) and it is well known that ultrafast electron injection in the time scale of a few tens of femtoseconds dominates the process with an additional slow picosecond timescale, the rate and amplitude of which depend on the sensitivity of the experimental conditions.

In contrast, SnO_2 and In_2O_3 showed a relatively slow injection process, thetimescale of which is about 10 ps. Further, ZnO showed a much slower process on a timescale of about 100 ps. Using Equation (10.1) with an effective

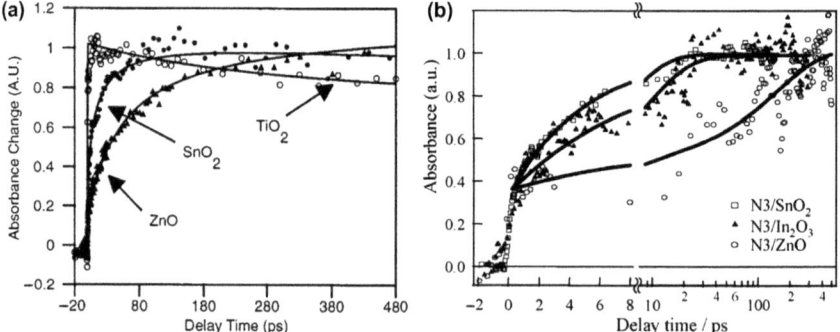

Figure 10.5   (a) Comparison of electron injection dynamics in RuN3-sensitized TiO_2, SnO_2 and ZnO thin films. All samples were excited at 400 nm and probed at 2150 cm^{-1} (TiO_2), 2066 cm^{-1} (SnO_2) and 1900 cm^{-1} (ZnO). Reprinted (adapted) with permission from Asbury *et al.*[18] Copyright 2001 American Chemical Society. (b) Rise profiles of the transient absorptions at 4000, 1700 and 1960 nm for N3/In_2O_3, N3/SnO_2 and N3/ZnO, respectively, after 540 nm excitation. Reprinted with permission from Furube *et al.*[9] Copyright © 2006 Elsevier.

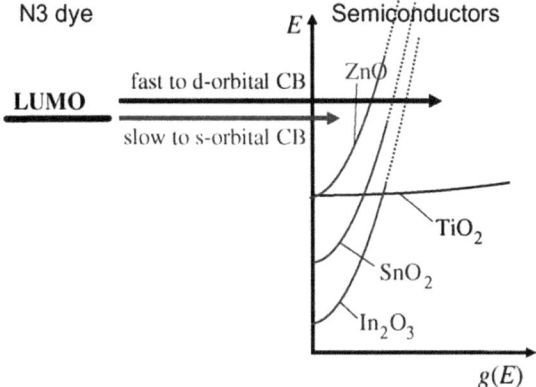

Figure 10.6 Energy scheme of electron injection for excited N3 dye on various metal oxide semiconductors: $g(E)$ is the DOS of the conduction band. The scheme is taken from Furube *et al.*[9]

mass (m^*) and the energy of the conduction band edge (E_{CB}), the DOS function, $g(E)$, for the four metal oxide materials, TiO_2, SnO_2, In_2O_3 and ZnO can be shown schematically in Figure 10.6.[9] Basically, electron transfer occurs isoenergetically from the LUMO level of N3. The amplitudes of $g(E)$ at the LUMO level show good agreement of order with the rate of electron injection. The faster the electron injection rate is, the more efficient the reaction yield can be expected, thinking of the injection as competing with all other deactivation processes. Therefore, TiO_2 is expected to be the best electron acceptor among these metal oxides. In fact, TiO_2 seems to be the most successful material for the electrode of dye-sensitized solar cells.

10.3.2 Dye Dependence

Although many kinds of sensitizer dyes are tested for the development in dye-sensitized solar cells, studies using time-resolved transient IR spectroscopy to understand the electron injection process have been limited so far. As mentioned earlier, Ru-complexes like N3 and N719 on TiO_2 are well studied and the injection kinetics is characterized by a rapid process within 100 fs and an additional slow process in the 10–100 ps timescale.[20,24–27,35,60,62] The former takes place from the lowest excited singlet state (including vibrationally hot states) and the latter from the lowest triplet state. As the electronically excited state relaxes in energy through intersystem crossing and vibrational cooling, the corresponding density of acceptor states become less from a certain point in the conduction band to the bottom of the conduction band or even to somewhere within the surface state levels.

The rate and relative amplitude of slow injection from the triplet state differ from report to report. Since some groups use only visible/NIR transient absorption spectroscopy, the oxidized Ru-complex, which appears at around 800 nm, cannot be spectrally distinguished very well from the overlapping triplet absorption, which has two broad peaks around 700 nm and 1200–1300 nm depending on the manner of protonation and the environmental conditions. Therefore, IR probe transient absorption has an advantage of not overlapping with other transient species.

In contrast to metal complexes, the lowest excited singlet state of organic dye molecules have a longer lifetime, typically on the nanosecond timescale. Therefore, the electron injection process is different from that of metal complex dyes. In the case of coumarin derivatives (called NKX-2311, NKX-2677 and C343), only femtosecond electron injection processes are observed.[64,71,72] Also, oligothiophen-functionalized carbazole dyes (called MK dye series)[73–75] and several porphyrin dyes[53,66,76] displayed only femtosecond injection. In spite of ultrafast injection, the electron injection yield cannot reach 100% because of the inhomogeneous nature of organic dyes on the TiO_2 surface. Dye aggregation, back electron transfer and some complex formation seem to be the reasons for such low electron injection efficiency in the case of organic dyes. Some characteristics of electron injection are summarized for Ru-complex and organic sensitizers in Figure 10.7.

Lian and co-workers performed extensive studies on Re-complex, such as the ligand effect and spacer length dependence, using IR probe transient absorption in order to understand the detailed mechanism of electron injection. How the driving force and electronic coupling strength affect the rate of electron injection is well summarized in their review articles.[18,42] The distance between the chromophore and the metal oxide surface is an important parameter varying electron injection rate. Lian and co-workers used Re-complex with different anchor lengths with and without 1 or 3 CH_2 units (named ReC0A, ReC1A and ReC3A, respectively).[38] They directly observed electron injection from the excited states of Re dyes into TiO_2 by simultaneously measuring the broad IR absorption of injected electrons and the vibrational spectrum of the Re complex in its ground, excited and oxidized states. The spectra are shown in Figure 10.8 (A)–(C). Electron injection from ReC0A to TiO_2 was found to occur on the <100-fs timescale judged by the prompt rise of broad absorption of injected electrons over the observation range from 2000–2120 cm^{-1} as well as another prompt rise of the peak of the oxidized state at 2095 cm^{-1}. On the other hand, the excited state vibrational spectrum at 2070 cm^{-1} evolved with a time

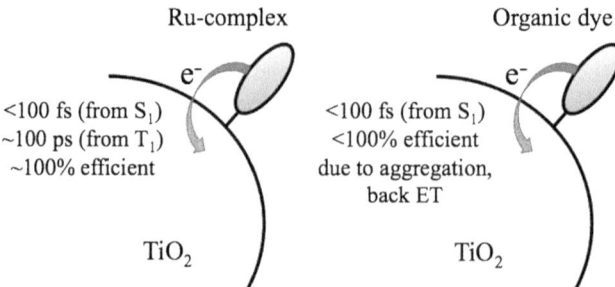

Figure 10.7   Some characteristics of electron injection for Ru-complex and organic sensitizers.

constant of a few picoseconds, indicating that injection occurred before complete vibrational relaxation in this system. They also found that the electron injection rate decreased by a factor of >200 from ReC0A to ReC1A. This decrease is much larger than the predicted change for non-adiabatic electron transfer processes, where the electron transfer rate should decrease exponentially with increasing length of CH_2 spacers. In contrast, they found that the electron injection rate decreased by a factor of 13.7 from ReC1A to ReC3A, which agreed qualitatively with the trend predicted for non-adiabatic electron transfer processes. This study shows that combined observation of injected electrons and vibrational bands of sensitizer is very efficient for obtaining detailed information.

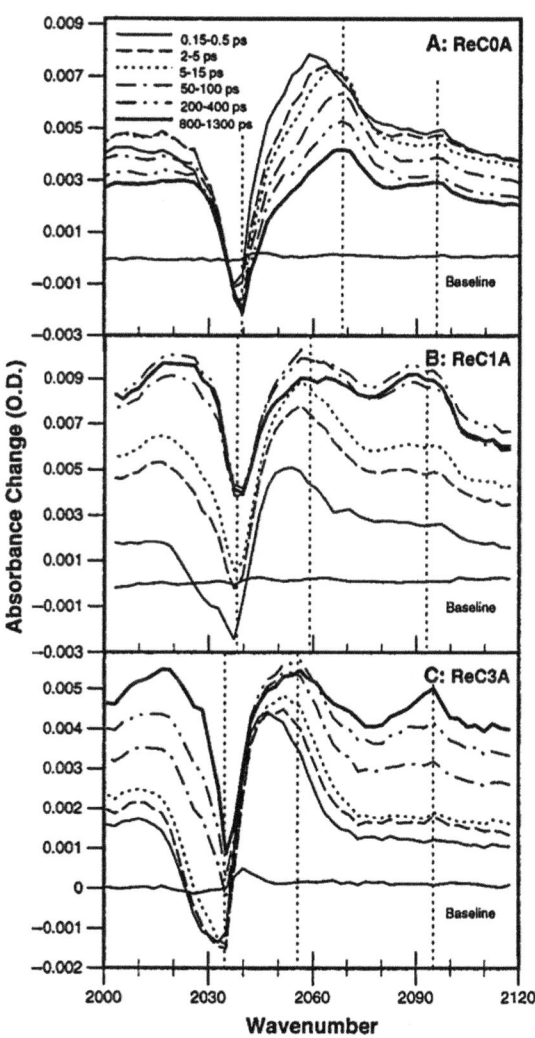

Figure 10.8 Transient IR difference spectra of (A) ReC0A, (B) ReC1A and (C) ReC3A on TiO_2 film.
The graph is taken from Asbury *et al.*[38]

10.3.3 Effect of Solvent

When dye-sensitized metal oxide is used in solar cells, liquid electrolyte usually is employed to relay the electrons from the counter electrode (usually Pt) to the oxidized dyes. Therefore, understanding the effect of solvent on the electron injection process is important. Not many studies are reported using IR probe transient absorption, probably because of the difficulty in experiment of making the solvent layer thin enough for the probe light to transmit. Lian and co-workers investigated photoinduced electron injection rates from Re-complex to TiO_2 nanocrystalline films in air, pH buffer, methanol, ethanol and dimethylformamide (with the thickness of $100\,\mu m$ or $<10\,\mu m$). Although the injection rates in these solvents were noticeably different, they pointed out that the presence of adsorbed water lowered the band edge of TiO_2 and reduced the rate differences more than expected.[49]

Polar solvents can change the level of the conduction band of metal oxide by forming surface dipoles. This should affect the energetics of the electron injection. Nowadays, ionic liquids have been attracting much interest as electrolyte solvents. They have recently been used to improve the dye-sensitized solar cell device lifetime because ionic liquids have negligible vapor pressure and will not evaporate between two electrodes of the cell. Dye-sensitized solar cells with ionic liquids, however, tend to give a lower incident photon to current efficiency (IPCE) by about 10–25% than those with acetonitrile used typically.[77] Although such low IPCE has been expected to be due to slow diffusion of redox ions in very viscous ionic liquids, a recent visible-to-NIR transient absorption study demonstrated that even the initial electron injection was significantly affected in an ionic liquid based electrolyte.[78] IR transient absorption would be a more suitable technique to observe selectively injected electrons and elucidate the effect of solvent on the interfacial electron transfer process. The presence of high-density charges provided by anions and cations of ionic liquid will have a big influence on the local environment around the dye on the surface of metal oxide.

10.3.4 Effect of Ions

When Li^+ ions, which are known to be potential-determining ions for semiconductor conduction and valence bands, are present in a dye-sensitized semiconductor film, the electrochemical potential of TiO_2 undergoes a positive shift due to adsorption of the positive Li^+ ions on the TiO_2 surface as well as to intercalation of Li^+ ions into the TiO_2 lattice.[79] This shift results in a large driving force for electron injection (due to the larger potential difference from the conduction-band edge to the LUMO level of the dye).

An example of transient IR absorption study for the electron injection from a photoexcited coumarin derivative (called NKX-2311) into nanocrystalline TiO_2 is shown in Figure 10.9, where the effect of Li^+ ions was examined by comparing films with and without Li^+ ions. Lithium iodide (LiI) dissolved in acetonitrile (0.1 mol L^{-1}) was used to examine the effect of Li^+ ion on the

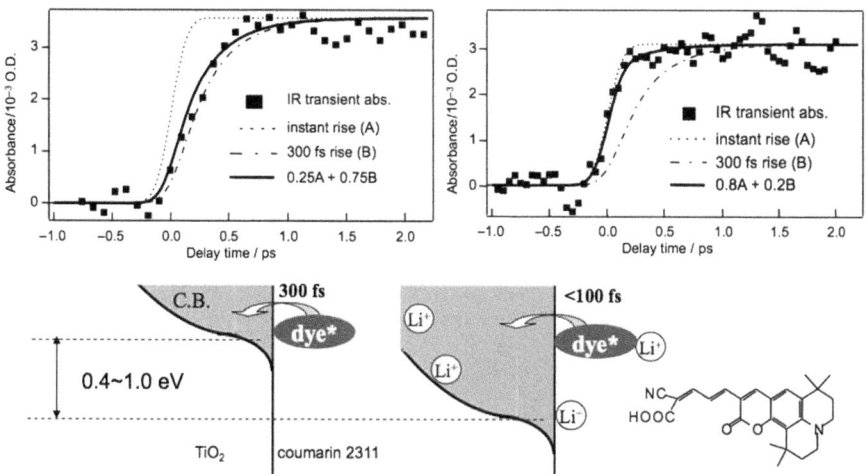

Figure 10.9 Transient absorption time profile at IR wavelengths of NKX-2311/TiO$_2$ in the absence (upper left) and presence (upper right) of Li + ions. Lines are simulation curves. Bottom: Schematic energy diagram explaining the observed electron injection dynamics in the absence and presence of Li$^+$ ions on the TiO$_2$ surface.

electron injection process. Just before the spectroscopic measurements, one or two drops of the solution were placed on the dye-sensitized film to immerse the film fully and the film was dried in the air. The schematic energy diagram in Figure 10.9 can be used to explain the observed electron injection dynamics in the absence and presence of Li$^+$ ions on the TiO$_2$ surface. The density of electron acceptor states is indicated by the grey area, which is composed of a conduction band with an $E^{1/2}$ shape, as expressed by the Equation (10.1) and a tail below the conduction-band edge due to surface trap states. In the absence of Li$^+$ ions, the LUMO of the NKX-2311 dye is located close to the conduction-band edge.[8] The majority of excited dye molecules underwent electron transfer into the conduction band of TiO$_2$ with a time constant of 300 or <100 fs. In the presence of Li$^+$ ions, rapid electron injection within 100 fs was mainly observed. Although interaction between Li$^+$ ions and the dye molecule was also seen, it was likely that the major role of Li$^+$ ions was to shift the conduction band downward to a considerable degree. The injection rate was accelerated by a factor of more than 3. Probably, as indicated in Figure 10.9, the density of states at the LUMO level increased.

Recently, Sunahara *et al.* have investigated more systematically the effect of Li$^+$ ions on electron injection into TiO$_2$ metal oxide nanocrystalline film.[80] One successful porphyrin sensitizer (called GD2) was used as the electron donor and two kinds of acetonitrile electrolytes (including LiI 0.7 M and I$_2$ 0.05 M (denoted as LiI0.7M), or LiI 0.1 M, DMPImI 0.6 M, tBP 0.5 M and I$_2$ 0.05 M (denoted as LiI0.1M)) were used to control the TiO$_2$ conduction band level. Actually, LiI0.7M gives a ~0.2 eV lower conduction band level than LiI0.1M. This larger driving force for electron injection resulted in (i) a larger amplitude

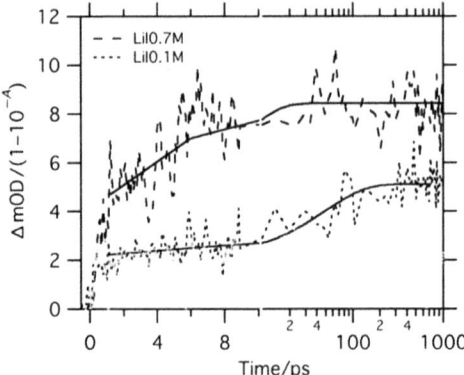

Figure 10.10 Transient absorption kinetics probed at 3440 nm for porphyrin dye (GD2) on TiO_2 with the acetonitrile electrolytes including LiI 0.7 M and I_2 0.05 M (denoted as LiI0.7M) and LiI 0.1 M, DMPImI 0.6 M, tBP 0.5 M and I_2 0.05 M (denoted as LiI0.1M). The bold lines are fits to single exponential functions. Transient absorption signals were corrected by the ground state absorption fraction $(1–10^{-A})$.

of ultrafast injection (<100 fs) by a factor of two, (ii) a faster time-constant of picosecond injection by a factor of ∼10 (from ∼50 ps to ∼6 ps) and (iii) a decrease in the amount to non-injecting dye (from ∼50% to ∼15%) as shown in Figure 10.10. Since IR probe transient absorption makes it easy to observe selectively generation of injected electrons without interference from other transient species, it was found that electron injection from the porphyrin dye was very inhomogeneous in nature so that a significant part of adsorbed dyes barely interact with the conduction band of TiO_2. Such inhomogeneity must be reduced by controlling the dye/TiO_2 interfacial properties in order to improve the solar cell performance.

10.3.5   Metal Nanoparticle to Metal Oxides

As described above, electron transfer at metal oxide semiconductor surfaces is very important for application of solar energy conversion. TiO_2 nanoparticles especially are known as a very efficient electron acceptors. Photo-induced electron transfer from dye molecules to the conduction band of TiO_2 occurs in the femtosecond time range in many cases and this can be understood by the large DOS in the TiO_2 conduction band. Study of these electron transfer dynamics by time-resolved spectroscopy gives important information from the viewpoint of both fundamental science and applications.

Recently, it was found that metal nanoparticles or nanostructures of gold also injected electrons into TiO_2 efficiently.[81–85] This reaction can be a good candidate for efficient solar energy conversion because of the high absorption property of gold nanoparticles, the tunability of the wavelength of the plasmon absorption in the near-IR region controlled by their shape and size, and their durability. IR-probe femtosecond transient absorption spectroscopy was

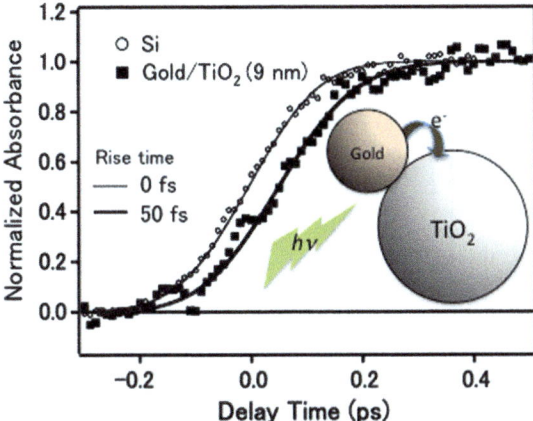

Figure 10.11   Transient absorption rise in the IR region of a gold/TiO$_2$ film and a silicon plate after visible light excitation.

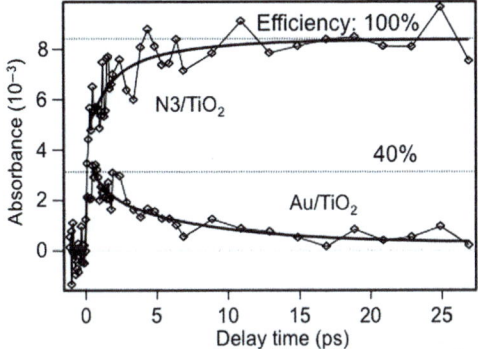

Figure 10.12   Transient absorption rise and decay profile in IR region of a gold/TiO$_2$ film along with that of an efficient Ru-complex dye (called N3) sensitized TiO$_2$ film. A 40% electron injection efficiency was estimated.

successfully utilized to reveal this reaction. Excitation around the plasmon band of spherical gold nanoparticles (10 nm diameter) attached to TiO$_2$ nanoparticles was revealed to result in ultrafast ($<$50 fs) and efficient ($>$40%) electron transfer into the conduction band of TiO$_2$.[59,61,62]

Figure 10.11 shows the IR transient absorption rise after excitation of the peak of the plasmon band (550 nm) with reference data for a silicon plate. The 50 fs rise time was elucidated, indicating that the electron transfer is a competing process with electron–electron scattering in gold, because it is known that after photoexcitation of gold nanoparticles, the electrons with a non-Fermi distribution relax by electron–electron ($<$100 fs), electron–phonon (1–10 ps) and phonon–phonon (ca. 100 ps) interactions.[86–88] By comparing the signal sizes with those obtained for Ru-complex dye (N3) sensitized TiO$_2$, the injection yield under plasmon band peak excitation was about 40% as shown in Figure 10.12.

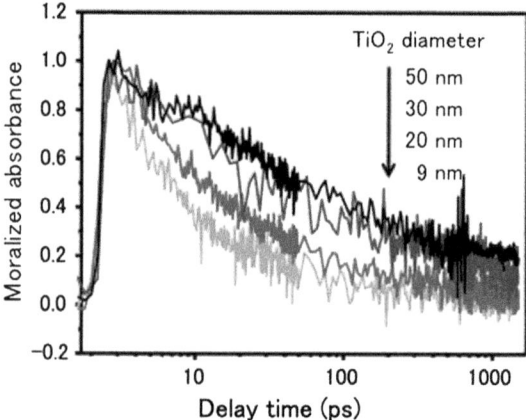

Figure 10.13 Transient absorption decay profiles in IR region of a gold/TiO$_2$ film as a function of the TiO$_2$ nanoparticle diameter.

The charge recombination time in the sub-nanosecond time region became longer for larger TiO$_2$ nanoparticles owing to the larger diffusion volume of injected electrons (Figure 10.13).[61] This is suggested for application, where long lifetimes of charge separation could be achieved. Because of the large number of injected electrons (roughly 10^3) per TiO$_2$ particle in this experiment, we expect the lifetime should be much longer under weak light irradiation, where a trap filling effect does not occur and the electron mobility is much smaller than under our experimental conditions.

10.4 Summary

Electron injection dynamics in the conduction band of metal oxide materials from dye molecules or metal nanoparticles, which is important when applied to sensitized solar cells, can be monitored in the infrared by ~100 fs time resolution. In this chapter, technical details of femtosecond visible-pump/IR-probe transient absorption spectroscopy and some typical spectroscopic data revealing the mechanism of electron injection process were described. A great advantage of this technique is that one can observe transient absorption of injected electrons easily because of the intense intraband transition of an electron at the bottom of or at the trap level just below the conduction band of the metal oxide that forms an electrode. In the case of dye-sensitized solar cells, the effects of metal oxide, dye, solvent and additive ions on the rate and efficiency of electron injection were discussed in detail. One recent discovery, plasmon-induced electron injection from a gold nanoparticle to a TiO$_2$ nanoparticle, was presented to show how femtosecond visible-pump/IR-probe transient absorption spectroscopy is useful in studying this kind of new charge transfer dynamics in a nanostructured system.

References

1. M. Matsumura, K. Mitsuda, N. Yoshizawa and H. Tsubomura, *Bull. Chem. Soc. Jpn.*, 1981, **54**, 692–695.
2. A. Hagfeldt, G. Boschloo, L. C. Sun, L. Kloo and H. Pettersson, *Chem. Rev.*, 2010, **110**, 6595–6663.
3. B. Oregan and M. Gratzel, *Nature*, 1991, **353**, 737–740.
4. A. Hagfeldt and M. Gratzel, *Chem. Rev.*, 1995, **95**, 49–68.
5. J. Moser and M. Gratzel, *J. Am. Chem. Soc.*, 1983, **105**, 6547–6555.
6. N. Sakai, N. Kawashima and T. N. Murakami, *Chem. Lett.*, 2011, **40**, 162–164.
7. K. Keis, E. Magnusson, H. Lindstrom, S. E. Lindquist and A. Hagfeldt, *Sol. Energy Mater. Sol. Cells*, 2002, **73**, 51–58.
8. R. Katoh, A. Furube, T. Yoshihara, K. Hara, G. Fujihashi, S. Takano, S. Murata, H. Arakawa and M. Tachiya, *J. Phys. Chem. B*, 2004, **108**, 4818–4822.
9. A. Furube, M. Murai, S. Watanabe, K. Hara, R. Katoh and M. Tachiya, *J. Photochem. Photobiol., A*, 2006, **182**, 273–279.
10. T. Yoshihara, R. Katoh, A. Furube, M. Murai, Y. Tamaki, K. Hara, S. Murata, H. Arakawa and M. Tachiya, *J. Phys. Chem. B*, 2004, **108**, 2643–2647.
11. A. Furube, R. Katoh, K. Hara and M. Tachiya, in *Ultrafast Phenomena in Semiconductors and Nanostrucutre Materials IX*, ed. K.-T. Tsen, J.-J. Song, H. Jian, Proc. of SPIE, 2005, 5725, SPIE, Bellingham, WA, 2005, 136–147.
12. N. A. Anderson, X. Ai and T. Q. Lian, *J. Phys. Chem. B*, 2003, **107**, 14414–14421.
13. Y. Fukai, Y. Kondo, S. Mori and E. Suzuki, *Electrochem. Commun.*, 2007, **9**, 1439–1443.
14. N. A. Anderson, X. Ai, D. T. Chen, D. L. Mohler and T. Q. Lian, *J. Phys. Chem. B*, 2003, **107**, 14231–14239.
15. X. Ai, N. A. Anderson, J. C. Guo and T. Q. Lian, *J. Phys. Chem. B*, 2005, **109**, 7088–7094.
16. X. Ai, N. Anderson, J. C. Guo, J. Kowalik, L. M. Tolbert and T. Q. Lian, *J. Phys. Chem. B*, 2006, **110**, 25496–25503.
17. J. C. Guo, D. Stockwell, X. Ai, C. X. She, N. A. Anderson and T. Q. Lian, *J. Phys. Chem. B*, 2006, **110**, 5238–5244.
18. J. B. Asbury, E. Hao, Y. Q. Wang, H. N. Ghosh and T. Q. Lian, *J. Phys. Chem. B*, 2001, **105**, 4545–4557.
19. J. Bisquert, F. Fabregat-Santiago, I. Mora-Sero, G. Garcia-Belmonte and S. Gimenez, *J. Phys. Chem. C*, 2009, **113**, 17278–17290.
20. Y. Tachibana, J. E. Moser, M. Gratzel, D. R. Klug and J. R. Durrant, *J. Phys. Chem.*, 1996, **100**, 20056–20062.
21. S. A. Haque, Y. Tachibana, R. L. Willis, J. E. Moser, M. Gratzel, D. R. Klug and J. R. Durrant, *J. Phys. Chem. B*, 2000, **104**, 538–547.
22. R. Huber, S. Sporlein, J. E. Moser, M. Gratzel and J. Wachtveitl, *J. Phys. Chem. B*, 2000, **104**, 8995–9003.

23. R. Huber, J. E. Moser, M. Gratzel and J. Wachtveitl, *J. Phys. Chem. B*, 2002, **106**, 6494–6499.
24. B. Wenger, M. Gratzel and J. E. Moser, *J. Am. Chem. Soc.*, 2005, **127**, 12150–12151.
25. G. Benko, J. Kallioinen, J. E. I. Korppi-Tommola, A. P. Yartsev and V. Sundstrom, *J. Am. Chem. Soc.*, 2002, **124**, 489–493.
26. J. Kallioinen, G. Benko, V. Sundstrom, J. E. I. Korppi-Tommola and A. P. Yartsev, *J. Phys. Chem. B*, 2002, **106**, 4396–4404.
27. J. Kallioinen, G. Benko, P. Myllyperkio, L. Khriachtchev, B. Skarman, R. Wallenberg, M. Tuomikoski, J. Korppi-Tommola, V. Sundstrom and A. P. Yartsev, *J. Phys. Chem. B*, 2004, **108**, 6365–6373.
28. M. Pellnor, P. Myllyperkio, J. Korppi-Tommola, A. Yartsev and V. Sundstrom, *Chem. Phys. Lett.*, 2008, **462**, 205–208.
29. T. Hannappel, B. Burfeindt, W. Storck and F. Willig, *J. Phys. Chem. B*, 1997, **101**, 6799–6802.
30. C. Zimmermann, F. Willig, S. Ramakrishna, B. Burfeindt, B. Pettinger, R. Eichberger and W. Storck, *J. Phys. Chem. B*, 2001, **105**, 9245–9253.
31. S. Iwai, K. Hara, S. Murata, R. Katoh, H. Sugihara and H. Arakawa, *J. Chem. Phys.*, 2000, **113**, 3366–3373.
32. R. J. Ellingson, J. B. Asbury, S. Ferrere, H. N. Ghosh, J. R. Sprague, T. Q. Lian and A. J. Nozik, *J. Phys. Chem. B*, 1998, **102**, 6455–6458.
33. H. N. Ghosh, J. B. Asbury and T. Q. Lian, *J. Phys. Chem. B*, 1998, **102**, 6482–6486.
34. H. N. Ghosh, J. B. Asbury, Y. X. Weng and T. Q. Lian, *J. Phys. Chem. B*, 1998, **102**, 10208–10215.
35. J. B. Asbury, R. J. Ellingson, H. N. Ghosh, S. Ferrere, A. J. Nozik and T. Q. Lian, *J. Phys. Chem. B*, 1999, **103**, 3110–3119.
36. J. B. Asbury, Y. Q. Wang and T. Q. Lian, *J. Phys. Chem. B*, 1999, **103**, 6643–6647.
37. R. J. Ellingson, J. B. Asbury, S. Ferrere, H. N. Ghosh, J. R. Sprague, T. Lian and A. J. Nozik, *Z. Phys. Chem-Int. J. Res. Phys. Chem. Chem. Phys.*, 1999, **212**, 77–84.
38. J. B. Asbury, E. C. Hao, Y. Q. Wang and T. Q. Lian, *J. Phys. Chem. B*, 2000, **104**, 11957–11964.
39. Y. X. Weng, Y. Q. Wang, J. B. Asbury, H. N. Ghosh and T. Q. Lian, *J. Phys. Chem. B*, 2000, **104**, 93–104.
40. J. B. Asbury, Y. Q. Wang, E. C. Hao, H. N. Ghosh and T. Q. Lian, *Res. Chem. Intermed.*, 2001, **27**, 393–406.
41. E. C. Hao, N. A. Anderson, J. B. Asbury and T. Q. Lian, *J. Phys. Chem. B*, 2002, **106**, 10191–10198.
42. J. B. Asbury, N. A. Anderson, E. C. Hao, X. Ai and T. Q. Lian, *J. Phys. Chem. B*, 2003, **107**, 7376–7386.
43. Y. H. Wang, K. Hang, N. A. Anderson and T. Q. Lian, *J. Phys. Chem. B*, 2003, **107**, 9434–9440.
44. X. Ai, J. C. Guo, N. A. Anderson and T. Q. Lian, *J. Phys. Chem. B*, 2004, **108**, 12795–12803.

45. N. A. Anderson and T. Lian, *Coord. Chem. Rev.*, 2004, **248**, 1231–1246.
46. N. A. Anderson and T. Q. Lian, *Annu. Rev. Phys. Chem*, 2005, **56**, 491–519.
47. J. C. Guo, C. X. She and T. Q. Lian, *J. Phys. Chem. B*, 2005, **109**, 7095–7102.
48. C. X. She, N. A. Anderson, J. C. Guo, F. Liu, W. H. Goh, D. T. Chen, D. L. Mohler, Z. Q. Tian, J. T. Hupp and T. Q. Lian, *J. Phys. Chem. B*, 2005, **109**, 19345–19355.
49. C. X. She, J. C. Guo and T. Q. Lian, *J. Phys. Chem. B*, 2007, **111**, 6903–6912.
50. L. Luo, C. J. Lin, C. Y. Tsai, H. P. Wu, L. L. Li, C. F. Lo, C. Y. Lin and E. W. G. Diau, *Phys. Chem. Chem. Phys.*, 2010, **12**, 1064–1071.
51. L. Y. Luo, C. W. Chang, C. Y. Lin and E. W. G. Diau, *Chem. Phys. Lett.*, 2006, **432**, 452–456.
52. L. Y. Luo, C. F. Lo, C. Y. Lin, I. J. Chang and E. W. G. Diau, *J. Phys. Chem. B*, 2006, **110**, 410–419.
53. C. W. Chang, L. Y. Luo, C. K. Chou, C. F. Lo, C. Y. Lin, C. S. Hung, Y. P. Lee and E. W. G. Diau, *J. Phys. Chem. C*, 2009, **113**, 11524–11531.
54. H. P. Lu, C. Y. Tsai, W. N. Yen, C. P. Hsieh, C. W. Lee, C. Y. Yeh and E. W. G. Diau, *J. Phys. Chem. C*, 2009, **113**, 20990–20997.
55. R. Katoh, A. Furube, A. V. Barzykin, H. Arakawa and M. Tachiya, *Coord. Chem. Rev*, 2004, **248**, 1195–1213.
56. A. Yamakata, T. Ishibashi and H. Onishi, *Chem. Phys. Lett.*, 2001, **333**, 271–277.
57. Y. Q. Wang, J. B. Asbury and T. Q. Lian, *J. Phys. Chem. A*, 2000, **104**, 4291–4299.
58. S. Cook, R. Katoh and A. Furube, *J. Phys. Chem. C*, 2009, **113**, 2547–2552.
59. L. Du, A. Furube, K. Hara, R. Katoh and M. Tachiya, *Thin Solid Films*, 2009, **518**, 861–864.
60. L. C. Du, A. Furube, K. Hara, R. Katoh and M. Tachiya, *J. Phys. Chem. C*, 2010, **114**, 8135–8143.
61. L. C. Du, A. Furube, K. Yamamoto, K. Hara, R. Katoh and M. Tachiya, *J. Phys. Chem. C*, 2009, **113**, 6454–6462.
62. A. Furube, L. Du, K. Hara, R. Katoh and M. Tachiya, *J. Am. Chem. Soc.*, 2007, **129**, 14852– +.
63. A. Furube, R. Katoh, K. Hara, S. Murata, H. Arakawa and M. Tachiya, *J. Phys. Chem. B*, 2003, **107**, 4162–4166.
64. A. Furube, R. Katoh, K. Hara, T. Sato, S. Murata, H. Arakawa and M. Tachiya, *J. Phys. Chem. B*, 2005, **109**, 16406–16414.
65. A. Furube, R. Katoh, T. Yoshihara, K. Hara, S. Murata, H. Arakawa and M. Tachiya, *J. Phys. Chem. B*, 2004, **108**, 12583–12592.
66. A. J. Mozer, M. J. Griffith, G. Tsekouras, P. Wagner, G. G. Wallace, S. Mori, K. Sunahara, M. Miyashita, J. C. Earles, K. C. Gordon, L. C. Du, R. Katoh, A. Furube and D. L. Officer, *J. Am. Chem. Soc.*, 2009, **131**, 15621– +.

67. Y. Tamaki, A. Furube, M. Murai, K. Hara, R. Katoh and M. Tachiya, *Phys. Chem. Chem. Phys.*, 2007, **9**, 1453–1460.

68. Y. Tamaki, K. Hara, R. Katoh, M. Tachiya and A. Furube, *J. Phys. Chem. C*, 2009, **113**, 11741–11746.

69. Z. F. Liu, M. Miyauchi, Y. Uemura, Y. Cui, K. Hara, Z. G. Zhao, K. Sunahara and A. Furube, *Appl. Phys. Lett.*, 2010, **96**.

70. S. Cook, R. Katoh and A. Furube, *Journal of Nanoelectronics and Optoelectronics*, 2010, **5**, 115–119.

71. D. Stockwell, Y. Yang, J. Huang, C. Anfuso, Z. Q. Huang and T. Q. Lian, *J. Phys. Chem. C*, 2010, **114**, 6560–6566.

72. K. Hara, Z. S. Wang, T. Sato, A. Furube, R. Katoh, H. Sugihara, Y. Dan-Oh, C. Kasada, A. Shinpo and S. Suga, *J. Phys. Chem. B*, 2005, **109**, 15476–15482.

73. Z. S. Wang, N. Koumura, Y. Cui, M. Takahashi, H. Sekiguchi, A. Mori, T. Kubo, A. Furube and K. Hara, *Chem. Mater.*, 2008, **20**, 3993–4003.

74. K. Hara, Z. S. Wang, Y. Cui, A. Furube and N. Koumura, *Energy Environ. Sci.*, 2009, **2**, 1109–1114.

75. X. H. Zhang, Z. S. Wang, Y. Cui, N. Koumura, A. Furube and K. Hara, *J. Phys. Chem. C*, 2009, **113**, 13409–13415.

76. H. Imahori, S. Kang, H. Hayashi, M. Haruta, H. Kurata, S. Isoda, S. E. Canton, Y. Infahsaeng, A. Kathiravan, T. Pascher, P. Chabera, A. P. Yartsev and V. Sundstrom, *J. Phys. Chem. A*, 2011, **115**, 3679–3690.

77. Z. S. Wang, N. Koumura, Y. Cui, M. Miyashita, S. Mori and K. Hara, *Chem. Mater.*, 2009, **21**, 2810–2816.

78. A. Furube, Z. S. Wang, K. Sunahara, K. Hara, R. Katoh and M. Tachiya, *J. Am. Chem. Soc.*, 2010, **132**, 6614– +.

79. D. F. Watson and G. J. Meyer, *Coord. Chem. Rev*, 2004, **248**, 1391–1406.

80. K. Sunahara, A. Furube, R. Katoh, S. Mori, M. J. Griffith, G. G. Wallace, P. Wagner, D. L. Officer and A. J. Mozer, *J. Phys. Chem. C*, 2011, **115**, 22084–22088.

81. Y. Tian and T. Tatsuma, *Chem. Commun.*, 2004, 1810–1811.

82. Y. Tian and T. Tatsuma, *J. Am. Chem. Soc.*, 2005, **127**, 7632–7637.

83. K. Yu, Y. Tian and T. Tatsuma, *Phys. Chem. Chem. Phys*, 2006, **8**, 5417–5420.

84. Y. Nishijima, K. Ueno, Y. Yokota, K. Murakoshi and H. Misawa, *J. Phys. Chem. Lett.*, 2010, **1**, 2031–2036.

85. E. Kowalska, O. O. P. Mahaney, R. Abe and B. Ohtani, *Phys. Chem. Chem. Phys*, 2010, **12**, 2344–2355.

86. S. Link and M. A. El-Sayed, *J. Phys. Chem. B*, 1999, **103**, 8410–8426.

87. S. Link and M. A. El-Sayed, *Int. Rev. Phys. Chem.*, 2000, **19**, 409–453.

88. S. Link, A. Furube, M. B. Mohamed, T. Asahi, H. Masuhara and M. A. El-Sayed, *J. Phys. Chem. B*, 2002, **106**, 945–955.

CHAPTER 11

Carrier Dynamics in Photovoltaic Structures and Materials Studied by Time-Resolved Terahertz Spectroscopy

ENRIQUE CÁNOVAS,*[a] JOEP PIJPERS,[a]
RONALD ULBRICHT[a] AND MISCHA BONN[a,b]

[a] FOM-Institute AMOLF, Science Park 104, 1098 XG Amsterdam, The Netherlands; [b] Max Planck Institute for Polymer Research, Ackermannweg 10, 55128 Mainz, Germany
*Email: canovas@amolf.nl

11.1 Introduction

Currently, the fossil-fuel based production of the world's energy demand is far from sustainable. The massive emission of greenhouse gases has been linked to global climate change, with possible detrimental effects such as an increase in climate-related civil conflicts,[1] rising sea levels and reduced agricultural production in certain regions. This predicament will only get worse, with emerging economies consuming increasing amounts of (fossil-fuel based) energy. Renewable energy sources will undoubtedly play a major role in meeting the energy demand of the growing world population and in reducing the impact of energy production on the environment.

RSC Energy and Environment Series No. 8
Solar Energy Conversion: Dynamics of Interfacial Electron and Excitation Transfer
Edited by Piotr Piotrowiak
© The Royal Society of Chemistry 2013
Published by the Royal Society of Chemistry, www.rsc.org

The direct conversion of light to electricity (photovoltaics, PV) is a clean and technically well-established[2] method of producing electricity, which has the potential to play a major role in the realization of more sustainable energy production within the next decades.[3] To realize this prospect, cost reduction of solar energy is imperative. Reducing the PV costs can be achieved by developing cheaper solar cell devices[4,5] and/or by reaching higher efficiencies through novel solar cell concepts.[6–8]

Generally, Si or GaAs highly crystalline bulk semiconductors, forming p–n junctions, are used as light absorbing materials in photovoltaic (PV) devices. The employment of high purity crystals is directly correlated with higher device efficiencies but unfortunately also with higher device costs. In order to reduce the costs in PV devices, solar cells based on polycrystalline materials and thin films have emerged. These technologies can be produced at competitive prices because of lower processing temperatures[3] and hence costs. Alternatively, even cheaper PV concepts have been introduced recently, including solar cells based on semiconducting polymers,[9,10] dye-sensitized solar cells[5,11] (DSSCs) and quantum dot sensitized solar cells[12,13] (QDSSCs).

The upper theoretical efficiency limit for the solar cell approaches described above is $\sim 30\%$[14] (for a single absorption threshold device with band gap of ~ 1.2 eV). This limit is set by two intrinsic energy loss mechanisms in solar cells: (i) their inability to absorb below band gap photons and (b) their inability to avoid energy dissipation of photogenerated carriers excited well above the conduction and valence band edges (thermalization processes). Tandem solar cell geometries[15] have proven to surpass the 30% efficiency threshold, but their manufacturing cost is as yet incompatible with mass production.[3] In parallel, a number of novel PV concepts have been proposed to avoid the loss mechanisms described above,[7,8] among them, "hot carrier solar cells" in which "hot" carrier extraction has to compete favourably with carrier thermalization (resulting in higher cell photovoltage[16]) and solar cells that are based on carrier multiplication (CM), the process in which the excess energy of hot carrier is used to excite additional carriers over the band gap (resulting in higher photocurrents).[17]

It is evident that a full understanding – and possible application – of "hot carrier" and CM concepts in novel PV devices requires detailed knowledge of the photophysical processes that occur upon absorption of a photon in a potential building block for the said PV devices. In this chapter, we present time-resolved terahertz spectroscopy (TRTS) as a powerful tool for studying carrier dynamics and carrier transfer processes in semiconductor architectures that are used as building blocks for conventional and novel PV structures. The control of creating and detecting freely propagating terahertz (THz) pulses allows time-resolved far IR studies to be performed with sub picosecond time resolution in a non-contact fashion. Furthermore, the ability to measure both the amplitude and the phase of the THz transient electric field permits one to obtain information about the complex dielectric function of the material in the THz frequency region, yielding the complex index of refraction, or, equivalently, the complex conductivity.

11.2 Time-Resolved Terahertz (THz) Spectroscopy (TRTS)

In this section, the basic characteristics of time-resolved THz spectroscopy (TRTS) are discussed. Generation and detection of THz radiation is briefly introduced and the layout and operation of a common THz setup is described. Also, theoretical models suited to describe the response of carriers in the THz frequency window are presented for three different semiconductor geometries that are relevant for conventional and novel PV applications: (Section 11.2.2.1) bulk semiconductors, (Section 11.2.2.2) polycrystals and mesoporous oxide films and (Section 11.2.2.3) semiconductor quantum dots (QDs).

11.2.1 Terahertz Generation, Detection and Time-Resolved Terahertz Spectroscopy Setup

The frequency of 1 THz (10^{12} Hz) corresponds to a photon energy of 4.2 meV or a wavelength of 300 μm. Accordingly, THz frequencies lie between the infrared and microwave frequency regions of the electromagnetic spectrum (Figure 11.1). The THz window is generally considered to range from 100 GHz to 10 THz. As shown in Figure 11.1, different photon frequencies are suitable for probing specific light–matter interactions that are related to various physical mechanisms. Among them, the THz window is particularly suited to studying electronic motion in conducting materials.

Historically, the application of THz radiation for spectroscopic purposes has long been hampered by a lack of suitable emitters and detectors for which reason the term "terahertz gap" had been coined. The reliable generation and

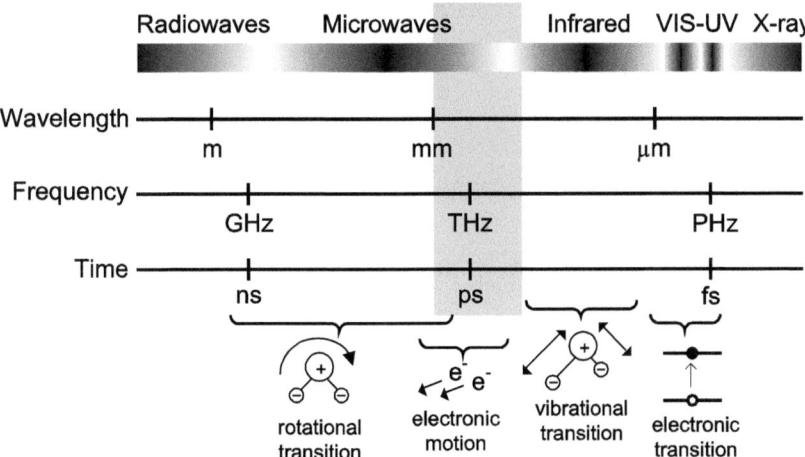

Figure 11.1 The electromagnetic spectrum from radio waves to the X-ray region. Rotational, vibrational and electronic transitions are shown along with electronic motion. The area accessible by THz spectroscopy is shown in shaded grey.

detection of freely propagating coherent THz pulses was achieved for the first time[18,19] in the late 1980s, opening the door for the employment of THz frequencies for spectroscopic characterization. After this technological breakthrough, methods for the emission and detection of THz radiation were optimized and employed in a pump–probe scheme for time-resolved studies in many different systems, including polar and non-polar liquids,[20–22] biological media,[23] semiconductors,[24–26] conducting polymers[27,28] and superconductors.[29] A particular feature of the generation and detection methods, which we will briefly introduce below, is that phase-stabilized single-cycle or few-cycle pulses are created which are detected in the time-domain, that is, the electric field of the THz pulse is recorded, thus providing direct access to the phase. As a consequence, not only the absorption properties of a material can be determined, but also the polarizability, that is, refraction.

In the following we will briefly introduce common methods to generate and detect THz radiation, drawing on the design of a typical TRTS setup, as used in our laboratory, of which the design is shown schematically in Figure 11.2. This setup is driven by an amplified Ti:Sapphire laser delivering an output of around

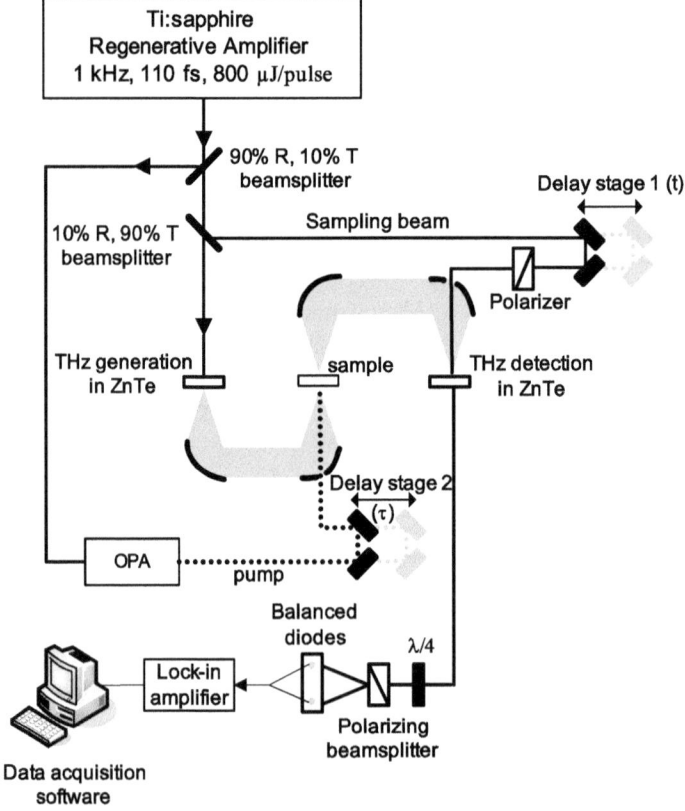

Figure 11.2 Schematic layout of a conventional time resolved THz spectroscopy setup.

0.8 W at a repetition rate of 1 kHz, with pulses centred around a wavelength of 800 nm and a duration of 110 fs.

The generation and detection of coherent THz radiation is commonly achieved in photoconductive antennae or by optical rectification in a nonlinear optical medium.[30] In both cases, the output of a femtosecond laser is split into two parts: the generation pulse and the detection (or gate) pulse. In the case of photoconductive antennae, the generation pulse is used to excite interband transitions resonantly in a voltage-biased semiconductor wafer (generally a low-temperature grown GaAs wafer with two metal strips deposited on top). Owing to the acceleration of the photoexcited electrons under the presence of the applied DC electric field, a THz electromagnetic transient is emitted that can propagate in free space. Probe bandwidths up to 5 THz can now routinely be generated by this method.[31,32] Photoconductive antennae are particularly suitable for ultrafast lasers such as Ti:Sapphire systems that deliver low-energy pulses at MHz repetition rates. For amplified laser systems that provide pulse energies in the μJ or mJ regime, like the one employed in our setup, optical rectification in a suitable nonlinear optical crystal is the most common choice.[33] In contrast to the resonant nature of the excitation process in photoconductive antennae, optical rectification is a non-resonant nonlinear optical process and the employed materials are thus capable of withstanding higher pump intensities without saturation or even crystal-damaging effects. Various THz spectral ranges are accessible by different crystals. Common choices are: $ZnTe$[34] (0–3 THz window), GaP[35] (2–7 THz window) and $GaSe$[36] (8–40 THz window) when pumped by a Ti:sapphire laser. Recently, ultrafast fibre lasers have been used as well.[37,38]

In our setup, 10% of the laser output is used as generation and detection beams. Most of this energy is split off and used as the generation beam by focusing it on a ZnTe crystal where THz radiation is generated by optical rectification. The diverging radiation is collimated and focused on the sample by a pair of off-axis parabolic mirrors. After transmission through the sample, the THz probe is again collimated and focused by another pair of parabolic mirrors on a second ZnTe crystal. Here, the THz beam is overlapped in space and time with the detection pulse that had been separated from the generation pulse before. As already mentioned, the THz pulse is detected in the time-domain by recording its electric field.

An example trace from our setup is shown in Figure 11.3 (left). The THz pulse resembles a single oscillation of an electric field. Fourier transforming the time-domain trace into the frequency domain yields the power spectrum and the frequency-dependent phase of the THz pulse, covering the spectral range from 0.2–2 THz (Figure 11.3 (right)). Similar to the two possibilities for generating THz, detection in the time-domain can be accomplished using either photoconductive antennae or nonlinear optical crystals. In the latter case, the mechanism is electro-optic sampling which is reminiscent of the Pockels effect but can also be seen as the inverse process of optical rectification.[39] Thus, a similar crystal as for the generation is usually used for detection. In electro-optic sampling, the detection pulse, which is much shorter that the

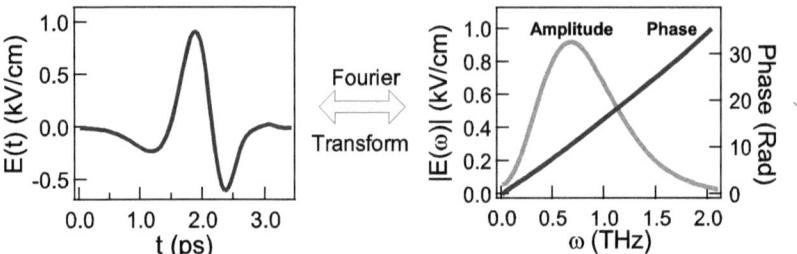

Figure 11.3   Typical waveform of a THz pulse propagating through air $E_{air}^{out}(t)$, measured by electro-optic sampling. Fourier-transforming the time-domain
waveform yields the spectral amplitude and the phase of the THz electric
field.

to-be-detected THz pulse (100 fs *vs.* 2 ps in our setup), is overlapped in time by
a certain portion of the THz electric field. Phase matching ensures that during
propagation of the two pulses through the detection crystal, the electric field of
the THz pulse experienced by the sampling pulse can be seen as constant. The
THz field induces a transient birefringence in the detection crystal which in turn
rotates the polarization of the detection beam. This change in polarization is
measured. Since it is proportional to the electric field strength it can be used the
retrieve the electric field at that particular point in time. The entire waveform of
the THz electromagnetic transient can then be resolved by varying the time
delay between the THz and the optical probe using a mechanical delay stage
(see Figure 11.2). The polarization of the detection beam can be measured
sensitively by splitting the *s* and *p* components of the detection pulse by a
Wollaston prism. The intensity differences of the two polarization components,
being proportional to the polarization rotation, are then detected by a pair of
balanced diodes.

Detection by means of a photoconductive antenna is done by monitoring the
changes in the antenna's photocurrent induced by the impinging THz field: as
in the generation case, the detection pulse excites charge carriers in the antenna.
However, now the carriers are not accelerated by an electrically applied DC
field but rather by the THz electric field. The measured photocurrent can then
be used to infer the THz electric field at that point in time.

In many THz experiments one wants to study the time evolution of the THz
response of a photoexcited system. For this purpose, in addition to the generation and detection pulses, a third beam (the pump or excitation pulse, see
Figure 11.2) is split off from the laser output and used to excite charge carriers
in the sample which are subsequently probed by the THz probe pulse. In our
setup, 90% of the output is used for this purpose. Variation of the time delay
between the two pulses by means of a second mechanical delay stage allows
study of the dynamics of the charge carriers after photoexcitation and in
principle following their relaxation back to equilibrium. The time resolution is
commonly in the sub-picosecond range and pump–probe delays, usually limited
by the length of the delay stage, of up to a few nanoseconds are possible.

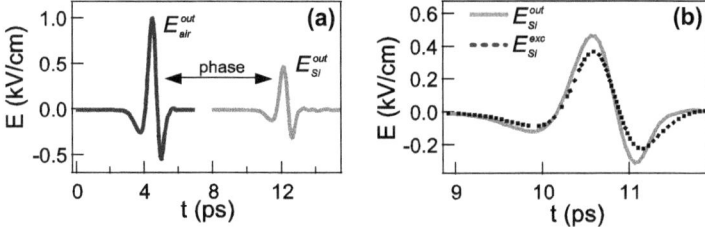

Figure 11.4   Measured waveforms in a typical THz-TDS experiment. (a) THz field
transmitted through air (black line) and unexcited silicon (grey line). The
scans contain both amplitude (absorption) and phase (refraction) information. (b) Transmitted THz fields through unexcited (grey solid line)
and excited (black dashed line) silicon, required to extract the real and
imaginary photoconductivity.

Note that TRTS is a contact-free technique, which is especially advantageous
when studying the photoconductivity in systems where it is difficult or even
impossible to apply electrical contacts. For a Ti:Sapphire laser system,
the output wavelength of 800 nm (1.55 eV) can be easily converted to 400 nm
(3.1 eV) and 266 nm (4.66 eV) by second and third harmonic generation,
respectively. In more sophisticated setups, optical parametric amplifiers (OPAs)
can be used to generate widely tunable pump energies, ranging from the
infrared to the ultraviolet.

The goal of THz spectroscopy is to measure the optical response of a material in the probe spectrum. As mentioned before, the phase-resolving nature
of the time-domain measurement allows the extraction of the extinction coefficient, that is, the imaginary part of the complex index of refraction n, but
also the refractive index, that is, the real part of n. This is nicely demonstrated
in Figure 11.4 (a)) which shows the recorded THz pulse with (right side) and
without (left side) sample inserted. The phase shift due to the inserted sample,
being proportional to the real part of the index of refraction and the sample
thickness, is clearly visible. The amplitude reduction is due to reflection losses
(and thus to the real part of n) and/or to absorption (and thus the imaginary
part of n). The analysis is, however, performed in the frequency domain[1] by
calculating the Fresnel coefficients, yielding the spectrally resolved complex
index of refraction $n(\omega)$, where ω is the angular frequency. This is done either
numerically or analytically for certain cases.[40,41] Such an analysis can be become quite complex, for instance in the case of samples containing various slabs
of materials.[42]

So far we have not considered the inclusion of the pump pulse that photo-
excites charge carriers: the method described above only yields the THz
response of the unexcited sample. In order to obtain the photoinduced change
in the refractive index of the material, one wishes to record the difference between the unexcited sample ($E^{out}(\omega)$ in Figure 11.4 b)) and the photoexcited

---

[1]Some methods however, which deal with intricate correlations in time-resolved measurements,
perform the analysis in the time-domain.

sample ($E^{exc}(\omega)$ in Figure 11.4 (b)). This is usually done by mechanically chopping the pump beam and recording the difference $\Delta E(\omega) = E^{out}(\omega)-E^{exc}(\omega)$ directly. From the ratio $\Delta E(\omega)/E^{out}(\omega)$, the Fresnel coefficients are calculated by taking into account the precise sample geometry.[40] Depending on the complexity of the sample and parameters, such as the penetration depth and material thickness, this has to be done numerically or can be performed in an analytical way with certain approximations.[40,43] Note that at early pump–probe delays (~ 1 ps) and in systems that rapidly change the photoinduced response (on timescales comparable to the probe duration, that is, ~ 1 ps for few-THz- bandwidth pulses), care has to be taken in the interpretation of the data. The time evolution of the system has then to be deconvoluted from the detector response by taking detailed grid scans and applying more elaborate algorithms involving 2D-Fourier transformation.[44,45]

It is common to express the THz response in terms of the complex dielectric constant $\varepsilon(\omega)$ or the conductivity $\sigma(\omega)$, rather than the complex index of refraction. The former is related to $n(\omega)$ via the relation: $n^2(\omega) = \varepsilon(\omega)$. The latter is most often used for photoexcited systems ("photoconductivity") and can be obtained from $\sigma(\omega) = -2\pi i \omega \varepsilon_0(\varepsilon_{exc}(\omega) - \varepsilon_{out}(\omega))$, where ε_0 is the free-space permittivity and $\varepsilon_{exc}(\omega)$ and $\varepsilon_{out}(\omega)$ are the dielectric constant of the excited and the unexcited sample, respectively. The real part of the conductivity is related to the absorption while the imaginary part is related to the polarizability.

11.2.2 Characteristic Terahertz Responses in Semiconductors

The properties of charge carriers in semiconductors depend on material morphology, temperature and material properties such as the crystal structure, band gap, dielectric function and carrier–lattice interaction. Much progress has been made in the past two decades in understanding the physics of elementary electronic excitations and carrier dynamics in the THz frequency range. A description of this work is outside the scope of the present chapter so we refer the reader to several reviews published on this topic.[46–48] In this section, we will only introduce different theoretical models which are suited to describing the THz response of charge carriers in semiconductor systems commonly employed in solar cell architectures.

11.2.2.1 Bulk Semiconductors

The majority of commercially available solar cells are based on highly crystalline bulk semiconductors. The conductivity of free charge carriers in many bulk semiconductors can be described by the Drude model. In this model, carriers are treated as an ideal gas of free particles interacting with the crystal lattice only by momentum-changing collisions. In the absence of an electric field, the movement of charge carriers has no preferential direction, exhibiting a random walk throughout the lattice (characterized by the mean free path in between collisions). Scattering from impurities and phonons are commonly the dominant processes that damp the motion of free carriers. At low temperatures, scattering from impurities is the dominating mechanism while at elevated temperatures carriers will predominantly scatter with phonons in the crystal lattice.

The carrier drift velocity $v_d(t)$ induced by a time-varying electric field $(E(t) = E_0 e^{i\omega t})$ and damped by scattering events (defined by a scattering time τ_s) can be obtained by solving the differential equation:

$$-\frac{e}{m^*}E(t) = \frac{\partial}{\partial t}v(t) + \frac{1}{\tau_s}v(t) \qquad (11.1)$$

resulting in:

$$v_d(t) = -\frac{e\tau_s}{m^*}\frac{1}{1 - i\omega\tau_s}E(t) = -\mu(\omega)E(t) \qquad (11.2)$$

where m^* is the carrier effective mass, e is the electron charge and $\mu(\omega)$ is the carrier mobility in the frequency domain. As shown in Equation (11.2), the mobility of carriers in semiconductors is defined by the carrier effective mass and the scattering time. While the former is largely an intrinsic material property, the latter is influenced by the quality of the sample (amount of defects per unit volume), the temperature (availability of phonons) and the carrier density (carrier–carrier interactions). From Equation (11.2) we can obtain the macroscopic electrical conductivity of the material as follows:

$$\sigma(\omega) = Ne\frac{v_d(t)}{E(t)} = Ne\mu(\omega) \qquad (11.3)$$

or equivalently:

$$\sigma(\omega) = \varepsilon_0\omega_p^2\tau_s\left(\frac{1}{1 + \omega^2\tau_s^2} + i\frac{\omega\tau_s}{1 + \omega^2\tau_s^2}\right) \qquad (11.4)$$

This relationship is called the Drude equation, where ε_0 is the vacuum permittivity and $\omega_p = \sqrt{Ne^2 / (m^*\varepsilon_0)}$ is the plasma frequency. Figure 11.5(a) plots the real and imaginary contributions to the Drude equation in the THz frequency window between 0 and 10 THz. One can see that the complex conductivity resembles a Lorentzian oscillator that is centred at zero frequency. The crossing of the imaginary part with the real conductivity at the full-width-half-maximum indicates the scattering rate, the inverse of the scattering time τ_s. This yields the carrier mobility μ through $\mu = e\tau_s / m^*$. The magnitude of the conductivity is determined by the plasma frequency, from which the charge carrier density can be inferred. Thus one is able to determine the two key parameters of conductivity (carrier density and mobility) independently, as opposed to, for instance, traditional photocurrent measurements where their product is measured. This can be particularly useful in materials with fast recombination channels such as trapping where a rapid reduction in charge density can give wrong mobility values. Figure 11.5(b) shows the conductivity of free carriers measured (squares) in a germanium wafer (under 400 nm pump excitation and at 50 ps pump–probe delay) in the THz probe window between

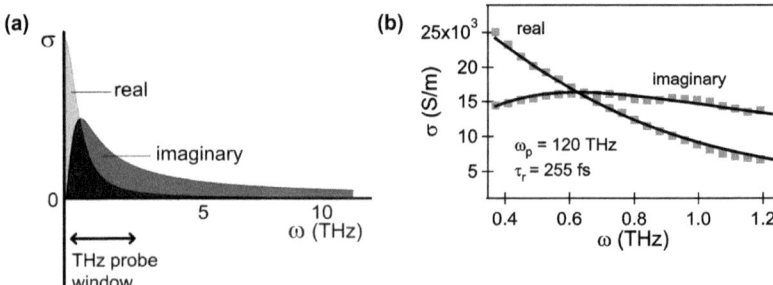

Figure 11.5 (a) Plot of the Drude equation as a function of the angular frequency. The resonance of free carriers is positioned at $\omega = 0$ and the frequency at which the real and imaginary conductivity cross, is determined by the scattering time: $\tau_r = 1/2\pi\omega_{cross}$. (b) Measured conductivity of free carriers in germanium (grey dots), after excitation with 400 nm pulses. Fitting the data (spheres) with the Drude equation (solid line) yields values for ω_p and τ_r.

0.4 and 1.2 THz and the fitting (lines) to the real and imaginary components in Equation (11.4). The fitting using the Drude equation describes the offset, the curvature and the amplitude of the two conductivity components perfectly and allows independent determination of the plasma frequency and carrier scattering time (shown in Figure 11.5(b)) for the sample under study.

In many situations, the Drude equation is a suitable description to model the electrical behaviour of carriers in a bulk semiconductor sample. However, it is worth noting that the Drude model as discussed above is based on a number of drastic simplifications, such as the neglect of Coulombic carrier–lattice interactions (polaron formation), carrier–carrier interactions and a distribution of scattering times owing to the occurrence of several scattering events (the Drude model only assumes one type of scattering process). Depending on the material, these effects might merely affect certain parameters (such as the effective mass in the case of polaron formation or the scattering time in the case of electron–hole interactions) while still leaving a Drude-like THz response.[49,50] In other cases such effects can significantly alter the nature of charge transport, for instance by changing from delocalized (Drude-like) to localized, as it is for instance found in polaronic hopping transport in organic semiconductors or other disordered systems.

11.2.2.2 *Polycrystals and Mesoporous Oxide Films*

Mesoporous oxide films, acting as electron acceptors, are an essential constituent of dye/QD-sensitized solar cells.[5] These films commonly consist of 10–50 nm diameter oxide particles (typical materials: TiO_2, SnO_2 and ZnO). Owing to sintering between the individual particles, the oxide network is a percolating pathway for electron transport. Although the individual building blocks of the films posses in principle bulk semiconductor character (*i.e.* the particle diameter is much larger than the exciton Bohr radius of the oxide material and no quantum confinement occurs), the THz conductivity of carriers within oxide mesoporous films does not follow the Drude model as in macroscopically-sized bulk semiconductors.[51-53] There are in principle two reasons for this. At first, the isotropic motion of charge carriers might be impeded by the presence of potential wells such as the particle boundaries in mesoporous oxide films. As a result, carriers do not scatter isotropically any more, as is assumed in the Drude model, but in preferential directions (away from the particle boundary). Often, an adaptation of the Drude model is used to take the effect of particle boundaries into account. This so-called Drude–Smith model[54] releases the constraint imposed by the Drude model of isotropic carrier scattering. In contrast, scattering events following preferential directions (*e.g.* from particle boundaries in the reverse direction) are included in the model. With this simple modification, the Drude–Smith model was shown to account successfully for the THz response of photogenerated carriers in, for example, polycrystalline silicon samples.[55,56] Recent Monte-Carlo simulations on microscopically-sized materials such as mesoporous oxide films have extended the capability to describe such systems in more detail than the Drude–Smith model by including more physical parameters.[57,58] It was shown that the precise nature of the conductivity response depends mainly on the mean free path and the thermal velocity of the charge carriers, the particle dimensions and the tunnel probability through the boundary.[59]

Even when preferential scattering events are negligible, which is the case when the particle dimensions are much larger than the carrier mean free path and the conductivity is thus locally Drude-like, the measured THz response differs from the Drude response. The reason is that it is not only the local dielectric response that determines the overall dielectric response, but also the dielectric contrast between the material of interest and its surroundings. When nano-structured semiconductors are not highly packed but instead are porous (like mesoporous oxide films), the connection between the local conductivities and the averaged conductivities measured in the far field requires the application of an effective medium theory. To estimate the conductivity of mesoporous films, it is necessary to define an effective dielectric function for a material consisting of two phases (*e.g.* semiconductor and air). A commonly used approach is the Maxwell–Garnett effective medium theory[60] which relates the dielectric function of the oxide particles, ε_p and of the surrounding medium, ε_m, to the *effective* dielectric function ε of the composite by:

$$\frac{\varepsilon - \varepsilon_m}{\varepsilon + 2\varepsilon_m} = f \frac{\varepsilon_p - \varepsilon_m}{\varepsilon_p + 2\varepsilon_m} \qquad (11.5)$$

where f is the filling fraction of the particles. The photoconductivity of carriers in a nanoporous medium can be obtained by inserting the Drude model (Equation (11.4) into Equation (11.5)), obtaining:

$$\frac{\varepsilon + {}^{i\sigma}/_{\omega\varepsilon_0} - \varepsilon_m}{\varepsilon + {}^{i\sigma}/_{\omega\varepsilon_0} + 2\varepsilon_m} = f \frac{\varepsilon_p + {}^{i\sigma_p}/_{\omega\varepsilon_0} - \varepsilon_m}{\varepsilon_p + {}^{i\sigma_p}/_{\omega\varepsilon_0} + 2\varepsilon_m} \qquad (11.6)$$

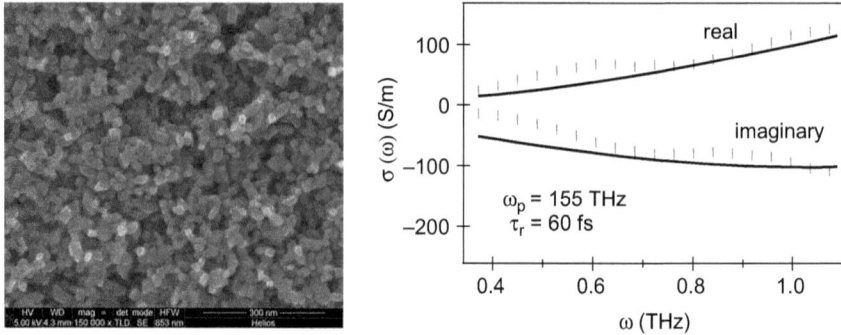

Figure 11.6   (a) SEM image of a nanostructured TiO_2 film (grain size ~ 40 nm). (b) THz conductivity of a nanostructured TiO_2 film (grey dots), measured after direct excitation with 266 nm light. The solid line is a fit to the Maxwell–Garnett effective medium theory (see text).

here, σ_p denotes the local (*i.e.* Drude) conductivity inside the particle and σ the measured averaged conductivity.

Figure 11.6(a) shows the SEM image of a sintered anatase TiO_2 film (particle diameter of 15 nm) and Figure 11.6(b) shows its characteristic conductivity response measured under a 266 nm pump. As one can see, the conductivity does not resemble the Drude model (as in Figure 11.5 (b)) but can be modeled using the effective medium approach. The solid line in Figure 11.6(b) represents the best fit to the Maxwell–Garnett effective medium theory (Equation (11.6)). Clearly, the combination of the Drude equation with effective medium theory represents a valid route for describing the conductivity in such complex systems, allowing the determination of the plasma frequency and carrier scattering time.

11.2.2.3   Semiconductor Quantum Dots (QDs)

The development of high quality semiconductor systems where quantum confinement occurs, manifested in the creation of discrete, "atom-like", electronic states rather than continuous bulk states, has boosted the interest of the PV research community in the implementation of solar cell geometries based on nanostructures. Particularly, colloidal quantum dots have emerged as potential candidates for developing inexpensive and highly efficiency concepts,[61] including tandem solar cells,[62] hot carrier solar cells[63] and cells based on carrier multiplication.[64]

The quantum confinement in QDs has a dramatic effect on their optoelectronic properties compared to their bulk counterparts. One particular feature is the tunability of the band gap via the QD size, that is, the band gap increases with decreasing QD size. At the same time, the electron and hole energy states become discrete (now commonly called intraexcitonic levels) and their relative energy separations increase with decreasing QD size. Typical energy level separations of tens of meV for hole levels and 100s of meV for electron energy levels are characteristic in most semiconductor QDs. Since now no continuous electronic states are available any more, electric conduction in the classical sense cannot occur, the local conductivity is thus not Drude-(or Drude–Smith)-like. In fact, the large level spacing (> 10 meV; 1 THz = 4 meV) implies that no real conductivity is measured but only an imaginary part that is increasing with frequency, as can be seen in Figure 11.7(b). This can be understood by realizing that the THz pulse is now probing intraexcitonic transitions non-resonantly. In the electromagnetic spectrum these intraexcitonic transitions can be seen as a series of Lorentzian resonances. This is schematically shown in Figure 11.7(a) where the Lorentzian peak of the lowest intraexcitonic level is plotted in terms of the conductivity. Since usually the THz probe bandwidth reaches up to only several THz (1 THz = 4 meV), its energy is insufficient to resonantly drive intraexcitonic transitions within QDs (the THz spectrum is indicated in Figure 11.7 by the double sided arrow). As a consequence, the absorption peak cannot be detected and no real photoconductivity is observed. In conventional, non-phase-resolved measurements this "would be it".

Owing to the time-domain approach in THz spectroscopy, one is however able to detect the phase-shift that is associated with a transition that is driven off-resonantly, which is manifested here as a negative imaginary conductivity component.[65–67] The THz probe thus only polarizes the exciton and this polarizability can be calculated. For this, one has, in principle, to sum over all intraexcitonic transitions although it turns out that generally the lowest energy transition accounts for the majority. The exciton polarizability is then given by:

$$\alpha_{exc}(\omega) = \alpha_{0,exc} \frac{\omega_0^2}{\omega_0^2 - \omega^2 - i\gamma\omega} \tag{11.7}$$

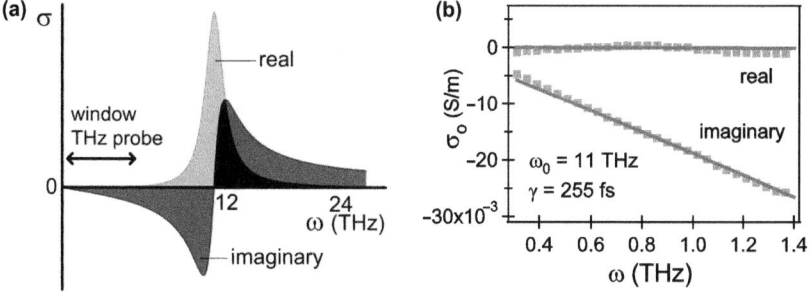

Figure 11.7   (a) shows a schematical representation of the conductivity of the lowest energetic intraband transition for a QD. In the frequency probe region of few THz, the light is not absorbed by carriers in QDs, giving no real photoconductivity. On the other hand, polarization effects play a role and are manifested in the imaginary component. The measured THz conductivity data of a colloidal suspension of InAs QDs after photo-excitation (800 nm pump) is shown in (b) and fitted by Equation (11.8). From the fitting an estimation of ω_0 and γ can be obtained.

where α_0 is the static (DC) polarizability associated with the intraexcitonic transition, ω_0 is the intraexcitonic transition energy defined by the spacing between the intraband energy level and the ground state and γ is the absorption line width.

Inserting this into an effective medium approach yields the measured conductivity σ_0:

$$\sigma_0(\omega) = \frac{-i\omega\varepsilon_0(\varepsilon + 2)^2\alpha_{exc}(\omega)N}{(9 - (\varepsilon + 2)\alpha_{exc}(\omega)N)} \tag{11.8}$$

where ε_0 is the vacuum permittivity, ε is the macroscopic dielectric constant, N is the number of probed QDs (considering one photogenerated electron–hole pair per QD) and α_{exc} is the QD polarizability of an excited QD. In this way the exciton polarizability can be inferred. As an example, the measured THz conductivity data for a colloidal suspension of InAs QDs after photoexcitation (800 nm pump) is shown in Figure 11.7(b) with the fit to Equation (11.8).

The exciton polarizability has been measured as a function of QD size for several materials, such as CdSe,[66] PbSe[67] and InAs.[68] It was found that the polarizability scales with the QD radius R between $R^{3.6}$ and R^4.

11.3 TRTS Carrier Dynamics Studies of High Efficiency Photovoltaic Concepts

As briefly commented in the introduction, the thermalization of carriers photoexcited well above the semiconductor band edges represents one of the intrinsic fundamental energy losses in solar cells. For this reason, novel PV concepts have been proposed to avoid this loss mechanism and hence to obtain device efficiencies beyond the Shockley and Queisser limit of $\sim 30\%$.[14] In one of these concepts, the so-called hot carrier solar cell,[16] thermalization losses in the absorber are avoided by extracting hot electrons (by means of suitable, energy-selective contacts) before they can thermalize. Theoretically, these devices are able to produce higher photovoltages and higher light-to-current conversion efficiencies ($>60\%$). The second proposed concept is based on a process called carrier multiplication (CM) (also known as impact ionization). In this process, the excess kinetic energy of hot electrons or holes is utilized to excite additional carriers over the band gap,[17] resulting in higher photocurrents.

Solar cells based on hot carrier extraction and CM rely on precise control of hot carrier relaxation were expected to be realized in nanostructured semiconductors (e.g. QDs) because of enhanced carrier–carrier interactions and discretized energy levels.[61] As will be shown below, TRTS is capable of probing charge carrier dynamics at early times after photoexcitation, including intraband relaxation and CM in bulk materials and quantum dots. As such, TRTS represents a powerful technique for evaluating novel semiconductor systems that may be used in the design of more efficient solar cells.

11.3.1 Relaxation of Hot Carriers in Semiconductor Quantum Dots

As we commented before, the search for suitable PV absorbers where carrier cooling is slowed down or inhibited can have important implications in the development of hot carrier solar cells.[16] Semiconductor QDs have been proposed as suitable candidates for hot carrier absorbers owing to their theoretically expected slow carrier cooling (resulting from the "phonon bottleneck" effect). The phonon bottleneck in QDs was postulated as a consequence of the discretization of energy levels in QDs. As the characteristic spacing between QD energy levels (hundreds of meVs) is large compared to the LO phonon frequency (~ 25 meV), electron thermalization via electron–phonon coupling in QDs will require multiphonon relaxation, which is a process with low probability. While many works have reported bottleneck effects in colloidal and solid state QDs, there are also many reports that report the absence of a phonon bottleneck.[69] Hence, the subject is still controversial and dependent on material selection and preparation (*e.g.* different surface chemistry defects). The controversy seems to have been influenced at least in part by the proper definition of what the bottleneck effect implies. In the limit of an ideally efficient bottleneck effect in QDs an infinite relaxation time for hot electrons within QDs can be expected (and monitored, for example, as total quenching of their luminescence). However, one can consider a less restricted definition implying that the bottleneck is manifested in QDs if slower carrier cooling is obtained compared with its bulk counterpart. For solar cell applications (hot carrier solar cells) a sufficient requirement for the "phonon bottleneck" to be useful is that the hot carrier relaxation time is long compared to the timescale necessary for carrier extraction.

A number of ultrafast carrier dynamics studies have been performed in colloidal dots.[70–72] In these works ultrafast electron relaxation times of hundreds of femtoseconds were reported and attributed to an Auger process for electron relaxation bypassing the bottleneck effect. TRTS has been successfully applied in combination time-resolved photoluminescence to study the phonon bottleneck effect in CdSe QDs.[73] In this work, thanks to the employment of THz frequencies, the authors were able to probe the rate of hole cooling following photoexcitation of the QDs. It was shown that the hole relaxation rate critically depends on the electron excess energy. These results, constituting a quantitative measurement of electron-to-hole energy transfer, proved that in colloidal CdSe QDs the phonon bottleneck effect can be bypassed by an Auger like recombination (as suggested before by other authors). In this process the excess electron kinetic energy after photoexcitation is given to a hole which can undergo phonon relaxation owing to the lower discretization of hole energy levels in CdSe QDs.

The previous results imply, in any case, that slow carrier cooling can be achieved in QD materials with strong discretization of energy levels for both electron and holes (such as, for example, lead salts[74]), in QDs surrounded by ligands acting as hole scavengers or potentially in heteronanocrystals (*e.g.* core–shell structures) where electron and hole are spatially separated and hence electron–hole energy transfer is inhibited.[75]

11.3.2 Carrier Multiplication (CM) in Semiconductors

Carrier multiplication (CM) is the process in which the absorption of a single high-energy photon results in the creation of multiple photogenerated electorn–hole pairs. In bulk material, CM is often referred to as impact ionization. The generally accepted model is that an initially generated "hot" carrier can relax to the valence or conduction band edge by two competitive processes of relaxation. First, hot carriers can relax via sequential emission of phonons (heat generation) but an alternative relaxation pathway for hot carrier relaxation is impact ionization. In this Auger process, the excess kinetic energy of the initially excited electron is employed to excite a second electron over the band gap, rather than being converted into heat. The rate of impact ionization is described by Fermi's golden rule and is determined by Coulombic coupling between initial (hot carrier) and final (bi-exciton) states and by the density of final states. In bulk materials, unfortunately, relaxation of hot carriers via phonon emission is faster than relaxation via impact ionization. As a result, CM has been shown to be relatively inefficient for bulk materials like silicon[76] and germanium[77] at visible photon energies and impact ionization does not contribute significantly to higher photocurrents in bulk semiconductor solar cells.

However, CM has been argued to be more efficient in QDs owing to quantum-confinement effects causing (i) a slowing of the phonon-mediated relaxation channel[61] and (ii) enhanced coulomb interactions,[64] resulting from forced overlap between wave functions and reduced dielectric screening at the QD surface.[78] The expectation of high CM factors (defined as the number of photogenerated excitons per absorbed photon) in QDs was fulfilled by a first report on efficient CM in PbSe QDs by the Klimov group.[64] After this initial report, several femtosecond spectroscopy studies have revealed high CM factors in PbS and PbSe,[64,79–83] PbTe,[84] CdSe,[85] Si,[86,87] and InAs[88,89] QDs. The highest CM factor was reported for PbSe QDs, with a demonstrated yield of up to seven carriers per photon.[81] More recently, however, the reported high CM factors in QDs have been questioned by several groups.[90–92] As will be shown in the next two sub-sections, TRTS studies have contributed to a better understanding of CM in QDs by assessing the CM factor in InAs QDs[88,93] and by determining the CM factor in bulk PbSe and PbS, two model materials in CM QD studies that allow a quantitative comparison between the occurrence of CM in QDs and bulk material.[94,95]

11.3.2.1 CM in Quantum Dots

CM in QDs is usually demonstrated by signatures of multi-exciton recombination (MER) (indicating the presence of multi-excitons) in the limit of low excitation densities (<<1 absorbed photon per QD). MER can be monitored by probing the transient population of excitons in transient absorption (TA) measurements[64,79,81,82,85,91] or by monitoring the radiative emission of (multi)excitons in time-resolved luminescence measurements.[92] TRTS can also be used to assess CM in QDs by monitoring the transient exciton population via the exciton polarizability (*i.e.* effectively probing the population of the $1S_{3/2}$ hole level, see Section 11.3.1). Figure 11.8(a) illustrates the nature of the signal in these different experimental approaches.

In a typical time-resolved CM experiment (TA, TRTS or time-resolved luminescence), QDs are excited by photons that have more than twice energy the QD energy gap. The reason that photons should have *more* energy than twice the QD energy gap is that the excess energy of the photon is usually distributed over the electron and the hole. Hence, a photon with exactly twice the energy of the QD 1S transition does not usually create hot carriers with sufficient energy for CM to occur. The demonstration of CM relies on the presence of a MER signature in the limit where at most one photon was absorbed per QD. This low excitation density limit ($\ll 1$ photon absorbed per QD, on average) is typically achieved by exciting the sample with very low pump fluence. It is crucial that the excitation fluence is sufficiently low, since multi-excitons can also be created by sequential absorption of multiple photons.[88]

Figure 11.8(b) illustrates two possible outcomes for a typical TRTS measurement to determine the occurrence of CM in QDs. In case the absorption of a single high-energy photon results in the formation of a single exciton (no CM, black dashed line), the signal is a step-like function since the radiative lifetime of single excitons is much larger ($\sim$ tens of ns to µs) than the time window of the ultrafast experiment. In contrast, in the case of efficient CM, the absorption of a single high-energy photon results in the generation of multi-excitons. Multi-excitons decay to the single exciton state on a timescale of tens to hundreds of picoseconds (grey solid line). The CM factor, η_{CM}, is determined by dividing the signal right after excitation (S_1) by the long lived signal (S_2).

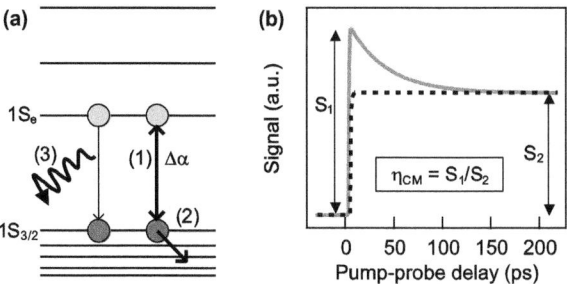

Figure 11.8 (a) Different experimental approaches to monitor multi-exciton recombination in CM studies: (1) TA, probing the bleach ($\Delta\alpha$) of the 1S exciton transition, (2) THz-TDS, probing the population of the $1S_{3/2}$ hole level via the exciton polarizability and (3) time-resolved luminescence probing the radiative emission originating from the $1S$ exciton transition. (b) Two possible outcomes of a CM experiment, as explained in the text.

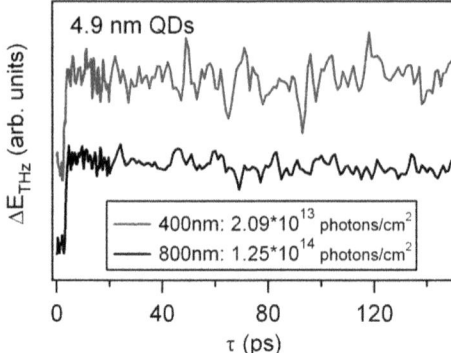

Figure 11.9 Comparison of transient TRTS data for 400 nm (grey line) and 800 nm (black line) excitation for 4.9 nm QDs. The 400 and 800 nm fluences are roughly equivalent, *i.e.* the average number of absorbed photons per QD is similar.
Reprinted with permission from Pijpers *et al.*[93] Copyright 2008 American Chemical Society.

TRTS has been employed to study CM InAs QDs of various sizes.[88,93] Figure 11.9 shows TRTS data for 4.9 nm InAs QDs excited by low-fluence 400 nm and 800 nm excitation pulses. The photon energy of a 400 nm photon corresponds to 2.74 times the energy gap, hence CM is in principle allowed from the point of view of energy conservation. It is apparent from the data that there is no significant bi-exciton decay visible for excitation at low 400 nm fluence, pointing to the absence of CM. The observation of no or negligible CM in InAs QDs is consistent with the report by Ben-Lulu *et al.*[91] but is in disagreement with other reports on CM in InAs QDs.[88,89]

These contradicting reports are representative of the controversy that emerged on the occurrence and efficiency of CM in QDs. The key question in this controversy was to what extent quantum-confinement effects lead to enhancement of CM in QDs. TRTS experiments contributed to answering this question by assessing CM factors in bulk materials, allowing a quantitative comparison between CM in bulk and QDs, as will be shown in the next section.

11.3.2.2   CM in Bulk Semiconductors

The contradicting literature values for CM factors in QDs (see previous section) resulted in a controversy about the efficiency of CM in QDs. In addition to discrepancies in experimental reports, tight binding calculations[96] suggested that CM factors in QDs were not only not enhanced relative to bulk, but were actually lower. Answering the key question in the controversy – whether the CM factor increases due to quantum-confinement effects– required a reliable comparison between CM factors in bulk and QDs. Remarkably, reliable numbers for CM factors in bulk materials were sparse. PbSe and PbS are arguably the most important materials in the CM discussion, since their small bulk band gap values result in an optimal energy gap for QDs of these materials to utilize the excess energy of visible photons of the solar spectrum. However, reports of bulk CM factors in PbSe and PbS were, respectively, absent or dated.[97]

Whereas CM in bulk materials is usually determined in photocurrent device measurements, that is, by collecting the carriers, CM in QDs is studied by (optical) spectroscopic measurements, in which the orbital occupation of the QDs is probed on ultrafast (picosecond) timescales. Hence, the commonly used experimental procedures to determine CM in QDs (ultrafast spectroscopy) and in bulk (device measurements) are rather different. While time-resolved optical and IR spectroscopies are ideally suited to probe carrier populations in colloidal QDs,[64,79,81] light of terahertz (THz) frequencies interacts strongly with free carriers in the bulk material and allows the direct characterization of carrier density and mobility.[98–100] From THz-time domain spectroscopy (TDS) experiments, one can quantitatively assess the number of photogenerated carriers in bulk semiconductors picoseconds after the light is absorbed. Additionally, as a result of the contact-free nature of the THz probe, it is possible to determine the CM factor in isolated samples of bulk semiconductors without the need to apply contacts, which is necessary in the device measurements. For these reasons, THz-TDS experiments have been employed to quantify CM in bulk PbSe and PbS on ultrafast timescales[94] in order to make a bulk-QD comparison in the context of the CM controversy. The CM factor in bulk PbS and PbSe was determined for excitation with various photon energies from the UV to the IR.

The reliable determination of CM factors in single crystalline PbSe and PbS films requires the accurate determination of two parameters: the number of absorbed photons per unit area and the number of generated carriers. In Pijpers *et al.*,[94] a homogeneous photon flux was achieved using a diffuser and the fluence at the sample position was determined with $\sim 3\%$ accuracy using five calibrated pinholes with increasing area. To calculate the number of absorbed photons per mm^2, the reflective losses at the sample surface were taken into account by using the frequency-dependent optical properties of bulk PbS and PbSe. Second, the number of generated electron–hole pairs was determined using TRTS measurements, by inferring the carrier density N from the Drude fits of photoconductivity. Figure 11.10 shows the complex conductivities $\sigma\omega$ for PbS ($E_{gap} = 0.42$ eV) following excitation with two different 266 nm (4.66 eV) fluences, as inferred from the Fourier transforms of the time-domain fields.[101]

The black dots in Figure 11.10 represent the photoconductivity that was measured at an excitation fluence of 15.2 nJ mm^{-2} 266 nm light (corresponding to 9.56×10^9 absorbed photons/mm^2). The data in Figure 11.10 (measured at a pump–probe delay of 10 ps) was fitted with the Drude expression (see Section 11.2.2.1), yielding a plasma frequency of $\omega_p = 3.30 \times 10^{14}$ Hz. From the value of ω_p, a carrier density of 6.85×10^{24} m^{-3} is calculated, given the electron and hole effective masses ($m_e \sim m_h \sim 0.20$ at room temperature[102]). Using the 266 nm penetration depth of PbS, the carrier density is converted into a sheet density of

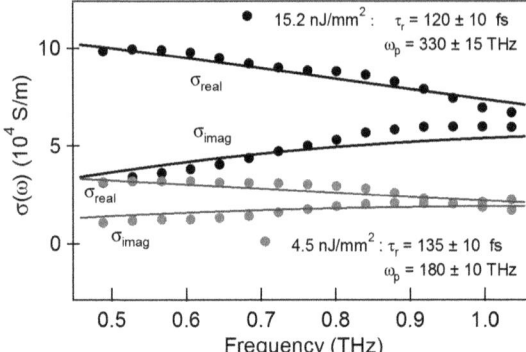

Figure 11.10   Complex, frequency dependent conductivities for PbS measured 10 ps after photoexcitation for two different 266 nm excitation fluences (black and grey dots). The data is described well by the Drude expression (solid lines) yielding the plasma frequency (directly related to density) and the carrier scattering time.
Reprinted with permission from Pijpers *et al.*[94] Copyright 2009 Nature Publishing Group.

2.92×10^{10} electron–hole pairs per mm^2. Here, it was taken into account that in PbS, both electrons and holes make comparable contributions to $\Delta E(t)$ because of their similar effective masses. Dividing the sheet density by the number of absorbed photons per mm^2 yields the CM factor η, the number of electron–hole pairs per absorbed photon. For the $15.2\,nJ\,mm^{-2}$ data in Figure 11.10, η was found to be 3.05 ± 0.15 electron–hole pairs per absorbed photon. This value clearly demonstrates the occurrence of CM in bulk PbS for excitation with 266 nm photons (having eleven times the energy of the PbS band gap). Performing the same procedure for the $4.5\ nJ\ mm^{-2}$ fluence data (grey dots in Figure 11.10), one finds a value for η of 3.06 ± 0.15 electron–hole pairs per absorbed photon. The good agreement between the found CM factors illustrates the validity of the procedure for a range of fluences. The same analysis for 800 nm excitation results in an average of 1.02 ± 0.05 generated electron–hole pairs for each absorbed photon, indicating that CM does not occur significantly for 800 nm excitation (3.7 times the band gap energy).

The CM factor for bulk PbSe is given in Figure 11.11 for a range of excitation wavelengths along with literature values for PbSe QDs. Figure 11.11 contains a selection of the literature that was published on CM in QDs between 2008 and 2011. Earlier QD reports are not included in Figure 11.11, since it has been identified that several challenges are associated with reliably measuring CM in QD colloidal suspensions.[88,91] These challenges include QD photocharging (which can be prevented by stirring the sample) and good control over the excitation fluence to prevent multi-photon excitation. It is evident in Figure 11.11 that the CM factors observed in these reliable QD reports are lower than or at most equal to our bulk values. This observation indicates

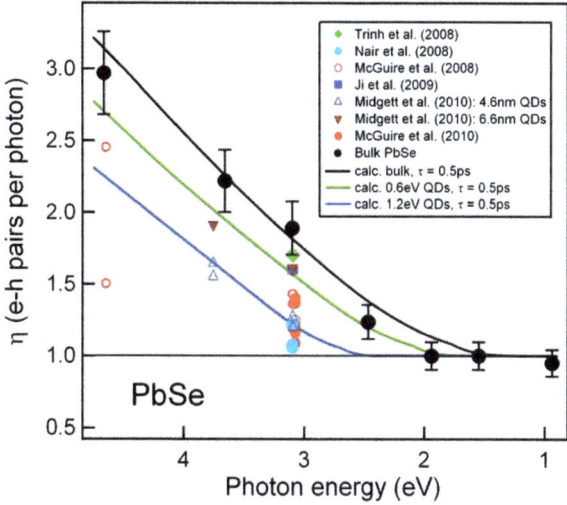

Figure 11.11 Comparison of CM in bulk PbSe (black dots) and PbSe QDs reported in the literature (different coloured markers). Tight binding calculations based on impact ionization for a range of phonon-assisted relaxation times (τ_h, dotted lines) yield CM factors that are in good agreement with our experimental results. Coloured solid lines represent tight binding calculations for PbSe QDs with 0.6 and 1.2 eV energy gap, for $\tau_h = 0.5$ ps. Error bars indicate 95% confidence level. Reprinted with permission from Pijpers *et al.*[94] Copyright 2009 Nature Publishing Group.

that quantum-confinement effects do not result in enhancement of the CM factor.

Although the number of generated excitons is smaller in QDs than in the corresponding bulk semiconductor at a given photon energy, this does not directly imply that there is no motivation to use QDs in solar cells for their CM abilities. For PV applications, the CM factor is not the quantity of interest because the energy of the excitons is not the same in QDs and in the bulk. Rather, a more relevant quantity is the energy efficiency, $\Phi(h\nu)$, defined as the ratio of the total excitonic energy (the number of excitons times the energy gap) and the photon energy $h\nu$. As such, the energy efficiency corresponds to the fraction of the photon energy that is transformed into excitons, that is, chemical energy, instead of heat after relaxation of the carriers. In Figure 11.12, the energy efficiency is plotted (by definition valued between 0 and 1) including (solid lines) and omitting (dashed lines) CM effects. At the gap energy, the energy efficiency is 1 since no energy is lost via "heat generation". For increasing photon energies, the energy efficiency decreases until CM sets in, so that energy efficiency increases again at higher energies. Interestingly, the energy efficiency above the gap is much larger in small QDs than in the bulk. For PbSe QDs with a gap of 1.2 eV, the energy efficiency is always larger than 50%. Also, the contribution of CM to the energy efficiency is largest for the

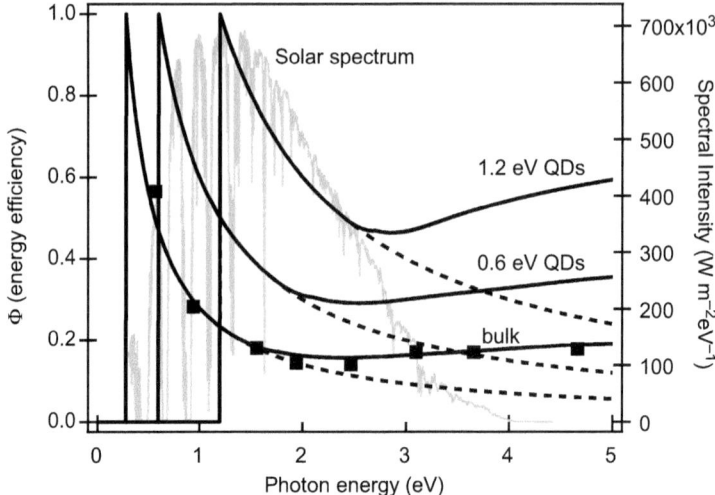

Figure 11.12   Energy efficiency of light absorption *versus* photon energy *hv*. Thick
solid lines: simulations ($\tau_h = 0.5$ ps) for bulk PbSe ($\varepsilon_g = 0.28$ eV) and
for two QDs ($\varepsilon_g = 0.6$ eV and $\varepsilon_g = 1.2$ eV). Dashed lines: simulations in
absence of CM, *i.e.* when excited carriers can only relax by emission of
phonons. Black squares: TRTS results for bulk PbSe. The grey shaded
region corresponds to the AM1.5 solar spectrum. Reprinted with
permission from Delerue *et al.*[95] Copyright 2010 American Physical
Society.

smallest QDs. Apparently, even if the number of generated excitons is smaller
in QDs than in the bulk, this is more than compensated by the increase in the
excitonic energy owing to the larger gap in QDs. Therefore strongly confined
QDs are energetically relatively efficient for photons with energy above
their gap.

Despite the observation that PbSe and PbS QDs exhibit higher energy efficiencies than bulk material, the added value of CM for photovoltaics seems
limited. This follows from averaging the calculated energy efficiency for the
solar spectrum. The values for the averaged energy efficiency, Φ^*, are obtained
by multiplying the energy-dependent energy efficiencies in Figure 11.12 with the
unconcentrated AM 1.5 reference solar spectrum using the following formula:

$$\Phi^* = \frac{\int I(hv) \cdot \Phi(hv) \cdot d(hv)}{\int I(hv) \cdot d(hv)} \tag{11.9}$$

where $I(hv)$ is the energy-dependent spectral intensity of solar light. In
this calculation, the assumption is made that each photon with energy
above the gap is absorbed and that each generated carrier contributes to the
photocurrent at the maximum voltage (not true for an operating solar cell). The
absolute increase in the averaged energy efficiency induced by CM is 9% for
bulk PbSe, but such small gap devices are inherently inefficient. For a realistic
gap of a PV device (~ 1.2 eV), the absolute CM-related gain in the light to

current conversion efficiency is limited to 2% when using PbSe QDs. This relatively small value is due to the fact that most of the photon flux in the solar spectrum is below the CM threshold (see Figure 11.12). Furthermore, it might be challenging to harvest the additional carriers from the 1.2 eV PbSe QDs because of the short lifetime ($\sim$ 50–100 ps) of the bi-exciton state in QDs.[103] Therefore, the benefit of CM in presently available QDs seems minor for PV applications, in spite of the relatively high energy efficiency of CM in PbSe QDs.

11.4 Interfacial Electron Transfer in Photovoltaic Structures Probed by Time-Resolved Terahertz Spectroscopy

A key feature of many emerging low cost thin film solar cells technologies is the excitonic character in their opto-electronical properties, which means that these are referred to sometimes as excitonic solar cells.[104] In excitonic solar cells, the absorption of a photon initially results in a neutrally charged bound electron–hole pair (*i.e.* an exciton) instead of free electrons and holes (as occurs in conventional p–n junctions). These excitons need to be dissociated into free carriers, so that an electrical current (consisting of free electrons and holes) can be generated in the solar converter. This particularity has important implications in the design and geometry of devices, which should include tailoring of the physicochemical interfacial charge separation processes taking place between donor and acceptor phases.

In this section, we present TRTS as a suitable tool to be employed for the study of exciton dissociation in excitonic solar cells. TRTS has been employed to study the timescale of charge separation in QD- and dye-senstized mesoporous oxide films. Finally, TRTS studies on carrier transport in QD super-lattices (also known as QD solids) are briefly discussed and connected to PV concepts.

11.4.1 Dye Sensitizing Mesoporous Oxide Films

In 1991 O'Reagan and Gratzel introduced the dye sensitized solar cell (DSSC)[5] as a low cost alternative PV concept to conventional p–n junction based solar cells. In DSSCs, a high surface area, dye-sensitized mesoporous oxide film (usually TiO_2) acts as anode. After photon absorption by the dye molecules, electrons are injected from the excited dye into the oxide acceptor and the remaining positively charged dye molecules are reduced by a suitable redox couple in the electrolyte, which is interpenetrated in the mesoporous oxide film. The oxidized redox species migrates towards the photocathode, where it is reduced by electrons that re-enter the system via the external circuit. The mechanism of charge separation in DSSCs differs from charge separation in conventional p–n junctions, where the space charge region generated between the n and p type semiconductors produces an electric field

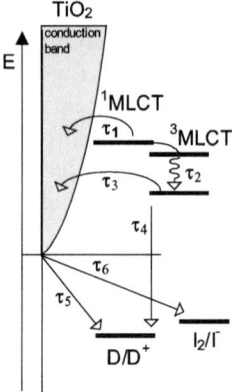

Figure 11.13 Schematic of the energy levels in a dye-sensitized TiO$_2$ anode. All possible dynamic processes after photoexcitation of the dye are indicated by arrows (see explanation in text).
Reprinted with permission from Pijpers *et al.*[120] Copyright 2011 American Chemical Society.

causing photogenerated electron–hole pairs to separate.[3] Instead, in DSSCs, charge separation occurs via exciton dissociation at the dye–oxide surface,[11,104] the driving force for which is the gain in chemical potential. Assessment of the timescale for exciton dissociation is crucial for a thorough understanding of a DSSC.[105] Obviously, the interfacial electron transfer from the dye into the oxide needs to occur on a faster timescale than the radiative lifetime of the excited dye molecule.

An illustration of the different dynamic processes in a dye-sensitized TiO$_2$ oxide film after photoexcitation is given in Figure 11.13. As a result of absorption of a visible photon by the dye, an electron is promoted from the ground state (D) to the singlet excited state (^{1}MLCT). Photoexcited electrons in the ^{1}MLCT state can be injected into the oxide (characterized by τ_1). Alternatively, the ^{1}MLCT state can undergo intersystem crossing to the excited manifold of the triplet ^{3}MLCT state, followed by intramolecular vibrational relaxation to the ^{3}MLCT ground state (characterized by τ_2). Significant injection from the singlet state into TiO$_2$ will only occur when τ_1 is small compared to intersystem crossing time and τ_2. In the situation where $\tau_1 > \tau_2$, the majority of the excited molecules in the ^{1}MLCT state will decay to the ^{3}MLCT state. From here, electrons can either be injected into TiO$_2$ or recombined in the ground state (characterized by τ_3 and τ_4, respectively). The injected electrons in TiO$_2$ will thermalize to the conduction band edge, from where they can recombine with the absorbed cationic dye molecule (D$^+$) or with the redox couple in the electrolyte (characterized by τ_5 and τ_6, respectively).

The general theory for predicting the rate of electron transfer between two chemical species was developed in the 1960s by Marcus,[106] amongst others. Later on this theory has been adapted to describe electron transfer between dyes and bulk semiconducting oxides,[107] yielding the following equation for the transfer rate constant (k_{ET}):

$$k_{ET} = \frac{2\pi}{\hbar} \int\limits_{-\infty}^{\infty} dE \rho(E) \frac{|H(E)|^2}{\sqrt{4\pi\lambda k_B T}} \exp\left(-\frac{(\lambda + \Delta G - E)^2}{4\lambda k_B T}\right) \qquad (11.10)$$

Three parameters determine k_{ET}: (i) the electronic coupling between electron donor and acceptor (represented by the electronic coupling matrix element, $|H(E)|^2$; $\rho(E)$ being the density of semiconductor states at energy E from the conduction band edge), (ii) the energy difference between donor and acceptor levels (defined as ΔG) and (iii) the reorganizational energy (λ) which accounts for energy fluctuations in the system due to the charge transfer process. The dynamics of electron injection from ruthenium-based dyes into mesoporous oxide materials have been studied by a number of experimental techniques. Commonly the adsorbed dyes are excited with visible (femtosecond) pump pulses with wavelengths ranging from 400–630 nm. In several works[108,109] the injection of electrons from excited dyes into oxides was monitored by probing the emergence of the spectral signature of the cationic dye (concurrently produced with injection) at near-infrared frequencies. Alternatively, mid-infrared frequencies (~ 5 µm wavelength) were used to probe the transient population of injected electrons in the oxide films directly.[110–112] Also, TRTS measurements have been performed on dye-sensitized TiO$_2$ films and injection dynamics were found to take place on a (sub)picosecond timescales,[113–116] in correspondence with reported electron injection timescales obtained by other femtosecond spectroscopic techniques. Similar to TA measurements with a mid-infrared probe,[107,117] TRTS provides direct evidence of injected electrons since ΔE_{THz} corresponds directly to the emerging population of free carriers in TiO$_2$, in contrast to TA measurements with a visible or near-infrared probes, where injection timescales are inferred from transient signatures originating from the emerging population of cationic adsorbed dye molecules.[108,109,118,119]

Figure 11.14 shows the carrier dynamics monitored by TRTS for TiO$_2$ films sensitized by N719 ruthenium based dyes.[120] The THz signal exhibits "biphasic" electron transfer kinetics which is typically reported for TiO$_2$ sensitized by ruthenium-based dyes.[108–111,118] The fast component has been attributed to injection from the ^{1}MLCT state (characterized by τ_1 in Figure 11.14) and the slower component has been assigned to injection from the ^{3}MLCT state (characterized by τ_3). Injection from the high-energy singlet state is only possible if it occurs on timescales faster than, or comparable to intersystem crossing from the singlet to the triplet state (characterized by τ_2 in Figure 11.13).

The data in Figure 11.14 can be described by the following equation:

$$N_e(t) = A(1 - \exp(-t/\tau_1)) + B(1 - \exp(-t/\tau_3)^\alpha) \qquad (11.11)$$

In this equation, $N_e(t)$ corresponds to the population of electrons transferred from the dye to the oxide as a function of pump–probe delay. A and B are the amplitudes of injection from the singlet and triplet state respectively and α is the

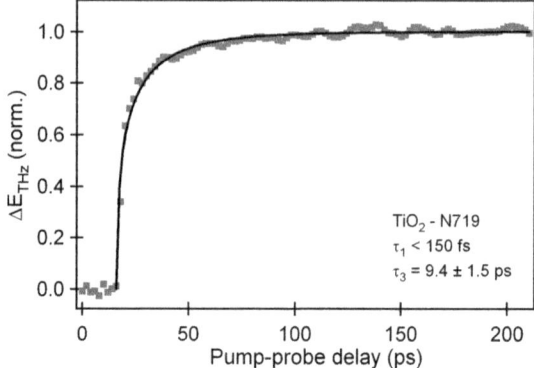

Figure 11.14   THz modulation – proportional to the product of electron density and mobility – for nanostructured TiO_2 films sensitized with N719, immersed in acetonitrile. The film was excited with 2 μJ mm^{-2} pulses, with a duration of 100 fs and centred at 590 nm. The data can be described by biphasic electron transfer kinetics (solid line).
Reprinted with permission from Pijpers *et al.*[120] Copyright 2011 American Chemical Society.

so-called stretch parameter. The use of the latter is necessary since electron injection does not follow single exponential kinetics, an observation which has been related to local inhomogeneities in the sensitized film[110,118,119](leading to local shifts of the conduction band edge, electronic coupling, *etc.*). Fitting the data in Figure 11.14 to Equation (11.11) yields values for τ_3 of 9.4 ± 1.5 ps (with $\alpha = 0.7$). An upper limit for the time constant τ_1 of 150 fs was obtained for this sample, which was limited by the instrument response function of the TRTS setup.

In many ultrafast studies, dye-sensitized oxide films are measured in inert solvents in absence of the electrolyte redox couple. In a complete DSSC, the dye-sensitized film is typically immersed in a redox electrolyte containing different chemical additives intended to optimize solar cell performance. In such complete solar cells, significant changes in the injection dynamics have been reported, indicating that the timescale of electron transfer between the dye and the oxide is retarded in the presence of the electrolyte species, resulting from a shift of the position of the conduction band (CB) edge energy level.[118,119] TRTS can monitor electron injection dynamics in the presence of the redox couple and/or electrolyte additives[120] with picosecond time resolution, posing a distinct advantage over, for example, time-resolved photon counting, where the time-resolution is ~ 250 ps.[119] Figure 11.15 compares the interfacial transfer kinetics in inert solution (acetonitrile), in the presence of the I^-_3/I^- redox couple that is typically used in DSSCs and in the presence of the recently introduced[121] T_2/T^- redox couple. Clearly, electron injection is strongly retarded in the presence of the T_2/T^- couple, which is useful information for designing a DSSC based on this redox couple.

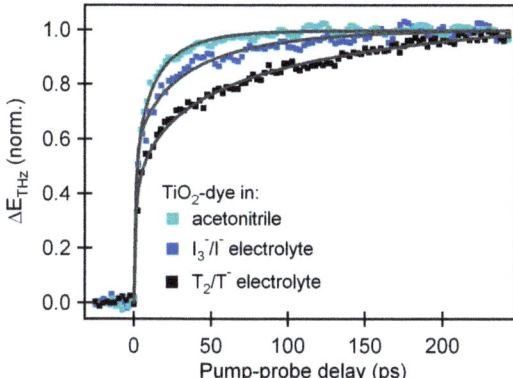

Figure 11.15   THz modulation – proportional to the product of electron density and mobility – for nanostructured TiO_2 films sensitized with N719 dye, immersed in acetonitrile (light blue dots), I_3^-/I^- in acetonitrile (blue dots, 0.7 M TBA I and 0.1 M I_2), and T_2/T^- in acetonitrile (black dots, 0.4 M T_2 and 0.1 M T^-).
Reprinted with permission from Pijpers *et al.*[120] Copyright 2011 American Chemical Society.

Future TRTS work can contribute to a better understanding of electron injection kinetics in a real device environment by monitoring carrier injection dynamics as a function of electrolyte composition, type of dye molecule and oxide material.

11.4.2   Quantum Dot Sensitizing Mesoporous Oxide Films

The QD sensitization of mesoporous oxides (on QD sensitized solar cells, QDSSCs) has been proposed as an alternative for dye sensitization with dyes.[122] The reason is the increased absorption cross section for QDs and the possibility of tuning the QD band gap by varying the size (allowing for readily matching the solar spectrum). However, these potential advantages have not yet been reflected in better efficiencies for QD based devices when compared with DSSCs. Slower electron injection, the non-homogeneous sensitization of the oxide matrix with QDs, the emergence of undesired QD surface defects and the incompatibility of QDs with many electrolyte environments are presumably the cause for the current low efficiencies of QDSSCs.[123]

Common techniques usually employed to monitor electron transfer processes on QD sensitizing oxides are photoluminescence (PL) and transient absorption (TA) spectroscopy (both probing carrier populations within the QDs). The reduction of QD luminescence[124] or increase in absorption[125] signals after their attachment to the oxide film is an indirect signature of efficient electron transfer between the species. However, the disappearance of electrons from the QD can also be explained by alternative recombination pathways (caused by changes in the QD surface chemistry after attachment) that are not present for dispersed QDs in a solvent. Indeed, TA and PL measurements on, for example,

CdSe-sensitized SiO_2[125] and PbSe-sensitized TiO_2[126] clearly show decay signals, often interpreted as electron injection, even though the alignment between the QD $1S_e$ level and the oxide conduction band edge is unfavourable for electron injection. TRTS, in contrast, is unique for probing electron transfer processes on these systems. With TRTS, it is possible selectively to excite the QDs and probe the emergence of free carriers in the oxide film as a function of pump–probe delay. Right after photoexcitation, the charge carriers are localized in the QD and there is no real conductivity (*i.e.* the THz field is not absorbed, see Section 11.2.2.3). After electrons have been injected into the mesoporous oxide film, the THz field interacts strongly with free electrons populating the oxide conduction band. Hence, the emergence of real conductivity in such TRTS experiments provides direct information about electron transfer rates on QDs sensitizing oxide surfaces.

Pijpers *et al.*[126] showed the usefulness of TRTS for demonstrating interfacial electron transfer in QD-sensitized oxide systems. Later on, Cánovas *et al.*[127] extended this work to explore the QD size-dependent interfacial electron transfer occurring between PbSe QDs and SnO_2 mesoporous oxide films. In this system, the timescale of electron transfer increased from ~ 100 ps for small ~ 2 nm diameter QDs, to ~ 1 ns for ~ 7 nm QDs (see Figure 11.16(a)). A simple

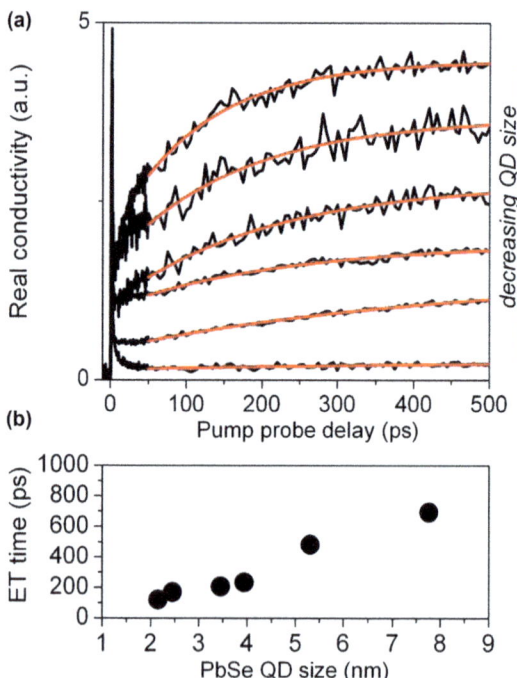

Figure 11.16   (a) QD size dependent electron transfer monitored by THz-TDS on PbSe QDs sensitizing SnO_2 films; the solid lines show a single-exponential fit to the conductivity traces. (b) Estimated electron transfer times (in ps) as a function of QD size.
Reprinted with permission from Cánovas *et al.*[127] Copyright 2011 American Chemical Society.

single-exponential ingrowth model (as in Equation (11.11) with $B=0$) reproduced the data remarkably well, indicating a high degree of homogeneity in the attached QD, in terms of both the average distance between the dot and the oxide (defined by the molecular bridge) and QD size (determining the relative band alignment between donor and acceptor). The retardation of electron transfer times for larger QDs can be explained by reduced quantum confinement and the concomitant smaller energy difference between the $1S_e$ QD level and the oxide CB edge (expressed as ΔG in Equation (11.10)).

The same TRTS measurements were used to estimate that the efficiency of electron transfer was only a few percent in their samples, which was likely to have been caused by the formation of QD agglomerates and QD induced surface traps at the oxide surface as a result of sensitization. These agglomerates and surface states presumably act as recombination centres. Fitting the data (in Figure 11.16(b)) according to the Marcus theory[106] (see Equation (11.10)), it was found that the reorganizational energy had a small value of $\lambda = 1 \pm 50$ meV, which might be characteristic for electron transer between nanocrystals and oxides,[125,128] suggesting that a barrierless electron transfer mechanism is typical for these systems.

Future directions of TRTS studies on carrier transfer rates in QD–oxide systems should focus on different QD–oxide material combinations, the dependence on donor–acceptor separation (by employing linker molecules of different length) and the effect of electrolyte composition on transfer rates. Such studies may further clarify the complex physicochemical processes governing carrier transfer in QDSSCs.

11.4.3 Quantum Dot Superlattices

Electronically coupled QD superlattices have been also considered for novel PV applications. Indeed, the best reported photocurrents on QD-based solar cells have been obtained for Schottky cells made of QD solids.[129] Also, QD superlattices have been employed as building blocks for tandem QD-based PV devices.[62] For all these reasons, it is hence of fundamental interest to monitor carrier transfer between QDs and across QD superlattices and TRTS represents an unique tool to do so.

TRTS is very sensitive to the degree of electronic coupling between nanocrystals.[130–133] In the case where there are strong overlapping of QD energy level wavefunctions, carriers populating the QDs can "hop" between dots, giving rise to a real component of the complex conductivity even at frequencies below the intraband transitions of isolated QDs.[132,134] In disordered superlattices of InP QDs, Beard *et al.*[130] reported an increase in the transient photoconductivity when the packaging of QDs is enhanced (the different separation among QDs was controlled by modifying the capping layer of the QD) and modelled the measured response using the Drude–Smith transport model (see Section 11.2.2.2). Beard *et al.* also found triexponential decay kinetics in these systems. Longer carrier lifetimes were observed when QDs are highly packed, which can be attributed to the increased tunnelling probability between closely spaced QDs, allowing carriers to escape from trap sites. Similar effects have also been reported for PbSe QD arrays.[133]

11.4.4   Exciton Dissociation in Semiconducting Polymers

Organic semiconductors have emerged as potential candidates for use in low-cost electronic devices, including solar cells.[135] They are commonly divided into two classes: small molecules that are typically deposited as crystalline material by evaporation and semiconducting organic polymers that are processed in solution. Optically excited excitons in organic semiconductors possess high binding energies that can reach values of several hundred meV. This is beneficial for application in light-emitting devices but detrimental in solar cells. Finding routes to dissociate excitons efficiently in order to generate a photocurrent is thus imperative. A popular approach in polymer-based solar cells is to blend them with electron-accepting molecules that scavenge the electrons from excitons that are photoexcited in the polymer. In this way, electrons and holes are unbound and thus free to diffuse and produce an electrical current. Obviously, the efficiency of this dissociation process depends on a number of parameters, which are related to the molecular structure of the constituents and the morphology of the blend. THz spectroscopy can give useful information about the early (ps) timescales after excitons have been excited by providing a means to obtain the yield of exciton and free-carrier generation. Put simply, one uses the fact that unbound charges generate real conductivity while excitons are characterized by the lack thereof and only give an imaginary contribution to the photoconductivity. Examinations are possible in great detail by careful analysis of the measured photoconductivities. Published works on pristine, that is, unblended polymer samples have investigated the effect of the type of polymer,[136] the polymer chain separation on the charge separation between polymer strands[137] and the influence of conjugation effects, molecular weight and film deposition conditions on the charge carrier mobility.[138,139] Films blended with fullerene derivatives have been investigated in several publications in which free carrier yields between a few and 30 times the yield in pristine films have been reported.[140–144]

11.5   Summary

In this chapter we have presented time-resolved THz spectroscopy as a powerful tool for the study of carrier dynamics in PV structures and materials. Optical and electrical information can be obtained in a non-contact fashion and resolved in time with sub-picosecond resolution. This versatility makes TRTS particularly suited to interrogate physical processes on nano-structured semiconductors, which nowadays constitutes the research forefront towards high efficiency and low cost PV devices.

Further development the TRTS technique itself, by obtaining increased THz probe bandwidths and intensities, as well as an increased stability on the THz sources, will undoubtedly extend the capabilities of TRTS for exploring novel materials and sophisticated device geometries, including those encountered in PV concepts where different material phases are mixed in nanometer scales.

References

1. S. M. Hsiang, K. C. Meng and M. A. Cane, *Nature*, 2011, **476**, 438.
2. A. Jäger-Waldau, *PV Status Report 2008*, Institute for Energy. Joint Research Centre (JRC), European Commission, 2008.
3. A. Luque and S. Hegedus, *Handbook of Photovoltaic Science and Engineering*, Wiley, Chichester, 2003.
4. T. Buonassisi, A. A. Istratov, M. A. Marcus, B. Lai, Z. Cai, S. M. Heald and E. R. Weber, *Nat. Mater.*, 2005, **4**, 676.
5. B. O'Regan and M. Grätzel, *Nature*, 1991, **353**, 737.
6. N. S. Lewis, G. Crabtree, A. J. Nozik, M. R. Wasielewski and P. Alivisatos, *Basic Research Needs for Solar Energy Utilization*, US Department of Energy, Office of Basic Science, 2005.
7. A. Martí and A. Luque, *Next Generation Photovoltaics: High Efficiency through Full Spectrum Utilization*, Institute of Physics Publishing, Bristol, UK, 2004.
8. M. A. Green, *Third Generation Photovoltaics*, Springer, Berlin, 2005.
9. C. Mayer, S. R. Scully, B. E. Hardin, M. W. Rowell and M. D. McGehee, *Mater. Today*, 2007, **10**, 28.
10. R. Kroon, M. Lenes, J. C. Hummelen, P. W. M. Blom and B. de Boer, *Polym. Rev.*, 2008, **48**, 531.
11. M. Gratzel, *Nature*, 2001, **414**, 338.
12. A. Zaban, I. Micic, B. A. Gregg and A. J. Nozik, *Langumir*, 1998, **14**, 3153.
13. R. Plass, S. Pelet, J. Krueger, M. Grätzel and U. Bach, *J. Phys. Chem. B*, 2002, **106**, 7578.
14. W. Shockley and H. J. Queisser, *J. Appl. Phys.*, 1961, **32**, 510.
15. R. R. King, D. C. Law, K. M. Edmondson, C. M. Fetzer, G. S. Kinsey, H. Yoon, R. A. Sherif and N. H. Karam, *Appl. Phys. Lett.*, 2007, **90**, 183516.
16. R. T. Ross and A. J. Nozik, *J. Appl. Phys.*, 1982, **53**, 3813.
17. J. H. Werner, S. Kolodinski and H. J. Queisser, *Phys. Rev. Lett.*, 1994, **72**, 3851.
18. M. van Exter, C. Fattinger and D. Grischkowsky, *Appl. Phys. Lett.*, 1989, **55**, 337.
19. C. Fattinger and D. Grischkowsky, *Appl. Phys. Lett.*, 1989, **54**, 490.
20. E. Knoesel, M. Bonn, J. Shan, F. Wang and T. F. Heinz, *J. Chem. Phys.*, 2004, **121**, 394.
21. E. Knoesel, M. Bonn, J. Shan and T. F. Heinz, *Phys. Rev. Lett.*, 2001, **86**, 340.
22. S. R. Keiding, *J. Phys. Chem. A*, 1997, **101**, 5250.
23. A. G. Markelz, A. Roitberg and E. J. Heilweil, *Chem. Phys. Lett.*, 2000, **320**, 42.

24. R. Huber, F. Tauser, A. Brodschelm, M. Bichler, G. Abstreiter and A. Leitenstorfer, *Nature*, 2001, **414**, 286.
25. R. A. Kaindl, M. A. Carnahan, D. Hagele, R. Lovenich and D. S. Chemla, *Nature*, 2003, **423**, 734.
26. T.-I. Jeon and D. Grischkowsky, *Phys. Rev. Lett.*, 1997, **78**, 1106.
27. E. Hendry, J. M. Schins, L. P. Candeias, L. D. A. Siebbeles and M. Bonn, *Phys. Rev. Lett.*, 2004, **92**, 196601.
28. T.-I. Jeon, D. Grischkowsky, A. K. Mukherjee and R. Menon, *Appl. Phys. Lett.*, 2000, **77**, 2452.
29. R. D. Averitt, A. I. Lobad, C. Kwon, S. A. Trugman, V. K. Thorsmølle and A. J. Taylor, *Phys. Rev. Lett.*, 2001, **87**, 017401.
30. K. Reimann, *Rep. Prog. Phys.*, 2007, **70**, 1597.
31. K. Sakai, *Terahertz Optoelectronics*, Springer, Berlin, 2005.
32. R. Cheville, *Terahertz Time-Domain Spectroscopy with Photoconductive Antennas, Terahertz Spectroscopy: Principles and Applications*, CRC, Boca Raton, 2008.
33. A. Nahata, A. S. Weling and T. F. Heinz, *Appl. Phys. Lett.*, 1996, **69**, 2321.
34. Q. Wu and X. C. Zhang, *Appl. Phys. Lett.*, 1997, **70**, 1784.
35. R. A. Kaindl, F. Eickemeyer, M. Woerner and T. Elsaesser, *Appl. Phys. Lett.*, 1999, **75**, 1060.
36. R. A. Huber, F. T. Brodschelm and A. Leitenstorfer, *Appl. Phys. Lett.*, 2000, **76**, 3191.
37. A. Sell, R. Scheu, A. Leitenstorfer and R. Huber, *Appl. Phys. Lett.*, 2008, **93**, 251107.
38. M. E. Fermann and I. Hartl, *IEEE J. Sel. Top. Quantum Electron.*, 2009, **15**, 191.
39. Q. Wu and X. C. Zhang, *Appl. Phys. Lett.*, 1995, **67**, 3523.
40. E. Knoesel, M. Bonn, J. Shan, F. Wang and T. F. Heinz, *J. Chem. Phys.*, 2004, **121**, 394.
41. L. Duvillaret, F. Garet and J. L Coutaz, *IEEE J. Sel. Top. Quantum Electron.*, 1996, **2**, 739.
42. R. A. Kaindl, D. Hagele, M. A. Carnahan and D. S. Chemla, *Phys. Rev. B*, 2009, **79**, 045320.
43. H. Nemec, F. Kadlec and P. Kuzel, *J. Chem. Phys.*, 2002, **117**, 8454.
44. H. Nemec, F. Kadlec, S. Surendran, P. Kuzel and P. Jungwirth, *J. Chem. Phys.*, 2005, **122**, 104503.
45. M. C. Beard, G. M. Turner and C. A. Schmuttenmaer, *Phys. Rev. B*, 2000, **62**, 15764.
46. R. Ulbricht, E. Hendry, J. Shan, T. F. Heinz and M. Bonn, *Rev. Mod. Phys.*, 2011, **83**, 543.
47. P. U. Jepsen, D. G. Cooke and M. Koch, *Laser Photonics Rev.*, 2011, **5**, 124.
48. C. A. Schmuttenmaer, *Chem. Rev.*, 2004, **104**, 1759.
49. E. Hendry, F. Wang, J. Shan, T. F. Heinz and M. Bonn, *Phys. Rev. B*, 2004, **69**, 081101.

50. E. Hendry, M. Koeberg, J. Pijpers and M. Bonn, *Phys. Rev. B*, 2007, **75**, 233202.
51. E. Hendry, M. Koeberg, B. O'Regan and M. Bonn, *Nano Lett*, 2006, **6**, 755.
52. H. K. Nienhuys and V. Sundstrom, *Appl. Phys. Lett.*, 2005, **87**, 012101.
53. G. M. Turner, M. C. Beard and C. A. Schmuttenmaer, *J. Phys. Chem. B*, 2002, **106**, 11716.
54. N. V. Smith, *Phys. Rev. B*, 2001, **64**, 155106.
55. L. Fekete, P. Kuzel, H. Nemec, F. Kadlec, A. Dejneka, J. Stuchlik and A. Fejfar, *Phys. Rev. B*, 2009, **79**, 115306.
56. D. G. Cooke, A. N. MacDonald, A. Hryciw, J. Wang, Q. Li, A. Meldrum and F. A. Hegmann, *Phys. Rev. B*, 2006, **73**, 193311.
57. H. Nemec, P. Kuzel, F. Kadlec, D. Fattakhova-Rohlfing, J. Szeifert, T. Bein, V. Kalousek and J. Rathousky, *Appl. Phys. Lett.*, 2010, **96**, 062103.
58. H. Nemec, P. Kuzel and V. Sundstrom, *J. Photochem. Photobiol. A: Chem.*, 2010, **215**, 123.
59. H. Nemec, P. Kuzel and V. Sundstrom, *Phys. Rev. B*, 2009, **79**, 115309.
60. C. F. Bohren and D. R. Huffman, *Absorption and Scattering of Light by Small Particles*, John Wiley & Sons: New York, 1983.
61. A. J. Nozik, *Physica E*, 2002, **14E**, 115.
62. X. Wang, G. Koleilat, J. Tang, H. Liu, I. J. Kramer, R. Debnath, L. Brzozowski, D. A. R. Barkhouse, L. Levina, S. Hoogland and E. H. Sargent, *Nat. Photonics*, 2011, **5**, 480–484.
63. W. A. Tisdale, K. J. Williams, B. A. Timp, D. J. Norris, E. S. Aydil and X. Y. Zhu, *Science*, 2010, **328**, 1543.
64. R. Schaller and V. Klimov, *Phys. Rev. Lett.*, 2004, **92**, 186601.
65. M. C. Beard, G. M. Turner and C. A. Schmuttenmaer, *Nano Lett.*, 2002, **2**, 983.
66. F. Wang, J. Shan, M. A. Islam, I. P. Herman, M. Bonn and T. F. Heinz, *Nature Mater.*, 2006, **5**, 861.
67. G. L. Dakovski, S. Lan, C. Xia and J. Shan, *J. Phys. Chem. C*, 2007, **111**, 5904.
68. J. J. H. Pijpers, M. T. W. Milder, C. Delerue and M. Bonn, *J. Phys. Chem. C*, 2010, **114**, 6318.
69. A. J. Nozik, *Annu. Rev. Phys. Chem.*, 2001, **52**, 193.
70. V. I. Klimov and D. W. McBranch, *Phys. Rev. Lett.*, 1998, **80**, 4028.
71. P. Guyot-Sionnest, M. Shim, C. Matranga and M. Hines, *Phys. Rev. B*, 1999, **60**, R2181.
72. V. I. Klimov, A. A. Mikhailovsky, D. W. McBranch, C. A. Leatherdale and M. G. Bawendi, *Phys. Rev. B*, 2000, **61**, R13349.
73. E. Hendry, M. Koeberg, F. Wang, H. Zhang, C. de Mello-Donegá, D. Vanmaekelbergh and M. Bonn, *Phys. Rev. Lett.*, 2006, **96**, 057408.
74. R. D. Schaller, J. M. Pietryga, S. V. Goupalov, M. A. Petruska, S. A. Ivanov and V. I. Klimov, *Phys. Rev. Lett.*, 2005, **95**, 196401.
75. A. Pandey and P. Guyot-Sionnest, *Science*, 2008, **322**, 929.

76. S. Kolodinski, J. H. Werner, T. Wittchen and H. J. Queisser, *Appl. Phys. Lett.*, 1993, **63**, 2405.
77. O. Christensen, *J. Appl. Phys.*, 1976, **47**, 689.
78. V. I. Klimov, *Ann. Rev. Phys. Chem*, 2007, **58**, 635.
79. R. J. Ellingson, M. C. Beard, J. C. Johnson, P. R. Yu, O. I. Micic, A. J. Nozik, A. Shabaev and A. L. Efros, *Nano Lett.*, 2005, **5**, 865.
80. R. D. Schaller, V. M. Agranovich and V. I. Klimov, *Nature Phys*, 2005, **1**, 189.
81. R. D. Schaller, M. Sykora, J. M. Pietryga and V. I. Klimov, *Nano Lett.*, 2006, **6**, 424.
82. M. T. Trinh, A. J. Houtepen, J. M. Schins, T. Hanrath, J. Piris, W. Knulst, A. Goossens and L. D. A. Siebbeles, *Nano Lett.*, 2008, **8**, 1713.
83. M. Ji, S. Park, S. T. Connor, T. Mokari, Y. Cui and K. J. Gaffney, *Nano Lett.*, 2009, **9**, 1217.
84. J. E. Murphy, M. C. Beard, A. G. Norman, S. P. Ahrenkiel, J. C. Johnson, P. R. Yu, O. I. Micic, R. J. Ellingson and A. J. Nozik, *J. Am. Chem. Soc.*, 2006, **128**, 3241.
85. R. D. Schaller, M. A. Petruska and V. I. Klimov, *Appl. Phys. Lett.*, 2005, **87**, 253102.
86. M. C. Beard, K. P. Knutsen, P. Yu, J. M. Luther, Q. Song, W. K. Metzger, R. J. Ellingson and A. J. Nozik, *Nano Lett.*, 2007, **7**, 2506.
87. D. Timmerman, I. Izeddin, P. Stallinga, I. N. Yassievich and T. Gregorkiewicz, *Nat Photonics*, 2008, **2**, 105.
88. J. J. H. Pijpers, E. Hendry, M. T. W. Milder, R. Fanciulli, J. Savolainen, J. L. Herek, D. Vanmaekelbergh, S. Ruhman, D. Mocatta, D. Oron, A. Aharoni, U. Banin and M. Bonn, *J. Phys. Chem. C*, 2007, **111**, 4146.
89. R. D. Schaller, J. M. Pietryga and V. I. Klimov, *Nano Lett.*, 2007, **7**, 3469.
90. G. Nair and M. G. Bawendi, *Phys. Rev. B*, 2007, **76**, 4.
91. M. Ben-Lulu, D. Mocatta, M. Bonn, U. Banin and S. Ruhman, *Nano Lett.*, 2008, **8**, 1207.
92. G. Nair, S. M. Geyer, L.-Y. Chang and M. G. Bawendi, *Phys. Rev. B*, 2008, **78**, 125325.
93. J. J. H. Pijpers, E. Hendry, M. T. W. Milder, R. Fanciulli, J. Savolainen, J. L. Herek, D. Vanmaekelbergh, S. Ruhman, D. Mocatta, D. Oron, A. Aharoni, U. Banin and M. Bonn, *J. Phys. Chem. C*, 2008, **112**, 4783.
94. J. J. H. Pijpers, R. Ulbricht, K. J. Tielrooij, A. Osherov, Y. Golan, C. Delerue, G. Allan and M. Bonn, *Nature Phys.*, 2009, **5**, 811.
95. C. Delerue, G. Allan, J. J. H. Pijpers and M. Bonn, *Phys. Rev. B*, 2010, **81**, 125306.
96. G. Allan and C. Delerue, *Phys. Rev. B*, 2006, **73**, 205423.
97. A. Smith and D. Dutton, *J. Opt. Soc. Am.*, 1958, **48**, 1007.
98. T.-I. Jeon and D. Grischkowsky, *Phys. Rev. Lett.*, 1997, **78**, 1106.
99. E. Hendry, M. Koeberg, J. Pijpers and M. Bonn, *Phys. Rev. B*, 2007, **75**, 233202.
100. M. C. Beard, G. M. Turner and C. A. Schmuttenmaer, *J. Phys. Chem. B*, 2002, **106**, 7146.

101. E. Hendry, M. Koeberg, J. Pijpers and M. Bonn, *Phys. Rev. B*, 2007, **75**, 233202.
102. C. F. Barton, *J. Appl. Phys.*, 1971, **42**, 445.
103. V. I. Klimov, J. A. McGuire, R. D. Schaller and V. I. Rupasov, *Phys. Rev. B*, 2008, **77**, 195324.
104. B. A. Gregg, *J. Phys. Chem. B*, 2003, **107**, 4688.
105. M. Gratzel, *Inorg. Chem.*, 2005, **44**, 6841.
106. R. A. Marcus, *J. Chem. Phys.*, 1965, **43**, 679.
107. J. B. Asbury, E. Hao, Y. Wang, H. N. Ghosh and T. Lian, *J. Phys.Chem. B*, 2001, **105**, 4545.
108. Y. Tachibana, J. E. Moser, M. Grätzel, D. R. Klug and J. R. Durrant, *J. Phys. Chem.*, 1996, **100**, 20056.
109. Y. Tachibana, M. K. Nazeeruddin, M. Grätzel, D. R. Klug and J. R. Durrant, *Chem. Phys.*, 2002, **285**, 127.
110. J. B. Asbury, E. Hao, Y. Wang, H. N. Ghosh and T. Lian, *J. Phys.Chem. B*, 2001, **105**, 4545.
111. J. B. Asbury, N. A. Anderson, E. Hao, X. Ai and T. Lian, *J. Phys.Chem. B*, 2003, **107**, 7376.
112. J. Huang, D. Stockwell, A. Boulesbaa, J. Guo and T. Lian, *J. Phys.Chem. C*, 2008, **112**, 5203.
113. G. M. Turner, M. C. Beard and C. A. Schmuttenmaer, *J. Phys.Chem. B*, 2002, **106**, 11716.
114. S. G. Abuabara, C. W. Cady, J. B. Baxter, C. A. Schmuttenmaer, R. H. Crabtree, G. W. Brudvig and V. S. Batista, *J. Phys.Chem. C*, 2007, **111**, 11982.
115. W. R. McNamara, R. C. Snoeberger, G. Li, J. M. Schleicher, C. W. Cady, M. Poyatos, C. A. Schmuttenmaer, R. H. Crabtree, G. W. Brudvig and V. S. Batista, *J. Am. Chem. Soc.*, 2008, **130**, 14329.
116. P. Tiwana, P. Parkinson, M. B. Johnston, H. J. Snaith and L. M. Herz, *J. Phys. Chem. C*, 2009, **114**, 1365.
117. J. Huang, D. Stockwell, A. Boulesbaa, J. Guo and T. Lian, *J. Phys.Chem. C*, 2008, **112**, 5203.
118. S. A. Haque, E. Palomares, B. M. Cho, A. N. M. Green, N. Hirata, D. R. Klug and J. R. Durrant, *J. Am. Chem. Soc.*, 2005, **127**, 3456.
119. S. E. Koops, B. C. O'Regan, P. R. F. Barnes and J. R. Durrant, *J. Am. Chem. Soc.*, 2009, **131**, 4808.
120. J. J. H. Pijpers, R. Ulbricht, S. Derossi, J. N. H. Reek and M. Bonn, *J. Phys. Chem. C*, 2011, **115**(5), 2578.
121. M. Wang, N. Chamberland, L. Breau, J.-E. Moser, R. Humphry-Baker, B. Marsan, S. M. Zakeeruddin and M. Gratzel, *Nat. Chem.*, 2010, **2**, 385.
122. I. Mora-Sero and J. Bisquert, *J. Phys. Chem. Lett.*, 2010, **1**(20), 3046 (and references therein).
123. G. Hodes, *J. Phys. Chem C*, 2008, **112**, 17778.
124. J. L. Blackburn, D.C. Selmarten and A. J. Nozik, *J. Phys Chem B*, 2003, **107**, 14154.
125. K. Tvrdy, P. A. Frantsuzov and P. V. Kamat, *PNAS*, 2011, **108**(1), 29.

126. J. J. H. Pijpers, R. Koole, W. H. Evers, A. J. Houtepen, S. Boehme, C. de Mello-Donega, D. Vanmaekelbergh and M. Bonn, *J. Phys. Chem. C*, 2010, **114**, 18866.

127. E. Cánovas, P. Moll, S. A. Jensen, Y. Gao, A. J. Houtepen, L. D. A. Siebbeles, S. Kinge and M. Bonn, *Nano Lett.*, 2011, **11**, 5234–5239.

128. G. D. Scholes, M. Jones and S. Kumar, *J. Phys. Chem. C*, 2007, **111**, 13777.

129. J. M. Luther, M. Law, M. C. Beard, Q. Song, M. O. Reese, R. J. Ellingson and A. J. Nozik, *Nano Lett.*, 2008, **8**(10), 3488.

130. M. C. Beard, G. M. Turner, J. E. Murphy, O. I. Micic, M. C. Hanna, A. J. Nozik and C. A. Schmuttenmaer, *Nano Lett.*, 2003, **3**, 1695.

131. S. A. Crooker, T. Barrick, J. A. Hollingsworth and V. I. Klimov, *Acta Phys. Pol. A*, 2003, **104**, 113.

132. D. G. Cooke, F. A. Hegmann, Y. I. Mazur, W.Q. Ma, X. Wang, Z. M. Wang, G. J. Salamo, M. Xiao, T. D. Mishima and M. B. Johnson, *Appl. Phys. Lett.*, 2004, **85**, 3839.

133. J. E. Murphy, M. C. Beard and A. J Nozik, *J. Phys. Chem. B*, 2006, **110**, 25455.

134. P. Boucaud, K. S. Gill, J. B. Williams, M. S. Sherwin, W.V. Schoenfeld and P. M. Petroff, *Appl. Phys. Lett.*, 2000, **77**, 510.

135. H. Hoppe and N. S. Sariciftci, *J. Mater. Res.*, 2004, **19**, 1924.

136. E. Hendry, M. Koeberg, J. M. Schins, L. D. A. Siebbeles and M. Bonn, *Chem. Phys. Lett.*, 2006, **432**, 441.

137. E. Hendry, M. Koeberg, J. M. Schins, H. K. Nienhuys, V. Sundstrom, L. D. A. Siebbeles and A. Bonn, *Phys. Rev. B*, 2005, **71**, 125201.

138. O. Esenturk, R. J. Kline, D. M. Deongchamp and E. J. Heilweil, *J. Phys. Chem. C*, 2008, **112**, 10587.

139. O. Esenturk, J. S. Melinger and E. J. Heilweil, *J. Appl. Phys.*, 2008, **103**, 023102.

140. X. Ai, M. C. Beard, K. P. Knutsen, S. E. Shaheen, G. Rumbles and R. J. Ellingson, *J. Phys.Chem. B*, 2006, **110**, 25462.

141. P. D. Cunningham and L. M. Hayden, *J. Phys.Chem C*, 2008, **112**, 7928.

142. E. Hendry, M. Koeberg, J. M. Schins, L. D. A. Siebbeles and M. Bonn, *Phys. Rev. B*, 2004, **70**, 033202.

143. P. Parkinson, J. Lloyd-Hughes, M. B. Johnston and L. M. Herz, *Phys. Rev. B*, 2008, **78**, 115321.

144. H. Nemec, H. K. Nienhuys, E. Perzon, F. L. Zhang, O. Inganas, P. Kuzel and V. Sundstrom, *Phys. Rev. B*, 2009, **79**, 245326.

CHAPTER 12

X-ray Transient Absorption Spectroscopy for Solar Energy Research

LIN X. CHEN

Chemical Sciences and Engineering Division, Argonne National Laboratory, Building 200, 9700 South Cass Avenue, Argonne, Illinois 60439; Department of Chemistry, Northwestern University, 2145 Sheridan Road, Evanston, Illinois 60208, USA
Email: lchen@anl.gov; l-chen@northwestern.edu

12.1 Introduction

The amount of energy from sunlight reaching the earth in one hour equals the total amount of energy consumption in the world for one year.[1] In spite of its tremendous potential as an abundant, renewable and clean energy resource, solar energy has only contributed a small portion of the renewable energy and represents a miniscule part of the total energy consumption. Solar energy utilization, almost entirely in the form of photovoltaic devices, faces great challenges in cost reduction, infrastructure renovation, energy storage and extension of device lifetime.[2,3] Although chemical bonds have been recognized as effective solar energy storage vehicles that lead to fuel generation through light driven water splitting and CO_2 reduction,[4–7] there has been minimal commercialization of photo-driven fuel production.[8] Factors behind the energy cost issues are device efficiency and the lifetime of performing their light to energy conversion functions,[8] many of which originate from our limited

RSC Energy and Environment Series No. 8
Solar Energy Conversion: Dynamics of Interfacial Electron and Excitation Transfer
Edited by Piotr Piotrowiak
© The Royal Society of Chemistry 2013
Published by the Royal Society of Chemistry, www.rsc.org

understanding of the light-matter interaction at a fundamental level in many emerging materials. For example, we still do not know how to predict the performance of a catalyst/photocatalyst in its ground state structure, because we often do not know how the useful photoinduced energy/electron transfer catalytic actions compete with wasteful exciton/excited state quenching via geminate recombination or internal conversion to heat. Frequently, redox reactions leading to solar fuel production may not happen as predicted and are difficult to characterize with well-established transient optical and magnetic resonance methods, because the former provide indirect structural information while the latter often does not have the time resolution needed for the initial steps of solar energy conversion. Transient Raman spectroscopy, especially recently developed ultrafast Raman spectroscopic methods, provides specific local structural information inferred from vibrational modes, such as C=O stretching.[9–11] In order to obtain direct and detailed structures with atomic resolution along the reaction coordinates, X-rays with wavelengths shorter than or comparable to the atomic scale are used to yield structural precision of hundredths of an Ångström or less. Several X-ray structural determination methods have been used, such as X-ray diffraction, scattering, imaging and spectroscopy, each of which is suitable for a particular set of samples and conditions. For many solar energy conversion systems involving metal centers as light absorbers or electron donors/acceptors, knowing the transient local structure of the metal center is particularly informative for our understanding of the reaction mechanism and structure/function correlation. In this chapter, one of the transient X-ray structural methods, X-ray transient absorption spectroscopy (XTA),[12–14] will be presented, with a focus on its application to solar energy research. A few examples from recent literature on XTA studies, in which snapshots of molecules or other moieties participating in solar energy conversion were taken, will be given. Because XTA is a relatively new technique, its future development and applications in solar energy research will be discussed in the final section.

12.2   X-ray Transient Absorption Spectroscopy (XTA): Capabilities and Development

12.2.1   Development of X-ray Transient Absorption Spectroscopy

12.2.1.1   X-ray Absorption Spectroscopy

X-ray absorption spectroscopy (XAS), including X-ray absorption near edge structure (XANES) and X-ray absorption fine structure (XAFS), is based on resonant absorption processes resulting from electronic transitions from core levels to vacant orbitals and to the continuum, respectively (Figure 12.1).[15–18] XANES measures dipole-mediated transitions from a deep core orbital to unoccupied orbitals and proceeds from an initial state $|i\rangle$ to a final state $|f\rangle$. The transitions from 1s, 2s, p, 3s, p, d, and so on, to higher states or continuum correspond to K-, L-, M-,...edge absorptions. The absorption $\mu(E)$ is

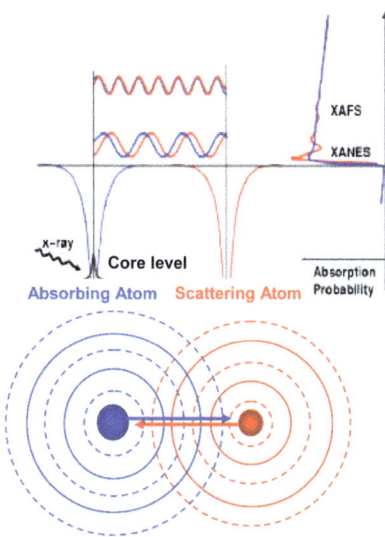

Figure 12.1 Illustration of XAS (after Matthew Newville, The University of Chicago).

proportional to $\sum |\langle f | e_x \cdot r_x | i \rangle|^2 \, \delta(E)$, where e_x and r_x are the X-ray electronic field vector and the transition dipole vector, respectively.[19] The pre-edge and edge features originate from electronic transitions from a core level to empty bound states,[20] which relate to the density of states and electronic configuration of the X-ray absorbing atom. For example, the pre-edge features observed in K-edge X-ray absorption of transition metal complexes with a center of symmetry are quadrupole-allowed with the intensities two orders of magnitude lower than the dipole-allowed transitions. However, these weak pre-edge features can gain extra intensity from mixing with the np character in a non-centrosymmetric coordination environment. Hence, the amplitudes and positions of these pre-edge features can be used to reveal both the energy levels of the vacant molecular orbitals (MO) and the coordination geometry. Because the X-ray photon energy required to eject a core level electron to the continuum depends on the charge of the X-ray absorbing atom, the transition edge in XANES spectra is sometimes the only unambiguous way to characterize the oxidation states of the metal during a chemical reaction.

XAFS are the oscillatory features in an X-ray absorption spectrum above the transition edge, originating from the interference between an outgoing photoelectron wave from the X-ray absorbing atom with back scattered photoelectron waves from neighboring atoms (Figure 12.1).[15–18] These oscillatory features have been successfully Fourier transformed into an atomic radial distribution centered at the X-ray absorbing atoms by Sayers, Stern and Lytle in their pioneering work of the 1970s[15–18] and expressed by the equation below:

$$\chi(k) = \sum_j N_j F_j(k) e^{-2\sigma_j^2 k^2} e^{-\frac{2R_j}{\lambda_j(k)}} \frac{\sin[2kR_j + \delta_j(k)]}{kR_j^2} \tag{12.1}$$

where j is the index for the neighboring atom shells around the X-ray absorbing atom, $F(k)$ is back-scattering amplitude, N is the coordination number, R is the average distance, σ is the Debye–Waller factor, λ is the the electron mean free path and δ, is the phase shift of the photoelectron wave. k is photoelectron wave vector, $k = [2m(E-E_0)/\hbar^2]^{1/2}$, where m is the electron mass and E_0 is the the threshold energy for the transition edge. XAS measurements can be carried out using a transmission or fluorescence detection mode. The latter is mostly for dilute samples based on the proportionality of the X-ray fluorescence signals to the absorbed X-ray photons. Experimental details of transmission and fluorescence detections can be found in a recent textbook by Grant Bunker,[21] as well as earlier comprehensive reviews.[19,20]

XAS provides precise local structures (*i.e.*, bond length precision to ± 0.01 Å) without the limitation of the sample form (*e.g.*, solids, amorphous powders, solution and gas phase), but its probing range is limited to a few neighboring atom shells around the X-ray absorbing atom and is dependent on the disorder in the samples. For a perfectly ordered sample with uniform distances for each neighboring shell, the structural range probed can be >5 Å from the central X-ray absorbing atom, whereas this range could be as small as the nearest neighboring shell if the distances diverge owing to a static or dynamic disorder in the sample. For many photochemical reactions in disordered media, XAS is a very useful structural tool for examining precise and detailed local structures around the X-ray absorbing atom, many of which are metal centers or other heavy atoms. The XAS in the soft X-ray regime (<3 keV), although it is also widely used and has advanced rapidly in recent years,[22] will not be covered in this review.

12.2.1.2 X-ray Transient Absorption Spectroscopy (XTA)

XTA is analogous to the optical transient absorption spectroscopy (OTA) (see the overview in Chapter 5); a laser pump pulse triggers photoexcitation or a photochemical reaction, mimicking light absorption of solar photons and an X-ray pulse (instead of the second laser pulse in OTA) probes electronic and geometric structures of the samples as the result of the first pump pulse (Figure 12.2). The time-resolution of the experiment with such a "pump–probe" scheme is determined by the pulse duration of the pump (the laser pulse) or the probe (the X-ray pulse), whichever is longer, convoluted with the instrumental response function (IRF). Hence, XTA differs from those early time-resolved XAS studies using synchrotron X-rays as a continuous wave (CW) source combined with lasers or steady-state light sources,[23–35] where the time resolution of the measurement was independent of the timing structure of the X-ray pulses.

Although both OTA and XAS techniques have been used for decades, XTA, viewed as the combination of the two, faces unique challenges as outlined below:

A. Discrepancy in Absorption Coefficients for Pump-Probe Photons. A key consideration for the laser pulse pump and X-ray pulse probe XTA is optimizing two different transitions with 2–3 orders of magnitude difference in

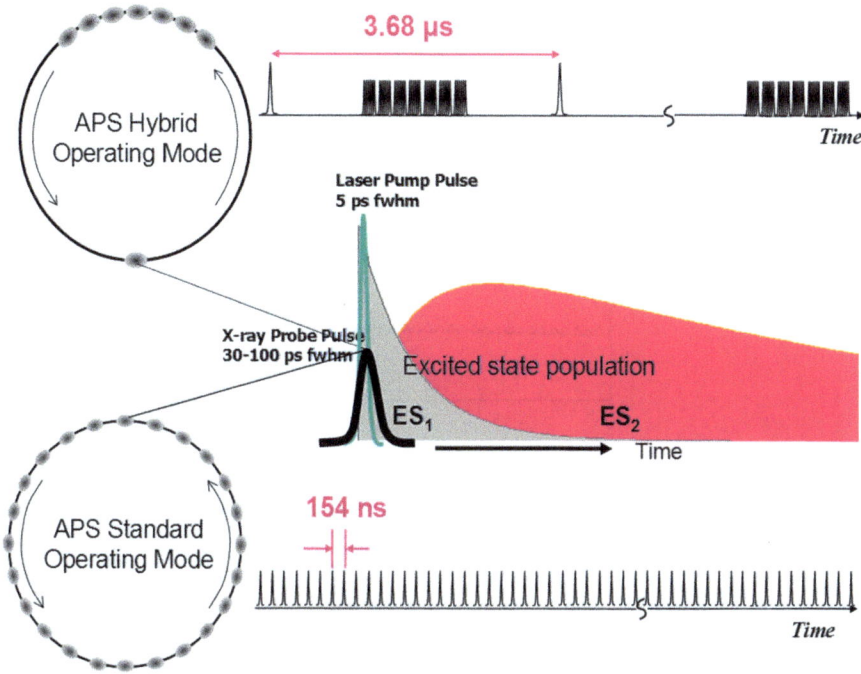

Figure 12.2 Timing sequence of XTA experiments at the Advanced Photon Source (APS). Two commonly provided modes for XTA are shown. The timing overlap shown in the middle indicates that the laser pulse first creates the first excited state ES_1 which decays with a delay time between the pump and probe pulses while the second excited state ES_2 (or a subsequent transient state) grows. The pump–probe delay time can be adjusted (see text) to probe the structures of the transient state at delay times according to their respective optimal concentrations.

energy and absorption efficiencies. The laser photons used to mimic the sunlight in photon–matter interactions have energies of a few eV corresponding to a wavelength range of 200–1000 nm, while hard X-ray photons used to probe transient structures in XTA measurements have energies of a few to a few tens of keV, corresponding to wavelengths in the range of 0.1 nm. The systems for efficient light harvesting and conversion in the solar spectrum should have absorption coefficients for sunlight wavelengths in a range of 10^3 to 10^6 $M^{-1} cm^{-1}$ or 20–400 cm^{-1} in a concentration range of 1–20 mM (Figure 12.3). In comparison, the X-ray photons in the probe pulse used to resolve molecular structures on the atomic scale have absorption coefficients in a range of <2 cm^{-1} throughout the same concentration range. Figure 12.3 also demonstrates that in dilute solution at concentrations of tens of mM, the majority of the X-ray photons are absorbed by the solvent. Therefore, the attenuation of the X-ray after the solution sample is nearly solute concentration independent in this concentration range.

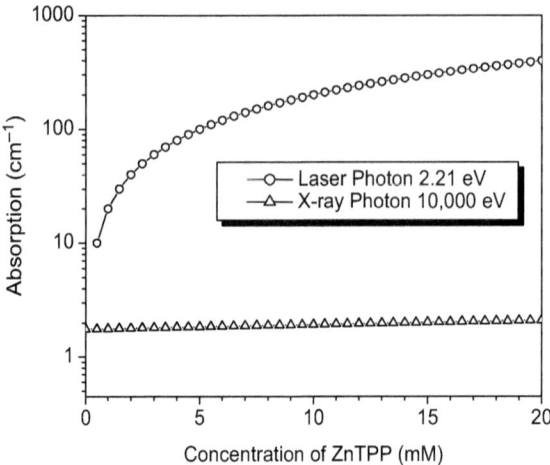

Figure 12.3   Absorption coefficients for laser (2.21 eV) and X-ray (10 keV) photons *vs.* zinc tetraphenylporphyrin concentration in toluene solution. The absorptions of light atoms, C, N and H are included in the X-ray absorption coefficients in unit length (cm⁻¹).

As a result, photons from UV/vis lasers have a much smaller penetration depth in the sample than photons of hard X-rays. A multidimensional parameter optimization approach needs to be carried out for the optimization of experimental conditions necessary to achieve simultaneously relatively high fractions of excited species (which favors dilute samples) and as high quality as possible XAS spectra (which favors high concentration samples). Analogous to UV/vis optical absorption, X-ray absorption is measured directly by $\mu(E)L = \ln(I_0/I)$, where I and I_0 are incident and transmitted X-ray intensities, respectively, $\mu(E)$ is the absorption coefficient for the X-ray photons at an energy E and L is the thickness of the sample. One can choose the concentration and the thickness at which a substantial fraction of excited state/transient species can be created in order to extract the structures from XAS spectra. In practice, conditions for each sample (*i.e.*, concentration, thickness and sizes) need to be evaluated before XTA experiments, with using the following criteria: (1) the fraction of the excited state created by the laser pump pulse should exceed 10% (which varies with the quality of the data and signal amplitude change between the photoexcited and ground states); (2) for transmission detected XTA signals, the transmission difference before and after the transition edge should exceed 10% of the amplitude of the step-function at the edge for the transmission detected XTA; (3) the number of X-ray fluorescence photon at the energy of interest, as well as contributions from other elastic scattering and Compton scattering, need to be included in fluorescence detector selection to make sure that the detector has a sufficient dynamics range as well as a fast response time to collect isolated pulsed signal without saturation; and (4) laser pulse intensity and beam size need to be optimized in order to optimize the penetration length discrepancy between the pump and probe beams. Several reviews have described the details of XTA experiments in transmission detection mode[14,36-38] and in fluorescence/emission mode.[12,13,39]

B. Determining Fractions of Transient Species. As in any spectroscopic method, the determination of a transient structure by the XTA method requires deconvolution of the signal of the transient species of interest from those of other species, such as any remaining ground state or photoproducts. In general, total signals detected via X-ray transmission or fluorescence can be expressed as:

$$I_{total}(E,t) = \sum_i f_i I_i(E,t) = \sum_i f_i I_{0i} e^{-k_i t} \qquad (12.2)$$

assuming all photoinitiated processes are first order with their rate constants of k_i and where i is the index of species in the sample, such as the ground state, first excited state, ..., and so on. f_i values are the initial fractions of the species generated by the laser pulse. Most of the XTA measurements carried out so far include two states, the ground/starting state and the excited/transient state. The certainty of f_i directly affects the accuracy of the transient structures determined by this method and hence is very important in data analysis. There are three main methods for extracting f_i values.

The first method extracts f_i based on optical transient absorption results using a laser only pump–probe approach, where the photoexcitation conditions are set to be as close as possible to those in the XTA measurements. The f_i values can be obtained by comparing ΔOD (change in optical density) and the total OD in the spectral region of the ground state bleach, as long as minimal interference from other spectral features is present in the same region.[12,40,41] This method appears to be accurate but sometimes is not feasible because of the interference caused by significant fractions of the excited state or other transient state absorption in the same spectral region.

The second approach is based on the multidimensional interpolation approach (MIA) developed by Smolentsev, Soldatov and co-workers[42-44] as well as Minuit XANes (MXAN) by Benfatto and Della Longa and co-workers [45-47] where the structural parameters, such as bond distances and angles as well as the fractions of the transient state are simultaneously used in iterative fitting processes to extract the excited state population and structures (see below).

The third method has been involved in steered parameter fitting employed by van der Veen et al.[48] where a local search of the optimal fitting results in a narrow range of f_i values is performed iteratively, yielding a set of structural parameters associated with a particular f_i value. In some special cases, characteristic XANES features, such as the sharp $1s \rightarrow 4p_z$ peak at the transition edges of the K-edge XANES spectra of the first row transition metals (see below), of either the ground or the transient state, can be used to extract the population fraction in the system by subtracting a fraction of the ground state spectrum until the characteristic feature disappears. As XTA measurements are used broadly in many different photoinitiated systems, various reaction pathways will be present, which complicates the data analysis. Various data analysis methods for multiple components, such as principle component analysis and singular vector decomposition, have been used to investigate the optimal method of extracting f_i.

C. Signal Detection in Dilute Samples with Low Concentration Transient Species. To accommodate XTA measurements of laboratory synthesized systems for solar fuel/electricity generation and catalysis with limited solubility and scarce quantity, we have been optimizing XTA experiments with X-ray fluorescence detection, which is the preferred detection scheme in the study samples with low concentrations (*i.e.*, $\sim 1\,mM$). In special cases, when the sample is abundant, highly soluble and has a very high yield of photo-excitation even at a high concentration (*i.e.*, $>10\,mM$), transmission XTA measurements can be carried out.[49–54] In addition, the recent development of catalysts with robust inorganic polyoxometalate-based water oxidation properties for hydrogen production[55–57] and other inorganic frameworks for water oxidation[58] call for the need to develop grazing incident (GI) XTA for heterogeneous catalytic systems on electrode surfaces.

In order to extract structural parameters from a mixture of the ground and excited state by XAS, sufficient laser photons must be available in each pulse, which limits the pump laser pulse repetition rate to a few kHz compared to a few MHz for X-ray pulses. Therefore, the pump–probe cycle limits usable X-ray probe pulses to one in every 600–6000 X-ray pulses in the pulse train, resulting in prolonged data acquisition time (*i.e.*, 4–40 hours per XAFS spectrum). However, this dire situation has been greatly improved by X-ray photon flux increases at beamlines with new insertion devices. In addition, high repetition rate lasers with fundamental or harmonic generation wavelengths have been used in XTA[59,60] and other pump–probe experiments which in some cases drastically shorten the data acquisition time when combined with a Kirkpatrick–Baez mirror[61] focused X-ray beam of a few to ten micrometer diameters. However, some systems with long-lived transient species or with absorption outside of the wavelength provided by these high repetition rate lasers cannot be studied.

In most solution samples with millimolar or lower concentrations used in solar energy conversion systems, solvent absorbs a major or significant portion of the X-ray photons, making the fraction of X-ray photons absorbed by the atom of interest even smaller. For example, in a sample containing 1 mM Ni in a toluene solution, the fraction of Ni absorption in the sample is only 0.6%, and 99.4% absorption of X-ray photons is due to C, O, N and H atoms in the system.[39] For most solar energy research samples, with a few exceptions, it is difficult to achieve $>10\,mM$ concentrations owing to sample scarcity and solubility. Under a few millimolar or lower concentrations, transmission detection will be difficult because the transition edge is very small on top of a huge transmitted X-ray signal as a background, resulting in a limited dynamic range for high quality XAFS signals. In the case where only XANES spectra are of interest, the detection selection limit can be more relaxed because XANES signals can be strong and have characteristic features. However, these results may only yield electronic configuration information and omit the nuclear geometry of the transient states.

For the fluorescence detection scheme, the incident intensity normalized signal can be expressed by Equation (12.3):

$$\frac{I_f}{I} = \frac{I_a}{I} \cdot \frac{\Omega}{4\pi} \cdot \eta \cdot \frac{\mu_k}{\mu_T} \cdot \eta_{\text{det}} \tag{12.3}$$

where I, I_f and I_a are intensities of incident, fluorescence and absorbed X-ray photons, respectively; $\Omega/4\pi$ is the solid angle covered by the detector, η is the quantum yield of the fluorescence, μ_k and μ_T are the absorption cross sections of the atoms of interest and of the whole sample, respectively. η_{det} is the detector efficiency, defined as the fraction of the fluorescence photons among the total photons entering the detector. As mentioned earlier, many photochemical processes occur in dilute solutions, so we use a dilute sample as an example for the calculation. For a 1 mM Ni-containing sample with 1 mm^2 area and 0.5 mm thickness, about 15% of incident X-ray photons will be absorbed. Depending on the geometry of the detection, the solid angle covered by the detector can be calculated and may vary from 1–20%. The fluorescence quantum yield η for Ni is 0.70 and the fraction of Ni absorption in the sample is 0.6%. With all the above factors considered, there would be ≫100 fluorescence photons/pulse entering the detector. In this case, the previous detection scheme with energy dispersive detectors coupled with single photon counting will not be feasible. Instead, we use a photomultiplier tube (PMT) or avalanche photodiode detectors (APD) operating in current mode and coupled with plastic scintillator and soller slits with z-1 filters.[62] This detection method can accommodate the intense light sources of today, maximize the signal-to-noise ratio and minimize the background signals from elastic scattering. The drawback of this method is that the energy resolution is only gained by physical methods such as the filters and slits rather than through built-in potential selection as used in solid state germanium detector arrays.

D. XTA Facility at the Advanced Photon Source. The development of the current XTA started when intense pulsed X-rays from third generation synchrotron sources, such as European Synchrotron Radiation Source (ESRF) in France, the Advanced Photon Source (APS) in the USA and Spring-8 in Japan, became available over last 15 years.[63–69] The first XTA experiment, which resolved a transient structure using the timing structure of X-ray pulses from a synchrotron source, was carried out in 2000 using X-ray pulse clusters with a 14-ns time span at the APS.[69] This was followed by later experiments using single X-ray pulses at the same facility and at the Advanced Light Source (ALS).[70] XTA measurements at a third generation synchrotron facility are becoming relatively straightforward in several synchrotron sources now. More recently, ultrafast X-ray free electron laser (XFEL) facilities are operational or under construction,[71] which advance the time-resolution for transient structural determination to sub-100 fs and shorter, comparable to what can be achieved by ultrafast lasers.

An XTA facility has been constructed at Beamline 11ID-D of the APS in Argonne National Laboratory (Figure 12.4).[13,39–41,62,69,72–74] The laser pump pulse uses the second or third harmonic output of a Nd:YLF regenerative amplified laser with 1–2 kHz repetition rate, giving 527 or 351 nm laser pulses with a temporal duration of 5 ps full width at half maximum (FWHM). The X-ray probe pulses were derived from electron bunches extracted from the storage ring with 80 ps fwhm and 6.5 MHz repetition rate. The laser pump, X-ray probe cycle was 1–2 kHz, limited by the laser repetition rate. The laser and X-ray pulses intersect at a flowing sample fluid with continuous purging of dry nitrogen gas and circulated using a peristaltic pump. The laser and X-ray beams are overlapped spatially and temporally on the flowing sample. The delay time between the laser and X-ray pulses was adjusted by a programmable delay line (PDL-100A-20NS, Colby Instruments) that varied the phase of the sinusoidal signals of the mode-lock driver of the laser relative to that of the RF signal of the storage ring with a precision of 500 fs. Two PMTs or APDs coupled by plastic scintillators were used at 90° angle on both sides to the incident X-ray beam to collect the X-ray fluorescence signals.

A Soller slit/Z-1 filter combination, which was custom-designed for the specific sample chamber configuration and the distance between the sample and the detector, was inserted between the sample fluid jet and the detectors. Two PMTs operating in current mode were used to acquire multiple photons from

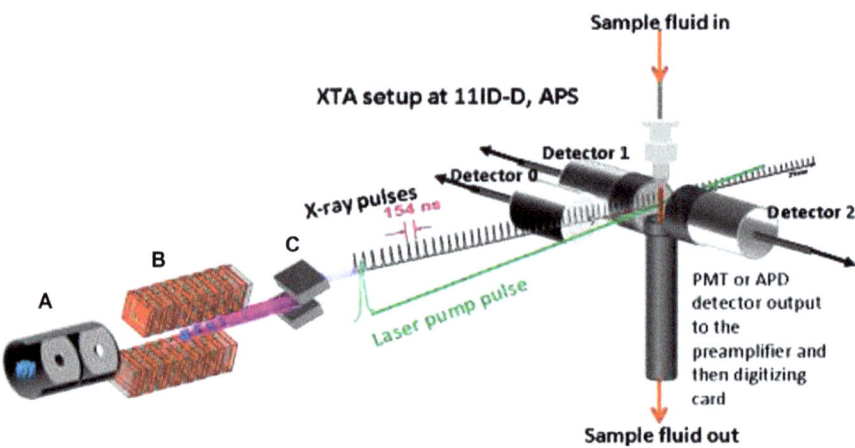

Figure 12.4 Schematic setup of the XTA facility at Beamline 11IDD at the APS. A, accelerator; B, undulator insertion device; C, monochromator. Detector 0 is for intensity normalization, Detector 1 and 2 are for X-ray fluorescence signals from the jet stream of the sample (red) and hence have the soller slit/z-1 filter/plastic scintilator/combo (dark cylinders) in front of the detector. The outputs of the detectors shown by black arrows are amplified and connected to the inputs of the digitizing board (from Agilent)

each X-ray pulse as the flux of the X-ray photons significantly exceeded the single photon counting limit. The outputs of the detectors were sent to two fast analyzer cards (Agilent) that were triggered at 1–2 kHz by a signal generated by a photo diode sampling the laser pulses. The analyzer card digitized the X-ray fluorescence signals as a function of time at 1ns/point after each trigger pulse. An additional detector of the same kind was placed upstream from the sample chamber to collect the elastic scattering signals from the air or a piece of thin film as the signal for the incident X-ray intensity normalization (Figure 12.4).

It is important to have the signals from detectors of the same type with the same response time to obtain the ratio of I_f/I_0 because the pulse-to-pulse noise would not be minimized if the two detectors have different response times. The use of the analyzer boards combined with software engineering (Attenkofer/Jennings, APS) also allows digitization of all X-ray pulses in the pulse train between the two consecutive laser pulses. For solar energy research, this feature is very important because the excited state could be generated by the laser pulse instantaneously while the catalytic intermediate could appear on a much slower timescale. Having all the X-ray pulses digitized gives the freedom to obtain data simultaneously on different timescales. If necessary, signals from later pulses can be binned to enhance the signal-to-noise ratio in an XTA spectrum. A procedure for *in situ* curve fitting (by G. Jennings, APS) was implemented to extract the signal intensity from the detectors. This method extracted the signal intensity from the integrated area of the pulsed signals by first curve fitting the pulse shape without and with X-ray photons as references, followed by scaling the pulse shape in real signals. The procedure removed the pulse shape dependence of the signal extracted based on the peak intensity or on the integrated area and made the signal processing much more tolerant to the electronic noise from the detectors.

12.3 Applications of XTA in Solar Energy Conversion Research: Examples

12.3.1 Metalloporphyrin Excited State Structural Dynamics (Homogeneous Electron/Energy Transfer)

Metalloporphyrins and derivatives play many important roles in solar energy conversion as light harvesting antennae and electron donors in natural photosynthesis, dye sensitizers in solar cells, photocatalysts for redox reactions, and so on.[75–79] Their intense absorption from the π–π^* transitions in the solar spectrum enables light harvesting, energy and electron transfer in natural and artificial photosynthesis.[79] Their roles in photocatalysis are facilitated by the oxidation state changes at the metal centers associated frequently with the coordination geometry changes and reversible ligation.[80] Although the excited and transient states of these molecules have been extensively studied, some portions of the structural–functional correlations of these states remain vague because most of the transient structural information obtained is indirect and dependent on theoretical predictions. As light harvesting antennae, metallo-porphyrins with long excited state lifetimes are used to facilitate energy transfer (EnT), such as that of Mg and Zn porphyrins.[81-83] As photocatalytic centers, metal center oxidation state and ligation state changes are key to their catalytic functions.[84-87] Fe, Co, Ru and several other open-shell metalloporphyrin derivatives have been studied as catalysts and can either harvest the light energy and directly drive the reactions, or accept energy from other sensitizers.[87]

As an example, we studied different parts of photochemical pathways of NiTMP [nickel(II)tetramesityl porphyrin] shown in Figure 12.5, mapped out by the earlier work of Holten *et al.* on NiTPP [nickel(II)tetraphenyl porphyrin], an almost identical nickel porphyrin.[88-90] Although NiTMP does not perform known catalytic functions, its interactions with light result in several typical photochemical processes, such as photoinduced changes in the 3d molecular orbital electron occupancy, molecular geometry and the Ni ligation state. Therefore, the example demonstrates the current capabilities of XTA at a synchrotron source in capturing these structural changes which can be applicable in solar energy conversion processes involving metal complexes.

The ground state NiTMP has two possible structures, a square-planar co-ordinated Ni in non-ligating solvent (*e.g.*, toluene) and a square-pyramidal- or octahedral-coordinated Ni in coordinating solvent (*e.g.*, pyridine, piperdine and pyrrolidine will bind Ni as axial ligands above and below the porphrin plane). Both forms of NiTMP can be excited by two $\pi-\pi^*$ transitions at Q- or B-band generating singlet S_1 or S_2 states in the square-planar form and triplet T_1 or T_2 states in the octahedral form. The transformations from one state to another take place with time constants ranging from less than 200 fs to tens of ns. The first pump–probe XTA measurement is shown in the purple box looking at the photodissociated NiTPP from doubly ligated

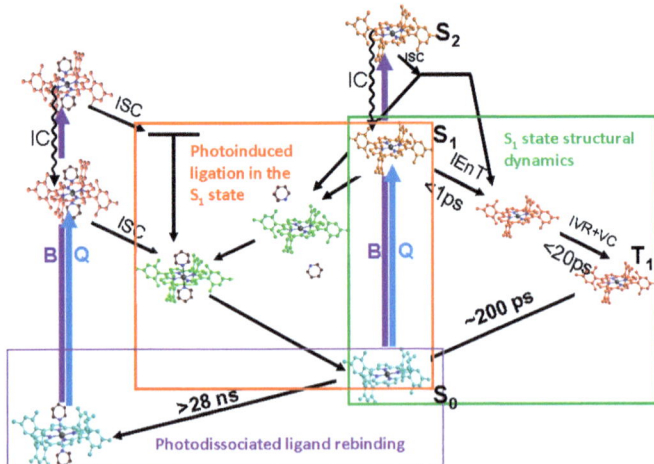

Figure 12.5 Photochemical pathways of nickel porphyrins studied by XTA. The processes are labeled as shown. B and Q stand for photoexcitation of B (or soret) around 420 nm and Q bands around 530 nm.

NiTPP-(piperidine)$_2$. At that initial stage of the APS operation, clusters of six X-ray pulses with a combined duration of 14 ns were used to probe the intermediate structure before the rebinding of the ligands.[69] By today's standards, the experiment was relatively straightforward with only nanosecond time resolution, however, the results demonstrated for the first time (1) the feasibility of XTA measurements at a synchrotron source for a dilute solution (1mM), (2) that structural determination with 100-ps resolution can be achieved using singlet X-ray probe pulses when they were available, (3) what the pros and cons are using the available solid state Ge detector array (Canberra) and (4) that the photodissociation intermediate is square-planar under the time resolution limit.

The next type of structural dynamics of nickel porphyrin by the XTA method to be explored was the excited state structural dynamics shown in the green box in Figure 12.5. The ground state (S$_0$) electronic configuration of square-planar Ni(II) (3d^8) porphyrin has only one empty 3d$_{x2-y2}$ MO and a fully occupied 3d$_{z2}$ MO.[91] Within 350 fs following the excitation of the Q-band, the S$_1$ state was believed to convert, through energy transfer, to an intermediate state T$_1'$ which then undergoes vibrational relaxation within 20 ps to a relaxed triplet state, T$_1$ with a presumed ^{3}(3d$_{x2-y2}$, 3d$_{z2}$) configuration, where the 3d$_{x2-y2}$ and 3d$_{z2}$ MOs are each singly occupied (Figure 12.6(A) and inset).[88,89,92,93] The T$_1$ state returns to the S$_0$ state in approximately 200 ps. Under the time-resolution limitation of the X-ray pulses from the APS, only the electronic configuration and geometry of the T$_1$ state can be investigated.

In the XANES region, especially the pre-edge region, a single peak due to the 1s → 3d$_{x2-y2}$ transition in the S$_0$ state and two peaks due to the 1s → 3d$_{x2-y2}$ and 1s → 3d$_{z2}$ transitions have been observed by the XTA spectra, confirming vacancies in both MOs and their energy separation of 2.2 eV, a result obtainable heretofore only via theoretical calculations. Because of the sensitivity of the XTA spectra to electronic configurations of metal centers, it directly measures excited state MO energy levels from the core level. XTA is particularly suitable for correlating molecular geometry with electronic configurations and MO energy levels when the excited state is optically "dark" or overwhelmed by other strongly absorbing or emitting states.

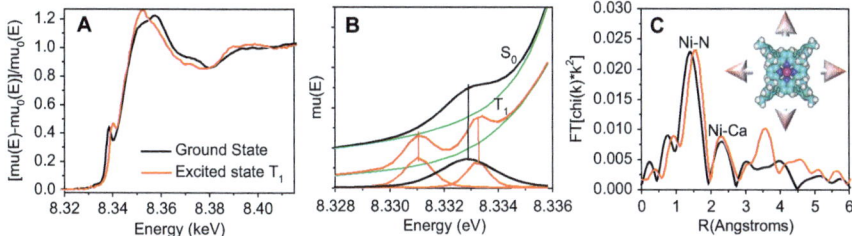

Figure 12.6 XTA spectra of NiTMP in toluene. (A) XANES spectra showing different regions with changes due to the formation of the T$_1$ state; (B) pre-edge region highlighted by a rectangle in (A), where the electron configuration for the S$_0$ and T$_1$ states are shown; and (C) XAFS region where the first and the second neighboring shells expand in the T1 state.

A noticeable difference in the bandwidths of the $1s \rightarrow 3d_{x2-y2}$ transition in the S_0 state and those of the $1s \rightarrow 3d_{x2-y2}$ and $1s \rightarrow 3d_{z2}$ transitions in the T_1 state was observed for the pre-edge peaks of the $1s \rightarrow 3d_{x2-y2}$ transition in the S_0 state and the $1s \rightarrow 3d_{x2-y2}$ and $1s \rightarrow 3d_{z2}$ transitions in the T_1 state (Figure 12.6(B)). The former has an apparent linewidth of $>2\,eV$ (FWHM), whereas the latter has two peaks with linewidths of $1.0–1.2\,eV$. The broader linewidth of the S_0 state is indicative of a wide distribution of unoccupied $3d_{x2-y2}$ MO energy due to multiple coexisting non-planar conformations in solution revealed by Raman spectroscopy and quantum mechanical calculations,[94–97] and now being observed by XTA and verified by density functional theory (DFT) calculations for a series of NiTMP conformations with saddled, twisted and domed distortions, with Ni–N distances adjusted accordingly. The results suggest that 3d MO energies with various conformers can vary by more than $1\,eV$ while the total energy of the molecules including surrounding solvent molecules corresponding to different conformers can still be less than kT. Because many solar energy conversion processes in transition metal complexes involve d-electrons, the results suggest the potential for influencing d-MO energy levels by altering the geometry of these complexes.

At the rising edge region from 8335–8343 eV (Figure 12.6), a distinctive peak at 8337.5 eV in the S_0 state was assigned as a dipole allowed $1s \rightarrow 4p_z$ transition,[98] which became 1.5 eV higher in the T_1 state than in the S_0 state. The "white line" peak (the sharp feature with the highest signal amplitude) region of 8348–8360 eV composed of the $1s \rightarrow 4p_{x,y}$ transitions and multiple scattering contributions, also shows significant changes upon $S_0 \rightarrow T_1$ transition. Because the $1s \rightarrow 4p_z$ transition peak energy increases with the number of the axial ligands and becomes less pronounced as it shifts towards to the white line peak, its amplitude can be used to monitor the ligation state of Ni. Ni(II) porphyrins are known to undergo double axial ligation in the S_0 state in the presence of ligating molecules (Figure 12.6(B)).

Ni(II) porphyrins have been shown to undergo wavelength-selected photo-induced ligation/deligation.[90,99] The dual ligated NiTMP has a $1s \rightarrow 4p_z$ feature shifted to a higher energy.[69,98] In the presence of pyridine, the photoexcitation induced a decrease in the sharp $1s \rightarrow 4p_z$ peak while a new peak appeared at a higher energy (Figure 12.7), similar that for the T_1 state of NiTMP in neat toluene without the ligation.[100] Therefore, the new absorption peak at higher energy is due to the formation of excited unligated NiTMP. Between 100 ps and 400 ps, the transient absorption intensity at 8339.5 eV is decreased and approaches zero after 600 ps, indicating the diminishing population of the unligated excited state after 600 ps. Meanwhile, the $1s \rightarrow 4p_z$ peak at 8337.5 eV reappeared with a reduced intensity. The time evolution of the $1s \rightarrow 4p_z$ peak is an indication of the branched pathways of the T_1 state in the presence of pyridine or other ligating molecules as mentioned earlier. Because the dual ligated NiTMP has shifted its $1s \rightarrow 4p_z$ peak to a higher energy under the white line transition, it will not contribute intensity at 8337.5 eV. Therefore, the reduced intensity at 8337.5 eV indicates that ligation has occurred and can be used to extract the fraction of the remaining unligated population. The ground

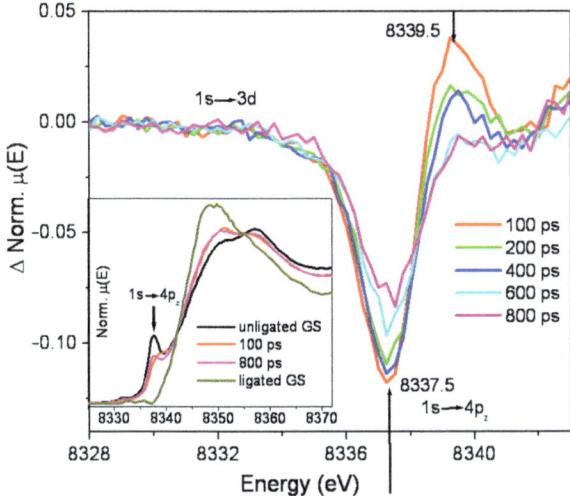

Figure 12.7 Difference transient XANES spectra (inset) in the 1s→4p$_z$ region of NiTMP in toluene:pyridine = 3:1 by volume ratio as a function of the pump–probe delay times.

state bleach of the unligated 1s→4p$_z$ transition at 8337.5 eV could last for several tens of nanoseconds because the photoinduced ligated NiTMP decays to the ground state with a time constant of 42 ns.

These results provide answers to the previous questions regarding the time sequence of the photoinduced ligation. The solvent ligation happens with a time constant of ~1 ns, long after the vibrational relaxation of the T$_1$ state. This time constant agrees with a solvent diffusion process. Because the ligation process is diffusion limited, its intrinsic diffusion rate determines the branching ratio of the T$_1$ state after its creation. The relatively long-lived ligated species also indicates that the ligated species is a triplet state that requires tens of nanoseconds to release the ligands and return to the S$_0$ state. This example therefore demonstrates the capability of XTA to monitor the ligation state of a metal center which is extremely important in studying photocatalytical reactions and intermediate catalyst–substrate interactions.

In the XAFS region (Figure 12.5(C)), the nuclear geometry changes in the same transition are captured, showing the lengthening of the average Ni–N and Ni–C$_{\beta\mu}$ α distances (without phase corrections) in the T$_1$ state, determined to be 0.08 Å and 0.07 Å, respectively, resulted from a porphyrin ring expansion. Because of the addition of one electron to the higher energy antibonding 3d$_{x2-y2}$ MO in the T$_1$ state, the Ni–N bond order is effectively reduced, resulting in longer Ni–N bond distances. As the porphyrin ring expands in the T$_1$ state, the electron density on Ni(II) by σ bonding and π-conjugation decreases while its effective charge increases compared to that of the S$_0$ state, which may consequently cause energy up-shift of the 1s→3d$_{x2-y2}$ and 1s→4p$_z$ transitions in the T$_1$ state (Figure 12.5(A)).

XTA studies on the S_0 state of NiTMP in solution revealed more significant 3d MO energy fluctuations than expected owing to the interconversion between different conformers, which could produce transient degeneracies between the two upper 3d MOs, a filled $3d_{z^2}$ and an empty $3d_{x^2-y^2}$ and enable $3d_{z^2}$ electrons to hop onto the empty $3d_{x^2-y^2}$, resulting in a vacancy in $3d_{z^2}$ MO to accommodate the axial ligation. Similarly, the photoexcitation of NiTMP also creates singly occupied $3d_{z^2}$ and $3d_{x^2-y^2}$ MOs in favour of the axial ligation. The photoexcitation shifted the ligation equilibrium $NiTMP^* + 2py \leftrightarrow NiTMP(py)_2^*$ to the right with a two-fold enhancement in the population of the ligated species compared to that in the ground state. Hence, these XTA studies provide a unified mechanism for axial ligation driven by the vacancy in the $3d_{z^2}$ MO in both the ground and excited state NiTMP which can be applicable to other metalloporphyrin photocatalysts or light sensitizers. The knowledge obtained here will have a general impact in designing photocatalysts for solar fuel generation, promoting binding to the metal center in a well-defined orientation of desirable functions.

12.3.2 Photoinduced Interfacial Electron Transfer (Heterogeneous Electron Transfer)

Since their discovery in early 1990s, dye sensitized solar cells (DSSCs) have been extensively investigated because of their potential in wide commercial applications for conversion of sunlight to electricity. The key interaction of light and matter in DSSCs is light induced interfacial charge transfer from transition metal complex dye sensitizers to semiconductor nanoparticles.[101–108] Extensive studies of energetics, kinetics and structural dynamics of the interfacial charge transfer processes have generated a wealth of information,[101–108] but structural evolution of the adsorbed dye sensitizer and the nanocrystal surface associated with the electron density shift during and after the interfacial charge injection remains vague and is actively pursued by several groups with theoretical modeling.[109–111] The example below demonstrates the current synchrotron XTA capabilities in solving transient electronic and geometric structures of a ruthenium complex dye sensitizer undergoing interfacial photoinduced charge separation, mimicking the electron injection process in DSSCs. This is a heterogeneous system where the hybrid material composed of RuN3 and TiO_2 nanoparticles is in a suspension that may be difficult to study by OTA because the decorated particles may form larger aggregates causing visible light scattering. XTA, on the other hand, is not sensitive to visible light transparency of the sample and hence is an ideal tool for studying such a system.

In a DSSC, $Ru^{II}(dcbpy)_2(NCS)_2$ [or RuN3 or N3, dcbpy = bis(4,4′-dicarboxy-2,2′-bipyridine] was attached to the surface of TiO_2 nanocrystals in a film as a working electrode. The metal-to-ligand-charge-transfer (MLCT) state of RuN3 injects one electron into the conduction band of TiO_2, leading to an interfacial charge separated state, $(TiO_2)_n^-/RuN3^+$ (Figure 12.8). The electron

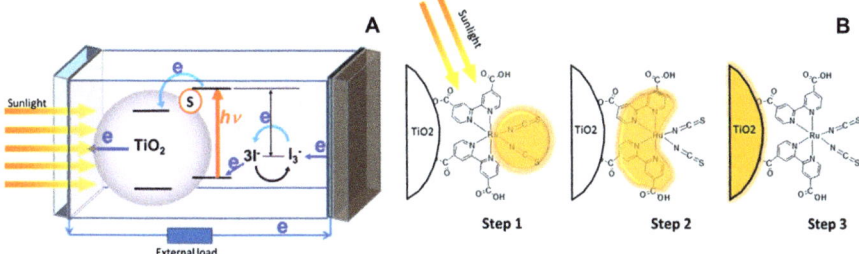

$$[Ru^{II}(bpy)_2(NCS)_2]^{+2}\ TiO_2 + hv \rightarrow [Ru^{II}(bpy)_2(NCS)_2]^{+2*}\ TiO_2 \rightarrow [Ru^{III}(bpy)_2(NCS)_2]^{+3}\ TiO_2\text{-}e$$

Figure 12.8   (A) DSSC pathways and (B) electron density (orange) distribution (1) before the excitation, (2) immediately after the excitation and (3) after electron injection (after Angelis and Fantacci).

injection dynamics of this process has been characterized by a dual-exponential function with time constants of <100 fs and 1 ps to 100 ps depending on the sample environment.[104,112,113] Because the current XTA time resolution at a synchrotron source is about 80 ps, only the RuN3 transient structure of the $(TiO_2)_n^-$/RuN3$^+$ state after photoinduced interfacial electron injection can be captured and the structure of the RuN3 MLCT state itself could only be resolved with fs X-ray pulses from X-ray free electron source, such as the Linear Coherent Light Source (LCLS).

Figure 12.9 displays the XANES spectra as well as difference XANES spectra at the Ru K-edge without and with the laser pump pulse. At the 50 ps delay time, the excited RuN3 molecules have already completed electron injection into the nanoparticles and are mostly in the oxidized form, $Ru^{III}N3^+$, according to OTA studies. The edge energy of the laser-on XANES spectrum is slightly shifted to a higher energy as expected for $Ru^{III}N3^+$ after the electron injection to the TiO_2 lattice.

The data analysis was complicated by simultaneously having two types of ligand (NCS and dcbpy) and two different states (the ground and charge separate states) in RuN3 at this delay time. Because the core–hole lifetime decreases as the atomic number of the element increases,[114] the bandwidth of a second and third row transition metal is broader than those of the first row transition metals. Consequently, the transition edges of high Z element appear to rise more slowly with an increase in the incident X-ray photon energy, which smears characteristic transition features. By taking the difference spectra between the laser pump without and with pulses, the changes caused by the generation of the transient species become more distinct, as seen in Figure 12.9. Therefore, we chose to use the multidimensional interpolation approach (MIA)[42] implemented in the FitIt code[43] and full multiple scattering (FMS) calculations of XAS using FEFF8.2[115] This method focuses on the difference spectrum which can be extracted precisely when the laser-on and laser-off spectra are collected strictly under the same detection and X-ray beam conditions.

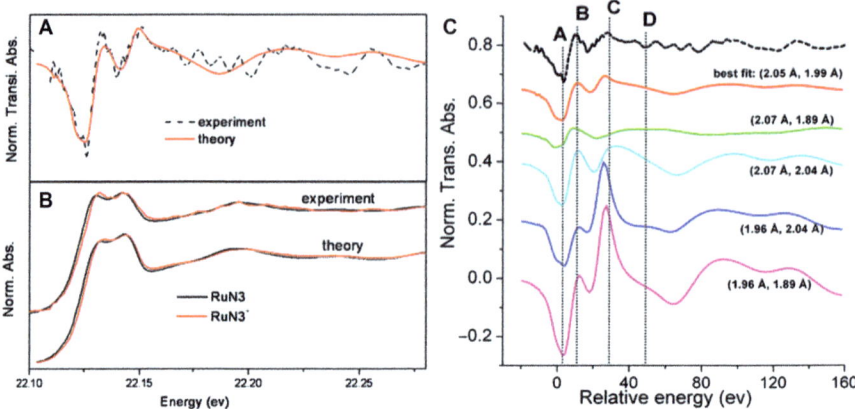

Figure 12.9   (A) Difference XANES spectrum taken at 50 ps delay from the pump laser pulse; (B) experimental and calculated XANES spectra for the ground state RuN3 and RuN3$^+$ after charge injection; (C) difference XANES spectra with minor variations in the structural parameters showing the sensitivity of the calculated spectra for the structures.

The data fitting includes the following steps: (1) calculating the ground state spectrum and optimizing non-structural parameters (*e.g.* transition edge position) if it is necessary; (2) calculating a series of transient state spectra by varying different structural parameters and constructing the interpolation polynomial; and (3) minimizing the discrepancy between the two sets of difference spectra, one corresponding to the experimental difference spectrum and the other corresponding to the theoretical difference spectrum in order to retrieve the best fitting parameters. The best fit can be found using Equation (12.4):

$$\min\left(\frac{1}{E_2 - E_1}\int_{E_1}^{E_2}\left[\left(\mu_{exp}^{laser\_on} - \mu_{exp}^{laser\_off}\right) - f\left(\mu_{theor}^{ts}(\delta p_1, \delta p_2) - \mu_{theor}^{gs}\right)\right]^2 dE\right)$$

$$(12.4)$$

where $\mu_{exp}^{laser\_on}$ and $\mu_{exp}^{laser\_off}$ are experimental absorption coefficients measured with and without the photoexcitation, respectively; μ_{theor}^{ts} and μ_{theor}^{gs} are the theoretically calculated absorption coefficients for the transient and ground states, respectively, f is the fraction of transient species (*i.e.*, the charge separated state, $(TiO_2)_n^-/RuN3^+$) and E_1 and E_2 define the energy range of spectral comparison.

By using MIA, XTA spectra were constructed from different structural parameters for RuN3 dye, which shows a high sensitivity in fitting over structural variations. A key in searching for the best structural parameters is to match the shape of the spectrum in the XANES region that account for the multiple scattering effect in the molecule. Extracting a correct fraction of the transient structure created by the pump laser pulse is often a key to determining correct structural parameters of the transient state. In this particular case, iterative fits with varying fraction and structural parameter were used. The results show that the conversion from the ground to the charge separate states causes different responses in the Ru–N bonds with different ligands, resulting in the average Ru–N(NCS) bond length being shortened by ~ 0.06 Å, from 2.05 Å to 1.99 Å, although the average Ru–N (dcbpy) bond length shows almost no changes. Other combinations of structural variations were used in analyses, but the results were significantly worse in the difference spectra, as shown in Figure 12.9(C).

These results can be rationalized by a recent time-dependent density functional theory (TDDFT) calculation on a RuN3–TiO$_2$(lattice) system which suggested two steps in photoinduced electron injection (Figure 12.8(B)), (1) shifting electron density from the highest occupied molecular orbital (HOMO), assigned as Ru t$_{2g}$ character with a sizable contribution from the NCS ligand orbitals, to the dcbpy ligands and (2) another electron density shift from dcbpy bridging ligands to the TiO$_2$ lattice.[109,110] Consequently, a net electron density loss at the Ru–NCS moiety and a net electron density gain at the surface of TiO$_2$ lattice take place, while the dcbpy ligands anchored on the TiO$_2$ act as an electron density relay with a minimal net change in the electron density. The example demonstrates the XTA capabilities for studying interfacial charge and energy transfer which are key processes in many DSSCs, hybrid solar conversion systems and catalytical systems. We learned from the results that anchoring the dye molecule in the charge injection direction aligned with the electron density gradient orientation was important to the electron injection efficiency and rate. Because high efficiency DSSCs require both fast electron injection and minimal germination charge recombination, it is not entirely clear whether little or no structural reorganization could accelerate the charge injection because certain structural reorganization could also prevent the germination recombination of the charges and enhance the device efficiency.

This work has extended the XTA studies from those of homogeneous systems to heterogeneous interfacial charge separation systems, which is a very active research area relevant to future developments of photovoltaic devices and heterogeneous catalysis. In order to study these surface specific interfacial light and matter interactions, solid film samples need to be studied by XTA, raising the challenge of minimizing the radiation damage of the stationary sample and circulating the film samples. Grazing angle incident XAS conducted in XTA will be pursued by additional low temperature studies.

On the scientific front of the DSSC research, alternative dyes with cheaper first row transition metals and other organic molecules have been investigated.[116–122] One type of alternative are copper(I) diimine complexes that have been extensively studied for their interesting photochemistry and structural dynamics.[123–130] These complexes have amazingly similar UV/vis absorption spectra to those of Ru(II)-trisbipyridyl complex and its derivatives and also undergo the MLCT transition where Cu(I) diimine becomes a Cu(II) diimine complex with one of the ligands becoming an anion. One such example is

$[Cu(I)(dmp)_2]^+ + h\nu \rightarrow [Cu(II)(dmp)(dmp)^-]^+$ (dmp = 2,9-dimethyl-1,10-phenan-throline) which shifts the electron density from the Cu(I) ($3d^{10}$) center to the ligands, resulting in a transient Cu(II) ($3d^9$) center.[72,131–148] What distinguishes Cu(I) diimine complexes from the Ru(II) complexes is their susceptibility to Jahn–Teller distortion as a result of the MLCT transition, which transforms the pseudo-tetrahedral coordination geometry of the Cu(I) center in the ground state to a flattened coordination geometry in the MLCT excited state. Meanwhile, the formation of an "exciplex" between the Cu(II) center in $[Cu(II)(dmp)(dmp)^-]^+$ and a solvent molecule lowers the energy of the MLCT state and shortens its lifetime from hundreds of ns to <1 ns (Figure 12.10).

In order to prolong the MLCT state lifetime and preserve the initial energy of the MLCT state, we have started structural dynamics studies on a series of Cu(I) diimine complexes with variable structural hindrances for the Jahn–Teller distortion and solvent accessibility.[149] One important part of this study is the direct structural determination of the MLCT state by XTA to identify the oxidation state of the copper center as well as its coordination geometry. Our studies found that the intersystem crossing (ISC) time constant in these complexes is strongly coordination symmetry dependent. The ISC time constant is <1ps if the MLCT state has two ligand planes orthogonal to each other, but is >10 ps if the relative orientation of the two ligand planes is far from orthogonal.[148,150] In the context of DSSC-related issues, the study intends to determine which structure, the one with orthogonal ligand planes or the one with flattened tetrahedral coordination geometry, favors electron injection into the semiconductor nanoparticle electrode and if we can use structural constraints to prevent Jahn–Teller distortion or solvent accessibility to the excited state and hence to modulate the excited state properties. Recently, we have found that one of the water soluble Cu(I) diimine complexes with sulfonic groups[151] performs photoinduced electron injection into TiO_2 nanoparticles. The structural dynamics of similar complexes hybrid with TiO_2 nanoparticles are under investigation from energetic, dynamic and structural control aspects.

12.4 Future Research and Development

A decade of development of XTA has transformed transient structural determination from a fantasy to somewhat routine measurements. However, its contributions to solar energy research as well as our fundamental understanding of photon–matter interactions need further advances in different areas as outlined below.

12.4.1 Visualization of Fundamental Events in Photon–Matter Interactions: Capturing the Transition States

Solar energy conversion processes start from photon–matter interactions and a photon triggered chemical reaction almost always generates changes in nuclear geometry of those involved. The availability of high repetition rate, intense and femtosecond hard X-ray pulses could enable the visualization of nuclear geometry evolution in many light-induced reactions in solar energy conversion processes, including "the transition state" (TS)[152] at the top of the potential barrier crossing trajectory between reactants and products. It is believed that a TS has "no lifetime" or a barrier crossing time of a fraction of the vibrational period. Almost all XTA experiments carried out so far have investigated transient structures that equilibrated in an excited state or transient state potential well rather than followed atomic or electron motions coherently. Hence, the TS has eluded experimental observation especially in the solution phase. Visualization of coherent vibrational motions leading to the TS in chemical reactions will transform our understanding of chemistry by providing information about the displacement pathways of all of the atoms involved, thereby replacing the simplistic descriptions now employed. Many important chemical reactions in solar energy conversion processes involve atomic displacements, such as natural and artificial photosynthesis and catalysis. Despite decades of study using spectroscopic probes, mechanisms of isomerization about a carbon–carbon double bond, and/or multiple double bonds, remain uncertain. Theoretical theoretical studies on barrier crossing and conical intersections for isolated small molecules have been carried out[153] which, combined with femtosecond structural studies, could provide new knowledge in chemical sciences. Some examples shown in the previous section are excellent candidates for such studies. To visualize the TS, the X-ray pulse duration must be shorter than the period of vibrational motion that forms the products.

Most XTA measurements conducted so far employed continuous energy tunability in about 1-keV range from synchrotron sources and collected X-ray absorption spectra step-by-step at each individual X-ray photon energy defined by the monochromator. The current femtosecond X-ray free electron laser (XFEL) sources provide either monochromatic or narrowly distributed energy spectra that are not continuously tunable for the step-by-step approach. For example, the Linear Coherent Light Source (LCLS) at the Stanford Linear Accelerator Center (SLAC) currently has about a 50-eV band width centered around 7.1 keV due to intrinsic bandwidth of self-amplified-spontaneous-emission (SASE).[154]

A preliminary XTA measurement at the X-ray pump–probe station on $[Fe(II)(bpy)_3]^{2+}$ in water demonstrated the feasibility of XANES measurements with $\sim 50\,eV$ spectral range, which is sufficient to cover the pre-edge, edge and above edge region (M. Cammarata, D. Fritz *et al.*, unpublished results). Combining the electronic configuration information obtained from the pre-edge/edge region with the structural information obtained by MIA[42–44] from the XANES region and the full XAFS spectra from synchrotron sources, it is possible to extract the dynamics of electronic transition and nuclear geometry changes simultaneously and to verify directly the Franck–Condon principle using the fs X-ray pulses from XFELs.

The first examples of such studies that are relevant to solar energy conversion are those of transition metal complexes functioning as light harvesters and catalysts. For example, the ground state metal to nitrogen stretching frequency is in a range of 500–700 cm^{-1},[155] corresponding to vibrational periods of about 50–70 fs. In more recent examples of platinum dimer complexes, coherent vibrational motions of Pt–Pt stretching have been observed,[156,157] which are in the range 100–250 cm^{-1}. Therefore, it is possible to follow structural reorganizations resulting from electron transfer directly and hence the reorganization energy if the metal–to–ligand bonds are directly involved. The internal reorganization energy appearing in the Marcus Equation of electron transfer[158–161] applied to organic and organometallic materials has been approximated by a C = C bond stretching frequency of around 1500 cm^{-1},[161] as a generic value, but the actual reorganization is vague owing to lack of the information on transient structures of excited state and charge separated state. Using XTA with fs time resolution, correlations in metal electronic configurations associated with photoinduced electron transfer and the reorganization of nuclear structures included implicitly in the Marcus Equation of electron transfer can be visualized for the first time. This information will help the design of artificial photosynthetic systems for solar fuel or electricity generation.

12.4.2   Detecting Transient Structures with Low Concentrations in Photocatalytic or Irreversible Processes and Rephasing the Coherence of Nuclear Motions in an Ensemble

Most current XTA studies are on the transient structures of excited states that are directly generated by photons through ground to excited state transitions. In solar energy conversion processes, such as photocatalysis, the most interesting structures to be determined are often those resulting from the excited state, such as a photocatalyst–substrate complex and other intermediate structures produced in subsequent transformations. In order to find the structure/property correlations of the catalysts, a detailed knowledge and direct observation of reaction trajectories is necessary.

As illustrated by Figure 12.11, a photoinitiated chemical reaction starts from the light absorption/excited state formation, followed by its transformation to other intermediates and finally the product. It is only possible to probe the structures of these transient species if they have synchronized or in-phase actions. When a laser pulse with 10$^{13-14}$ photons and a bandwidth much smaller than the vibronic level separation strikes a sample, it creates a sub-set of excited state molecules with similar energetics, or more intuitively, at the same point of the multi-dimensional potential landscape (Figure 12.11). This sub-set of the excited state population initially with coherent vibrational motions will proceed to pass multiple potential barriers and then to different intermediate states. As the barrier crossing occurs, the coherence of molecular motions will be lost because the probability of barrier crossing is highly dependent upon the trajectory. A slight deviation of the trajectory could result in success or failure of the barrier crossing. Owing to the nature of the reaction dynamics understood so far, capturing the transition state structures beyond the first barrier crossing will be very challenging, the timescales for the barrier crossing (*i.e.*, a few to a

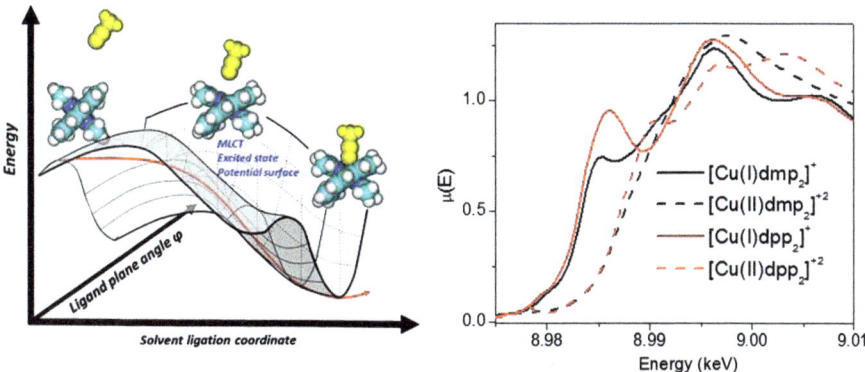

Figure 12.10 (A) MLCT state structural dynamics of $[Cu(I)(dmp)_2]^+ + h\nu \rightarrow$ $[Cu(II)(dmp)(dmp)^-]^+$ as functions of two key reaction coordinates, the angle between the two ligand planes and the solvent ligation distance; (B) structural dependence of XANES spectra for $[Cu(I)(dmp)_2]^+$ and $[Cu(I)(dpp)_2]^+$ (dpp = 2,9-dipenyl-1,10-phenanthroline) and their corresponding Cu(II) species generated by bulk electrolysis. The significant differences in the Cu(I) and Cu(II) spectra provide bases for monitoring the oxidation state and geometry during photoinduced interfacial charge transfer in DSSC composed of the derivatives of these complexes.

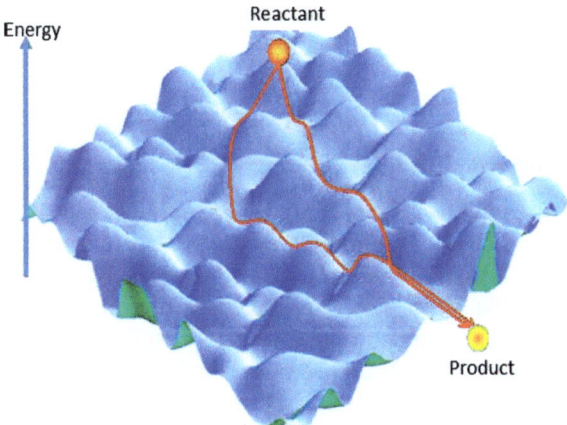

Figure 12.11 Illustration of the potential energy surface and possible trajectories from the same starting point to the final product. The coexisting pathways can destroy the coherence in the trajectory along the reaction coordinates and complicate the interpretation of XTA data.

few tens of fs) is many orders of magnitude shorter than the average time for the molecular trajectory from one thermally equilibrated intermediate to another (*e.g.*, ns–ms). In order to deal with multiple transient species on multiple timescales and gradual dephasing of the coherent motions, the signal-to-noise ratio of the data must be high in order to implement principle component analysis or singular value decomposition in data analyses. Meanwhile, rephasing molecules at a certain time after the excitation remains a challenge.

12.4.3   Theoretical Modeling

Advances in XTA enable direct comparisons between transient structures obtained from experimental observations and theoretical calculations, which simultaneously provide guidance and pose challenges for theoretical modeling. A few requirements in theoretical modeling that only reflect the authors' opinion are briefly outlined below.

12.4.3.1   *Excited State Structural Dynamic Calculations for Transition Metal Complexes Involved in Solar Energy Conversion*

Increasing numbers of groups are now engaging in calculations of excited state structural dynamics from first principles in isolated molecular systems and with relatively small molecules of light elements.[162–165] However, only very limited calculations of excited state structural dynamics with heavy elements are available; these calculations provide not only excited state structures but also dynamics, as well as potential energy surfaces around the metal centers, enabling accurate XANES calculations. This is not uniquely for XTA analyses, but for the whole field of XAS including steady-state measurements.

12.4.3.2   *Extracting Multiple Transient Structures in XTA Data Analyses*

As shown in Equation (12.2), the accuracy of simultaneously and independently obtaining fractions of the transient species and their corresponding transient spectra can be challenging. Current approaches using MIA and MXAN have accomplished the first steps. The requirement is to combine these approaches with accurate potential energy surfaces around metal centers, which by itself is already difficult. Since MXAN and MIA or similar approaches take advantage of the structurally sensitive multiple scattering features in a relatively narrow energy window and with varying combined nuclear coordinates, they succeeded in extracting multiple species and their fractions simultaneously. However, the current development has not been incorporated in high level calculations for both electronic and nuclear structural contributions. This remains a challenge for extracting accurate structures when the two effects interfere.

12.4.3.3   *Normal Mode Analysis Coupled Structural Analysis*

As the time resolution improves, experimental observation approaches direct visualization of nuclear motions during the photoexcitation, which provides the basis for structural control of light–matter interactions in solar energy conversion. The excited state to final photochemical product trajectories are often initiated by vibrational motions. When photons add energy to molecules, their energy will be dissipated among different vibrational modes. Therefore, carrying out a normal mode analysis (NMA) of the excited state will provide a guide to how the nuclear motions were launched at the very beginning of the photon–matter interaction. In data fitting with MIA and similar methods, it is important to choose the correct nuclear coordinates that are adjusted to generate theoretically calculated spectra that fit the experimental results. So far the choice of the combined coordinates has been mainly based on chemical intuition. If the NMA can be conducted for the excited states, one can gain a much better initial selection of the structural variation along the normal modes, which may consequently generate the correct answers sooner with better fits to the experimental results.

12.4.3.4   Ensemble Calculations of Structural Trajectories: Coupling Molecular Dynamics Simulation in Data Analysis

Solar energy conversion functions often depend on dynamic structures in solution or on a supporting matrix where a transiently appearing dynamic structure could evolve into a precursor for catalytic intermediates. Such dynamic structures are implicitly depicted by the Debye–Weller factor in the conventional XAS data analysis in Equation (12.1), without specific description of the structural origin. In many homogeneous photochemical reactions, metal complexes interact with solvent molecules to form transient dynamic solvated structures, such as dynamic bonding between the catalyst molecule and the solvent or substrate molecules. These dynamic structures may well be the precursor or transition states in catalytic reactions, but were unfortunately obscured in the conventional data analysis.

  Molecular dynamics (MD) simulation for many years has been used as a "partner" in analyzing XAFS data for studying the atomic radial distribution function of materials without long range order.[166,167] Unfortunately XAFS is not sensitive enough to the bond angles and thus the MD-XAFS combination is not capable of extracting the full three-dimensional (3D) molecular structure, even though bond angles are very important for gaining deep insight into the "structure–function" relationship of catalytic systems. XANES analysis using a MIA or MXAN scheme has proven to be sensitive enough to extract full 3D local structural parameters around active metal centers in complex systems.[44,168] Recently, MD simulations were applied in advanced XANES analysis of the 3D local atomic structure of a metal ion in water[169,170] using the XANES analysis program MXAN.[45,46] However, the full molecular potential XANES simulation (*i.e.* simulations beyond muffin-tin approximation[171]) used were computationally very expensive, limiting the calculations to a small number of atoms surrounding the metal ion. Hence, a more effective approach is needed to deal with the relatively large number of atoms in multi-metallic catalytic metal complexes for solar energy conversion.

12.5 Summary

Advances in third generation synchrotron sources for XTA over the past decade have significantly enhanced our understanding of photon–matter interactions and the structural origins of photochemical properties of excited state molecules relevant to photoinduced homogeneous and heterogeneous electron and energy transfer processes. As the technique develops, higher quality data and better theoretical modeling are crucial in advancing XTA to become a common method for solving transient structures along excited state trajectories during photochemical processes involving multiple temporal and spatial scales. Emerging ultrafast X-ray sources with pulse durations comparable to ultrafast lasers open new opportunities to study molecular structural dynamics in real time with atomic resolution during fundamental chemical events, such as bond breaking. Making molecular movies is no longer a fantasy, but a reality. Therefore, we expect many future breakthroughs in solving molecular and electronic reorganization during solar energy conversion, which will provide guidance for the design of efficient catalysts for solar fuel generation and materials for efficient photovoltaic generation of electricity.

Acknowledgements

The experimental work was funded by the Division of Chemical Sciences, Geosciences and Biosciences, Office of Basic Energy Sciences of the US Department of Energy through Grant DE-AC02-06CH11357. The authors would like to thank Drs. Klaus Attenkofer, Guy Jennings and Xiaoyi Zhang from the Advanced Photon Source for their long term collaboration and postdoctoral research associates and students who have contributed to the development of XTA method and samples, and Drs. Wighard Jäger, Tao Liu, George B. Shaw, Erik C. Wasinger, Jenny V. Lockard, Andrew B. Stickrath, Michael R. Harpham, Jier Huang, Mr. Michael W. Mara, Ms. Nosheen Gothard and Megan Shelby. The author is thankful for collaboration with colleagues at Argonne, Drs. David J. Gosztola, Jan Hessler, Di-Jia Liu, David M. Tiede and the Solar Energy Conversion (Photosynthesis) Group. Dr. Grigory Smolentsev and Prof. Alexander Soldatov have collaborated in implementing MIA into XTA data analyses. LXC is grateful for collaboration with Profs. Gerald J. Meyer, Jonathan S. Lindsey, Michael D. Hopkins, Michael R. Wasielewski, F. Castellano, Jean-Pierre Sauvage, Fraser Stoddart, M. Benfatto, S. Della Longa and Daniel G. Nocera and their groups in various institutions. The long term support from the Sector 11 & 12 staff at the Advanced Photon Source has been invaluable to the work presented here. Use of the Advanced Photon Source was supported by the US Department of Energy, Office of Science, Office of Basic Energy Sciences, under Contract No. DE-AC02-06CH11357. The author would like to acknowledge the assistance received from Brian Roczynski, Michael Mara and Michael Harpham in proof reading the manuscript.

References

1. R. E. Blankenship, D. M. Tiede, J. Barber, G. W. Brudvig, G. R. Fleming, M. Ghirardi, M. R. Gunner, W. Junge, D. Kramer, A. Melis, T. A. Moore, C. C. Moser, D. G. Nocera, A. J. Nozik, D. R. Ort, W. W. Parson, R. C. Prince and R. T. Sayre, *Science*, 2011, **332**, 805–809.
2. N. S. Lewis and G. Crabtree, Basic Research Needs for Solar Energy Utilization, *Basic Energy Sciences*, US Department of Energy, 2005.
3. N. S. Lewis and D. G. Nocera, *Proc. Natl. Acad. Sci. U. S. A*, 2007, **104**, 20142–20142.
4. H. G. Park and J. K. Holt, *Energy Environ. Sci.*, 2011, **3**, 1028–1036.
5. H. Dau and I. Zaharieva, *Acc. Chem. Res.*, 2009, **42**, 1861–1870.
6. D. Gust, T. A. Moore and A. L. Moore, *Acc. Chem. Res.*, 2009, **42**, 1890–1898.
7. D. G. Nocera, *Inorg. Chem.*, 2009, **48**, 10001–10017.
8. T. R. Cook, D. K. Dogutan, S. Y. Reece, Y. Surendranath, T. S. Teets and D. G. Nocera, *Chem. Rev.*, 2010, **110**, 6474–6502.
9. D. W. McCamant, P. Kukura and R. A. Mathies, *J. Phys. Chem. A*, 2003, **107**, 8208–8214.
10. P. Kukura, D. W. McCamant and R. A. Mathies, *Ann. Rev. Phys. Chem.*, 2007, **58**, 461–488.
11. E. T. J. Nibbering, H. Fidder and E. Pines, *Ann. Rev. Phys. Chem.*, 2005, **56**, 337–367.
12. L. X. Chen, *Ann. Rev. Phys. Chem.*, 2005, **56**, 221–254.
13. L. X. Chen, *Angew. Chem., Int. Ed.* 2004, **43**, 2886–2905.
14. C. Bressler and M. Chergui, *Chem. Rev.*, 2004, **104**, 1781–1812.
15. D. E. Sayers, E. A. Stern and F. Lytle, *Phys. Rev. Lett.*, 1971, **27**, 1204–1207.
16. F. W. Lytle, D. E. Sayers and E. A. Stern, *Phys. Rev. B*, 1975, **11**, 4825–4835.
17. D. E. Sayers, F. W. Lytle, M. Weissbluth and P. Pianetta, *J. Chem. Phys.*, 1975, **62**, 2514–2515.
18. E. A. Stern, D. E. Sayers and F. W. Lytle, *Phys. Rev. B*, 1975, **11**, 4836–4846.
19. B. K. Teo, D. C. Joy, Eds, EXAFS [Extended X-Ray Absorption Fine Structure] *Spectroscopy: Techniques and Applications*, Springer-Verlag, New York, LLC, 1981.
20. D. C. Koningsberg and R. Prins, X-ray Absorption:Principles, Applications, Techniques of EXAFS, *SEXAFS and XANES*, John Wiley & Sons, New York, 1988.
21. G. Bunker, *Introduction to XAFS: A Practical Guide to X-ray Absorption Fine Structure Spectroscopy*, Cambridge University Press, Cambridge, 2010.
22. J. Stöhr, NEXAFS Spectroscopy, Springer-Verlag, New York, LLC, 2010.
23. B. Chance, R. Fischetti and L. Powers, *Biochemistry*, 1983, **22**, 3820–3829.
24. D. M. Mills, A. Lewis, A. Harootunian, J. Huang and B. Smith, *Science*, 1984, **223**, 811–813.
25. B. Chance, C. Kumar, Z. R. Korszun, V. Legallis, W. Pennie, J. Sorge and S. Khalid, *Nucl. Instrum. Methods Phys. Res., Sect. A*, 1984, **222**, 180–184.

26. J. Z. Tischler, B. C. Larson and D. M. Mills, *Mater. Res. Soc. Symp. Proc.*, 1985, **35**, 119–124.

27. T. Matsushita, H. Oyanagi, S. Saigo, U. Kaminaga, H. Hashimoto, H. Kihara, N. Yoshida and M. Fujimoto, *Jpn. J. Appl. Phys., Part 2*, 1986, **25**, L523–L525.

28. M. R. Chance, M. D. Wirt, E. M. Scheuring, L. M. Miller, A. Xie and D. E. Sidelinger, *Rev. Sci. Instr.*, 1993, **64**, 2035–206.

29. L. X. Chen, M. K. Bowman, P. A. Montano and J. R. Norris, *Mater. Res. Soc. Symp. Proc.*, 1993, **307**, 45–50.

30. L. X. Chen, M. K. Bowman, P. A. Montano and J. R. Norris, *J. Am. Chem. Soc.*, 1993, **115**, 4373–4374.

31. L. X. Chen, M. K. Bowman, Z. Wang, P. A. Montano and J. R. Norris, *J. Phys. Chem.*, 1994, **98**, 9457–9464.

32. L. X. Chen, Z. Wang, J. K. Burdett, P. A. Montano and J. R. Norris, *J. Phys. Chem.*, 1995, **99**, 7958–7964.

33. L. X. Chen, M. R. Wasielewski, T. Rajh, P. L. Lee, M. C. Thurnauer and P. A. Montano, *J. Phys. IV*, 1997, **7**, 569–572.

34. L. X. Chen, P. L. Lee, D. Gosztola, W. A. Svec, P. A. Montano and M. R. Wasielewski, *J. Phys. Chem. B*, 1999, **103**, 3270–3274.

35. L. X. Chen, P. L. Lee, D. Gosztola, W. A. Svec and M. R. Wasielewski, *J. Synchrotron Radiat.*, 1999, **6**, 403–405.

36. C. Bressler, R. Abela and M. Chergui, *Z. Kristallogr.*, 2008, **223**, 307–321.

37. M. Chergui, *Acta Crystallogr. A*, 2010, **66**, 229–239.

38. C. Bressler and M. Chergui, *Ann. Rev. Phys. Chem.*, 2010, **61**, 263–282.

39. G. Jennings, W. J. H. Jaeger and L. X. Chen, *Rev. Sci. Inst*, 2002, **72**, 362–368.

40. L. X. Chen, X. Y. Zhang, E. C. Wasinger, K. Attenkofer, G. Jennings, A. Z. Muresan and J. S. Lindsey, *J. Am. Chem. Soc.*, 2007, **129**, 9616–9618.

41. L. X. Chen, G. B. Shaw, I. Novozhilova, T. Liu, G. Jennings, K. Attenkofer, G. J. Meyer and P. Coppens, *J. Am. Chem. Soc.*, 2003, **125**, 7022–7034.

42. G. Smolentsev and A. Soldatov, *J. Synchrotron Radiat.*, 2006, **13**, 19–29.

43. G. Smolentsev and A. V. Soldatov, *Comput. Mater. Sci.*, 2007, **39**, 569–574.

44. G. Smolentsev, A. V. Soldatov and L. X. Chen, *J. Phys. Chem. A*, 2008, **112**, 5363–5367.

45. M. Benfatto and S. Della Longa, *J. Synchrotron Radiat.*, 2001, **8**, 1087–1094.

46. M. Benfatto, S. Della Longa, K. Hatada, K. Hayakawa, W. Gawelda, C. Bressler and M. Chergui, *J. Phys. Chem., B*, 2006, **110**, 14035–14039.

47. S. Della-Longa, L. X. Chen, P. Frank, K. Hayakawa, K. Hatada and M. Benfatto, *Inorg. Chem.*, 2009, **48**, 3934–3942.

48. R. M. van der Veen, C. J. Milne, A. El Nahhas, F. A. Lima, V. T. Pham, J. Best, J. A. Weinstein, C. N. Borca, R. Abela, C. Bressler and M. Chergui, *Angew. Chem., Int. Ed.*, 2009, **48**, 2711–2714.

49. M. Saes, C. Bressler, R. Abela, D. Grolimund, S. L. Johnson, P. A. Heimann and M. Chergui, *Phys. Rev. Lett.*, 2003, **90**, 047403–1.
50. W. Gawelda, M. Johnson, F. M. F. de Groot, R. Abela, C. Bressler and M. Chergui, *J. Am. Chem. Soc.*, 2006, **128**, 5001–5009.
51. W. Gawelda, V. T. Pham, M. Benfatto, Y. Zaushitsyn, M. Kaiser, D. Grolimund, S. L. Johnson, R. Abela, A. Hauser, C. Bressler and M. Chergui, *Phys. Rev. Lett.*, 2007, **98**, 05740.
52. V. T. Pham, W. Gawelda, Y. Zaushitsyn, M. Kaiser, D. Grolimund, S. L. Johnson, R. Abela, C. Bressler and M. Chergui, Observation of the solvent shell reorganization around photoexcited atomic solutes by picosecond X-ray absorption spectroscopy, *J. Am. Chem. Soc.*, 2007, **129**, 1530–1531.
53. M. Khalil, M. A. Marcus, A. L. Smeigh, J. K. McCusker, H. H. W. Chong and R. W. Schoenlein, *J. Phys. Chem. A*, 2006, **110**, 38–44.
54. A. Cavalleri, M. Rini, H. H. W. Chong, S. Fourmaux, T. E. Glover, P. A. Heimann, J. C. Kieffer and R. W. Schoenlein, *Phys. Rev. Lett.*, 2005, **95**, 067405.
55. Q. S. Yin, J. M. Tan, C. Besson, Y. V. Geletii, D. G. Musaev, A. E. Kuznetsov, Z. Luo, K. I. Hardcastle and C. L. Hill, *Science*, 2010, **328**, 342–345.
56. Y. V. Geletii, B. Botar, P. Koegerler, D. A. Hillesheim, D. G. Musaev and C. L. Hill, *Angew. Chem., Int. Ed.*, 2008, **47**, 3896–3899.
57. Y. V. Geletii, C. Besson, Y. Hou, Q. S. Yin, D. G. Musaev, D. Quinonero, R. Cao, K. I. Hardcastle, A. Proust, P. Kogerler and C. L. Hill, *J. Am. Chem. Soc.*, 2009, **131**, 17360–17370.
58. D. A. Lutterman, Y. Surendranath and D. G. Nocera, *J. Am. Chem. Soc.*, 2009, **131**, 3838–3839.
59. A. M. March, A. B. Stickrath, G. Doumy, E. P. Kanter, B. Krässig, S. H. Southworth, K. Attenkofer, C. A. Kurtz, L. X. Chen and L. Young, *Rev. Sci. Inst.*, 2011, **82**, 073110.
60. F. A. Lima, C. J. Milne, D. C. V. Amarasinghe, M. H. Rittmann-Frank, R. M. van der Veen, M. Reinhard, V. T. Pham, S. Karlsson, S. L. Johnson, D. Grolimund, C. Borca, T. Huthwelker, M. Janousch, F. van Mourik, R. Abela and M. Chergui, *Rev. Sci. Instr.*, 2011, **82**, 126106.
61. P. Kirkpatrick and A. V. Baez, *J. Opt. Soc. Am.*, 1948, **38**, 766–773.
62. X. Zhang, G. Smolentsev, J. Guo, K. Attenkofer, C. Kurtz, G. Jennings, J. V. Lockard, A. B. Stickrath and L. X. Chen, *J. Phys. Chem. Lett.*, 2011, **2**, 628–632.
63. C. P. J. Barty, M. Ben-Nun, T. Guo, F. Raksi, C. Rose-Petruck, J. Squier, K. R. Wilson, V. V. Yakovlev, P. M. Weber, Z. Jiang, A. Ikhlef and J.-C. Kieffer, *Oxford Ser. Synchrotron Radiat.*, 1997, **2**, 44–70.
64. R. W. Schoenlein, S. Chattopadhyay, H. H. W. Chong, T. E. Glover, P. A. Heimann, C. V. Shank, A. A. Zholents and M. S. Zolotorev, *Science*, 2000, **287**, 2237–2240.

65. V. Srajer, S. Crosson, M. Schmidt, J. Key, F. Schotte, S. Anderson, B. Perman, R. Zhong, T. Y. Teng, D. Bourgeois, M. Wulff and K. Moffat, *J. Synchrotron Radiat.*, 2000, **7**, 236–244.

66. M. F. DeCamp, D. A. Reis, P. H. Bucksbaum, B. Adams, J. M. Caraher, R. Clarke, C. W. S. Conover, E. M. Dufresne, R. Merlin, V. Stoica and J. K. Wahlstrand, *Nature*, 2001, **413**, 825–828.

67. R. Neutze, R. Wouts, S. Techert, J. Davidsson, M. Kocsis, A. Kirrander, F. Schotte and M. Wulff, *Phys. Rev. Lett.*, 2001, **87**, 195508/1–195508/4.

68. S. Techert, F. Schotte and M. Wulff, *Phys. Rev. Lett.*, 2001, **86**, 2030–2033.

69. L. X. Chen, W. J. H. Jager, G. Jennings, D. J. Gosztola, A. Munkholm and J. P. Hessler, *Science*, 2001, **292**, 262–264.

70. M. Saes, C. Bressler, R. Abela, D. Grolimund, S. L. Johnson, P. A. Heimann and M. Chergui, *Phys. Rev. Lett.*, 2003, **90**, 047403/1–047403/4.

71. http://www.xfel.eu/, and http://www.riken.jp/XFEL, http://www-ssrl.slac.stanford.edu/lcls/.

72. L. X. Chen, G. Jennings, T. Liu, D. J. Gosztola, J. P. Hessler, D. V. Scaltrito and G. J. Meyer, *J. Am. Chem. Soc.*, 2002, **124**, 10861–10867.

73. L. X. Chen, G. B. Shaw, T. Liu, G. Jennings and K. Attenkofer, *Chem. Phys.*, 2004, **299**, 215–223.

74. L. X. Chen, *Ann. Rev. Phys. Chem.*, 2005, **56**, 221–254.

75. D. Gust, T. A. Moore and A. L. Moore, *Acc. Chem. Res.*, 2001, **34**, 40–48.

76. J. H. Alstrum-Acevedo, M. K. Brennaman and T. J. Meyer, *Inorg. Chem.*, 2005, **44**, 6802–6827.

77. T. Hasobe, H. Imahori, P. V. Kamat, T. K. Ahn, S. K. Kim, D. Kim, A. Fujimoto, T. Hirakawa and S. Fukuzumi, *J. Am. Chem. Soc.*, 2005, **127**, 1216–1228.

78. D. Holten, D. F. Bocian and J. S. Lindsey, *Acc. Chem. Res.*, 2002, **35**, 57–69.

79. M. R. Wasielewski, *Chem. Rev.*, 1992, **92**, 435–61.

80. J. Rosenthal, T. D. Luckett, J. M. Hodgkiss and D. G. Nocera, *J. Am. Chem. Soc.*, 2006, **128**, 6546–6547.

81. V. S. Y. Lin, S. G. Dimagno and M. J. Therien, *Science*, 1994, **264**, 1105–1111.

82. D. M. Guldi, *Chem. Soc. Rev.*, 2002, **31**, 22–36.

83. D. Holten, D. F. Bocian and J. S. Lindsey, *Acc. Chem. Res.*, 2002, **35**, 57–69.

84. J. Costamagna, G. Ferraudi, J. Canales and J. Vargas, *Coord. Chem. Rev.*, 1996, **148**, 221–248.

85. R. Gerdes, D. Wohrle, W. Spiller, G. Schneider, G. Schnurpfeil and G. Schulz-Ekloff, *J. Photochem. Photobiol., A*, 1997, **111**, 65–74.

86. A. Maldotti, A. Molinari and R. Amadelli, *Chem. Rev.*, 2002, **102**, 3811–3836.

87. M. J. Esswein and D. G. Nocera, Hydrogen production by molecular photocatalysis, *Chem. Rev.*, 2007, **107**, 4022–4047.
88. D. Kim, C. Kirmaier and D. Holten, *Chem. Phys.*, 1983, **75**, 305–322.
89. J. Rodriguez and D. Holten, *J. Chem. Phys.*, 1989, **91**, 3525–3531.
90. J. Rodriguez and D. Holten, *J. Chem. Phys.*, 1990, **92**, 5944–5950.
91. C. J. Ballhausen, Introduction to Ligand Field Theory, McGraw-Hill, New York, 1962.
92. S. Gentemann, N. Y. Nelson, L. Jaquinod, D. J. Nurco, S. H. Leung, C. J. Medforth, K. M. Smith, J. Fajer and D. Holten, *J. Phys. Chem. B*, 1997, **101**, 1247–1254.
93. H. S. Eom, S. C. Jeoung, D. Kim, J.-H. Ha and Y.-R. Kim, *J. Phys. Chem. A*, 1997, **101**, 3661–3669.
94. C. M. Drain, S. Gentemann, J. A. Roberts, N. Y. Nelson, C. J. Medforth, S. Jia, M. C. Simpson, K. M. Smith, J. Fajer, J. A. Shelnutt and D. Holten, *J. Am. Chem. Soc.*, 1998, **120**, 3781–3791.
95. E. W. Findsen, J. A. Shelnutt, J. M. Friedman and M. R. Ondrias, *Chem. Phys. Lett.*, 1986, **126**, 465–71.
96. C. J. Medforth, M. O. Senge, K. M. Smith, L. D. Sparks and J. A. Shelnutt, *J. Am. Chem. Soc.*, 1992, **114**, 9859–9869.
97. X.-Z. Song, W. Jentzen, S.-L. Jia, L. Jaquinod, D. J. Nurco, C. J. Medforth, K. M. Smith and J. A. Shelnutt, *J. Am. Chem. Soc.*, 1996, **118**, 12975–12988.
98. L.-S. Kau, D. J. Spira-Solomon, J. E. Penner-Hahn, K. O. Hodgson and E. I. Solomon, *J. Am. Chem. Soc.*, 1987, **109**, 6433–6442.
99. J. L. Retsek, C. M. Drain, C. Kirmaier, D. J. Nurco, C. J. Medforth, K. M. Smith, I. V. Sazanovich, V. S. Chirvony, J. Fajer and D. Holten, *J. Am. Chem. Soc.*, 2003, **125**, 9787–9800.
100. L. X. Chen, X. Zhang, E. C. Wasinger, K. Attenkofer, G. Jennings, A. Muresan and J. S. Lindsey, *J. Am. Chem. Soc.*, 2007, **129**, 9616–9618.
101. A. Staniszewski, S. Ardo, Y. L. Sun, F. N. Castellano and G. J. Meyer, *J. Am. Chem. Soc.*, 2008, **130**, 11586–+.
102. B. O'Regan and M. Gratzel, *Nature*, 1991, **353**, 737–740.
103. A. Hagfeldt and M. Gratzel, *Acc. Chem. Res.*, 2000, **33**, 269–277.
104. N. A. Anderson and T. Q. Lian, *Ann. Rev. Phys. Chem.*, 2005, **56**, 491–519.
105. D. P. Hagberg, J. H. Yum, H. Lee, F. De Angelis, T. Marinado, K. M. Karlsson, R. Humphry-Baker, L. C. Sun, A. Hagfeldt, M. Gratzel and M. K. Nazeeruddin, *J. Am. Chem. Soc.*, 2008, **130**, 6259–6266.
106. T. W. Hamann, R. A. Jensen, A. B. F. Martinson, H. Van Ryswyk and J. T. Hupp, *Energy Environ. Sci.*, 2008, **1**, 66–78.
107. S. Ardo and G. J. Meyer, *Chem. Soc. Rev.*, 2009, **38**, 115–164.
108. S. Ardo, Y. Sun, A. Staniszewski, F. N. Castellano and G. J. Meyer, *J. Am. Chem. Soc.*, 2010, **132**, 6696–6709.
109. F. De Angelis, R. Car and T. G. Spiro, *J. Am. Chem. Soc.*, 2003, **125**, 15710–15711.
110. F. De Angelis, S. Fantacci, A. Selloni, M. K. Nazeeruddin and M. Gratzel, *J. Phys. Chem., C*, 2010, **114**, 6054–6061.

111. P. Persson and M. J. Lundqvist, *J. Phys. Chem., B*, 2005, **109**, 11918–11924.
112. G. Benko, J. Kallioinen, J. E. I. Korppi-Tommola, A. P. Yartsev and V. Sundstrom, *J. Am. Chem. Soc.*, 2002, **124**, 489–493.
113. D. F. Watson and G. J. Meyer, *Ann. Rev. Phys. Chem.*, 2005, **56**, 119–156.
114. Y. Ma, C. T. Chen, G. Meigs, K. Randall and F. Sette, *Phys. Rev. A.*, 1991, **44**, 1848–1858.
115. A. L. Ankudinov, B. Ravel, J. J. Rehr and S. D. Conradson, *Phys. Rev. B: Condens. Matter Mater. Phys.*, 1998, **58**, 7565–7576.
116. W. M. Campbell, A. K. Burrell, D. L. Officer and K. W. Jolley, *Coord. Chem. Rev.*, 2004, **248**, 1363–1379.
117. W. M. Campbell, K. W. Jolley, P. Wagner, K. Wagner, P. J. Walsh, K. C. Gordon, L. Schmidt-Mende, M. K. Nazeeruddin, Q. Wang, M. Gratzel and D. L. Officer, *J. Phys. Chem., C*, 2007, **111**, 11760–11762.
118. A. Kay and M. Gratzel, *J. Phys. Chem.*, 1993, **97**, 6272–6277.
119. B. A. Gregg, *J. Phys. Chem., B*, 2003, **107**, 4688–4698.
120. K. Hara, Z. S. Wang, T. Sato, A. Furube, R. Katoh, H. Sugihara, Y. Dan-Oh, C. Kasada, A. Shinpo and S. Suga, *J. Phys. Chem., B*, 2005, **109**, 15476–15482.
121. H. Imahori, T. Umeyama and S. Ito, *Acc. Chem. Res.*, 2009, **42**, 1809–1818.
122. J. E. Monat and J. K. McCusker, *J. Am. Chem. Soc.*, 2000, **122**, 4092–4097.
123. N. Alonsovante, J. F. Nierengarten and J. P. Sauvage, *J. Chem. Soc., Dalton Trans.*, 1994, 1649–1654.
124. T. Bessho, E. C. Constable, M. Graetzel, A. H. Redondo, C. E. Housecroft, W. Kylberg, M. K. Nazeeruddin, M. Neuburger and S. Schaffner, *Chem. Commun.*, 2008, 3717–3719.
125. E. C. Constable, A. H. Redondo, C. E. Housecroft, M. Neuburger and S. Schaffner, *Dalton Trans.*, 2009, 6634–6644.
126. B. A. Gandhi, O. Green and J. N. Burstyn, *Inorg. Chem.*, 2007, **46**, 3816–3825.
127. S. Hattori, Y. Wada, S. Yanagida and S. Fukuzumi, *J. Am. Chem. Soc.*, 2005, **127**, 9648–9654.
128. N. Robertson, *Chemsuschem*, 2008, **1**, 977–979.
129. S. Sakaki, T. Kuroki and T. Hamada, *J. Chem. Soc., Dalton Trans.*, 2002, 840–842.
130. A. Suzuki, K. Kobayashi, T. Oku and K. Kikuchi, *Mater. Chem. Phys.*, 2010, **129**, 236–241.
131. D. R. McMillin, M. T. Buckner and B. T. Ahn, *Inorg. Chem.*, 1977, **16**, 943–945.
132. G. Blasse and D. R. McMillin, *Chem. Phys. Lett.*, 1980, **70**, 1–3.
133. D. R. McMillin and M. C. Morris, *Proc. Natl. Acad. Sci. U. S. A.*, 1981, **78**, 6567–6570.
134. J. R. Kirchhoff, R. E. Gamache, M. W. Blaskie, A. A. Del Paggio, R. K. Lengel and D. R. McMillin, *Inorg. Chem.*, 1983, **22**, 2380–2384.

135. D. R. McMillin, J. R. Kirchhoff and K. V. Goodwin, *Coord. Chem. Rev.*, 1985, **64**, 83–92.
136. S. E. J. Bell and J. J. McGarvey, *Chem. Phys. Lett.*, 1986, **124**, 336–340.
137. C. E. A. Palmer and D. R. McMillin, *Inorg. Chem.*, 1987, **26**, 3837–3840.
138. F. N. Castellano, M. Ruthkosky and G. J. Meyer, *Inorg. Chem.*, 1995, **34**, 3 4.
139. K. L. Cunningham, C. R. Hecker and D. R. McMillin, *Inorg. Chim. Acta*, 1996, **242**, 143–147.
140. M. Ruthkosky, F. N. Castellano and G. J. Meyer, *Inorg. Chem.*, 1996, **35**, 6406–6412.
141. M. Anpo, M. Matsuoka, K. Hanou, H. Mishima, H. Yamashita and H. H. Patterson, *Coord. Chem. Rev.*, 1998, **171**, 175–184.
142. K. L. Cunningham and D. R. McMillin, *Inorg. Chem.*, 1998, **37**, 4114–4119.
143. M. Ruthkosky, C. A. Kelly, F. N. Castellano and G. J. Meyer, *Coord. Chem. Rev.*, 1998, **171**, 309–322.
144. C. T. Cunningham, J. J. Moore, K. L. H. Cunningham, P. E. Fanwick and D. R. McMillin, *Inorg. Chem.*, 2000, **39**, 3638–3644.
145. D. V. Scaltrito, D. W. Thompson, J. A. O'Callaghan and G. J. Meyer, *Coord. Chem. Rev.*, 2000, **208**, 243–266.
146. N. Armaroli, G. Accorsi, J. P. Gisselbrecht, M. Gross, J. F. Eckert and J. F. Nierengarten, *New J. Chem.*, 2003, **27**, 1470–1478.
147. L. X. Chen, G. B. Shaw, I. Novozhilova, T. Liu, G. Jennings, K. Attenkofer, G. J. Meyer and P. Coppens, *J. Am. Chem. Soc.*, 2003, **125**, 7022–7034.
148. G. B. Shaw, C. D. Grant, E. W. Castner, G. J. Meyer and L. X. Chen, *J. Am. Chem. Soc.*, 2007, **129**, 2147–2160.
149. J. V. Lockard, S. Kabehie, G. Smolentsev, A. Soldatov, J. I. Zink and L. X. Chen, *J. Phys. Chem. B*, 2010, **114**, 14521–14527.
150. M. Iwamura, S. Takeuchi and T. Tahara, *J. Am. Chem. Soc.*, 2007, **129**, 5248–5256.
151. N. A. Vante, V. Ern, P. Chartier, C. O. Dietrich-Buchecker, D. R. McMillin, P. A. Marnot and J. P. Sauvage, *Nouv. J. Chim.*, 1983, **7**, 3–5.
152. D. G. Truhlar, B. C. Garrett and S. J. Klippenstein, *J. Phys. Chem.*, 1996, **100**, 12771–12800.
153. Domcke, W., Yarkony, D. R., Koppel, H. *Conical Intersections: Theory, Computation and Experiment*, World Scientific, 2011.
154. Pellegrini, C., Stöhr, J. In *Beam Line*, Rutherford, A., Ed., Stanford Linear Accelerator Center, 2002, Vol. 32, pp 32.
155. D. C. Bradley and M. H. Gitlitz, *Nature*, 1968, **218**, 353–354.
156. S. Cho, M. W. Mara, X. Wang, J. V. Lockard, A. A. Rachford, F. N. Castellano and L. X. Chen, *J. Phys. Chem. A*, 2011, **115**, 3990–3996.
157. R. M. van der Veen, A. Cannizzo, F. van Mourik, A. Vlcek and M. Chergui, *J. Am. Chem. Soc.*, 2011, **133**, 305–315.

158. R. A. Marcus and N. Sutin, *Biochim. Biophys. Acta*, 1985, **811**, 265–322.
159. R. A. Marcus, *Rev. Mod. Phys.*, 1993, **65**, 599–610.
160. D. M. Adams, L. Brus, C. E. D. Chidsey, S. Creager, C. Creutz, C. R. Kagan, P. V. Kamat, M. Lieberman, S. Lindsay, R. A. Marcus, R. M. Metzger, M. E. Michel-Beyerle, J. R. Miller, M. D. Newton, D. R. Rolison, O. Sankey, K. S. Schanze, J. Yardley and X. Y. Zhu, *J. Phys. Chem., B*, 2003, **107**, 6668–6697.
161. P. F. Barbara, T. J. Meyer and M. A. Ratner, *J. Phys. Chem.*, 1996, **100**, 13148–13168.
162. L. Serranoandres, M. Merchan, I. Nebotgil, R. Lindh and B. O. Roos, *J. Chem. Phys.*, 1993, **98**, 3151–3162.
163. F. Furche and R. Ahlrichs, *J. Chem. Phys.*, 2002, **117**, 7433–7447.
164. Z. R. Grabowski, K. Rotkiewicz and W. Rettig, *Chem. Rev.*, 2003, **103**, 3899–4031.
165. A. Dreuw and M. Head-Gordon, *J. Am. Chem. Soc.*, 2004, **126**, 4007–4016.
166. B. J. Palmer, D. M. Pfund and J. L. Fulton, *J. Phys. Chem.*, 1996, **100**, 13393–13398.
167. S. W. Wallen, B. J. Palmer, D. M. Pfund, J. L. Fulton, M. Newville, Y. Ma and E. A. Stern, *J. Phys. Chem. A*, 1997, **101**, 9632–9640.
168. L. X. Chen, X. Zhang, E. C. Wasinger, J. V. Lockard, A. B. Stickrath, M. W. Mara, K. Attenkofer, G. Jennings, G. Smolentsev and A. Soldatov, *Chem. Sci.*, 2010, **1**, 642–650.
169. P. D'Angelo, V. Migliorati, G. Mancini and G. Chillemi, *J. Phys. Chem. A*, 2008, **112**, 11833–11841.
170. P. D'Angelo, A. Zitolo, V. Migliorati, G. Mancini, I. Persson and G. Chillemi, *Inorg. Chem.*, 2009, **48**, 10239–10248.
171. G. Smolentsev, A. V. Soldatov and M. C. Feiters, *Phys. Rev. B*, 2007, **75**, 144106–1.

Subject Index

adsorbate anchoring, 95–96
adsorbate ions, 9
advanced photon source, 345–347
alizarin, 126
alkali atoms, chemisorption of, 251
amplified spontaneous emission (ASE), 208
atomic force microscopy (AFM), 216, 276
Auger recombination processes, 211
Auger theory, 68–69
autocorrelation function (ACF), 50–52

benchmark redox couples, 23
bioinspired high-potential porphyrin photoanodes, 5
bremsstrahlung radiation, 270, 271
bulk hetero-junction (BHJ), 78
    polymers and, 104–105
bulk semiconductors, 308–310, 318–323
Butler-Volmer equation, 5

carrier multiplication (CM), 316–323
cathodoluminescence (CL), 273
central processing unit (CPU), 119
charge separated (CS), 191
chemical vapor deposition (CVD), 216, 217
circular dichroism in angular distributions (CDAD), 248
coarse grained, and finite element methods, 81
computational materials science, 79
conduction band (CB), 137, 141, 326

conductor-like polarizable continuum model (C-PCM), 123
configuration interactions (CI), 43
coumarin dyes, 10
coupling molecular dynamics simulation, in data analysis, 361
crystalline solar cells (CSC), 112
Cs σ-resonance, 252–253
cylindrical silver nanowires, 119

DDA see discrete dipole approximation (DDA)
delocalized bands, 246–250
density functional theory computational methods, 26
density functional theory (DFT), 3, 40, 80
density of states (DOS), 11
dephasing mechanism, 42
destiny of initial states (DOS), 228
DFT see density functional theory (DFT)
dipole transition moment, 128
direct mechanism, 42
direct recombination, 3
discrete dipole approximation (DDA), 120–122
disordered interfaces, functional control of, 102–103
Drude-Smith model, 311
DSSCs see dye-sensitized solar cells (DSSCs)
dye-sensitized metal oxide, 292
dye-sensitized semiconductor surfaces, 94

dye-sensitized solar cells (DSSCs), 2,
    78, 281–282, 323–324

effective mass theory (EMT), 38
electron-cation (EC), 152
electron dynamics, 101
electron equilibration method
    (EEM), 87
electronic coherences, 167
electronic interactions, 96
electron injection, 142
electron-phonon coupling, 63–64
electron-phonon relaxation, 69
electron transfer rates, 147
electrostatic interactions, 87
excitation intensity, 211
excited state absorption (ESA), 165
exciton-biexciton transitions, 195
external quantum efficiency (EQE), 148

Fano interference, 254
fast Fourier transform (FFT)
    algorithm, 176
FDTD *see* finite-difference
    time-domain (FDTD)
FEM *see* finite-element method
    (FEM)
fewest switches SH (FSSH), 40
finite-difference time-domain
    (FDTD), 116–117, 242
finite-element method (FEM), 81,
    117–120
fluctuation-induced tunneling
    conduction (FITC) model, 3, 13,
    16–18
focused ion beam (FIB), 242
Fourier transforming, 182
free-induction decay (FID), 241
full multiple scattering (FMS), 353
full width at half maximum (FWHM),
    346

gadolinium gallium garnet (GGG),
    206
Gaussian function, 188
ground state bleach (GSB), 165
group velocity dispersion (GVD), 207

Hamiltonian matrix, 193
Hartree-Fock (HF) method, 39
heterogeneous electron transfer
    see photoinduced interfacial
    electron transfer
high efficiency photovoltaic
    concepts, carrier dynamics studies
    of, 314
        semiconductor quantum dots,
            315–316
        semiconductors, CM in,
            316–323
highest occupied molecular orbital
    (HOMO), 183, 283
highly non-exponential slow electron
    injection, 140
homogeneous electron/energy
    transfer *see* metalloporphyrin
    excited state structural dynamics

impact ionization (II), 41–42
inhomogeneous/homogeneous
    broadenings, 196
interfacial charge transfer excitations,
    97
interfacial electronic coupling,
    97–99
interfacial electron transfer,
    dye-sensitized solar cell (DSC)
        electron-cation charge
            recombination, 143–148
        electron-cation interactions,
            151–157
        electron-cation recombination,
            148–151
        electron injection from
            sensitizer, 136–143
        Grätzel solar cells, 135–136
interfacial electron transfer (IET)
        excitations at interfaces
            molecular excitations,
                91–92
            nanomaterials,
                92–94
        interfacial interactions
            adsorbate anchoring,
                95–96

dye-sensitized
semiconductor surfaces,
94
electronic interactions, 96
materials modelling, 79
modelling methods, 79–81
multiscale modelling,
81–90
modelling complex materials
disordered interfaces,
functional control of,
102–103
mesoscopic charge
transport, 103–104
polymers and bulk
hetero-junctions,
104–105
semiconductor
hetero-structures and
hybrid systems, 105
surface electron transfer
electron dynamics, 101
interfacial charge transfer
excitations, 97
interfacial electronic
coupling, 97–99
mechanisms, 96–97
spacer groups, 99–101
interligand electron transfer (ILET),
140
internal conversion (IC), 139
intersystem crossing (ISC), 138
intraband optical transition, 285
intramolecular vibrational relaxation
(IVR), 139
ions effect, 292–294

Kerr gated fluorescence microscopy
(KGFM), 205–213
KGFM *see* Kerr gated fluorescence
microscopy (KGFM)

laser intensity fluctuations, 214
ligand charge transfer state (MLCT),
137
ligand-to-ligand charge transfer
(LLCT), 124

linear coherent light source (LCLS),
357
linear response time-dependent
density functional theory
(LR-TDDFT), 123–125
Liouville pathways, 197
lithium iodide (LiI), 292
localized electronic states, 250–256
localized surface plasmon (LSP), 264
local oscillator (LO), 168
lower incident photon to current
efficiency (IPCE), 292
lowest unoccupied molecular orbital
(LUMO), 183, 282
LR-TDDFT *see* linear response time-
dependent density functional
theory (LR-TDDFT)

Mach-Zehnder interferometer (MZI),
238
Maxwell-Boltzmann distribution, 155
MEG *see* multiexciton generation
(MEG)
mercury-cadmium-telluride (MCT),
287
mesoscopic charge transport, 103–104
metalloporphyrin excited state
structural dynamics, 347–352
metal nanoparticle, 294–296
metal oxides, interfacial electron
injection mechanism into, 287
dye dependence, 289–291
ions effect, 292–294
metal nanoparticle to, 294–296
semiconductor dependence,
288–289
solvent effect, 292
metal oxide semiconductor
dependence, 288–289
metal-to-ligand charge transfer
(MLCT), 137, 284
MMS *see* multiscale Maxwell-
Schrödinger scheme (MMS)
modelling methods, 79–81
molecular dynamics (MD), 361
simulations, 81
molecular excitations, 91–92

Monte Carlo method, 155
multidimensional interpolation
    approach (MIA), 353
multiexciton generation (MEG),
    43–46
        with dopants/defects and
            charging, 46–50
multi-hierarchy methods, 90
multiphoton photoemission, 229
multiple exciton generation
        dephasing mechanism, 42
        direct mechanism, 42
        impact ionization process, 41–42
multiple trap and release (MTR), 12
multiscale Maxwell-Schrödinger
    scheme (MMS), 128–129
multiscale modelling
        multi-hierarchy methods, 90
        multiscale challenges, 82–83
        quantum mechanical–molecular
            dynamics (QM–MD), 85–88
        strategies, 83–85
        surface reactions, 88–90

NAMD *see* nonadiabatic molecular
    dynamics (NAMD)
nanoplasmonic optics, 244–246
nanoporous metal oxide film, 14
nanoporous metal oxides, charge
    transport in, 12–13
        dark AC conductivity, power
            law dependence of, 19–20
        experimental methods, 20–21
        fluctuation-induced tunneling
            conductivity, 13–18
nanostructured TiO_2, 93
nanowires (NW), 198
near-field scanning optical
    microscopy (NSOM), 204
neutral density filter (NDF), 286
nonadiabatic molecular dynamics
    (NAMD), 40, 60–61
        simulation details, 61
non-bonded interactions, 87
non-collinear optical parametric
    amplifier (NOPA), 168

nonlinear optical spectroscopy,
    226
normal hydrogen electrode (NHE), 7
normal mode analysis (NMA),
    361–362
nuclear magnetic resonance (NMR),
    180

optical absorption spectra, 48
optical parametric amplifier (OPA),
    285
optical parametric oscillator (OPO),
    215
optical response function, 50–52

parameter-space analysis, 3
phase wrapping, 179, 181
phonon dephasing
        multiple exciton generation/
            fission and luminescence and,
            54–57
        in PbSe quantum dots, 52–54
        temperature dependence of, 54
phonon induced dephasing, 50
phonon motions induce fluctuations,
    62
photoabsorbers, inverse design of,
    10–12
photoanodes, 11
photocatalytic cells, computational
    modeling of
        nanoporous metal oxides,
            charge transport in,
            12–21
        photoabsorbers, inverse design
            of, 10–12
        photoelectrochemical device
            modeling, 3–10
        redox potentials, calculations of,
            21–29
photoelectrochemical device modeling
        bioinspired high-potential
            porphyrin photoanodes,
            5–10
        modeling current-voltage
            characteristics, 3–5

photoelectron momentum mapping
delocalized bands, 246–250
localized electronic states,
250–256
photo-induced electron transfer, 294
photo-induced interfacial electron
transfer, 352–356
photoluminescence (PL), 327
photon echo (PE) spectroscopy, 165
photon-matter interactions,
visualization of fundamental events
in, 356–358
2-photon tandem Z-scheme, 10
photovoltaic structures probed,
interfacial electron transfer in
dye sensitizing mesoporous
oxide films, 323–327
QD sensitizing mesoporous
oxide films, 327–329
QD superlattices, 329–330
semiconducting polymers,
exciton dissociation in, 330
plasmon-enhanced solar chemistry
continuum models, 115–116
DDA, 120–122
FDTD, 116–117
FEM, 117–120
hybrid approaches
hybrid RT-TDDFT/
FDTD approach,
129–131
MMS, 128–129
many-body theories
hybrid approaches,
127–128
LR-TDDFT, 123–125
RT-TDDFT, 125–127
plasmonic nanoparticles, 120
plasmonic phenomena, TR-PEEM
imaging of, 234–236
localized SP modes, 237–241
nanoplasmonic optics, 244–246
SPP dynamics, 241–244
polar solvents, 292
polycrystals, and mesoporous oxide
films, 310–312

porphyrin-based photoanodes, 9
Zn-porphyrin sensitizers,
142, 150
PPTAM, *see* pump probe transient
absorption microscopy (PPTAM)
propagating surface plasmon (PSP),
264
pump-probe photons
absorption coefficients
discrepancy for, 340–343
pump probe signal, 214
pump probe transient absorption
microscopy (PPTAM), 213–218

quantum chemistry, 80–81
quantum dots (QDs), 212,
312–314, 316–318, 327–329,
329–330
quantum interference, 253
quantum mechanical–molecular
dynamics (QM–MD), 85–88
quantum mechanical (QM), 79

random phase approximation (RPA),
128
real-time time-dependent density
functional theory (RT-TDDFT),
125–127
redox potentials, calculations of,
21–23
density functional theory
computational methods, 26
method benchmark results,
26–27
methodology and benchmark
results, 23–26
reference couple, choice of,
27–28
solvent polarity and supporting
electrolyte, 28–29
reflection high energy electron
diffraction (RHEED), 233
RMSD *see* root-mean-squared
deviation (RMSD)
root-mean-squared deviation
(RMSD), 15, 18

RT-TDDFT *see* real-time
   time-dependent density functional
   theory (RT-TDDFT)
RuN3, stimulated emission decay of,
   138

saturated calomel electrode (SCE), 23,
   24, 27
scanning electron microscope (SEM)
   images, 13, 18
scanning probe microscopy (SPM),
   261, 262
scanning tunneling luminescence
   (STL), 262
scanning tunneling microscopy
   (STM)
      light emission from
         experimental
            considerations, 265–266
         mechanism, 262–264
         surface plasmons in,
            264–265
      nanostructures, light emission
         from
            individual atoms/
               molecules and surface
               states, 266–268
            nanocrystals, 271–275
            thin film and organic
               semiconductors,
               268–271
         photon emission, time-resolved
            studies of, 275–276
second harmonic generation (SHG),
   204
self-assembled monolayers (SAMs),
   274
semiconducting polymers, exciton
   dissociation in, 330
semiconductor dependence, 288–289
semiconductor hetero-structures, and
   hybrid systems, 105
semiconductor quantum dots, 46
      Auger and phonon-assisted
         Auger processes, time domain
         ab initio study of, 66–67

Auger studies, 69–70
Auger theory, 68–69
electron phonon relaxation
   charge carriers,
      phonon-assisted
      relaxation of, 61–63
   ligands saturate dangling
      bonds and accelerate
      electron-phonon
      relaxation, 66–67
   NAMD, 60–61
   TDDFT, 58–59
   temperature dependence,
      63–66
MEG, 41–42
phonon induced dephasing
   multiple exciton
      generation/fission and
      luminescence, 54–57
   optical response function,
      50–52
   phonon dephasing in PbSe
      quantum dots, 52–54
   temperature dependence
      of, 54
SAC-CI, 42–43
   MEG, 43–50
theoretical approaches, 38–39
   DFT, 40
   electron correlation,
      incorporation of, 39–40
   Hartree–Fock Method, 39
   nonadiabatic molecular
      dynamics, 40
   TDDFT, 40
semiconductor Quantum Dots (QDs),
   312–314
semiconductors, characteristic
   terahertz responses in
      bulk semiconductors, 308–310
      polycrystals and mesoporous
         oxide films, 311–312
      semiconductor QD, 312–314
simple equivalent circuit, 4
single wall carbon nanotubes
   (SWCNTs), 215–216

single wall nanotubes (SWNTs), 215

Slater-type atomic orbitals (STO), 12

solvent effects, 292

spacer groups, 99–101

spectral interferometry, 170–171

SPP *see* surface plasmon polariton (SPP)

standard hydrogen electrode (SHE), 23, 27

Stanford Linear Accelerator Center (SLAC), 357

stimulated emission, 138

stimulated emission (SE), 165

sum frequency generation (SFG), 143, 204

surface plasmon polariton (SPP), 113, 235, 241–244

surface reactions, 88–90

symmetric interference distributions, 255

symmetry adapted cluster-configuration interaction (SAC-CI), 42–43

systematic inverse design methodology, 10

terahertz generation, and detection, 303–308

terahertz (THz), 151, 152

three-pulse photon echo peak shift (3PEPS), 166

time-correlated-single photon-counting (TCSPC), 204

time-dependent density functional theory (TDDFT), 58–59, 355

time domain density functional theory (TDDFT), 40

time-ordering arguments, 173–178

time resolved infrared spectroscopy experimental techniques, 286–287

    metal oxides, interfacial electron injection mechanism into, 287

    dye dependence, 289–291

    ions effect, 292–294

    metal nanoparticle to, 294–296

semiconductor dependence, 288–289

solvent effect, 292

time-resolved multiphoton photoemission spectroscopy (TR-MPP), 226

time-resolved photoemission electron microscopy (TR-PEEM), 234–236

time-resolved terahertz (THz) spectroscopy (TRTS), 151

    high efficiency photovoltaic concepts, carrier dynamics studies of, 314

    semiconductor quantum dots, 315–316

    semiconductors, CM in, 316–323

photovoltaic structures probed, interfacial electron transfer in dye sensitizing mesoporous oxide films, 323–327

    quantum dot sensitizing mesoporous oxide films, 327–329

    quantum dot superlattices, 329–330

    semiconducting polymers, exciton dissociation in, 330

semiconductors, characteristic terahertz responses in bulk semiconductors, 308–310

    polycrystals and mesoporous oxide films, 311–312

    semiconductor quantum dots, 312–314

terahertz generation and detection, 303–308

time-resolved ultrafast multiphoton photoemission (TRUMP), 227

tip-induced surface plasmons (TISPs), 262, 264
Ti:sapphire laser system, 284
transient absorption (TA), 144, 327
    decay, 296
    traces, 215
transient far-infrared conductivity spectra, 154
transient species, determining fractions of, 343–344
transition dipole moment, 186
transition metal complexes, excited state structural dynamic calculations for, 360
triplet electron injection, 142
TR-PEEM *see* time-resolved photoemission electron microscopy (TR-PEEM)
tunneling current density, 16
tunneling luminescence spectroscopy, 273
two-dimensional photon echo spectroscopy
    CdSe NCs
        2D spectra, simulating, 187–188
        exciton-biexciton transitions, determination of, 186–187
        lowest exciton and biexciton states in, 183–185
    CdTe/CdSe core/shell nanocrystals, 189–195
    electronic coherences, 167–168
    fundamentals of, 164–165
    semiconductor nanocrystals and quantum dot-based solar cells, 161–163
    solvation dynamics, 165–167
    spectral interferometry, 170–171
        background subtraction, 172–173
        collected signal, 171–172

phasing 2DPE spectra, 178–182
        time-ordering arguments, 173–178
    wedge calibration, 169–170

ultrafast intraband charge-phonon relaxation, 62
ultrafast multiphoton photoemission microscopy
    interfacial electron dynamics, frequency *vs.* time domain measurements of, 228–229
    photoelectron momentum mapping
        delocalized bands, 246–250
        localized electronic states, 250–256
    plasmonic phenomena, TR-PEEM imaging of, 234–236
        localized SP modes, 237–241
        nanoplasmonic optics, 244–246
        SPP dynamics, 241–244
    time-resolved photoemission electron microscopy, 231–234
    time-resolved ultrafast multiphoton photoemission, 229–231
ultrafast optical imaging
    KGFM
        background and experimental considerations, 205–208
        in $CdS_x Se_{1-x}$ nanobelts, 208–209
        quantum dots, fluorescence dynamics of, 212–213
    PPTAM
        background and experimental considerations, 213–215

exciton and charge carrier
dynamics studies,
215–218
single molecule femtosecond
microscopy, 218–220
spatial super-resolution,
221–222
sub 100 femtosecond time
resolution, 220–221
ultra high vacuum (UHV), 265

vacuum ultraviolet (VUV), 234
variable range hopping (VRH), 12

water oxidation, 1, 2

X-ray absorption spectroscopy
(XAS), 338–340
X-ray diffraction (XRD), 13, 18
X-ray free electron laser (XFEL),
345, 357
X-ray transient absorption
spectroscopy
capabilities and development
advanced photon source,
345–347
dilute samples with low
concentration, signal
detection in, 344–346
pump-probe photons,
absorption coefficients
discrepancy for,
340–343

transient species,
determining fractions
of, 343–344
X-ray absorption
spectroscopy (XAS),
338–340
future research and
development
detecting transient
structures with low
concentrations,
358–360
photon-matter
interactions,
visualization of
fundamental events in,
356–358
theoretical modeling,
360–361
multiple transient structures in,
360
in solar energy conversion
research
metalloporphyrin
excited state
structural dynamics,
347–352
photoinduced interfacial
electron transfer,
352–356

yttrium aluminum garnet (YAG),
206